5th International Conference on Mechanical Engineering and Automation Science (ICMEAS 2019)

IOP Conference Series: Materials Science and Engineering
Volume 692

Wuhan, China
10 - 12 October 2019

ISBN: 978-1-7138-0984-5
ISSN: 1757-8981

Printed from e-media with permission by:

Curran Associates, Inc.
57 Morehouse Lane
Red Hook, NY 12571

CURRAN ASSOCIATES INC. proceedings.com

Some format issues inherent in the e-media version may also appear in this print version.

Institute of Physics
Dirac House, Temple Back
Bristol BS1 6BE UK

Phone: 44 1 17 929 7481
Fax: 44 1 17 920 0979

techtracking@iop.org

Additional copies of this publication are available from:

Curran Associates, Inc.
57 Morehouse Lane
Red Hook, NY 12571 USA
Phone: 845-758-0400
Fax: 845-758-2633
Email: curran@proceedings.com
Web: www.proceedings.com

TABLE OF CONTENTS

MANUFACTURING ENGINEERING

MATERIALS ENGINEERING

MECHANICAL ENGINEERING

DESIGN ENGINEERING

IOP Conf. Series: Materials Science and Engineering **692** (2019) 011001 doi:10.1088/1757-899X/692/1/011001

PREFACE

2019 5th International Conference on Mechanical Engineering and Automation Science (ICMEAS 2019) was held in China University of Geosciences (Wuhan), Wuhan, China during Oct. 10-12, 2019. ICMEAS 2019 was Co-organized by China University of Geosciences (Wuhan) and Hong Kong Society of Mechanical Engineers (HKSME). The conference provides a useful and wide platform both for display the latest research and for exchange of research results and thoughts in Mechanical Engineering and Automation Science and other topics. The participants of the conference were from almost every part of the world, with background of either academia or industry, even well-known enterprise. The success and prosperity of the conference is reflected high level of the papers received.

The proceedings are a compilation of the accepted papers and represent an interesting outcome of the conference. This book covers 4 chapters: Manufacturing Engineering; Materials Engineering; Mechanical Engineering; Design Engineering.

We would like to acknowledge all of those who supported ICMEAS 2019. Each individual and institutional help were very important for the success of this conference. Especially we would like to thank the organizing committee for their valuable advices in the organization and helpful peer review of the papers.

We sincerely hope that ICMEAS 2019 will be a forum for excellent discussions that will put forward new ideas and promote collaborative researches. We are sure that the proceedings will serve as an important research source of references and the knowledge, which will lead to not only scientific and engineering progress but also other new products and processes.

Prof. Huafeng Ding, China University of Geosciences (Wuhan), China
Prof. Ian McAndrew, Capitol Technology University, USA

COMMITTEES

Conference Committee Co-chairs

Prof. Du Ruxu, South China University of Technology, China

Prof. Huafeng Ding, China University of Geosciences(Wuhan), China

Prof. Mohd Hamdi Bin Abd Shukor, University of Malaya, Malaysia

Program Committee

Prof. Stefan Dimo, University of Birmingham, UK

Prof. Yong Zhao, Nazarbayev University, Kazakhstan

Local Committee

Hong Xu, China University of Geosciences(Wuhan), China

Advisory Commitees

Prof. Hui Ding, Harbin Institute of Technology, China

Prof. Enrico Vezzetti, Politecnico di Torino, Italy

Prof. Egon Müller, Technische Universität Chemnitz, Germany

International Technical Committee

Prof. Tang Sai Hong, Universiti Putra Malaysia, Malaysia

Prof. Spilios D. Fassois, University of Patras, Greece

Prof. Shinn-Liang Chang, National Formosa University, Taiwan, China

Prof. Sandy To,Hong Kong Polytechnic University, Hong Kong, China

Prof. M. Chandrasekaran, Vels University, India

Prof. Josua Meyer, University of Pretoria, South Africa

Prof. Harijono Djojodihardjo, UPM, Malaysia

Prof. David Barton, University of Leed, UK

Prof. C. P. Jiang, National Formosa University, Taiwan, China

Prof. Antonis I.G. Vardulakis, Aristotle University of Thessaloniki, Greece

IOP Conf. Series: Materials Science and Engineering **692** (2019) 011003 doi:10.1088/1757-899X/692/1/011003

Peer review statement

All papers published in this volume of *IOP Conference Series: Materials Science and Engineering* have been peer reviewed through processes administered by the proceedings Editors. Reviews were conducted by expert referees to the professional and scientific standards expected of a proceedings journal published by IOP Publishing.

IOP Conf. Series: Materials Science and Engineering **692** (2019) 012001 doi:10.1088/1757-899X/692/1/012001

Research on the Robot's Automatic Road Detection Technology with SRPS

Zhang Jianzhong[1] and Xu Qiang[1]

[1]School of Mechanical Engineering, Hefei University, Hefei 230601, China

Email:zjzfirst@sohu.com

Abstract. Road detection technology has been widely used in many fields, such as vehicle's assisted driving, robot autonomous navigation, military defense, public security and so on. In this paper, the road detection method based on sensitive region pixel sequence (SRPS) solution is studied. Firstly, the recent advances of the road detection method based on machine vision are summarized. Secondly, analyses the principle of road detection technology based on sensitive region pixel sequence (SRPS) solution to improve road detection efficiency and accuracy of the vision method. Finally, experimental results show that the sensitive region pixel sequence (SRPS) solution has an important significance to reduce the calculation cost of the robot's road detection system.

1. Introduction

Since the 1980s, automatic road detection has been a core technology for autonomous driving. In recent years, some institutes and enterprises carried out a detailed study of road detection technology, and obtained some certain achievements [1]. With the development of machine vision algorithm and artificial intelligence technology, vision measurement technology has become more widely applied in the research on robotics. The vision detection technology in the early study is limited by the level of hardware. With the development of computer hardware, the research of vision detection technology and its application system have gained great achievements, such as ALVINN system designed by Carnegie Mellon University, VAMP system designed by Germany Federation of Munich National Defense University, GOLD system designed by Italy Parma University and so on. Furthermore, it is easy for vision detection method to subdivide and judge the road type by deep learning. Researchers in China have begun the study of driverless cars too. In 2005, the first city driverless car was developed at the Shanghai Jiao Tong University [2].

The road information acquisition methods mainly include the laser radar detection and the vision detection. With the laser radar detection method, the accurate three-dimensional information of the road can be obtained, which is more suitable for the judgment of the moving object, but the cost of this method is high. With the vision detection method, the road conditions are to be observed through the camera, which has the characteristics of low cost and high precision.

In recent years, the vision detection method has become a key technology in the road detection system, a variety of different road models are identified by image processing and understanding. [3]Vision detection method can be divided into monocular vision, binocular vision and multi-vision. [4]

1) With monocular vision method, the road's image is acquired with a single camera, and the road's image depth information is lost. It is difficult for the automatic pilot robot or car to make the precise judgment of the road with the monocular vision method.

2) Binocular vision is a method of obtaining the object's three-dimensional geometric information by two cameras based on the disparity principle. The two images of the same object from different locations are obtained, and calculate the locations of the corresponding points in the image.In the binocular vision system; the two cameras can help the automatic pilot robot or car to get more environmental information of the road's.

3) With multi-vision method, the road's information is acquired with more than two cameras.The acquisition of the road's position information is more abundant, but sometimes becomes more prone to ambiguity. [5,6]

In the existing literature, many researchers have done a lot of research on the road detection technology based on machine vision. Azevedo, V B [7] proposed a vision approach for real-time mapping of traversable road surfaces with a stereo camera. The road detection system based on machine vision should solve a complex pattern recognition problem, which has become a concerning focus in the research on the automatic pilot robot or car.

The current road detection system has solved many problems of road recognition, still with some defects. A common road detection system does not exist. The research on the road detection technology with adaptive and self-learning ability is a difficult task, too. [8]

One of the important challenges is that the heavy calculation load leads to the poor real-time performance of the road detection system.The complexity of traffic environmental information reduces the calculation precision and efficiency of the road detection. Further, the widely-used machine learning method learns from the training samples, and the increasing time cost of training samples will reduce the learning accuracy consequently, which leads to the poor adaptability of self-learning to the changing road environment.

To solve this problem, it's necessary to reduce calculation cost of the road detection system. In the course of this research, the SRPS solution has been put into application.

2. Principle of road detection technology based on SRPS

In order to improve the speed and efficiency of image matching, multi-resolution image matching algorithm has been proposed. Crowley [9] proposed a multi-resolution image pyramid for the first time.In 2004, David G Lowe [10,11] proposed a local feature description SIFT operator of the multi-resolution image.

In principle, multi-resolution image matching algorithm can be divided into two stages: multi-scale spatial decomposition stage and reconstruction stage.

1) In multi-scale spatial image decomposition stage, the image is down sampled with vertical direction and horizontal direction, and an image pyramid is formed. The spatial resolution at higher level will have a smaller pixel size, otherwise, the reverse (shown in **Figure 1**).

In 1988, Mallat proposed the orthogonal wavelet construction method and the fast orthogonal wavelet transform algorithm[12,13].

Suppose $f(t) \in L^2(R)$ and $\{C_n^0\}_{n\in Z} \in \ell^2$, make $f(t) = \sum_{n\in Z} C_n^0 \phi_0, n(t)$, the decomposition algorithm for Mallat is described as flow.

$$C_n^j = 2^{-1/2} \sum_{k\in Z} h_{2k-n} C_k^{j-1} \tag{1}$$

$$D_n^j = 2^{-1/2} \sum_{k\in Z} g_{2k-n} C_k^{j-1} \tag{2}$$

Using the decomposition of the low frequency coefficient C_n^j and the high-frequency coefficient D_n^j, the multi-resolution representation of image information is performed.

2)In the step of the image reconstruction, the image reconstruction is done, which is the reverse process of the multi-resolution image decomposition.

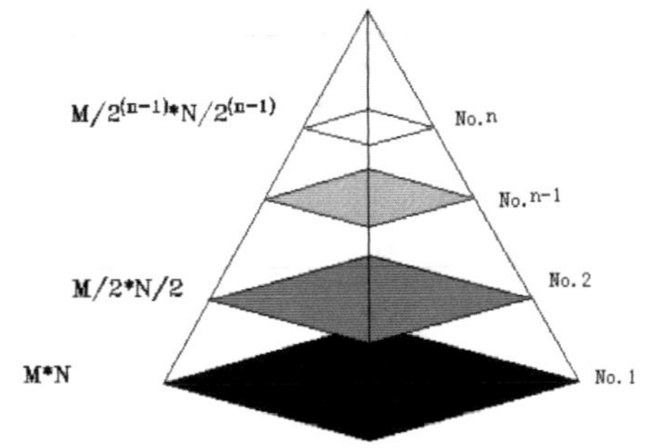

Figure 1. Multi-resolution image pyramid

Image fitting analysis in the road detection system is feasible in principle and has the characteristics of higher speed in computation, but this method has been lack of the ability in judging the road type. By contrast, image matching analysis has the characteristics of higher precision and good computational stability.On the basis of the study,the SRPS solution is studied.SRPS means the sensitive region pixel sequence of an image.The main two steps of the SRPS solution are as follows

Firstly, obtain the SRPS's feature pixels of a road image by the image segmentation at different higher levels,and create a data set corresponding to the sensitive region pixel sequence.

Secondly,calculate the correlation with a single-dimensional arrays corresponding to the SRPS at different higher levels.

An effective choice is to reduce the dimension of the road image,and change the road image gray value to one-dimensional data. The strength S calculation of the one-dimensional data can be defined as

$$S = a * b^T \tag{3}$$

Where, vector a is a one-dimensional pixel gray value of the road template, vector b is the one-dimensional pixel gray value of the real road image at the designated address corresponding to the one-dimensional array.

3. Experiment

In order to verify the feasibility of the SRPS solution, the experiment with SRPS solution has been conducted. The experiment goal is to detect the blind road. With the development of barrier-free design and upgrading of the attention of the community for the blind groups, the blind road has become an important barrier-free facility, which specifically helps the blind people to walk. The blind road generally consists of two kinds of blind road, including the marching blind road and the prompt blind road. The former is made up of the guide bar bricks, which can guide the blind to walk forward; the latter is made up of the bricks with a dot, which can remind the blind people of the road obstacles ahead.

Take the blind road's picture and calculate the SPRS feature pixels in the selected ROI region.The effect of a blind road's multi-resolution image with neighborhood average filter is shown in **Figure 2**.In this way, the blind road line is fitted with SRPS solution and the location of the road line is found, which is shown in **Figure 3**.The original blind road image is shown in **Figure 3**(a). The feasible path of the road obtained by the SRPS solution is shown in **Figure 3**(b).

IOP Conf. Series: Materials Science and Engineering **692** (2019) 012001 doi:10.1088/1757-899X/692/1/012001

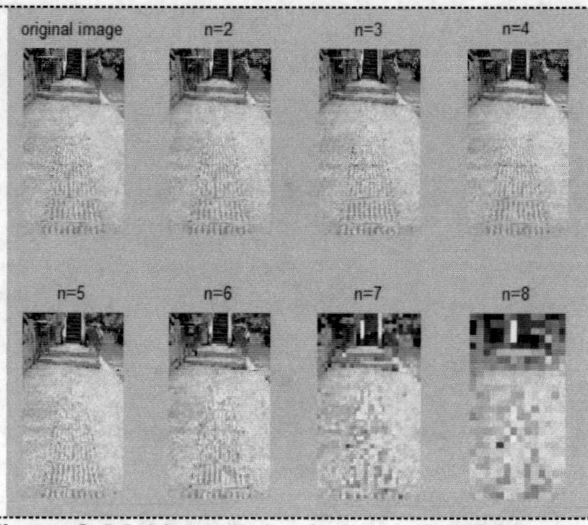

Figure 2. Multi-resolution images of a blind road

(a) Original image

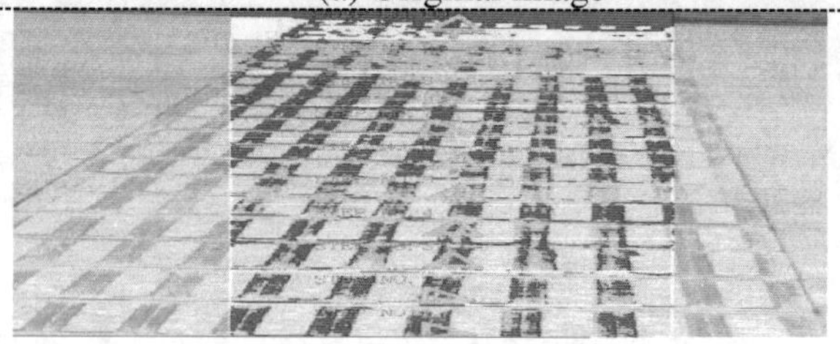

(b) Detected blind road

Figure 3. The blind road line's fitting result with SRPS solution

Besides this, the blind road line's matching result is shown in **Figure 4**.Five matching instances are obtained, and the feasible path obtained by the SRPS solution is accurate.

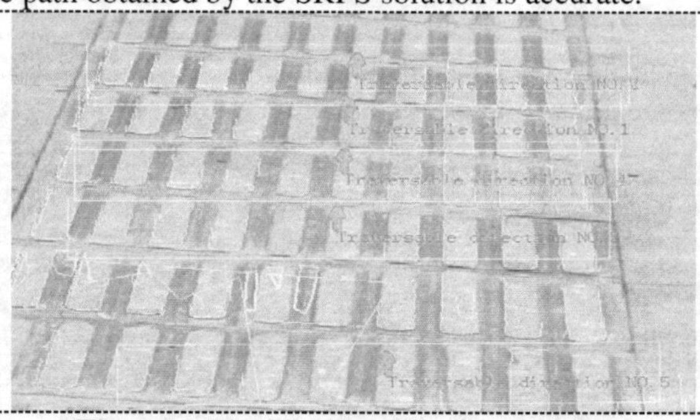

Figure 4. The blind road line's matching result with the SRPS solution

Comparing the traditional method, the total time of the road image processing with the SRPS solution is significantly reduced, which saves about 1/3 of the time of the traditional method's.

The experiment proves the feasibility of the SRPS solution. The further experiment on the SRPS solution will need to be the focus of our future research.

4. Conclusion

This paper does a principle analysis of the road detection method based on SRPS solution in order to improve the efficiency and accuracy of the vision detection method.The proposed solution has an important significance to reduce calculation cost of the road detection system, which is helpful to enhance the early-warning performance for the automatic pilot robot.Experimental results show that the SRPS solution has a positive significance for the improvement on the automatic road detection technology.

Acknowledgements

This work is supported by Provincial (Key) Natural Science Research Project of Anhui Colleges(KJ2017A537, KJ2017A538).

References

[1] G Wei 2014 *The Road Recognition Technology Research of Mobile Robot.*PhD Dissertation, Hebei University of Technology

[2] Weimin L,Chiying L 2013 Autonomous cross-floor navigation of a stair-climbing mobile robot using wireless and vision sensor. *2013 44th Int. Symp. on Robotics*, DOI:10.1109/ISR. 2013.6695649

[3] Steger C, Ulrich M, Wiedemann C 2008 *Machine Vision Algorthms and Applications.* Y. Shao-rong. Beijing, Tsinghua University Press

[4] Ni DJ,Song A,Tian L, Xu XN,Chen DF 2015 A walking assistant robotic system for the visually impaired based on computer vision and tactile perception. *Int. J. of Social Robotics*, **7**(5) pp 617-628

[5] Z Jigui, Y Jing 2011 *Principle and Method of Vision Measurement*, Beijing, Mechanical Industry Press

[6] Slavkovikj V,Verstockt S, De Neve W,Van Hoecke S, Van de Walle R 2014 Image-based road type classification *.Int. Conf. on Pattern Recognition* pp 2359-2364.DOI: 10.1109/ICPR

[7] Azevedo V B, De Souza A F , Veronese L P,Badue C,Berger M *2013* Real-time road surface mapping using stereo matching, v-disparity and machine learning. *The 2013 Int. Joint Conf. on Neural Networks* (IJCNN).DOI:10.1109/IJCNN.2013.6707066

[8] Rosenfeld A, Thurston M 1971 Edge and curve detection for visual scene analysis. *IEEE Trans. Computer*, **20** pp 562-569

[9] J L Crowley 1981 *A Representation for Visual Information* .PhD thesis, Carnegie-Mellon University,Robotics Institute, Pittsburgh, Pennsylvania

[10] David G Lowe 1999 Object recognition from local scale-invariant features. *Int. Conf. on Computer Vision*, Corfu Greece

[11] David G Lowe 2004 Distinctive image features from scale invariant keypoints . *Int. Journal of Computer Vision*, **60**(2) pp 91-110

[12] Gu Y J, Jin Z 2010 Grass detection based on color features.*Proc. of Chinese Conf. on Pattern Recognition.* Piscataway, USA, IEEE pp 1-5

[13] D Mingfang, W Junzheng, L Jing, C Guangtao, F Jianjun,C Haiqing 2014 Adaptive outdoor road detection for autonomous mobile robot.*J. of Image and Graphics*, **19**(7) pp 1046-1053

IOP Conf. Series: Materials Science and Engineering **692** (2019) 012002 doi:10.1088/1757-899X/692/1/012002

Injection Molding Cycle Time Reduction for Automobile Headlamp Shell in Moldflow

Tiomela Fouodji Cedric[1] and Huang Xiao Hua[2,*]

[1] Mechanical engineering, Nanjing University of Science and Technology, Nanjing, Jiangsu, 210094, China

[2] Mechanical engineering, Nanjing University of Science and Technology, Nanjing, Jiangsu, 210094, China

E-mail: *michhxh@163.com ; snoopytfc@yahoo.fr

Abstract. In the injection molding cycle, the cooling time consumes more than 70% of the molding process, optimizing the molding cycle time can be done through the selection and the design of appropriate cooling system that will be able to achieve equal distribution of the temperature along the part and quickly dissipate the heat. This paper presents an analytic study on the simulation of cool + fill + pack + warp in Autodesk Moldflow 2012 for the automobile headlamp shell, the simulation is made on of four different cooling systems, after comparison of the obtained results, the Spiral Zigzag Conformal Cooling Channel (SZCC) has the lowest time required for the part to reach the ejection temperature, with 61.22 seconds, it also shows an improvement on the volumetric shrinkage with the part's deflexion that have been reduced from 2.15 mm on the first design to 2.12 mm, making Spiral Zigzag Conformal cooling the best solution for this model.

1. Introduction

Injection molding process is among the largest mass production processing for plastic parts in the world and has an approximation of 32wt% of all plastic, making it second place to the extrusion process, which consumes approximately 36%wt[1]. The injection machine takes the granulate polymers, heats them until molten, then force it in the cavities through the nozzle and the sprue. After solidification and cooling, the molten granulate takes the shape of the mold before been ejected from the mold[2,3]. The injection molding process is generally achieved in three stages which are: the filling + packing stage, the cooling stage, and the ejection stage. The cooling stage is the most important among the three since it directly influences the product's quality and the productivity rate [4]. Traditionally, the cooling channels have been designed by drilling straight holes inside the mold, these straight holes also called conventional cooling channels are not uniform to the shape of the part as seen in figure 1a, because of that, the cooling process would take long and obviously the injection molding cycle time will increase while the production rate will decrease. It has been noticed that decreasing the cooling cycle could come out with other undesired defects such as excessive shrinkage and warpage on the product[5]. In order to prevent those defects, a new technic of designing cooling channels has been introduced by plastic molding researchers, it's called conformal cooling channels, and the cooling channel is designed as to conform with the contour of the part. And therefore, makes the cooling uniform in all areas along the part, figure 1b clearly illustrates the design. Conformal cooling channels have contributed in a great way in optimizing the cycle time. Wang et al. developed the spiral conformal cooling channel[6] and they have come to a conclusion that their design made of spiral curves do not reduce the rate of the coolant flow, which is good but it mostly achieves uniform mold cooling.

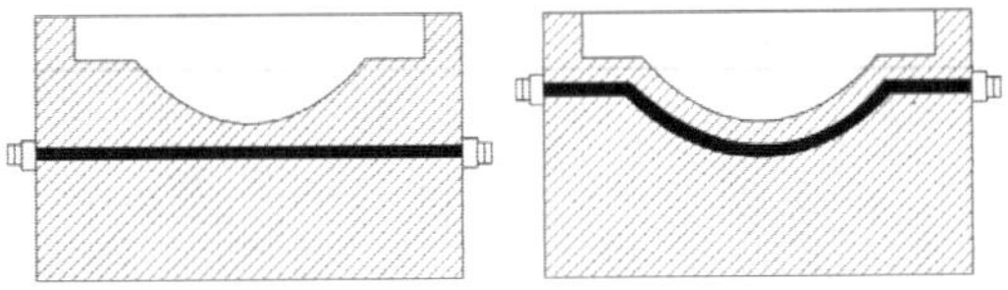

(a) (b)

Figure 1. (a) straight drilled cooling channel. (b) conformal cooling channel.

Konsulova Bakalova has used thermal simulation software to compare conformal cooling channels with circular and elliptical cross-sections, and concluded that the cooling time has been reduced and the appearance of the part has been improved for the design with conformal cooling[7]. In general, there are four different designs for conformal cooling respectively named the zigzag conformal cooling channels[8], the spiral conformal cooling channels[6,8] the scaffold conformal cooling channels[9] and the Voronoi diagram cooling channels[10]. Each design is chosen according to the need of the mold designer and mainly conditioned by the structure of the part[4].

In this work, we solve a practical problem, by running the cool + fill + pack + warp analysis in Moldflow simulation, we evaluate and compare four different designs of cooling systems of the headlamp shell for which the initial cooling system design was made with baffles cooling channels despite the variation of the thickness in the regions with ribs, the objective is to reduce the molding cycle time and to maintain or increase the quality of the part.

2. Part Design and Cooling System Design

The headlamp shell is meant to be used on the left and right front side of the car to assembly the lightening element such as bulbs, reflector, bezel and lens. The model was designed with Unigraphics NX8 (figure 2) and the parameters are 358 mm in length, 318 mm in width, 452 mm in height, the thickness varies from 2 mm in the flat areas to 3 mm in the areas with strengthening ribs.

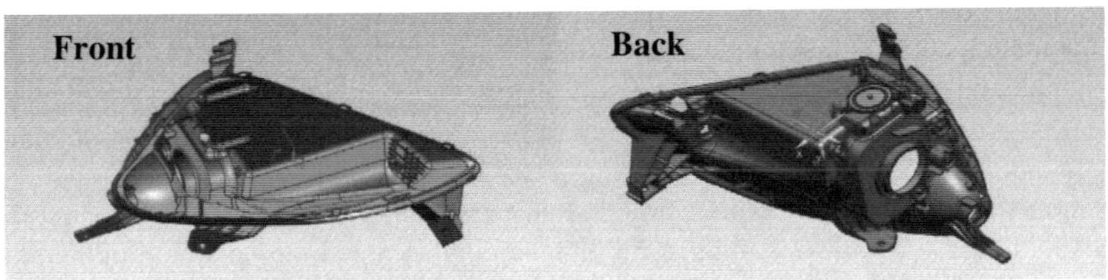

Figure 2. 3D model of the headlamp shell.

The location of the cooling lines is critical to achieving efficient cooling, reducing the defects in the geometry and improving the productivity of the product. The cooling channels should be larger enough to have reasonably low-pressure drop but not so large that excessive flow rate is needed to obtain maximum cooling efficiency via turbulent flow[11]. The first design is the Baffle Cooling Channels (BC) with a diameter of 18 mm, the second is the Baffle Zigzag Conformal Cooling Channels (BZCC), the respective diameters are 18 for baffle and 12mm for the zigzag conformal cooling, the third design is the Baffle Spiral Conformal Cooling Channels (BSCC) with the same dimensions as in the case of (BZCC), the last one is the Spiral Zigzag Conformal Cooling Channels (SZCC) with the diameter of 12 mm on both sides of the mold, Figure 3 shows the design of different cooling systems.

IOP Conf. Series: Materials Science and Engineering **692** (2019) 012002 doi:10.1088/1757-899X/692/1/012002

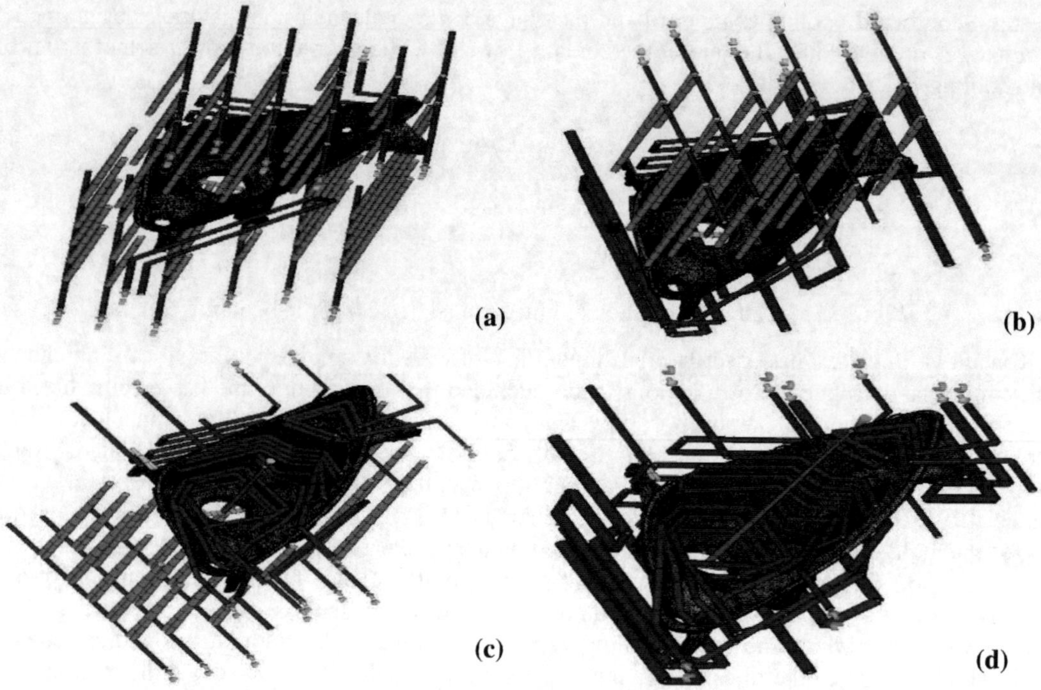

Figure 3. (a) Baffle Cooling Channels (BC); (b) Baffle Zigzag Conformal Cooling Channels (BZCC); (c) Baffle Spiral Conformal Cooling Channels (BSCC); (d) Spiral Zigzag Conformal Cooling Channels (SZCC).

3. Moldflow Analysis

Moldflow has three different types of meshing known as the midplane, the dual domain and the 3D mesh, before starting the analysis, we must understand difference and choose the one that will be suitable for the model we intend to analyse[12]. For this analysis, the 3D mesh type will be used because our model doesn't only have complex geometry but it also has thickness variance from 2 to 3mm.

3.1. Material selection

The headlamp shell except vibrations, does not sustain to any kind of stress or choc during its use since it is fixed on the car, the thermoplastic material that is going is the polypropylene (pp) because it's the cheapest among thermoplastic material and can resist to the vibration caused by the engine of the car. The model weights 534g and the thermal and mechanical properties of pp are listed in the Table 1.

Table 1. Thermal and mechanical properties.

Property	Value
Density (g/cm^3)	0.9
Melt temperature (°C)	240
Thermal conductivity (10^{-4}cal/sec cm°C)	2.8
Heat capacity (cal/g°C)	0.9

3.2. Gating system design

The gate is known as the last passageway where the molten material coming out of the barrel of injection machine should go through to fill the cavities, in order to obtain optimum filling during the process, in this case, the edge gate is the appropriate one due to its capacity of providing wide molding window for robust model, it also insure low shear rates. We choose two injection points located at the edge of the model in the hole where the bulb is supposed to be mounted, it is a cold gate and the geometry is rectangular with 8mm width and 3mm height.

3.3. Runner system design

For the runner system, we have previously compared the half-circular and the trapezoidal cold runner during the design and it has been noticed that the trapezoidal cold runner is best for this analysis since the molten plastic flows quickly, and the turbulent flow that has a tendency to increase stresses and sink marks is avoided. The dimensions of the runner are: top width 12mm, bottom width 8mm, and the height is 8mm.

3.4. Sprue design

We choose hot sprue for this model because, from the position where the runner is located to the surface of the fixed plate, the distance is more than 800 mm making it difficult to use cold sprue because the temperature of the molten material would drop before reaching the runner and the mold will not be completely filled.

4. Results and discussions

The time to reach the ejection temperature representing the overall time needed for the part to be filled cooled and ejected from the mold and the deflection will be discussed here. The results from figure 4 (a) shows that the time to reach the ejection temperature for the Baffle Cooling Channels (BC) is 74.7 seconds. Meanwhile the use of the Baffle Zigzag Conformal Cooling Channels (BZCC) decreases the time to reach the ejection temperature to 71.84 seconds as shown by figure 4 (b). By comparing these two cases, the BZCC provides 3.89% faster Cooling than the BC. The simulation results also show that Baffle Spiral Conformal Cooling Channels (BSCC), figure 4 (c) requires 65.75 seconds to reach the ejection temperature, whereas the Spiral Zigzag Conformal cooling channel (SZCC), figure 4 (d) only needs 61.22 seconds to reach ejection temperature. From these results, it's well known that the BC has to lowest performance in terms of the cooling system and the SZCC has the highest performance among the four cases and is 18.11% faster compared to the BC.

The second parameter of comparison is the deflection, which is conditioned by the volumetric shrinkage and temperature variance during the cooling phase, the cooling lines should be designed as making the temperature of the walls of the mold as uniform as possible in order to reduce the deflection along the axis[13]. Figure 5 presents the simulation results of the deflection for which BC is 2.15 mm, whereas BZCC, BSCC and SZCC respectively show 2.77 mm, 2.65 mm and 2.12 mm. Thus, the SZCC has the minimum deflection among the three cases. The simulation results obtained from Autodesk Moldflow Advisor 2012 are compiled in table 2. From the values obtained in figures 6 and 7 we conclude that SZCC is the most efficient cooling system design for our part, it has the shortest time to reach the ejection temperature, it doesn't have the lowest percentage of volumetric shrinkage compared to the BZCC but has the minimum deflection dimension, which makes it the optimum cooling design for this model.

Table 2. Summary of the results obtained from the analysis.

Cooling channel	Time to reach ejection Temperature seconds	Volumetric Shrinkage %	Deflection mm
BC	74.75	17.73	2.15
BZCC	71.84	17.71	2.77
BSCC	65.75	18.57	2.67

IOP Conf. Series: Materials Science and Engineering **692** (2019) 012002 doi:10.1088/1757-899X/692/1/012002

SZCC	61.22	17.79	2.12

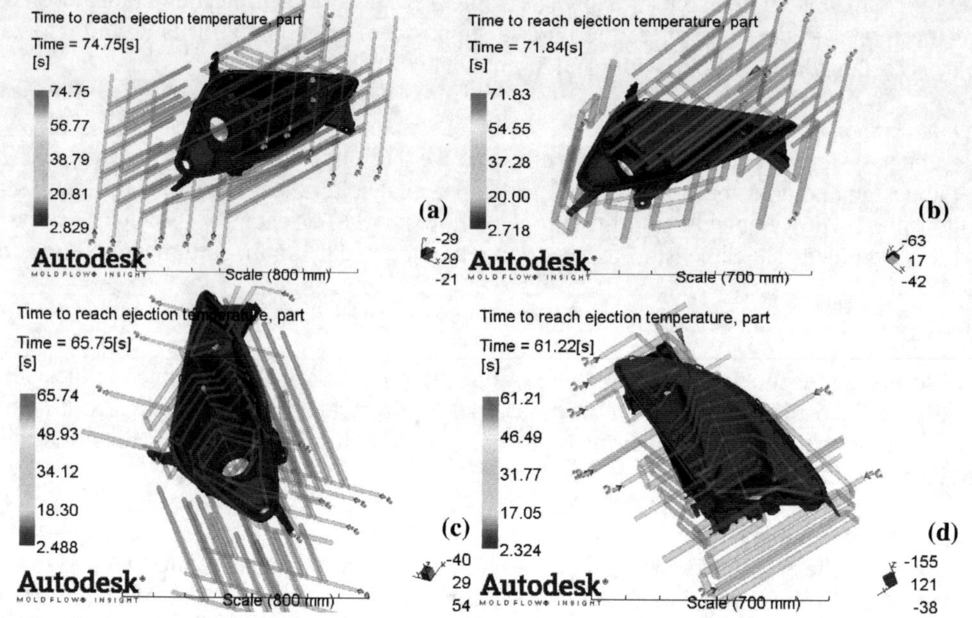

Figure 4. Time to reach ejection temperature with (a) Baffle Cooling Channels (BC); (b) Baffle Zigzag Conformal Cooling Channels (BZCC); (c) Baffle Spiral Conformal Cooling Channels (BSCC); (d) Spiral Zigzag Conformal Cooling Channels (SZCC).

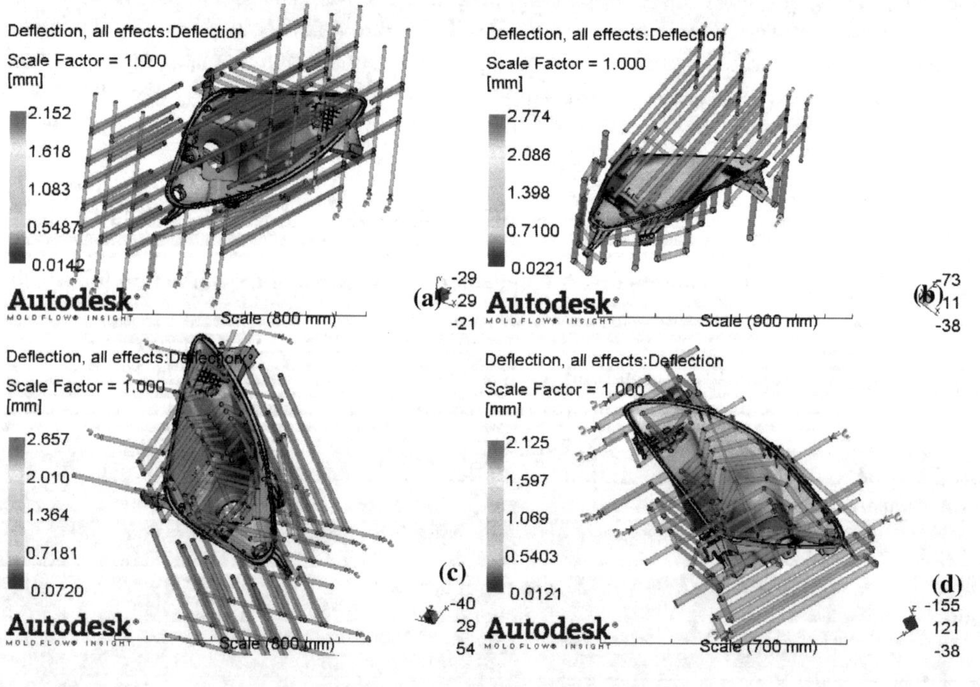

Figure 5. Deflection with (a) Baffle Cooling Channels (BC); (b) Baffle Zigzag Conformal Cooling Channels (BZCC); (c) Baffle Spiral Conformal Cooling Channels (BSCC); (d) Spiral Zigzag Conformal Cooling Channels (SZCC).

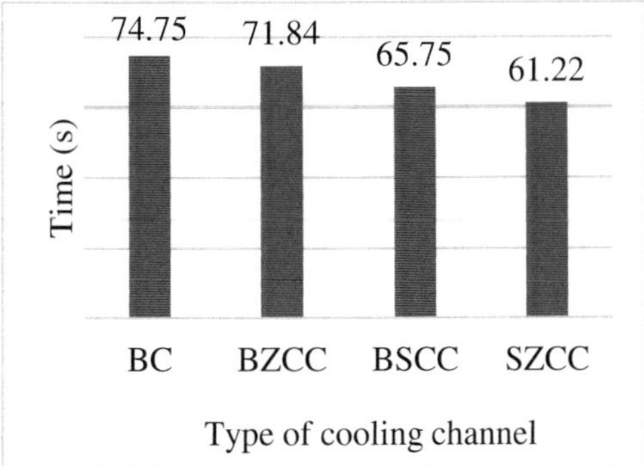

Figure 6. Comparison of time to reach ejection temperature.

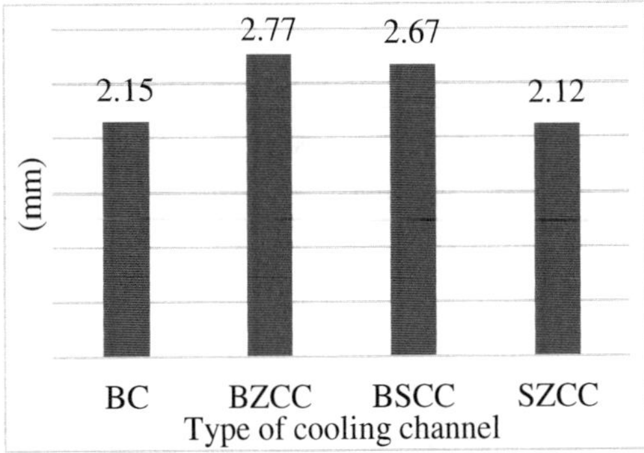

Figure 7. comparison of the part's deflection.

5. Conclusion

This study introduces the simulation analysis in Autodesk Moldflow of four different cooling channels respectively named, Baffle Cooling Channels (BC), Baffle Zigzag Conformal Cooling Channels (BZCC), Baffle Spiral Conformal cooling channel (BSCC) and the Spiral Zigzag Conformal Cooling Channels (SZCC), all designed for the cooling of the automobile headlamp shell. The cool + fill + pack + warp analysis was performed in Moldflow 2012. The time to reach ejection temperature that was previously 74.7s for BC has been gradually improved to 61.2s for the SZCC, the volumetric shrinkage and the part deflection have not been improved as much as expected but still the values are in the range of acceptance because the automobile headlamp shell does not require high dimension accuracy and high aesthetic due to its use condition. our main goal was then to reduce the molding cycle time and if possible, decrease the volumetric shrinkage and improve the dimension accuracy of the model, but nevertheless, the simulation shows that with the Spiral and Zigzag conformal cooling channels the production rate has been improved by 18.12%. The SZCC design, it's then the appropriate solution for this study. However, manufacturing the conformal cooling channels could be expensive and challenging since it requires some special tooling and technic of non-conventional machining. But nevertheless, conformal cooling designs can help make great profit and time saving during mass production cycle.

Acknowledgments

This project is a part of design and manufacturing of injection mold sponsored by VISION (Yangzhou) AUTO PARTS IND CO., LTD. The authors of this work are thankful to Mr. Li Wei and Mr. Li Kun Peng for their technical support and collaboration.

References

[1] Rosato D V and Rosato M 2000 *Injection Molding Handbook*. SPB Publishing the Springer p 246

[2] Kazmer, David O 2016 *Injection mold design engineering*. Carl Hanser Verlag GmbH Co KG

[3] Isayev, Avraam. 1991 *Modeling of polymer processing*: *recent developments* 312

[4] Park, Hong-Seok, and Xuan-Phuong Dang 2017 Development of a smart plastic injection mold with conformal cooling channels *J. Procedia Manufacturing* chapter 10 48-59

[5] Shoemaker J 2006 *Moldflow Design Guide*: A Resource for plastics Engineers, Hanser Publishers vol 10

[6] Wang Y, Yu K M, Wang C C L 2015 Spiral and conformal cooling in plastic injection molding *J* Computer-Aided Design chapter 63 1-11

[7] Dimla E 2015 Design Considerations of Conformal Cooling Channels in Injection Molding Tools Design *J. Thermal Engineering* chapter 1 627-635

[8] Park H S, Pham N H 2009 Design of conformal cooling channels for an automotive part *J Automotive Technology* chapter 10 87-93

[9] Au K M and Yu K M 2007 A scaffolding architecture for conformal cooling design in rapid plastic injection molding *J. Advanced Manufacturing Technology* chap 34 496-515

[10] Wang Y, Yu K M, Wang C C L, Zhang Y 2011 Automatic design of conformal cooling circuits for rapid tooling *J. Computer Aided Design* chapter 43 1001-1010

[11] Khan, Muhammad, Kamran Afaq S, Nizar Ullah Khan, and Saboor Ahmad 2014 Cycle time reduction in injection molding process by selection of robust cooling channel design *J.* ISRN Mechanical Engineering vol 2014 1-8

[12] Oliveira, André Antunes 2016 Structural Analysis of Components Obtained by the Injection Molding Process

[13] Wang Y, Yan Z and Shan X 2018 Optimization of Process Parameters for Vertical-Faced Polypropylene Bottle Injection Molding *J.* Advances in Materials Science and Engineering vol 2018 1-9

IOP Conf. Series: Materials Science and Engineering **692** (2019) 012003 doi:10.1088/1757-899X/692/1/012003

Service Time Performance Analysis of Improved Automated Restaurant by Layout Reconfiguration and Conveyor System

W J Chin[1], M Lim[1], J C E Yong[1,*], A A M Al-Talib[1] and K H Chaw[2]

[1] UCSI University, Jalan Menara Gading, UCSI Heights (Taman Connaught), Cheras 56000 Kuala Lumpur, Malaysia
[2] MODU System Malaysia, Jalan Astana 1, Bandar Bukit Raja, 41050 Klang, Selangor

[*]jonathanyong@ucsiuniveristy.edu.my

Abstract. Achieving effective service time in restaurants hinges on implementing and integrating innovation in design and technology. A combination of restaurant layout configuration and conveyor belt system integration is proposed to enhance the turnaround service time performance. Emphasis on the layout reconfiguration with optimized dining capacity and conveyor system for semi-automated function are applied to existing restaurants by heuristic method. Then operation is simulated by agent-based approach in Anylogic simulator. Results yielded an improved in the total turnaround service time by 2.75% to 5%, which correspond to an approximate 78% reduction on a food-to-table service time. The enhanced performance shows the effectiveness of reconfigured layout to service more customers without compromising delivery time. While conveyor belt mechanism provides the greatest food delivery time and manpower reduction, with a considerable payback period of 6 months. The findings and modelling will aid the implementation of semi-automated restaurant as a stepping stone to one fully automated.

1. Background

Restaurants have been an integral part of society for many centuries, with their history dating all the way back to the eighteenth century in Paris when they were in the form of taverns, inns, traiteurs, cook shops, and boarding houses [1]. Fast forward to present day society, restaurants have become a saturated profit-based & market-driven by the forces of urbanization, globalisation and digitalization. The restaurant industry of today, like any other service industry, relies heavily on customer satisfaction to determine its success and any feedback is immediately amplified and spread to thousands if not tens of thousands in seconds, highlighting the need and importance of keeping customers happy [2,3].

Based on the research done by previous researchers, one of the major factors affecting customer satisfaction in restaurants is service time. [4,5,6]. Therefore, it can be inferred that even after flourishing for centuries, the service time component still remains to be an issue for the restaurant industry of today. The service time component in a restaurant can be broken down into various individual components; these include the time to be seated, time for order to be taken and also time for food and drinks to be delivered. In order to improve the service time in restaurants, layout re-configuration of the restaurant as well as the implementation of semi-automation features such as conveyor systems may be applied to help increase the efficiency of service.

Several previous researches that featured different restaurant layout reconfiguration models were studied. Among the notable models is the theoretical Takashi model or the queuing theory which applied two-dimensional cellular automata and multi-agent simulation to aim for reducing number of waiting customers and average number of seat sharing ratio among customers. The 4-seated tables in the restaurant were reduced and 1-seated tables were increased to improve seating efficiency since the probability of customers coming alone is higher than that of them coming in a group. The research succeeded in reducing the average number of waiting customers by 72.91% [7]. Next, the empirical Nobutada model which aimed to decrease the required moving distance of kitchen staff members by adjusting number of cooking machines in the restaurant's kitchen and arranging closer to each other and succeeded with a reduction of 9.30% [8]. Besides, the empirical Takeshi model which aimed to shorten cooking time using the similar approach, which was also used in the Nobutada model, accomplished a reduction of 8.80% [9].

On the other hand, with the rise of the 4th industrial revolution, automation has been slowly but surely creeping into many industries. Some of the earlier incorporation of automation into the restaurant industry started off by implementing automated food ordering systems, using software systems that allowed customers to browse the menu and order via their own mobile phones [10]. Subsequently, these systems were upgraded by integrating built in touch panels at every table and by using more advanced technology such as wireless mesh networking technology in order to smoothen communication between stations [11]. In accordance with the continued advancement of technology, automation in restaurants gradually broke free from the ordering process limitations and moved into the actual serving and billing process by combining robots and high-level programming to successfully step into the era of full automation [12].

2. Methodology

There will be two selected cases of different restaurants with their layouts featured for improvements in this research. The methodology for the two cases will be focused on layout reconfiguration and implementation of automation technologies respectively. The research of case 1 is divided into two aspects: the restaurant layout reconfiguration based off of researched algorithms and then the simulation in AnyLogic. As for case 2, the two aspects are implementation of conveyor system based off of actual data collected through survey and then simulation using AnyLogic as well.

2.1. Conceptual framework

Ideal Table Mix is a technique used to calculate the optimal mix of tables with different seating sizes in an assigned dining area. From the original research that has proposed this technique, several layout rearrangement methods were compared and the Ideal Table Mix was proven to be superior to the other proposed methods in terms of customers lost daily and generated restaurant revenue. The technique utilises restaurant parameters like seating areas in a dining area, probability weightage of different number of customer parties and tables' seating sizes to calculate the optimum amount of tables in an assigned area [13].

For the introduction of automation, a simple conveyor system that connects the kitchen to all tables is proposed. This improvement was proposed because one of the leading causes of delay in service time is attributed to server movement [8]. By substituting the manual service for kitchen to table orders with an automated food delivery conveyor system, the service time can be greatly reduced.

2.2. Simulation in AnyLogic

Both the original and reconfigured layout floor plan of MODU automated restaurant were imported into AnyLogic and necessary blocks and components were drawn on the floor plans and model spaces of the software to simulate restaurant operations on them. The customers, waiters and food were assigned as three different agents and the Markov chain of blocks that simulate the actions of them in the simulation model will be elaborated.

In both of the layouts, the customers will enter the restaurant, queue up at the order counter to check the menu or order for food, then they will move to the seats in the different dining areas, one where the waiters will serve the food and one where food are delivered through conveyor belts. Then, the customers will wait for waiters to approach them for ordering or fulfilling extra needs. After consuming their food in the restaurant for a specific time, they will leave the restaurant. The chain of blocks responsible for simulating the customers' actions is shown in figure 1.

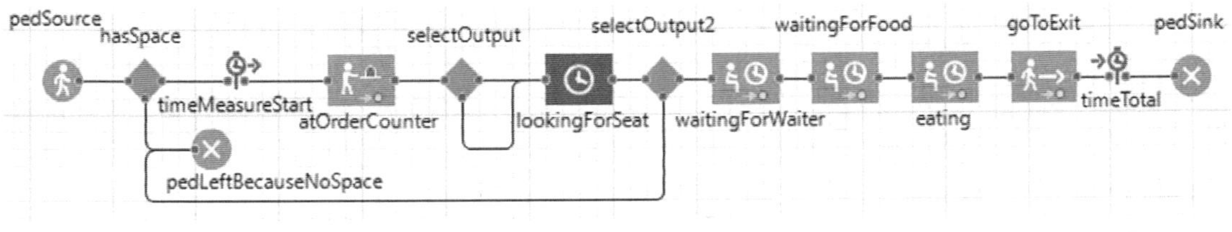

Figure 1. The Markov chain of blocks for different agents' actions.

The waiters however, will wait in the kitchen area until a customer sits down and need them to take their orders or have their food available to be delivered by the waiters. The waiters will return to the kitchen if there are no available works in the dining area. The chain of blocks that simulates the actions of the waiters is also shown in figure 1.

Lastly, the chain of blocks that simulates the food's actions is responsible for distributing the food into two routes, one which makes the food to be sent to customers through the conveyor and one which makes the waiters deliver them. This particular chain of blocks is shown in figure 1 as well.

Both of the simulations of the original and reconfigured layout of MODU restaurant have the similar blocks as discussed with some tweaks in their properties to ensure a well-controlled comparison.

3. Results

The operation of two different cases of selected restaurant layouts were simulated in AnyLogic and the results will be discussed below.

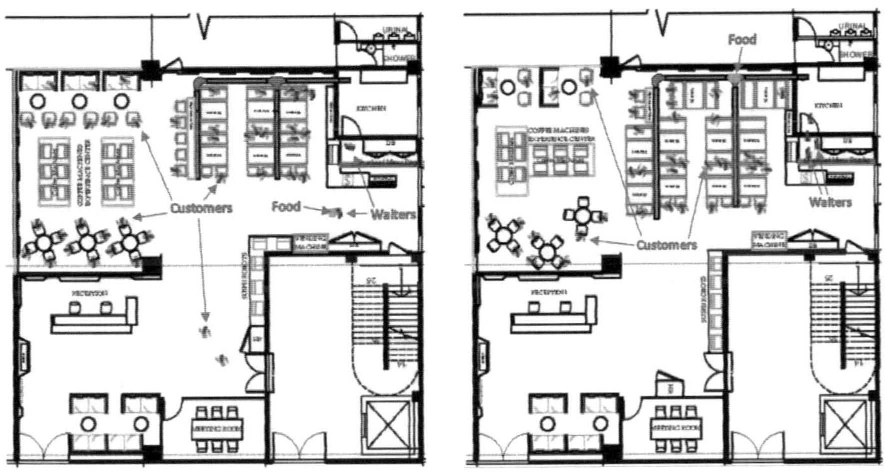

(a) Original layout (b) Reconfigured layout

Figure 2. Situations in original and reconfigured MODU restaurant layout during lunch hours.

3.1. CASE 1: Conveyor system with reconfiguration of layout

The simulation of the MODU restaurant's operation was run for 10 hours with a customer arrival rate ranging from 45 per hour for the peak lunch hours, 10 per hour for the less busy hours, 45 per hour for the peak dinner hours and 35 per hour for the less-busy dinner hours. After the simulation was run, the movement and egress of the agents on the layout was observed. Figure 2 shows the situation in the restaurant with both the original and reconfigured layouts during peak lunch hour from 12pm to 2pm.

Several time-measuring blocks were also placed in specific position in the Markov chain to measure different times in the model during simulation and to generate time-distribution graphs for easier results comparison of the original and reconfigured layouts. The resulting time distribution graphs that show the total time the customers spent in the restaurant for both layouts are shown in figure 3.

Ratio of customer frequency *Ratio of customer frequency*

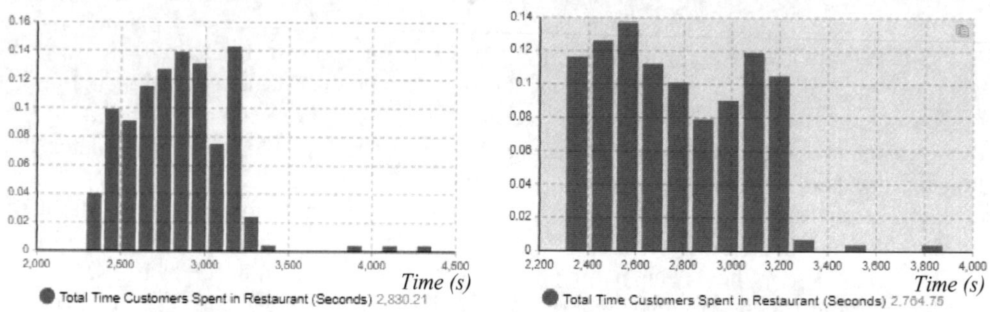

(a) Original layout (b) Reconfigured layout

Figure 3. Comparison of histogram of total time customers spent in restaurant of both layouts.

Lastly, the simulated resulting times waited by the customers before their food arrive after their orders were taken for both the conveyor-delivering area and waiter-serving area for both layouts are shown in table 1.

Table 1. Comparison of times waiting for food in different areas of both layouts.

Time distribution criteria	Original layout's time	Reconfigured layout's time
Time waiting for food in conveyor-delivering area	5 m 9 s	5 m 12 s
Time waiting for food in waiter-serving area	5 m 44 s	5 m 40 s

3.2. CASE 2: Conveyor system implementation without layout reconfiguration

The simulation for Sushi Jiro, Midvalley Megamall was carried out twice, the first time to simulated normal operating conditions while the second simulation incorporated a food delivery conveyor system that connected the kitchen to all tables in the restaurant. The conveyor was designed above the existing sushi belt and had a width of 30cm while operating at a speed of 0.8m/s. The restaurant layout with the conveyor system can be seen below in figure 4.

Similar to the layout reconfiguration simulation, time measuring blocks were used to measure the time starting from the moment the food is prepared and released from the kitchen to the moment it reaches the customers table. The resulting comparison between frequency-time graphs for original operating conditions and with food delivery conveyor system can be seen in figure 5 below.

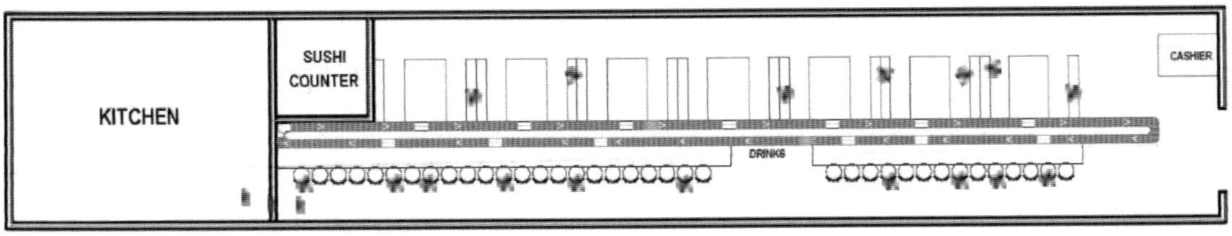

Figure 4. Simulated conditions for layout with conveyor system (bottom) of Sushi Jiro, Malaysia

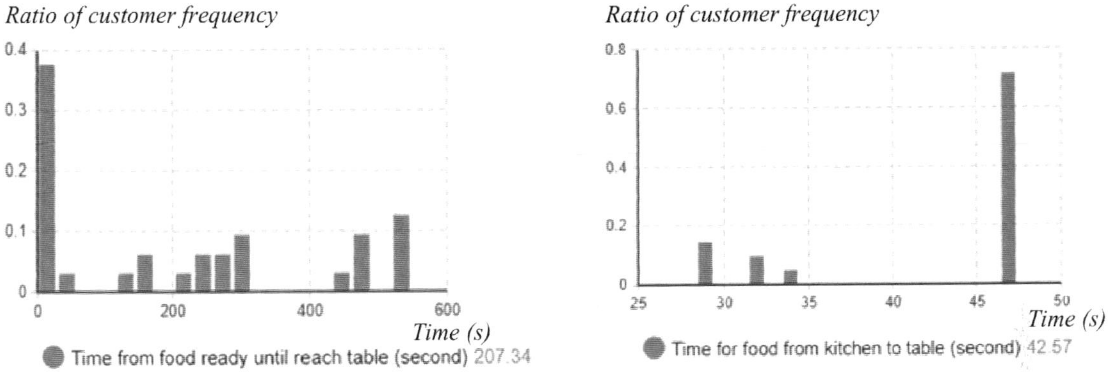

(a) Original operating conditions (b) Food delivery with conveyor system

Figure 5. Comparison of service time between original operating conditions and with addition of food delivery conveyor system.

The total dining time of customers in the restaurant was also measured in the simulation using time measuring blocks and the results were similarly compared between the original model with the model that had the additional food delivery conveyor system. The comparison between the two frequency-time graphs can be seen in figure 6 below.

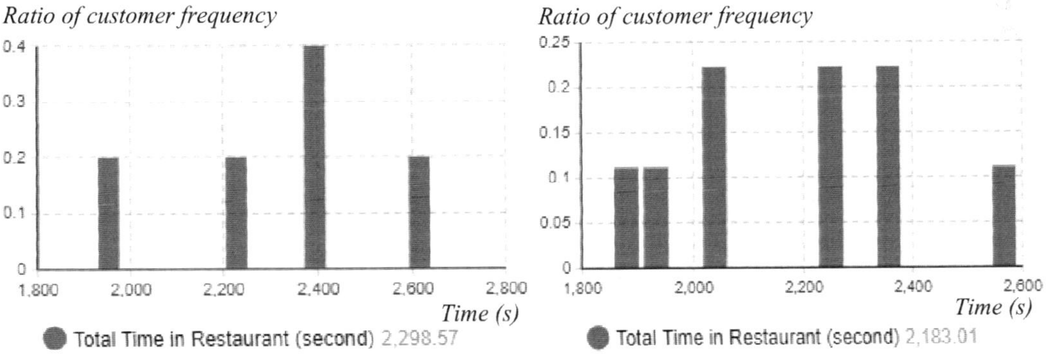

(a) Original operating conditions (b) Food delivery with conveyor system

Figure 6. Comparison of total dining time between original operating conditions and with addition of food delivery conveyor system.

From the graphs shown in figure 5 and figure 6, it can clearly be seen that the addition of the food delivery conveyor system greatly reduces the service time for kitchen to table process. The service time is reduced from an average of 207.3 seconds to 42.57 seconds and shows a reduction of almost

78%. The average total dining time of customers has also been reduced from 38.3 minutes to 36.3 minutes which translates into an overall average reduction of 5%.

Therefore, based on the results obtained from this simulation, it can be observed that the implementation of the food delivery conveyor system does indeed manage to achieve the objective of this research. Not only does this modification reduce the service time, it manages to do so by a great degree, largely improving the efficiency of service in a restaurant and showcasing the potential of automation.

4. Discussion

4.1 Improvement on Service Time

From case 1 which is was based on layout reconfiguration, improvements can be seen from both the average total dwell time of customers in the restaurant and the time customers spent waiting for food in the waiter-serving area. The amounts of reductions achieved are 2.75% and 1.17% respectively.

For case 2 which focused on implementation of the conveyor system, it can be seen that the improvement in service time obtained is significant as it is a 78% reduction when compared to the original. However, this degree of improvement is directly related to the limitations in the original layout of the site which contributes greatly to the delay in service time. Therefore, the same improvement when implemented in different settings might not bring about as great a degree of improvement as can be seen from the results of this study.

4.2 Justification of the distribution

For case 1, from the graph shown in figure 3 which was generated from the simulation of the reconfigured layout, what could be observed is its uneven distribution. This can be justified by the varying arrival rate of customers throughout the 10-hours simulation which was set according to a schedule. With the number of waiters remaining same throughout the simulation, the serving time fluctuates to the varying arrival rate.

For case 2, based on the graphs generated through simulation, it can be seen that for the original operating conditions, the distribution of service time for kitchen to table process is concentrated to be within 100 seconds especially at the start of the simulation. However, as the simulation continues, the tasks the servers have to perform start to pile up and the delay gradually starts to lengthen, ultimately increasing to above 8 minutes. After implementation of the conveyor system, as the conveyor is able to deliver multiple orders at the same time, the service time is limited to be within 1 minute which represents the furthest distance the conveyor has to deliver. For the total dining time of the customers, the distribution is normal for both with and without implementation of conveyor system.

4.3 Payback period

Cost analysis was only carried out during case 2 and therefore only results of case 2 will be shown. By estimating the cost of the conveyor system to be at RM1000 per metre and the wage of servers to be at RM 10 per hour working 12 hour shifts daily, the cost of implementing the conveyor system which spans 45 metres can be recouped in half a year by reducing the number of servers from 3 to 1.

5. Conclusion

From case 1 which focused on layout reconfiguration, the average total time the customers had to spent in the restaurant was reduced by 2.75% and the time customers had to wait before their food arrives after their orders were taken in the waiter-serving area was reduced by 1.17%. From case 2 which focused on implementation of automation, the average total dining time had been reduced by 5% while the service time for kitchen order to be served was reduced by a large margin of 78%. In conclusion, it can be seen from the results of the simulations that while both re-configuration of restaurant layouts and incorporation of semi-automation technology in the form of conveyor systems did both achieve reduction of service time in the restaurants, it is apparent that the usage of semi-automation technology is more efficient as it brings about greater impact while taking less effort to implement.

Acknowledgments
The authors would like to express special thanks to MODU Sdn Bhd for providing their automated restaurant showcase unit for this research.

References
[1] Kiefer, N. (2002). Economics and the origin of the restaurant. *The Cornell Hotel and Restaurant Administration Quarterly, 43(4), 58–64.* doi:10.1016/s0010-8804(02)80042-8
[2] Kivela, J. and Kagi, J., 2009. Applying six sigma in food service organizations. *Turizam: međunarodni znanstveno-stručni časopis*, *56*(4), pp.319-337.
[3] Algassem, F.S. 2016 Integration of Lean Six Sigma with Multi Agent Systems in the Food Distribution Industry in Small to Medium Enterprises. *Thesis: College of Engineering, Design and Physical Sciences, Brunel University, London.*
[4] Budhathoki N 2018 Customer Satisfaction in Restaurant Services. *Bachelor's Thesis, International Sales and Marketing, Haaga-Helia, University of Applied Sciences.*
[5] Phan, T.A. and Nguyen, T.H.T. (2016). An Analysis of Factors Impact on Customer Satisfaction in Vietnam Restaurants; Case of Fast Food Restaurants. *International Journal of Business and Management Review,* 4(6), 1-17.
[6] Chahal, D.K. and Naail, M.K. (2017). The Determinants of Customer Satisfaction and Behavioral Intention among Restaurants in Klang Valley Area of Malaysia: A Conceptual Study. *Journal of Social Sciences and Humanities,* 12(2), 307-317.
[7] Tanizaki, T., & Shimmura, T. (2017). Modeling and Analysis Method of Restaurant Service Process. *Procedia CIRP, 62, 84–89.* doi:10.1016/j.procir.2016.06.076
[8] Shimmura, T., Fujii, N., & Kaihara, T. (2017). Staff Motion Reduction at a Japanese Restaurant by Kitchen Layout Redesign after Kitchen Simulation. Procedia *CIRP, 62, 106–110.* doi:10.1016/j.procir.2016.07.007
[9] Shimmura, T., Takahashi, T., Oura, S., Asakawa, T., Kaihara, T., Fujii, N., & Nonaka, T. (2016). Enhancing Kitchen Layout and Training to Improve Management and Employee Satisfaction at a Multiproduct Japanese Cuisine Restaurant. *Serviceology for Smart Service System*, 139–145. doi:10.1007/978-4-431-56074-6_16
[10] Tanpure, S.S., Shidankar, P.R. and Joshi, M.M., 2013. Automated food ordering system with real-time customer feedback. *International Journal of Advanced Research in Computer Science and Software Engineering*, *3*(2).
[11] Nawaz, A. and Jalil, F. (2015). Touch Panel Based Modern Restaurants Automation using Zigbee Technology. *Conference: 4th International Conference on Aerospace Science & Engineering (ICASE), 2-4 Sept 2015, Islamabad, Pakistan. doi:10.21090/ijaerd.020466*
[12] Shiny, J.S., Ashok Kumar, M., Nanthagopal, V. and Raguram, R. (2017). Automation of Restaurant (Ordering, Serving, Billing). *International Journal of Advanced Research in Electrical, Electronics and Instrumentation Engineering,* 6(3), 1532-1539.
[13] Thompson, G. M. (2011). Inaccuracy of the "Naïve Table Mix" Calculations. *Cornell Hospitality Quarterly, 52(3), 241–252.* doi:10.1177/1938965511398261

Design of Storage and Monitoring System for Medical Dangerous Chemicals Based on Wireless Sensor Network

Tao Zhang[1], Ren-Ren Xu[1], Ya-Wei Zhang[1] and De-Qiang Meng[1]

[1] Tianjin University of Technology and Education, Section of State-owned Assets Management, Tianjin 300222, China

Corresponding author. Tao Zhang's E-mail addresses: 632779578@qq.com

Abstract. Medical hazardous chemicals storage system is a special storage system. Its management basically relies on getting rid of a lot of manpower and increasing the safety and reliability of management. Using information management is a necessary way to improve storage efficiency, reduce storage cost and improve storage security. This design adopts wireless sensor network technology according to the characteristics of storage system of medical dangerous chemicals (Wireless Sensor Networks, WSNs). The wireless sensor network node design scheme is proposed. Based on IEEE802.15.4/ZigBee protocol, the wireless sensor network node is designed and implemented by PIC18F4620 single chip microcomputer and wireless radio frequency chip CC2420. Realize low cost to improve the security and real-time of monitoring, realize information sharing and network remote monitoring.

1. Introduction

Domestic major hospitals and medical institutions have emergency rescue plans for medical hazardous chemicals accidents, and regularly organize drills to strengthen the effective control of hazardous chemical accidents and prevent accidents, so medical hazardous chemicals storage management has been paid attention to by medical institutions at all levels[1,2]. At present, most of the storage problems of dangerous chemicals in medical institutions are simply relying on people, which not only has a high management cost, but also has a great relationship with the quality and experience of managers, which makes the management process have some potential safety hazards. In order to solve the security problem of medical dangerous chemicals storage, the use of modern information technology to achieve storage management, wireless sensor network with its miniaturization, low cost, flexibility provides a reliable management scheme for tracking, monitoring and management of medical dangerous chemicals storage[3].

This test is based on the design of wireless sensor network node system based on IEEE 802.15.4 and ZigBee protocol. The experiment is mainly based on ZigBee/IEEE802.15.4 wireless communication protocol[4,5]. The radio frequency chip CC2420 and the micro-control chip PIC18F4620 are selected to design the radio frequency transceiver controlled by the micro-controller to realize the networking communication between nodes. This test only monitors the temperature information of the monitoring environment. MCU processes the information, WSN carries out data communication and uploads it to PC.

In this design, the sink node receives the temperature data from the terminal node after completing the generation and management of the network, and transfers the data to the PC through the serial port.

IOP Conf. Series: Materials Science and Engineering **692** (2019) 012004 doi:10.1088/1757-899X/692/1/012004

2. Hardware and software part design of wireless sensor networks

Wireless sensor network node consists of data processing unit, radio frequency transceiver unit, sensor unit and power supply.

In this design, the radio frequency transceiver CC2420, which meets the IEEE802.15.4 standard of 2.4 GHz, is selected from TI company for comprehensive functions and costs. DS18B20 temperature sensor is selected to collect data from analog storage environment. PIC18F4620 and CC2420 modules are selected to process and send data[6].

The design of transmitting and receiving module includes power module design, interface module design and reset circuit design. The difference between them is that ordinary nodes need to implement data collection functions, so they should include the design of the data acquisition module; the aggregation node needs to implement control and display data on the computer[7]. The design block diagram is shown in Figure 1 and Figure 2.

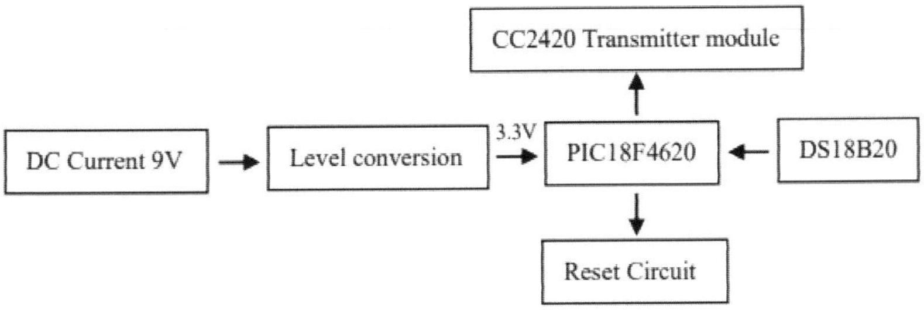

Figure 1. Overall block diagram of the transmitting module.

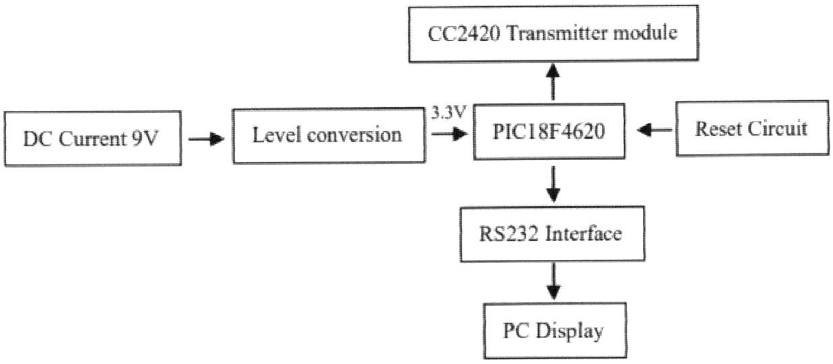

Figure 2. Overall block diagram of receiving module.

The software design of the wireless sensor network node selects the MAPLAB IDE integrated development environment, and selects the C18 compiler for simulation debugging. Temperature monitoring by temperature sensor-DS18B20, because DS18B20 adopts "first-line bus" protocol mode, PIC18F4620 single-chip hardware does not support single bus protocol, and software method is used to simulate single-bus protocol timing to complete access to DS18B20 chip [8].

In this design, the aggregation node creates a network, allows ordinary nodes to join, and forms a network. The data of the common node is obtained by pressing the aggregation node button [9]. The flow chart of the aggregation node is shown in Figure 3.

IOP Conf. Series: Materials Science and Engineering **692** (2019) 012004 doi:10.1088/1757-899X/692/1/012004

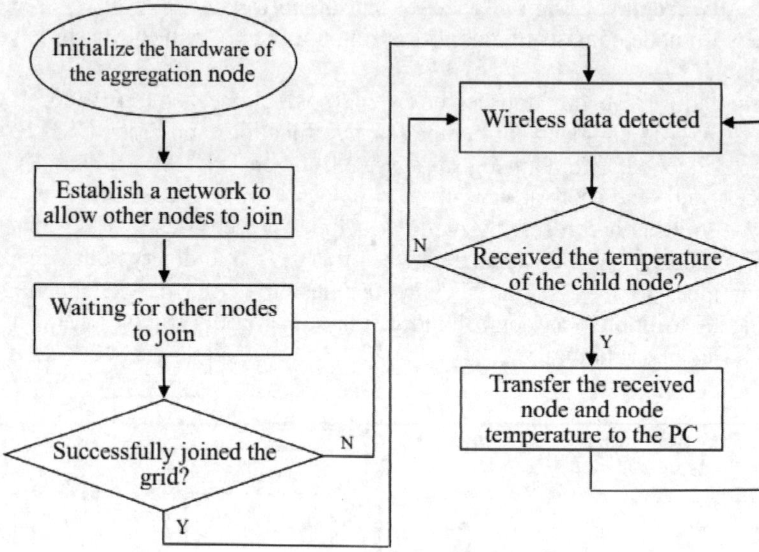

Figure 3. Wireless temperature measurement system aggregation node program flow chart.

The aggregation node will first detect whether it has formed a network. If it does not form a network, it will automatically establish a new network and allow other devices to join the network [10]. The ordinary node transmits the collected data to the aggregation node, and simultaneously receives the data and performs corresponding operations. After the network node is powered on, scan all available channels to find the neighbouring aggregation node and apply to join the network [11]. In order to reduce power consumption, the design uses a timed wake-up method to connect to the aggregation node to receive or transmit data, and the idle time is in the sleep mode [12]. In this test, the common node collects the temperature value, and then combines the temperature sensor data into a data packet and sends it to the receiving end.

3. System Testing

Run the aggregation node to run, connect the node to the serial port on the PC through the RS-232 connector, and use HyperTerminal to communicate with the aggregation node. Powering up the aggregation node, in the HyperTerminal window, displaying the message: Microchip ZigBee(TM) Stack – v1.0-3.5, the ZigBee Coordinator aggregation node will automatically try to find available wireless channels to form a network. If successful, the following message will be displayed:

Trying to start network..., PAN #### started successfully. #### is the network PAN ID of the group. When other nodes access the network, a message is displayed: Joining permitted. Power on the data collection node and display the message: Microchip ZigBee(TM) Stack – v1.0—3.5, ZigBee RFD, RFD attempts to access the network. Access network display message:

Trying to join network as a new device..., Network(s) found. Trying to join ####.

Join successful!

The aggregation node displays the following message to confirm the new node access: Node #### just joined, ####Assign a short address to the new node. The aggregation node sends a data request to the ordinary node, requests the data of the ordinary node, and sends the success display information: Node #### sending message. Message sent successfully. When receiving the data of the ordinary node, the following message will be realized: Received DS18B20 = #### °C. The result graph of the wireless sensor network node is shown in Figure 4 and Figure 5.

ZigBee Coordinator

Transceiver-CC2420

Hardware initialized
ZigBee initialized

Trying to start network...
PAN 0179 started successfully
Turning on joining...
Joining status changed.
Node 1001 just joined.
Requesting temperature...

Received
DS18B20=29.6250 C

Figure 4. Result Diagram of Convergence Node of Wireless Temperature Measurement System.

Microchip ZigBee(TM) Stack – v1.0

ZigBee RFD

Transceiver-CC2420

Trying to join network as a new device...
EA Error finding network. Trying again...
Trying to join network as a new device...
Network(s) found. Trying to join 2001.
Join successful!
Join successful!
Join successful!
Requesting data...
DS18B20=29.7500 C
DS18B20=29.8125 C
DS18B20=29.8750 C

Figure 5. Result Diagram of Common Nodes of Wireless Temperature Measurement System.

The research results show that the wireless sensor network based on ZigBee protocol integrates the characteristics of ZigBee technology, such as flexible networking, wide coverage and good security. The storage and monitoring system of dangerous chemicals based on wireless sensor network has strong applicability. By adding different types of sensors in the system, it can achieve medical application. The storage of dangerous chemicals requires various functions of data detection, so as to meet the requirements of saving manpower, reducing costs and improving the safety of management.

Acknowledgement

Fund projects: school-level scientific research projects (Item Number KJ1901)

References

[1] Xu G C and Li D G 2018 Talking about the management of dangerous chemicals *Modern Property (Mid-term Journal)* pp 26-7 [CrossRef]

[2] Qin S and Li J 2019 Application of wireless sensor network in environmental monitoring *Environment and development* **31** pp 173+5 [CrossRef]

[3] Xu L M and Xie B 2017 Standardized construction of medical compressed air system in new hospitals *Medical Gas Engineering* **2** pp 22-5 [CrossRef]

[4] He P 2017 IEEE 802.15.4 wireless sensor network protocol interference testing *Computer and Digital Engineering* **45** pp 2251-5 [CrossRef]

[5] Chen Z C 2013 Research on wireless sensor networks based on IEEE 802.15.4 *South China University of Technology* [CrossRef]

[6] Li X H, Qin X, He C J and Hu P P 2015 Design of storage monitoring system based on ZigBee wireless sensor network *Shandong Industrial Technology* pp 139-40 [CrossRef]

[7] Zhang Y L 2008 Design of wireless sensor network node based on radio frequency chip CC2420 *Metallurgical Branch of China Metrology Association, Metallurgical Automation magazine* Papers collection of 2008 meeting of metallurgical branch of China metrology association *Metallurgical Branch of China Metrology Association, Magazine of Metallurgical Automation: Magazine of Metallurgical Automation* p 3 [CrossRef]

[8] Li X X 2016 Analysis on the safety management of medical gas *Western Traditional Chinese Medicine* **29** pp 23-5 [CrossRef]

[9] Wang L D, Li S and Zhu G 2018 The wireless calibration system of environment temperature and humidity based on ZigBee technology *Measurement and Testing Technology* pp 17-9 [CrossRef]
[10] Ma K Z 2019 Research of key issues in wireless sensor network applications *Electronic Testing* pp 76-7 [CrossRef]
[11] Liu J 2010 Research on key technologies of storage monitoring and management system based wireless sensor network *China's Circulation Economy* **24** pp 17-9 [CrossRef]
[12] Li Z H 2018 Design and implementation of wireless communication network node based on ZigBee *China New Communications* **20** p 6 [CrossRef]

Planning Path Strategy and Simulation of Infusion Service Robot Based on Ant Colony Algorithm

Tao Zhang[1], De-Qiang Meng[1] and Su-Dong Zhao[1]

[1] Tianjin University of Technology and Education, Section of State-owned Assets Management, Tianjin 300222, China

Corresponding author. Tao Zhang's E-mail addresses: 632779578@qq.com

Abstract. The infusion service robot was taken as the research object, and the algorithm of path planning for infusion service robot was simulated by the sensor network system in this experiment. The ant colony algorithm was selected based on the sensor network node optimization deployment of the mobile service robot. Based on sensor network for hospital environmental information collection, combined with sensor network node information and RSSI algorithm based on MATLAB to achieve positioning, and then complete the simulation study of infusion service robot navigation. The ant colony algorithm was verified as the basic method and feasibility of applying the path strategy of the robot platform. During the testing process, some conditions and algorithms were simplified to avoid the complexity of the problem.

1. Introduction

The goal of "focusing on the development of the top ten iconic products of surgical robots and intelligent nursing robots in the fields of helping the elderly, helping the disabled, family services, and medical rehabilitation" was proposed by the *Robot Industry Development Plan (2016-2020)* ,which made the infusion service robots are valued and quickly promoted[1,2]. At present, the path planning for developing mainstream mobile service robots on the market to achieve positioning and navigation through GPS, ultrasonic or laser positioning. The advantage of GPS is that the robot hardware is simple, but it is limited by the large distance error of the civilian GPS positioning, the GPS technology has not been corrected, resulting in low accuracy, which can't meet the normal work of infusion service robots in hospitals and other places. Therefore, it is necessary to pay a high cost to realize the high-precision positioning and navigation function of the GPS technology[3,4]. Ultrasonic waves or lasers would cause certain interference to the medical electronic equipment detection data of the hospital during the work process.

Aiming at the shortcomings of traditional robot path planning and navigation technology in hospitals or medical institutions, the proposed scheme based on sensor network collects information on hospital environment, combined with information of sensor network nodes and RSSI algorithm to realize the positioning of infusion service robots. The infusion service robot can ensure high precision and reduce the production cost in the path planning strategy through the ant colony algorithm.

In this paper, the simulation research on the path planning strategy of the infusion service robot was carried out. The information acquisition of the sensor network system was used, and the ant colony algorithm was used to plan the path. The feasibility of the application was verified by MATLAB simulation.

IOP Conf. Series: Materials Science and Engineering **692** (2019) 012005 doi:10.1088/1757-899X/692/1/012005

2. Test Content and Environment Simulation

During the tests, the infusion service robot's motion environment was a set of mutually separated regular areas with dynamic topological characteristics, simulated in the actual situation of the hospital, such as the obstacles in the designated area (for example, people, beds, Long benches, etc.), as well as hospital inpatient corridors (about 1 meter in width). At the same time, the robot's location of interest to the room would generally have special markings (such as two-dimensional code, etc.) as a road sign during navigation in the above environment[5]. Therefore, some existing road signs were used to realize cooperation with wireless sensor network nodes, and the positioning navigation operation of the infusion service robot was completed together.

The infusion service robot and its surrounding exercise environment of this study have the following characteristics.

i. The wireless sensor network node in the infusion service robot motion environment is pre-deployed and the specific location is clear;
ii. The infusion service robot is also equipped with a wireless sensor network node and moves dynamically with the robot;
iii. Clock synchronization between wireless sensor network nodes;
iv. The infusion service robot does not need a pre-known environmental map, and does not need to be equipped with a positioning device such as GPS, and the surrounding environment of the robot may be dynamically changed.

Sensor network deployment is the precondition of navigation for infusion service robot. The sensor node placement algorithm of the virtual diamond grid (VRGSD, Virtual Rhomb Grid-based Sensor Deployment algorithm) was used as sensor network deployment algorithm[6]. The diamond mesh not only made full use of the sensor's sensing and communication capabilities, but also ensured complete seamless communication and coverage within the sensor area. The minimum number of nodes required in a completely seamless coverage sensor area is calculated by the following formula[7]:

$$N = \frac{F}{\delta} = \frac{F}{\frac{3\sqrt{3}}{2} r^2} \tag{1}$$

F: the area of the sensor area

δ : the effective coverage area of each node

R: the sensing or communication radius of the sensor node.

The data structure of the VRGSD algorithm and the pseudo code of the implementation are simulated by MATLAB, and 20 sensor nodes are randomly generated. The coordinates of the simulation result are: C= [91;13 2;5 3;18 4;14 5;18 6;17 7;10 8;3 9;4 10;6 11;12 12;16 13;15 14;8 15;20 16;7 17;11 18;2 19;120].

Randomly arranged in the sensor area, each node has a radius of emission of 5 units (r=5) and a sensor area of 30×40 units. Figure 1 shows the simulation results.

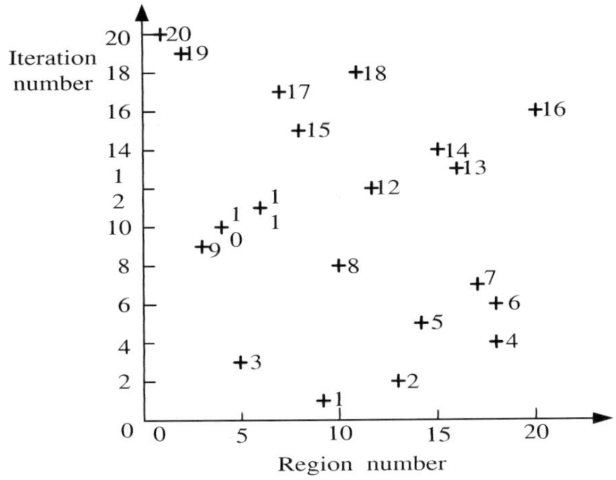

Figure 1. Simulation results of the VRGSD algorithm.

Ant colony algorithm is applied to let robots deploy 20 nodes of sensor networks to find the shortest path. The algorithm can be described as follows: m ants are assumed to be randomly placed on a fully connected graph with n nodes, $\tau_{ij}(t)$ indicates the amount of pheromone on the i, j line at time t, $\eta_{i,j}(t)$ is heuristic information related to the problem, Set here $\eta_{i,j}(t)=1/di$, j. At the initial moment, the amount of pheromone on each path is equal, assume $\tau_{ij}(0)=C$ (C is a constant). The ant k (k=1, 2, ..., m) determines the next transfer to which position according to the amount of pheromone on each path during the movement. The equation of $p_{i,j}^{k}(t)$ is represented the probability that ant k is transferred from position i to position j at time t, as shown by the calculation formula 2[8].

$$p_{i,j}^{k}(t)=\begin{cases}\dfrac{\left[\tau_{i,j}(t)\right]^{\alpha}\left[\eta_{i,j}(t)\right]^{\beta}}{\left[\sum\limits_{s\notin tabu_k}\left[\tau_{i,s}(t)\right]^{\alpha}\left[\eta_{i,j}(t)\right]^{\beta}\right]}, & j\notin tabu_k\\ 0, & j\in tabu_k\end{cases} \tag{2}$$

In Equation 2, the taboo tabuk records the location where the ant has traversed, α , β is a parameter that adjusts the relative importance of pheromone strength τ and heuristic information η. After n moments, the ants complete a tour, and the amount of pheromone on each path is adjusted according to formula 3.

$$\tau_{ij}(t+n)=(1-\rho)\tau_{ij}(t)+\sum_{k=1}^{m}\Delta\tau_{i,j}^{k}(t+n) \tag{3}$$

In the above formula: $\rho\in[0,1]$ represents the volatilization coefficient of the pheromone, $\Delta\tau_{i,j}^{k}$ represents the amount of pheromone left by the ant k on the path (i,j) in this cycle; $\Delta\tau_{i,j}$ represents the amount of pheromone left by all the ants on the path (i,j) in this cycle. The flow chart of the ant colony algorithm is shown in Figure 2.

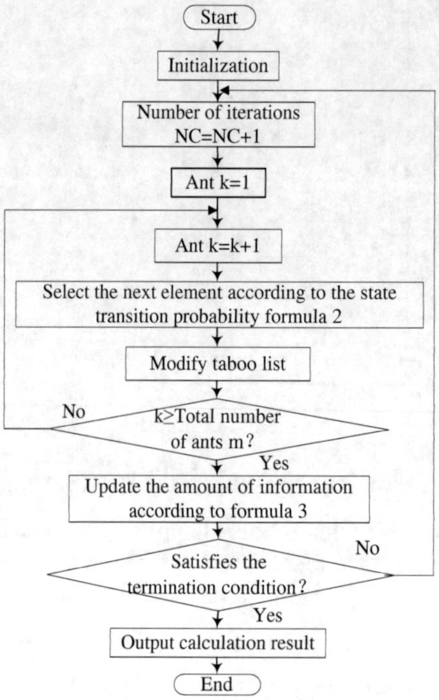

Figure 2. Ant colony algorithm flow chart.

For the 20 nodes of the simulation, the shortest path is sought according to the ant colony algorithm, and the sensor network node is deployed based on the infusion service robot. The run function is as follows[9]:

Function[R_best,L_best,L_ave,Shortest_Route,Shortest_Length]=ACATSP(C,NC_max,m,Alpha,Beta,Rho,Q)

Among them C: Coordinates of n locations, n×2 matrix; NC max: maximum number of iterations; m: the number of ants; Alpha: parameters characterizing the importance of the pheromone; Beta: parameters characterizing the importance of the heuristic factor; Rho: Pheromone evaporation coefficient; Q: pheromone increases the intensity coefficient; R_best: the best route of each generation; L_best: the length of the best route of each generation.

[R_best,L_best,L_ave,Shortest_Route,Shortest_Length]=ACATSP(C, 55, 20, 4, 0.3, 0.4, 0.7). The simulation results are derived: Shortest_Route = 13 14 7 6 4 2 5 8 11 17 19 20 12 15 18 9 10 3 1 16; Shortest_Length =116.0402.

The choice of the robot path should take into account the communication radius of the sensor node and the position of the obstacle to ensure communication with all nodes. The RSSI (received signal strength indicator) ranging positioning technology is applied to the infusion service robot navigation to provide accurate location information for the mobile service robot[10]. The infusion service robot may receive signals from n (n≥3) beacon nodes at the same time, and convert the real-time measurement received signal strength into distance to calculate its own position. The test wireless communication uses the CC2430 chip produced by Chipcon company, and its channel propagation attenuation model as follows[11]:

$$p = p_0 - 10r\lg(\frac{d}{d_0}) + \xi \tag{4}$$

Among them: p_0 is the signal strength from the transmitting node, the value of this p_0 is 41.1 dB; d_0 is the distance between the transmitting node and the receiving node, value d_0 = 1 m; r is a parameter related to the signal propagation environment, and the value is r=3; ξ is a Gaussian distributed random noise with a mean of 0, and the standard deviation ranges from 4 to 10. Test the measured value of the received signal strength of the two nodes and the theoretical value. The measured value and the theoretical deviation are smaller in the range of 0 to 15 m. When the distance between the two nodes is greater than 15 m, the measured value and the theoretical deviation are larger.

A maneuvering statistical model is established for the infusion service robot, and the current statistical model is used to define the state vector as:

$$X_k = [x_{k,} y_{k,} v_{xk,} a_{xk,} a_{yk}]^T \tag{5}$$

The discrete equation of state is:

$$X_{k+1} = \varphi(k+1,k)X_k + U(k) + W(k) \tag{6}$$

Among them,

$$\varphi(k+1,k) = \begin{bmatrix} I_2 & TI_2 & \frac{1}{\alpha^2}(-1+\alpha T+e^{-\alpha T})I_2 \\ O_2 & I_2 & \frac{1}{\alpha}(1-e^{-\alpha T}I_2) \\ O_2 & TI_2 & e^{\alpha T}I_2 \end{bmatrix}$$ is a state transition matrix;

T is the sampling period;

$$U(k) = \begin{bmatrix} \frac{1}{\alpha}(-T+\frac{\alpha}{2}T^2+\frac{1-e^{-\alpha T}}{\alpha})\bar{a}_x \\ (T-\frac{1-e^{-\alpha T}}{\alpha})\bar{a}_x \\ (1-e^{-\alpha T})\bar{a}_x \\ \frac{1}{\alpha}(-T+\frac{\alpha}{2}T^2+\frac{1-e^{-\alpha T}}{\alpha})\bar{a}_y \\ (T-\frac{1-e^{-\alpha T}}{\alpha})\bar{a}_y \\ (1-e^{-\alpha T})\bar{a}_y \end{bmatrix}$$ is the input matrix;

$\bar{a}_x$, $\bar{a}_y$: the average value of maneuvering acceleration

W (k): a discrete time noise sequence, variance matrix is

$$Q_k = w\left[w(k)W^T(k)\right] = \begin{bmatrix} 2\alpha\sigma^2_{ax}q^3 & O_3 \\ O_3 & 2\alpha\sigma^2_{ay}q^3 \end{bmatrix}$$

α : the reciprocal of maneuver time

q^3 : a constant matrix related to α

σ^2_{ax} and σ^2_{ay} are the mobile service robot acceleration variance, maneuver variance, respectively

The adaptive adjustment is:

1. The current acceleration is positive

$$\sigma^2_{ax}(k-1) = \frac{4-\pi}{\pi}\left[a_{x,max} - a_{x,k-1}\right], \quad \sigma^2_{ay}(k-1) = \frac{4-\pi}{\pi}\left[a_{y,max} - a_{y,k-1}\right]$$

2. The current acceleration is negative

$$\sigma^2_{ax}(k-1)=\frac{4-\pi}{\pi}\left[a_{x,-\max}-a_{x,k-1}\right],\quad \sigma^2_{ay}(k-1)=\frac{4-\pi}{\pi}\left[a_{y,-\max}-a_{y,k-1}\right]$$

$a_{x,\max}$, $a_{y,\max}$, $a_{x,-\max}$, $a_{y,-\max}$ represents the maximum and minimum accelerations that the mobile service robot may have in the x and y directions, respectively

$a_{x,k-1}$ and $a_{y,k-1}$ are the filtered output values at time k-1.

The target area is covered with wireless sensor network nodes, but is limited by the node transmit power, which causes the infusion service robot to communicate with only some of the nodes at a certain time. To simplify the problem, suppose that the robot can communicate with the beacon nodes B1 (x_1, y_1), B2 (x_2, y_2) and B3 (x_3, y_3) at the time of k, the system measurement equation is shown below:

$$Z_k = H(X_k) + V(k) \tag{7}$$

Among them, $Z_k = \begin{bmatrix} p_{1,k} & p_{2,k} & p_{3,k} \end{bmatrix}^T$, $H(X_k)=\begin{bmatrix} h_1 \\ h_2 \\ h_3 \end{bmatrix} = \begin{bmatrix} p_0 - 30\lg(\frac{d_{1,k}}{d_0}) \\ p_0 - 30\lg(\frac{d_{2,k}}{d_0}) \\ p_0 - 30\lg(\frac{d_{3,k}}{d_0}) \end{bmatrix}$,

$d_{i,k} = \sqrt{(x_k - x_i)^2 + (y_k - y_i)^2 + (z_k - z_i)^2}\ (i = 1,2,3)$ is to measure noise, and its variance matrix is R[12].

3. Simulation and conclusion of the infusion service robot planning path

MATLAB software is used for simulation, the red triangle in Figure 3 is the simulation infusion service robot. The green path in the figure is the path we originally envisioned, and the blue star is the sensor network node distributed in the environment. During the simulation process, the node would turn red after being used. According to the information of the sensor network node, combined with the ant colony algorithm calculation to achieve the role of positioning, and then complete the path planning test.

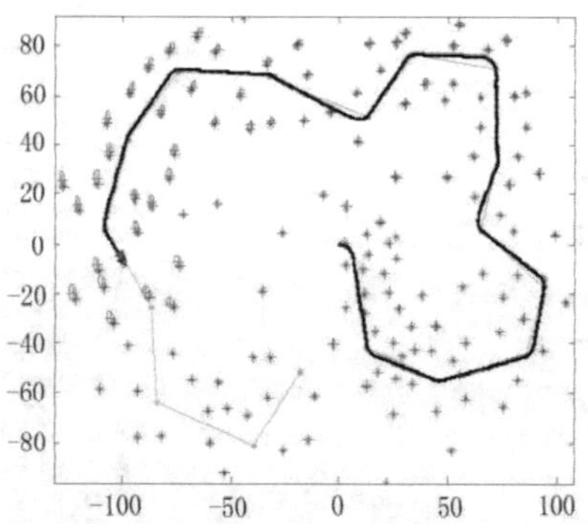

Figure 3. Effect diagram of robot-assisted positioning and navigation simulation process based on sensor network.

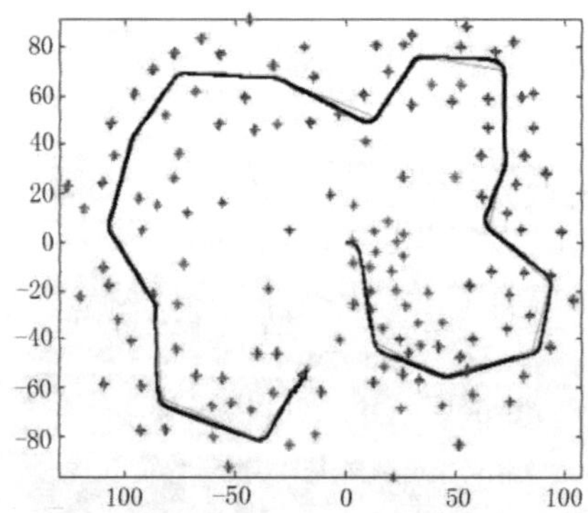

Figure 4. Robot-assisted positioning and navigation simulation results based on sensor network.

Figure 4 shows the black route as the simulation path. The difference between the comparison and the previously set path is very small. From the result, the simulation effect is better, indicating that the robot has completed the positioning and navigation very excellent. The information of the hospital environment is collected through the sensor network, and the information and RSSI algorithm of the sensor network node are used to realize the positioning and navigation of the infusion service robot. The ant colony algorithm is used to determine the path planning route of the infusion service robot. Combined with the complex and diverse environment of major hospitals and medical institutions, the feasibility and basic methods of applying the simulation results to the practical application of infusion service robots need further research.

Acknowledgement
Fund Project: School-level Scientific Research Project (Project No. KJ1901)

References

[1] Wu Z L 2019 Automatic route generation method based on ant colony algorithm *Ship Science and Technology* **41** pp 43-5 [CrossRef]
[2] Cai G B 2013 Discussion on the design and safety of hospital medical equipment use environment *Medical Equipment* **26** pp 48-50 [CrossRef]
[3] Zhang W and Feng L 2018 Simulation research on ground target navigation and positioning of mobile service robot *Computer Simulation* **35** pp 231-234+292 [CrossRef]
[4] Pan F 2018 Research on characteristics of laser interference infrared/television imaging system *Xi'an University of Electronic Science and Technology* [CrossRef]
[5] Li Y J 2019 Research on tourism route planning based on ant colony algorithm *Computer Knowledge and Technology* pp 1-4 [CrossRef]
[6] Li Z and Gu T W 2019 Review on the application of ant colony algorithm in power system *New industrialization* **9** pp108-13 [CrossRef]
[7] Tu L J, Li L S and Lin G X 2019 Research on path planning of orchard mobile robot in dynamic environment *Mechanical Design* **36** pp 75-9 [CrossRef]
[8] Zhang W and Feng L 2018 Simulation research on navigation target positioning of mobile robots *Computer Simulation* **35** pp 231-234+292 [CrossRef]
[9] Ding W, Sun H and Zeng J H 2006 Overview of mobile service robot navigation based on multisensory information fusion *Sensors and Microsystems* pp 1-3 [CrossRef]
[10] Cao R Y, Li S C, Ji Y H, Xu H B, Zhang M and Li M Z 2019 Multi-machine collaborative work task planning based on ant colony algorithm *Journal of Agricultural Machinery* **50** pp 34-9 [CrossRef]
[11] Zhao J W, Wang H Y, Tang B and Yang Z Y 2016 Research on GPS trajectory generation and location based on mobile robot *Industry and Mine Automation* **42** pp 17-9 [CrossRef]
[12] Lu J X, Li J and Zhou Y 2019 Research and application of ant colony algorithm in an IoT medical box system *Electronic Products World* **26** pp 80-84+88 [CrossRef]

Manufacturing of Electrical Dryer Machine for Food and Fruit Products

Adhan Efendi[1,*], Aditya Nugraha[2] and Ridwan Baharta[3]

[1,2] Subang State Polytechnic
[3] Lampung State Polytechnic

[*] adhan.efendi@gmail.com

Abstract. Indonesia is a country with a tropical climate that has two seasons, namely the dry season and the rainy season. In the dry season, sun drying is the most widely used method for agricultural products. In the rainy season, people who open businesses in the food processing sector have difficulty predicting uncertain weather. The community needs solar thermal energy to dry food ingredients in the form of opak, tubers, fruits, etc. In the rainy season, the value of production has decreased. This study aims to create a drying machine that can help people to dry various types of food and fruit products. The research was carried out in several stages, namely field observation, literature study, design, manufacturing, and performance testing. The study states that: (1) the manufacturing process of electrical machine dryer was done in three stages: the initial stage, advanced stage, and the final stage. (2) the trial of an electrical dryer machine was used to dry opak and banana. Data obtained in the form of opak before drying weighs 3.4 ounces and after drying to 2.2 ounces. Whereas the banana before drying weighs 7.3 ounces and after drying to 4.5 ounces; (3) electrical dryer machine is very helpful in the process of drying food processed products in Subang Regency.
Keywords: manufacturing, electrical dryer machine, opak and banana

1. Introduction

The food industry plays an important role in people's lives to increase market demand for food in the business. The market demand for food needs to be accompanied by an increase in the quality and quantity of food so that food needs in the community can be fulfilled. Technological developments make the food processing industry begin to shift from traditional systems to modern systems. A modern system that is more effective and efficient is intended to increase productivity without reducing the quality of product results. Small or micro-scale industries managed by community groups generally still use traditional systems in the production process. This traditional system has resulted in a decline in production effectiveness so that micro industries are slow to develop globally.

Indonesia is a country with a tropical climate that has two seasons, namely the dry season and the rainy season. Sun-drying is the most widely used method for agricultural products in many countries [1]. In the rainy season, the people who open businesses in the food processing sector have difficulty predicting uncertain weather. The community needs solar thermal energy to dry food ingredients in the form of opak, tubers, fruits, and others. In the rainy season, it is not often that the value of production has decreased.

Based on the background above, it is necessary to have an electrical dryer machine that can help the community in solving the problem. The presence of this machine is expected to help the community especially those engaged in food or food processing businesses so that in the rainy season the community can still dry semi-finished or raw products to return to the process and the products processed to be more effective.

2. Study Literature

2.1 Dryer Machine

Drying is among the most ancient and pre-eminent physical methods of food preservation [2]. The drying process is intended to reduce the water content contained in a product to be dried. According to [3], the drying method is selected based on the particular characteristics of the products and socio-economic considerations. Based on the opinion above, it can be concluded that a drying machine is a machine used to reduce the water content in a product to be dried. The dried product is generally in the form of food, fruit, or vegetables. Drying machines have many types depending on their uses, for example, aisle type dryer, cabinet type, electrical, and others. Added by [4], dryers are divided into three types, namely: based on type of food, type of heater, and type of heating technique.

2.2 Food and Vegetables

[3] Fruits and vegetables are important sources of essential dietary nutrients such as vitamins, minerals, and fiber. Since the moisture content of fresh fruits and vegetables is more than 80%, they are classified as highly perishable commodities. Added by [5] that damage to fruits and vegetables is caused by high levels of humidity at harvest. Dried vegetables are generally stable after processing and during storage due to low water activity, which is reduced during the drying process to a level that does not support the growth of microorganisms [6].

Fruits and vegetables are seasonal and due to their low shelf life after harvest they are sold in the markets at very low prices. [7] This is due to a large number of crop yields and unprocessed harvests by farmers. [8] Added that one of the preservation of post-harvest vegetables is to use the drying process. Based on the opinion above, it can be concluded that food, fruit, and vegetables are dried and processed into high-selling products. Clarified by [9], keeping the product fresh is the best way to maintain its nutritional value, but most storage techniques require low temperatures, which are difficult to maintain throughout the distribution chain. Foods such as fruit and vegetables have a high water content of more than 80% which makes it very vulnerable to bacteria that cause decay. Dehydration keeps food in a stable and safe condition by reducing water [10].

2.3 Design and Manufacturing of Electrical Dryer Machine

Autodesk Inventor is a software that is used to design 3D types of Computer Aided Drawing (CAD). In language, manufacturing comes from Latin, namely *manus* means the hand and *factus* mean to make. So, manufacturing is the process of making products using hands. Manufacturing is converting raw materials into finished materials as desired by considering aspects of needs, technology, and economics. In the context of this research, manufacturing is intended as a machine manufacturing process starting from the preparation of tools and materials, measurement, cutting of materials, assembly, and finishing.

3. Method

This research was carried out in the Subang Polytechnic Manufacturing Laboratory. The time needed to carry out this research is 5 months. The stages of this study are field observations, literature studies, design making, machine manufacturing, and performance testing. Research plan is shown in figure 1.

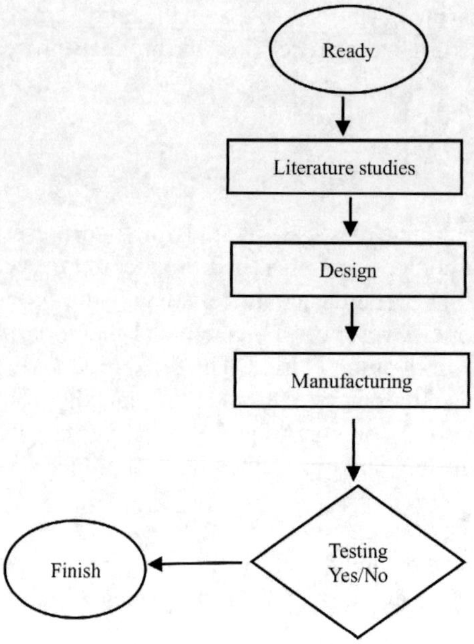

Figure 1. Research Plan

4. Results and Discussion

4.1 Literature Studies

The literature study is carried out by examining various literature books relating to the welding process. Besides, the more important literature study is reviewing international and national scientific journals that have been published. Literature studies are conducted by searching for and analyzing the work of experts.

4.2 Design of Electrical Dryer Machine

After the research team collected supporting data, the research team then designed an electrical dryer machine. Machine design uses a 3D inventor application which is then validated by experts. Validation is intended to see the feasibility of the design. Dryer design is shown in figure 2.

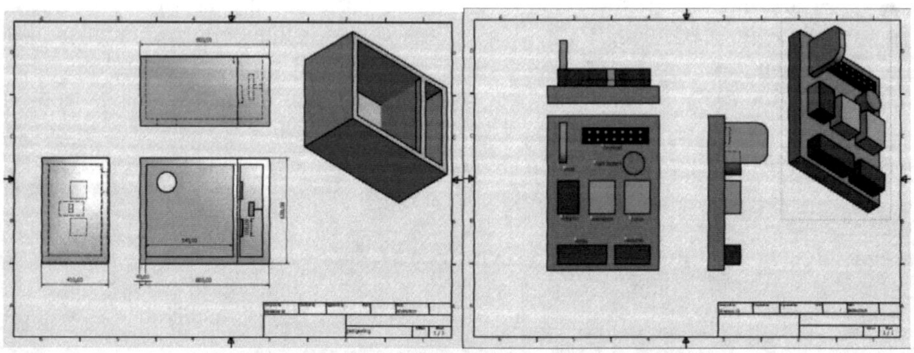

Figure 2. Electrical Dryer Machine Design

4.3 Manufacturing of Electrical Dryer Machine

This activity is a follow-up activity after the drying machine design is completed. Some of the main components used to make tunnel dryer machines are: heating components, fans, hollow iron ST37, iron elbows, plates, ram, arduino, etc. The making of the machine is made through three stages, namely initial stage, advanced stage, and final stage of machine manufacturing. It is supported by [11] that in the manufacturing process the standard factor of personal protective equipment (PPE) must be observed following standards to avoid work accidents.

4.3.1 Initial stage of dryer machine manufacturing. Machine manufacturing is following the design made by the design team. Then followed up by the manufacturing team, by making the frame and the outer cover. Initial dryer process is shown in figure 3.

Figure 3. Initial stage of dryer machine manufacturing

4.3.2 Advanced stage of dryer machine manufacturing. Activities in the advanced stage focus on the insulation inside the machine by installing heat shock (Styrofoam, plate and aluminum foil) and installation of the main components of which are in the drying chamber (heating elements, blowers, and sensors). Advanced dryer process is shown in figure 4.

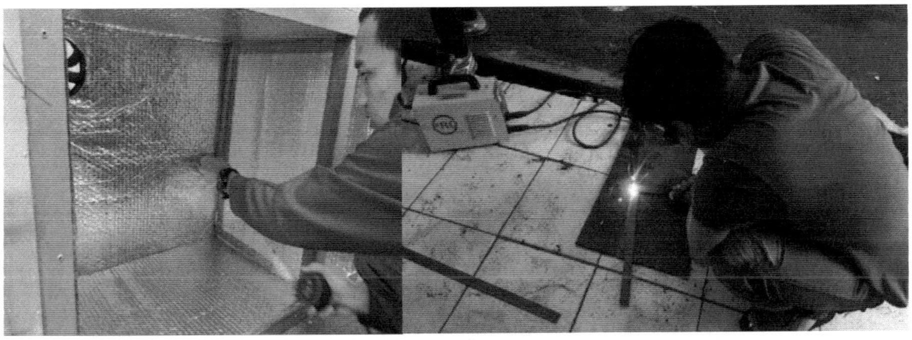

Figure 4. Advanced Stage of Dryer Machine Manufacturing

4.3.3 Final stage of dryer machine manufacturing. The activity at the final stage of the machine which has been done is completed, then followed up by the automation and electricity systems. This activity is in the form of installation of electrical, cabling, and electrical components. The arduino microcontroller automation system sketch was made in the final stages by adjusting construction to improve engine performance. Final stage process is shown in figure 5.

IOP Conf. Series: Materials Science and Engineering **692** (2019) 012006 doi:10.1088/1757-899X/692/1/012006

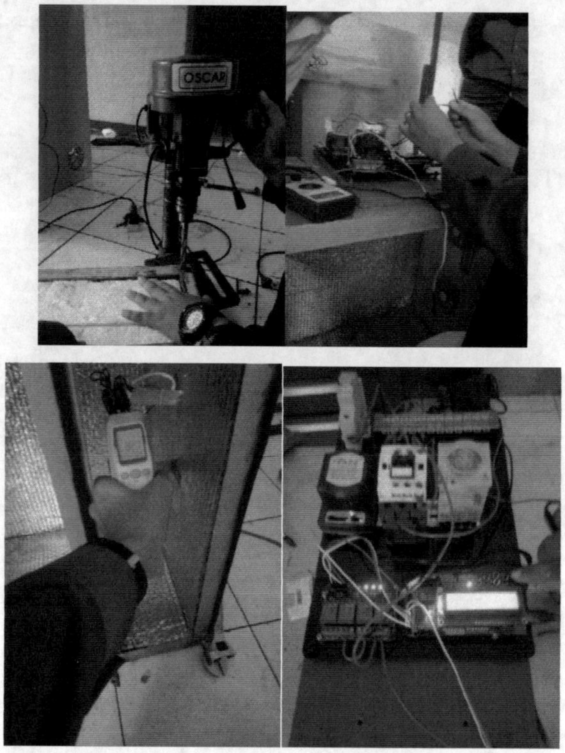

Figure 5. Final Stage Dryer Machine Manufacturing

4.4 Electrical Dryer Machine Testing

This activity aims to look at engine performance, engine durability, component durability, product quality, and energy efficiency. The testing phase is carried out using two types of foodstuffs namely opaque and banana. The data obtained in the form of opaque before drying weighs 3.4 ounces and after drying becomes 2.2 ounces. Whereas the banana before drying weighs 7.3 ounces and after drying it becomes 4.5 ounces. The above trial results are also supported by the opinion [5] For all the tomato sizes and at all airflow rate levels, the gram weight of the tomato decreased with an increase in drying time. Also for all sizes at all drying time levels, gram weight decreased with an increase in air flow rate. It is made clear by [1] that the drying tests conducted have shown that the combined drying method seems as well solution suited to the drying of food products, this is a scenario of drying oriented with fixed setpoints (drying air temperature and velocity), that are well-chosen according to the quality of the product to dry. The electrical dryer machine is shown in figure 6.

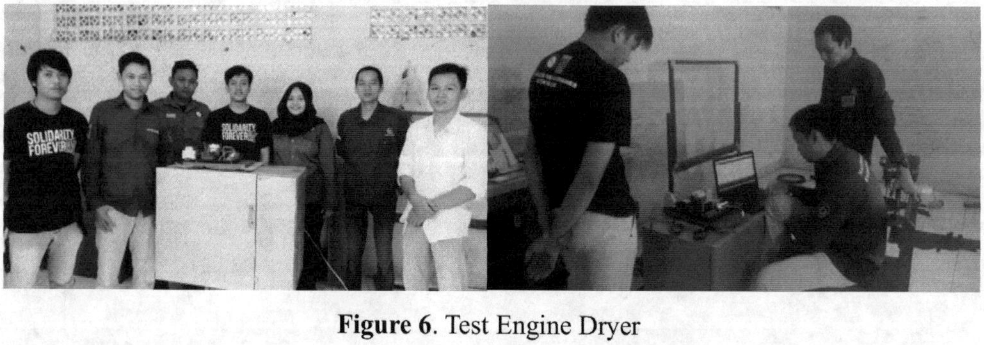

Figure 6. Test Engine Dryer

5. Conclusion

The study states that: (1) the manufacturing process of electrical machine dryer was done in three stages: the initial stage, advanced stage, and the final stage. (2) the trial of an electrical dryer machine was used to dry opak and banana. Data obtained in the form of opak before drying weighs 3.4 ounces and after drying to 2.2 ounces. Whereas the banana before drying weighs 7.3 ounces and after drying to 4.5 ounces; (3) electrical dryer machine is very helpful in the process of drying food processed products in Subang Regency.

Acknowledgments
Thank you to all the elements involved in making this research, and thanks to the Directorate General of Research and Development Strengthening Overseas Seminar Assistance Program, Kemenristekdikti as funders so that this research can be promoted

Bibliography

[1] H. Oueslati and S. Ben Mabrouk, "Design and Installation of a Solar-Gas Tunnel Dryer," no. May 2015, 2014.
[2] J. A. Moses, "Novel Drying Techniques for the Food Industry," no. September, 2014.
[3] V. Changrue and V. G. S. Raghavan, "Stewart Postharvest Review," no. January 2015, 2006.
[4] *Handbook of Industrial Dryer*. Taylor & Francis Group,LLC.
[5] J. C. Ehiem, S. V Irtwange, and S. E. Obetta, "Design and Development of An Industrial Fruit and Vegetable Dryer," vol. 1, no. 2, pp. 44–53, 2009.
[6] S. Mounir, A. S. Mujumdar, B. Bhandari, and Z. Fang, *Advances in Drying Science and Technology : Handbook of Drying of Vegetables and Vegetable Products*, no. November. 2017.
[7] *PR R OJ JE E CT T R RE EP P OR RT T ON DEHYDRATED FRUITS & VEGETABLES (SOLAR DRYER*. North Eastern Development Finance Corporation Ltd, 2017.
[8] A. Mohammed, "Design and construction of a Vegetable Drier," pp. 88–94, 2013.
[9] S. Pragati and B. Preeti, "Technological Revolution in Drying of Fruit and Vegetables," no. February, 2016.
[10] D. Maisnam, P. Rasane, A. Dey, S. Kaur, and C. Sarma, "Recent advances in conventional drying of foods .," vol. 1, no. 1, 2016.
[11] A. Efendi and D. Komarudin, "Evaluation of the Application of Occupational Safety and Health (OSH) at the Subang State Polytechnic Laboratory," Automotive Experiences, vol. 2, no. 1, pp. 9–14, 2019.

IOP Conf. Series: Materials Science and Engineering **692** (2019) 012007 doi:10.1088/1757-899X/692/1/012007

Application of Non-mold Casting Technology in the Tractor Diesel Engine Cylinder Block

Xing-hai SHAO*, Feng-jun LI, Hai-duo Xu, Qi AN, Ling-fang XIANG, Zhi-ming Feng and Le-hua LV

First Tractor Company Limited, Luoyang 471004, Henan, China

*yud1@163.com

Abstract. This paper researched the application of non-mold casting technology for reducing the development cycle and cost of tractor diesel engine cylinder body. The detail of the each process method has explained which based on the non-mold casting process of a diesel engine cylinder block. The results showed that rapid casting of diesel engine cylinder block can be realized by the non-mold casting composite forming technology, combined with CAD/CAE technology and virtual reality technology. The advantages of this method have fast speed, low cost, green, and lightweight, that is to say, piece manufacturing cycle was shortened by more than 50%, single piece manufacturing cost was reduced by more than 81%, and 8-9 molds were saved.

1. Introduction

With the development of machinery manufacturing, the speed of product development competition has become the focus of the market. In this case, the ability of autonomous rapid product development has become the strength foundation of the competition in the mechanical manufacturing industry [1]-[2]. Non-mold casting technology is a method to obtain castings by using the manufacturing technology of adding materials or reducing materials with the help of data model without direct molding of molds [3][4][5]. Because this technology directly eliminates the development and manufacturing process of the mold, the manufacturing cycle and manufacturing cost are greatly reduced, providing strong technical support for the rapid development and manufacturing of new products.

At present, in the trial production of new products of tractor diesel engine cylinder block, the traditional molded casting method is still adopted for the manufacture of blank parts, that is, the design and manufacture of molds first, and then the sand core is manufactured, and the blank parts are obtained after the core forming. In this traditional method, the development and manufacture of moulds take up most of the time of the whole production process. The application of non-mold casting technology to the trial production of new products of tractor diesel engine cylinder block can effectively solve the above problems.

2. Non-mold casting process for diesel engine cylinder blocks

Please follow these instructions as carefully as possible so all articles within a conference have the same style to the title page. This paragraph follows a section title so it should not be indented.

This paper expounds the features of complex structure and low yield of diesel engine cylinder block and in combination with the application of non-mold casting technology at home and abroad, and design the process flow of non-mold casting to blank parts, as shown in figure 1. In the process, the main processes are: ①use UG and other software to build 3d models of products. ② use CAE simulation software and virtual reality technology to design and optimize the casting process of complex cylinder parts. ③After the reverse calculation and design the casting mold. ④The mold forming of the cylinder body sand core is carried out, and to be assembled . ⑤The blank is formed by casting.

In the process of non - mold forming of sand core of cylinder body, Use the CAMTC series triaxial and five-axis forming equipment independently developed by the China Academy of Machinery Science and Technology. The forming process of furan self - hard resin sand core is completed by casting technology.

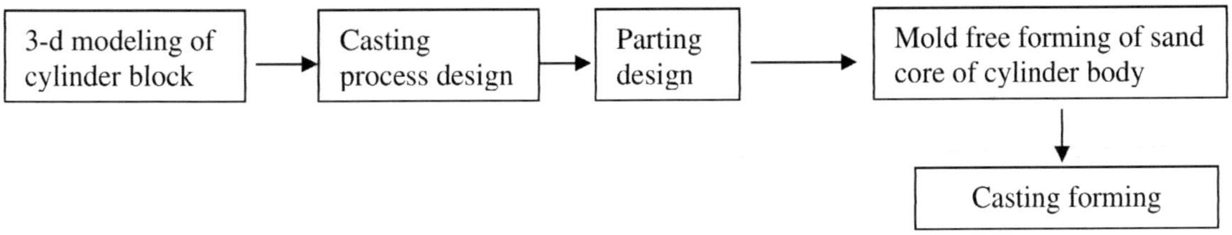

Figure 1. Flow chart of casting process for a diesel engine cylinder block.

3. Non - mold casting of diesel engine cylinder block

3.1. 3-d modeling of cylinder block

UG software is used to model a diesel engine cylinder parts and analyzed its structure characteristics. 3d model of cylinder parts and structural features of parts: The product is a four cylinder diesel engine cylinder block that is beyond to the engine part, the dimensions is 600mm*500mm*450mm, material: HT250, the total casting weight is 173kg.Inner structure of the casting is complex and thin, the thinnest wall thickness is only 5mm,expecially the water sleeve core of the cylinder body, which is extremely thin here and is prone to defects such as partial core and broken core, resulting in extremely high reject rate of the products.

After establish the 3-dmodeling of cylinder block, fill the hole structure that needs to be processed in the later stage, offset the machining surface on the processing surface to retain the machining allowance of 4-6mm, and scale it by 1% scale according to the grey iron material, so as to obtain the 3d model of the casting of the cylinder block.

Based on the 3-d modeling, use the CAD technology, CAE casting simulation system and virtual reality technology were used to design and optimize the casting process of the cylinder block.

3.2. Optimization of cylinder block casting process design

According to the analysis of the structural features of the cylinder block, two casting techniques of side lateral pouring and bottom vertically pouring were determined: side lateral pouring design and two exhaust risers were designed, with dimensions of 45 x 10mm; The bottom vertically pouring vertical casting design has multiple air holes. Two technological design schemes of cylinder block are designed using UG software on the casting 3d model.

The feasibility of the above two technological schemes was verified by CAE casting simulation system. Through the simulation of casting stress, filling and solidification, the possible deformation, porosity and shrinkage cavity and other defects of the metal liquid during filling and solidification in the resin sand mold are predicted. The simulation results of cylinder block should be analyzed on the flow field of metal liquid at the core of the lateral pouring process. Figure 2 shows the defects result of the partial solidification and shrinkage porosity. A dynamic virtual reality visual simulation model is rendered according to the simulation data of cylinder casting to reproduce the filling and solidification process of liquid metal. The virtual reality visualization model is displayed in the virtual reality lab in 3D visualization, and the process review and optimization is conducted by technicians such as casting process, smelting and modeling, as shown in figure 3. The above methods can perform more accurate process virtualization verification. The results of CAE simulation and process review showed that both of the two processes of the cylinder block could guarantee the steady filling of the metal liquid, and the shrinkage porosity and shrinkage hole defect in the body met the casting quality requirements. The simulation results showed that the process was qualified. The following two casting techniques are designed for rapid casting of cylinder block blank.

Figure 2. Simulation results of solidification of process cylinder head

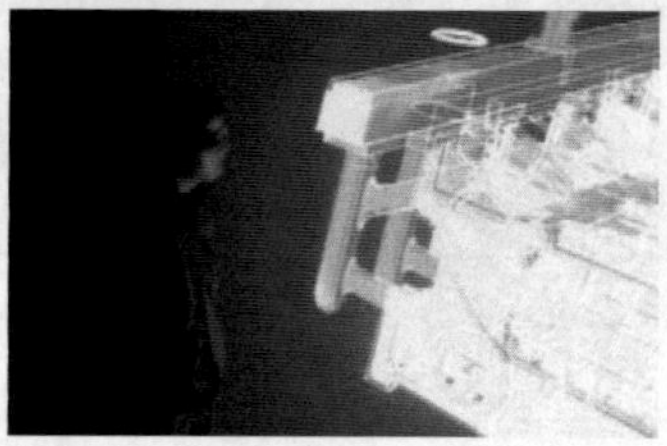

Figure 3. 3d visualization of simulation results of cylinder casting in a simulated real environment

3.3. Mold parting design and virtual assembly technology

After determining the two casting processes of the cylinder block, the appropriate amount of sand is selected. By using UG software, combining the existing processing equipment and features of five axes and triaxial, the fractal design is carried out on the basis of the original three-dimensional model. The classification scheme is shown in figure. 4. According to the plan, the cylinder shape core is mainly divided into water jacket core, four cylinder cylinders and multiple external dies.

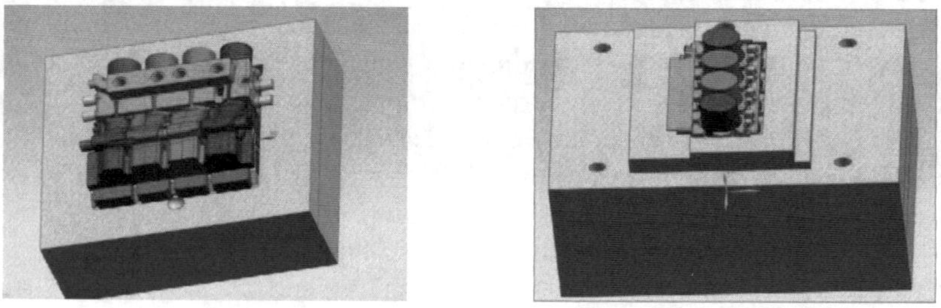

a) side lateral pouring b) bottom vertically pouring

Figure 4. Casting mold design scheme of cylinder block

In the virtual assembly environment of UG, or through the human-computer interaction and virtual scene technology of the virtual reality system, the virtual assembly and visualization of the partial parts of the cylinder body shape are realized. Figure 5 is the virtual assembly verification between the cylinder body shape cores in the virtual reality environment. Through verification, all the designed fractal schemes can realize assembly, in which the assembly of side injection leveling scheme should first assemble the four cylinder barrel cores and the water jacket cores before the overall lower core. 4 cylinders should be numbered to prevent confusion.

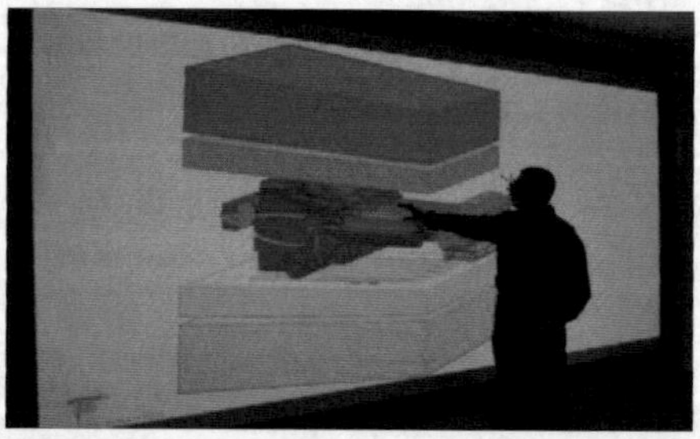

Figure 5 Virtual assembly process of the cylinder body core in virtual reality environment

3.4. Non-mold casting

Triaxial and five - axis equipment is used to fabricate cylinder sand core. According to the structural characteristics of the sand core of the cylinder body, the water sleeve core and cylinder barrel core are processed by five-axis linkage method, while the other three shafts are processed on one or both sides.

3.4.1 Design of machining tool path. On the basis of the fractal design scheme, the process USES UG software to design the blank and machining tool path respectively on the water jacket core, cylinder barrel core and external mould parts. The processing method is to choose the Cavity to be machined, and the processing order is generally M16 Cavity-Milling→M8 Rest-Milling→B6 Zlevel-Profile, Less use of external mold for bevel and curved surfaceM16 Cavity-Milling→M8 Rest-Milling or M16 Cavity-Milling, The processing mode can be completed. After the design of machining tool path for each part is completed, it is processed into PTP file respectively, and the corresponding naming is conducted, which is imported into the secretnon-mold molding machine. Figure 6 shows the design of machining cutter for part of the cylinder block.

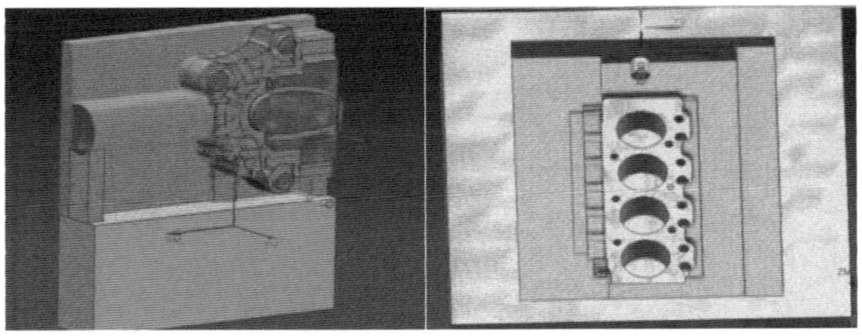

Figure 6 Part of the cylinder shape core processing design

3.4.2 Manufacturing of sand billets. The size of different sand cores of the cylinder body varies greatly, so it is necessary to make multiple sand billets. The sand cores of the cylinder body adopt furan self-hardening resin sand. In order to improve the production efficiency, the manufacture of sand billet adopts the method of artificial manufacturing sand billet: the sand is mixed through the sand mixing machine, the simple sand mould is made by using wood or aluminum alloy plate, and the sand lump is made by manual filling, tamping and scraping, and then the mold is unformed after hardening.

3.4.3Non-mold digitizing. After making the sand billet, the magnetic block is clamped and fixed on four sides in the digitalnon-mold precision forming machine for triaxial machining or five-axis machining. The process USES compressed air to blow out the sand. Figure 7 shows the five-axis machining process of the cylinder.

Figure 7 Five-axis machining of cylinder body cylinder core

3.4.4 Repair and size detection. After the core processing is completed, repair and size detection is carried out. The modification is mainly to remove the base and connection structure of the five-axis processing of sand core, such as water jacket and cylinder core. Some of the rectangular structures will not be able to process also need to use tools to make the processed rounded corners into right angles by hand. The precision of the processed sand core is tested on the three-coordinate measuring instrument. The accuracy of the sand core prepared by the above method is about 0.3mm.

IOP Conf. Series: Materials Science and Engineering **692** (2019) 012007 doi:10.1088/1757-899X/692/1/012007

3.5. Assemble and casting forming

According to the classification process and the assembly plan, the two plans of the cylinder block and sand core are combined and assembled: after the finished outer mold, water sleeve core and cylinder barrel are painted and dried, the assembly is carried out in accordance with the assembly sequence, among which the horizontal placement process needs to integrate the water jacket core and cylinder barrel core into the lower core. During the assembly process, the mud bar is used to measure the wall thickness. After passing the measurement, all parts are assembled together, and the casing is used to increase the height of casting and air outlet. The iron is pressed and then cast into shape. The assembly and molding process is shown in figure 8 and figure 9.

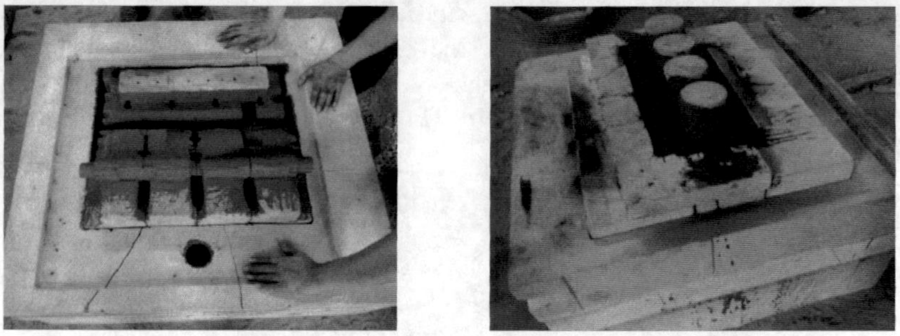

a) side lateral pouring b) bottom vertically pouring

Figure 8 Core assembly of body type core

Figure 9. The cast shape of the cylinder block and the state to be poured after pressing the iron

Two blank parts of the cylinder block are obtained after casting cooling, sand removal, shot blasting, casting risers removal and grinding, as shown in figure 10. The results of performance testing and anatomic verification of the castings showed that the sand core was floated or shifted by side lateral pouring , forming waste. Qualified casting was obtained by bottom vertically pouring.

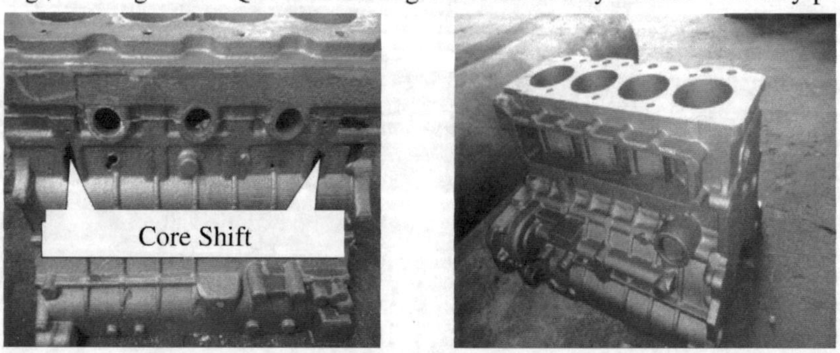

a) side lateral pouring b) bottom vertically pouring

Figure 10. Cylinder block blank

4. Conclusion

Through applied research, it is feasible to test the new product of cylinder block and to verify the different technology quickly by the method of non-mold casting technology. The manufacturing of cylinder block with non-mold casting technology has the following advantages:

1) Speed is fast, Replacc the traditional mould-casting method with the non-mold casting method, and the manufacturing cycle of individual parts is reduced by more than 50%. The single part of cylinder is manufactured by non-mold casting, wood and metal, and the expected cycle is 16 days, 32 days and 90 days respectively.

2) The cost is low, Replace the traditional mould-casting method with the non-mold casting method, and the cost of individual parts is reduced by more than 81%. The single part of cylinder is manufactured by non-mold casting, wood and metal, and the cost is 15,000yuan, 79,000 yuan and 8,035,000yuan.

3) Greenization and lightweight. Non-mold casting technology saves 8-9 sets of molds and realizes energy saving and emission reduction. Due to the fact that non-mold casting of does not need to consider the draft angle, it can reduce the machining of the casting and provide technical support for the lightweight design of the cylinder block.

Reference

[1] Shao Xinghai, Cai Anke, Guo Yahui, etc. The research of Rapid Sand Casting Process on Tractor Complex Casting [J].FOUNDRY, 2017, 66(10): 1049-1052.

[2] Shao Xinghai, Xu Haiduo, Cai Anke, etc. The research of Rapid Casting Process on Tractor Small Casting [J].FOUNDRY, 2018, 67(4): 295-298.

[3] Chen shaokai, Shan zhongde, Liu feng, etc. Application Research on Patternless Casting CNC Manufacturing Technology [J]. FOUNDRY EQUIPMENT AND TECHNOLOGY, 2010, (2) : 1-2.

[4] Zhongde Shan,Xiaoli Dong,Xinya Li, etc. NC Milling Technology of Patternless Sand Mold[C]. The 3rd International Conference on Rapid Prototyping and Manufacturing and The 2nd International Conference for Bio-manufacturing (ICRPM-BM 2008 Beijing) , 2008:350-353.

[5] Fonderie Saguenay Adds Patternless Casting[J].Modern Casting , 2011,101(7):13-13.

IOP Conf. Series: Materials Science and Engineering **692** (2019) 012008 doi:10.1088/1757-899X/692/1/012008

Adaptive Force Control for Robotic Grinding of Complex Blades

Pokuang Zhou, Yuming Zhou, Qianlong Xie* and Huan Zhao

State key Laboratory of Digital Manufacturing Equipment and Technology, Huazhong University of Science and Technology, Wuhan 430074, China.

*xieqianlong@hust.edu.cn

Abstract. Aiming at solving the robotic grinding problem of aircraft engine blades with complex surface, this paper proposed an adaptive force control method using fuzzy PID algorithm, which can ensure that blade surface is completely polished to eliminate residual texture and harmful stress concentration. First, an intelligent digital compliant grinder with active and passive compliance was developed. Second, a fuzzy PID adaptive algorithm was proposed, which can stabilize the contact force between the grinder and the blade. Third, a protection scheme was proposed to prevent excessive grinding. In the polishing process, the robot arm was fitted with the blade. According to the surface profile of the blade, the robot arm moved in a specific trajectory, keeping the blade in contact with the grinder. During grinding, some parameters, such as the abrasive belt velocity, the abrasive belt type and contact force, could affect the quality of the grinding process. Therefore, appropriate parameters were selected. The experimental results showed that the actual force fluctuates within 0.5N. In addition, the original surface roughness was 1.672μm, which was reduced to 0.054μm by adaptive force control method, comparing to 0.589μm with the non-force control. Not only is the contact force more precisely controlled, but the grinding accuracy is greatly improved.

1. Introduction

Robots have been widely applied in machining parts with complex surfaces. For the complex curved plane's engine blades, it is one of the key components, besides its quality and machining accuracy greatly affect the performance of the engine. Because of the intricacy of the aero-engine blades structure, when the engine works and blades are under enormous pressure, so it is easy for the blades to deform. In addition, a not quality surface and remaining texture lines on blades, which do not meet the requirements, may make wicked effect to the movement of air inside the engine. In addition, it leads to large fluctuations in machining accuracy of intake and exhaust edges of the blades [1-2].

There are several methods of blades grinding, such as manual grinding, abrasive flow grinding, and multi-axis Computerized Numerical Control machine grinding [3]. Abrasive flow machining (AFM) is a new type of finishing method using extruding abrasive media through parts' surfaces [4]. But AFM has a disadvantage is that the shape error cannot be corrected. The effects of manual grinding depend on the worker's experience and the tiny particles during grinding are harmful to health when inhaled by workers. To enhance both the processing efficiency and the quality, recently, industrial robots are diffusely applied in the blades grinding field [5]. For polishing blades, the advantages of robotic grinding are less costing, better agility and multiformity than a multiaxial machine tool. In the robots polishing intricate curved blade process, one of the difficulties to ensure the quality of blade surface is the fluctuation of contact force, which has a tremendous negative impact. For the better processing quality, some researchers put forward that the active compliance control and the passive compliance control should both be applied [6]. In addition, some researchers propose that limiting the fluctuation of contact force can effectively improve the blade processing quality [7]. The researchers proposed disparate methods of force control to precisely control the contact force. Dai et al. [8] put forward a parametric modelling controller to adjust the normal grinding force precisely. Liu et al. [9] propose a neoteric scheme for robot grinding, which uses passive force control to substantially reduce the drastic mechanical shock and applies active force control to acquire expectable force accuracy. All these methods have a common point, that is, the industrial robot whose instinct is low positioning accuracy is used in the active force control, therefore, it is difficult to guarantee the precision of force control. The researchers used a new approach to estimate and compensate industrial robot positioning errors. Specifically, the control method used an intelligent digital flexor grinder to directly control the contact force, instead of applying the active force of the robot.

As for the adaptive force control method, the conventional proportional–integral–derivative (PID) controllers stay to be the most prevalently used in the industrial processes [10]. In practice, lots of physical systems have inbred uncertainty traits such as non-linearities and high order. [11]. Hence, the famous method was proposed. Ziegler–Nichols method, supplies a systematic tuning method for the PID parameters which has good load disturbance reduction, but has unsatisfying weaknesses, such as, a long settling time and large overshoot [12]. For improving systems' overshoot, rise time, and integral of the absolute error. Theoretically, some researchers come up with it is suitable to utilize of fuzzy logic, combining features on the basis of the experiences of specialists with regard to PID gain scheduling. Practically, it has been used with great successful in industry applications. Fuzzy PID controllers have been displayed and inspected, and their adequate performance in several plants has been exhibited. Fuzzy PID controller is often deemed as a succedaneum to classical PID controllers in high non linearity and intricate situations. Fuzzy logic has a superiority of simplicity as well as precision with its simulation results.

This paper uses an unconventional blade grinding method, and applies a combination of intelligent digital compliant grinder including the flexible grinder and the robot arm. To be more specific, the robot arm loading fixture is used to hold the clamping parts, and artificially planning the unique and corresponding movement path of the robot arm according to the contour of the external surface of the grinding part. The flexible grinder has the active compliance control at the control strategy level and the passive compliance control is the spring as a compliant mechanical component, and the grinder with force sensor is carried by a motor to move with one degree of freedom, which is controlled with a closed loop of current. Moreover, the novel intelligent algorithms precisely control the contact force between the polishing tool and the workpiece during the grinding process, and a protection in the control program reduce the risk of excessive grinding during the beginning of the actual grinding process. All of those ensure better surface quality of the blades.

The paper is organized as follows. In Section II, the original position-based impedance control, an optimization method for the controller parameters and fuzzy-PID algorithm are detailed introduced. Section III discusses the intelligent digital compliant grinder, experiment procedures, the experimental program design, and an innovative method to prevent excessive grinding. Experiment results and analysis are provided in Section IV. Section V gives the conclusion.

2. Fuzzy - proportional integral differential (PID) controller design and preventing excessive grinding

2.1 Force control analysis

Due to the complex shape of aero-engine blade and the high quality of requirement for its machining surface, the contact force of grinding blade must be strictly controlled during mechanical grinding and polishing. Based on this, this paper combines the fuzzy theory with the PID control algorithm, a self-tuning fuzzy-PID controller formed out. Through the fuzzy-PID control, the reference position trajectory that adjusts the movement of the grinding machine is designed to generate the required contact force, in the absence of environmental stiffness and location information. In this scheme, adaptation allows automatic acquire modulation to provide a unitive property despite changes in environment parameters. The holistic structure of the robot grinding controller is revealed in figure 1, where e is the deviation, ec is the deviation change rate, F_a is the actual grinding force measured by the force sensor, as well as F_d is the required grinding force.

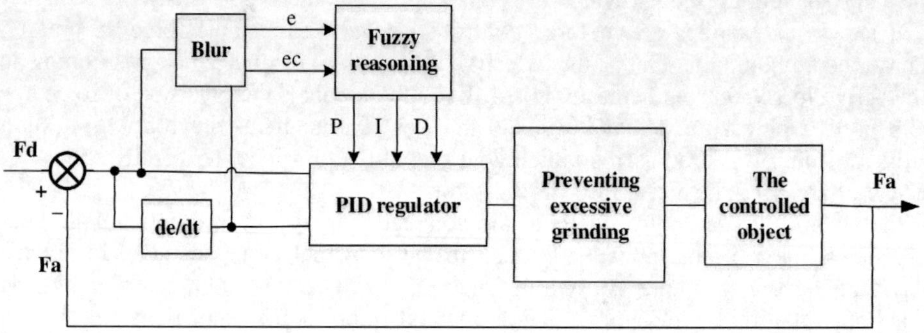

Figure 1. Holistic structure of the fuzzy-PID controller.

The PID control algorithm realized by computer is adopted, and its discrete PID control law is as follows:

$$F(n) = K_P e(n) + K_I \sum_{i=0}^{n-1} e(i) + K_D[e(n) - e(n-1)] \tag{1}$$

At the Nth sampling time, $F(n)$ is the output of the controller; $e(n)$ is the deviation signal as input of the controller; K_P, K_I and K_D are the proportional, integral, and differential coefficients, respectively.

Taking deviation e and deviation rate ec as inputs, which is able to keep self-tuning of the PID parameters of deviation e and deviation rate ec at reasonable values under different circumstances. By using fuzzy theory to correct PID parameters online, a self-tuning fuzzy-PID controller is constructed.

2.2 Fuzzy-PID parameter control algorithm

In the fuzzy structure, there are two inputs, more precisely, the deviation e and the deviation rate ec of the force signal. Equations (2) and (3) are used to periodically update the values of the deviation e and the deviation rate ec, where F_d represents the ideal contact force and F_a represents the actual measured force transmitted by tetrad force sensors.

$$e = F_d - F_a \tag{2}$$

$$ec = e(i) - e(i-1) \tag{3}$$

ΔK_P, ΔK_I and ΔK_D are the three outputs of the fuzzy force controller, and automatically adjust the value of PID parameters in accordance with the state of the force deviations. The real-time value can be calculated by using equation (4). Where K_{P0}, K_{I0} and K_{D0} are the initial design values of the proportional integral differential parameters, which is 0.5, 0.2 and 0.01 respectively according to the experiment experience.

$$\begin{cases} K_P = K_{P0} + \Delta K_P \\ K_I = K_{I0} + \Delta K_I \\ K_D = K_{D0} + \Delta K_D \end{cases} \tag{4}$$

Firstly, Fuzzification converts the input and output variables into corresponding fuzzy quantities. Input and output variables were divided into seven language labels: NB(negative big), NM(negative medium), NS(negative small), ZO(zero), PS(positive small), PM(positive middle), PB(positive big).

IOP Conf. Series: Materials Science and Engineering **692** (2019) 012008 doi:10.1088/1757-899X/692/1/012008

Meanwhile, in the discrete domain, their variation ranges were divided into seven levels: -3, -2, -1, 0, 1, 2, 3. Their actual variables are shown in Table 1. The relationship between variables in different fields is established by using the scaling factor, and the scaling factor is defined by using equation (5).

$$\begin{cases} K = \dfrac{W_h - W_l}{Z_h - Z_l} \\ K' = \dfrac{Z_h - Z_l}{W_h - W_l} \end{cases} \tag{5}$$

where K and K' represent the scale factors of the input and output, respectively, $[W_h, W_l]$ represents the range of variables in the discrete domain, and $[Z_h, Z_l]$ represents the range of the actual variables. Fuzzy reasoning is a process of establishing fuzzy rules. Fuzzy reasoning is a process of establishing fuzzy rules. Firstly, it is necessary to convert all variables into membership functions of fuzzy sets. The membership function is a curve that defines how each variable is mapped to a membership value between 0 and 1[13]. It is reasonable to choose the common triangle membership function, which has the characteristics of simple calculation and good performance [14]. As shown in Figure 2, it is a membership function of the reference fuzzy set.

Table 1. The actual parameter variables.

	e	ec	$\Delta K_m (m = P, I, D)$
Change block	[-10,10]	[-10,10]	[-0.7,0.7]
The scaling factor	0.35	0.35	0.23

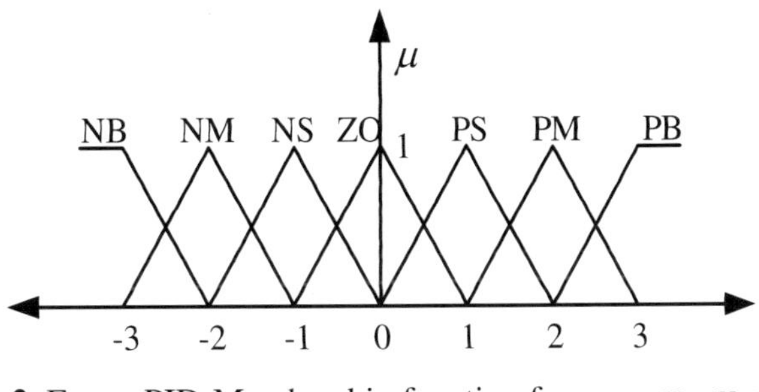

Figure 2. Fuzzy-PID-Membership function for e, ec, K_P, K_I and K_D.

The fuzzy rules are constructed by using the if-then statement, and the 49 rules are refined to form a fuzzy rule base of K_P, K_I and K_D .In the PID controller, the K_P value determines the response speed of the system, the K_I value is used to eliminate the steady-state error of the system, and the K_D value improves the dynamic characteristics of the system, eliminates oscillations, and reduces the influence of system inertia. In the early stage of regulation, a large K_P value ensures the system response speed, a small K_I value or even zero value to avoid system integration saturation, and a large K_D value to avoid overshoot; In the middle stage of regulation, small K_P, K_D and large K_I values are selected to ensure the response speed and stability of the system. In the later stage of regulation, in order to reduce the static error and improve the control accuracy of the system, the values of K_P and K_I should increase. At the same time, in order to compensate for the extended regulation time caused by the large K_D value in the early stage of regulation, the K_D value was supposed to be large. According to the above description, this section formulated the fuzzy rule base of K_P, K_I and K_D, as shown in sequence in Table 2.

Table 2. The K_P, K_I and K_D rule base for the fuzzy combiner.

(a)

	ec						
e	NB	NM	NS	ZO	PS	PM	PB
NB	PB	PB	PM	PM	PS	ZO	ZO
NM	PB	PB	PM	PS	PS	ZO	NS
NS	PM	PM	PM	PS	ZO	NS	NS
ZO	PM	PM	PS	ZO	NS	NM	NM
PS	PS	PS	ZO	NS	NS	NM	NM
PM	PS	ZO	NS	NM	NM	NM	NB
PB	ZO	ZO	NM	NM	NM	NB	NB

(b)

	ec						
e	NB	NM	NS	ZO	PS	PM	PB
NB	NB	NB	NM	NM	NS	ZO	ZO
NM	NB	NB	NM	NS	NS	ZO	ZO
NS	NB	NM	NS	NS	ZO	PS	PS
ZO	NM	NM	NS	ZO	PS	PM	PM
PS	NM	NS	ZO	PS	PS	PM	PB
PM	ZO	ZO	PS	PS	PM	PB	PB
PB	ZO	ZO	PS	PM	PM	PB	PB

(c)

	ec						
e	NB	NM	NS	ZO	PS	PM	PB
NB	PS	NS	NB	NB	NB	NM	PS
NM	PS	NS	NB	NM	NM	NS	ZO
NS	ZO	NS	NM	NM	NS	NS	ZO
ZO	ZO	NS	NS	NS	NS	NS	ZO
PS	ZO	ZO	ZO	ZO	ZO	ZO	ZO
PM	PB	NS	PS	PS	PS	PS	PB
PB	PB	PM	PM	PM	PS	PS	PB

Finally, the center of gravity method is selected to defuzzify and get the values of ΔK_P, ΔK_I and ΔK_D ,as shown in equation (6) , where $\mu(e)$ and $\mu(ec)$ respectively represent the degree of membership corresponding to the deviation e and ec, and $\mu(K_P)$, $\mu(K_I)$ and $\mu(K_D)$ respectively represent the membership degree corresponding to K_P, K_I and K_D .

$$\begin{cases} \Delta K_P = \dfrac{\sum_{i=0}^{1}\mu_m(e)\times\left[\sum_{j=0}^{1}\mu_j(ec)\times\mu_{ij}(K_P)\right]}{\sum_{i=0}^{1}\mu_i(e)\times\sum_{j=0}^{1}\mu_j(ec)} \\ \Delta K_I = \dfrac{\sum_{i=0}^{1}\mu_i(e)\times\left[\sum_{j=0}^{1}\mu_j(ec)\times\mu_{ij}(K_I)\right]}{\sum_{i=0}^{1}\mu_i(e)\times\sum_{j=0}^{1}\mu_j(ec)} \\ \Delta K_D = \dfrac{\sum_{i=0}^{1}\mu_i(e)\times\left[\sum_{j=0}^{1}\mu_j(ec)\times\mu_{ij}(K_D)\right]}{\sum_{i=0}^{1}\mu_i(e)\times\sum_{j=0}^{1}\mu_j(ec)} \end{cases} \tag{6}$$

The ΔK_P, ΔK_I and ΔK_D values obtained are sent to the PID controller. Equations (1), (4) and (5) are used to establish a robot control model to polish the blades.

2.3 Excessive grinding preventing

In the experiment of polishing the flat plate, there would be too much grinding at the position where the flat plate and the grinding belt just contacted, for the following reasons. According to the principle of fuzzy PID algorithm, when the blade and belt are not in contact at the beginning of grinding, grinding force is set and the force error is maximized, and the proportional coefficient calculated by fuzzy-PID algorithm is also relatively large, which further makes the motor speed very high. This condition can cause the initial contact position to be excessive ground. Excessive grinding is abominable and must be avoided, so a novelty means was proposed. According to the classification standard of contact force, the polishing process can be divided into three parts: Firstly, before polishing, the loading part of the robot arm moves towards the polishing machine, and the contact force is 0N, which may lead to excessive grinding. Secondly, during the polishing process, the contact force fluctuates between 3N and 7N.

Thirdly, after grinding, the blade will return to its original position with a contact force of 0, so it is impossible for the blade to contact the belt, and excessive grinding does not occur. According to the difference between the actual force F_a and the setting force F_d, which is used as the judgment basis. And the limiting motor voltage module is increased. This paper designed a protection module in the program. The logic is as shown in Figure 3:

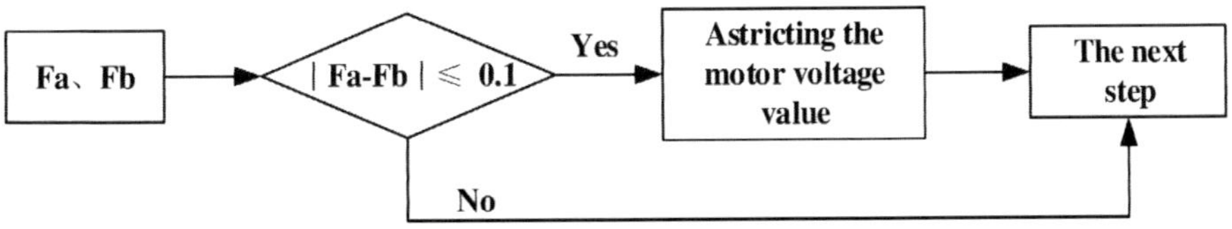

Figure 3. Excessive grinding preventing

3.Experimental validation

This section is not only going to introduce the laboratory equipment, such as an industrial robot and a sand belt grinding mechanism, but also puts forward the experimental process. On the basis of the correlative experience of the belt grinding of Computerized Numerical Control machine, advisable the polishing parameters was chosen, for instance, the abrasive belt type, the polishing contact force and the abrasive belt velocity. All these contribute to the grinding of blades that meet quality requirements.

3.1 The experimental platform

To carry out experimental analysis of the grinding blade processes, a setup for the robotic grinding was designed and applied as shown in Figure 4.

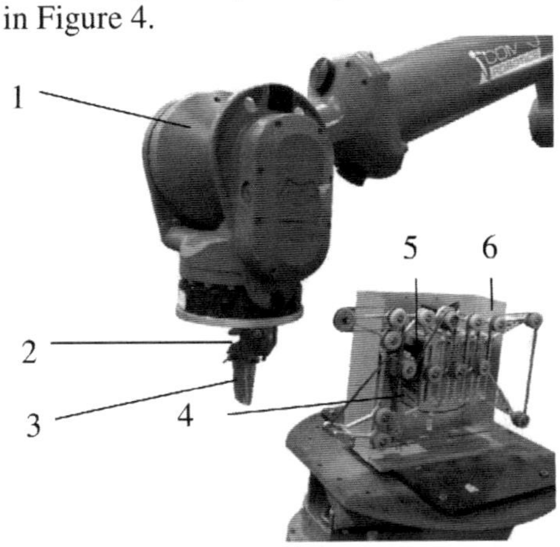

Figure 4. Robotic grinding setup;

1 - robot; 2 - fixture; 3 -blade ,4 –tension wheel;5 - force sensor; 6 –sand belt grinding mechanism

The robotic arm is equipped with a clamp attached to the engine blades. The robot arm moves in a set path that conforms to the blade surface profile during grinding. Abrasive belt grinding machine with flexible mechanism, for example, tensioning wheel with spring. A powerful sensor on the grinding device feeds back the contact force.

3.2 Plate grinding and overgrinding preventing

First of all, it is reasonable to perform experiments on a simple flat plate. Refer to previous experiments of the belt grinding of CNC machine, the material of engine blade is superalloy material, and we choose advisable the polishing parameters, for instance, the abrasive belt type, using alumina stacked abrasive belt (KK718X), the polishing contact force as 4.8N and the abrasive belt velocity as 2800 rpm. Experiments verified the procedures for preventing excessive grinding. In Figure 5, a superalloy plate with excessive grinding was shown. In the next subsection, the pictures of the blades without excessive grinding were shown.

Figure 5. The plate with excessive grinding

3.3 Blade grinding and experimental results

After finishing the grinding test of the above flat plate, the experience was summarized and the grinding experiment of the blade was carried out. Set the trajectory of the robot arm, in addition, used the same belt type, polishing contact force, and belt speed as the plant experiment did. Made sure that other influencing factors remain the same, two control groups were done. One group did not exert force control, the other group exerted force control fuzzy PID algorithm. Experimental results were obtained, as shown in the Figure 6.

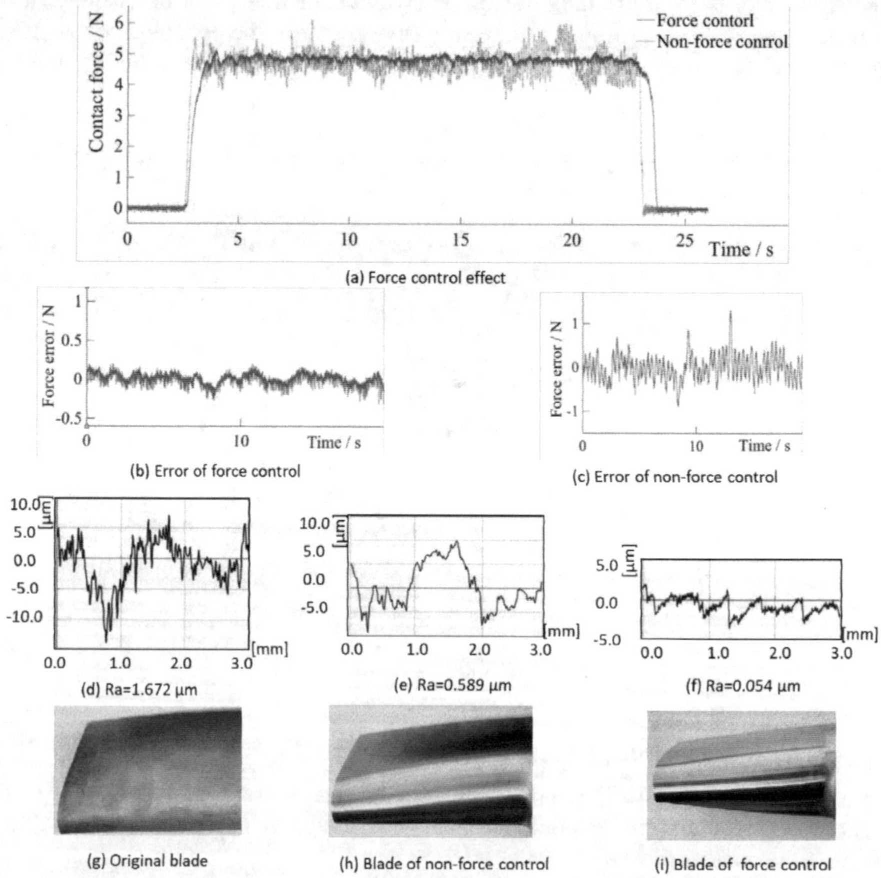

Figure 6. Comparison of experimental results

In the experimental group without force control, the surface roughness was improved, but it could not meet the requirements of blade surface roughness. In the experimental group with fuzzy PID algorithm, the roughness is greatly improved, specifically, the surface roughness is 0.054μm, and the blade surface has benign consistency, meeting the requirements of grinding.

4.Conclusion

This paper introduces an innovative intelligent digital compliant grinder, which applies fuzzy-PID algorithm for active force control to ensure constant contact force during the polishing process in robotic polishing. Those works enormously lessen the milling mask of the blade and boost the surface quality. Firstly, this paper did experiments with the superalloy plate, employing optimal polishing parameters, and put forward a program method to prevent excessive grinding. Secondly, this study conducted robotic blade polishing experiment, and verified the availability of the proposed method. The original surface roughness of engine blade was 1.672μm, when the active contact force under fuzzy PID algorithm was applied, surface roughness was reduced to 0.054μm, comparing to the surface roughness 0.589μm by the non-force control, and the actual force fluctuates were within 0.5N. The results demonstrate that the machining consistency and surface roughness are remarkably improved with the proposed the robotic intricate blade polishing algorithm.

Acknowledgement

This work was supported by the National Key Research and Development Program of China under Grant No.2017YFB1301501.

Reference

[1] Gao H , Zhao Z , Sun Y W . Recent Development of the Aero-Engine Impeller and Blade Surface Polishing Technology[J]. Advanced Materials Research, 2010, 135:7-12.
[2] Tian F , Lv C , Li Z , et al. Modeling and control of robotic automatic polishing for curved surfaces[J]. CIRP Journal of Manufacturing Science and Technology, 2016, 14:55-64.
[3] Liao L, Xi F, Liu K. Adaptive Control of Pressure Tracking for Polishing Process. ASME. J. Manuf. Sci. Eng. 2010;132(1):011015-011015-12.
[4] Haibo W, Hang G, et al. Development of novel guar gum hydrogel based media for abrasive flow machining: Shear-thickening behavior and finishing performance. (Elsevier: International Journal of Mechanical Science), 2019, Vol 157-158, pp 758-772
[5] Zhong Z W. Recent Advances in Polishing of Advanced Materials[J]. Advanced Manufacturing Processes, 2008, 23(5):8.
[6] Fengjie Tian , Chong Lv, et al. Modeling and control of robotic automatic polishing for curved Surfaces. Vol 14 (Elsevier: CIRP Journal of Manufacturing Science and Technology), 2016, pp 55-64.
[7] Duan Jihao,Shi Yaoyao,Li Xiaobiao,et al.Adaptive polishing for blisk by flexible grinding head[J].Acta Aeronautica et Astronautica Sinica,2011,32(5):934-940
[8] Antnio Lopes, Almeida F . A forceimpedance controlled industrial robot using an active robotic auxiliary device[J]. Robotics and Computer Integrated Manufacturing, 2008, 24(3):299-309.
[9] Dai H , Yuen K M , Elbestawi M A . Parametric modelling and control of the robotic grinding process[J]. International Journal of Advanced Manufacturing Technology, 1993, 8(3):182-192.
[10] Zhang J , Liu G , Zang X , et al. A hybrid passive/active force control scheme for robotic belt grinding system[C]// 2016 IEEE International Conference on Mechatronics and Automation. IEEE, 2016.
[11] D Puangdownreong, T Kulworawanichpong, S Sujitjorn. Input weighting optimization for PID controllers based on the adaptive tabu search. 2004 IEEE Region 10 Conference D. 2004: 451-454.
[12] Adnan Jabbar Attiya, Yang Wenyu, et al. Fuzzy-PID Controller of Robotic Grinding Force Servo System, Telekomnika Indonesian Journal of Electrical Enjineering.2015;15:515-525.
[13] Ocampo-Duque, W, et al. "Assessing water quality in rivers with fuzzy inference systems: A case study", Environment International, (Elsevier: Environment International)2006; Vol 32, pp 733-742.
[14] Hong-Liang Gao, Xi-Sheng Zhan, Yi-Ran Yuan, Zi-Jie Pan, Guo-Long Yuan. "Mitigation of low frequency oscillations in power systems based on Mamdani fuzzy inference", Transactions of the Institute of Measurement and Control, 2019.

Reliability Analysis of Belt Conveyor Based on Fault Data

Meiyan Li[1], Yingqian SUN[1,*] and Chuan Luo[1]

[1]Mining and Safety Engineering College, Shandong University of Science and Technology, Qianwangang Road, Qing Dao Shandong Province, 266590, China

[*]1932001736@qq.com

Abstract. In order to improve the accuracy of the reliable analysis of the belt conveyor, based on the belt conveyor fault data, firstly fit the probability density distribution function of the belt conveyor fault interval time. According to the characteristics of the fault data, it is inferred that it may obey the Weibull three-parameter distribution. The Weibull distribution model is obtained by combining the least squares method with the correlation coefficient method, and the accuracy of the model is verified by d test. Finally, the reliability of the belt conveyor is analysed, and the measures to improve the reliability of the belt conveyor are given, which provides a theoretical basis for the daily maintenance of the belt conveyor.

Keywords. Reliability Analysis; Belt Conveyor; Fault Data; Weibull Three-Parameter Distribution

1. Introduction

As a key equipment in coal mine production and transportation systems, mining belt conveyors will bring significant economic losses when they fail. Therefore, how to carry out reliability analysis on the belt conveyor and master the operation status of the belt conveyor is very important for the mining enterprise. At present, many scholars at home and abroad study the reliability analysis of belt conveyor mainly from the methods of combining qualitative analysis and quantitative analysis and quantitative analysis alone. The research of qualitative and quantitative combination mainly uses fault tree, FMECA and Bayesian network. Ren Zhongquan, Jing Hui, Xie Xin, Han Weihong use the fault tree model to analyze the reliability of the belt conveyor and find out the weak links in equipment maintenance [1-4]. Li Peng [5] uses the FMECA method to obtain the weak links in the maintenance process of the belt conveyor. Zhang Dawei [6] combined the fault tree with the Bayesian network to establish a fault management knowledge model for the belt conveyor, providing a complete set of belt conveyor fault knowledge model for maintenance. Although a combination of qualitative and quantitative analysis is used, errors in qualitative analysis can lead to inaccuracies in subsequent quantitative analysis results. In response to this problem, scholars have separately studied the reliability of belt conveyors from a quantitative perspective, mainly using mathematical statistics. Agnieszka Blokus Roszkowska [7] uses a mathematical approach to reliability analysis of belt conveyor systems by establishing a reliability function. Ovidiu-Bogdan Tomuş [8] uses a two-parameter Weibull statistic method to perform a comprehensive reliability analysis of the belt conveyor system based on the fault data. However, the following problems also exist in the quantitative analysis process:

(1) When scholars conduct quantitative analysis, it is considered that there is a linear relationship

between the fault data, and the two-parameter Weibull distribution model is used for analysis. However, the fault data is often nonlinear, which will cause large errors in parameter estimation. (2) Whether the fault data belongs to the Weibull distribution needs to be verified.

In response to the above problems, this paper analyses reliability from a quantitative perspective. First analyse the fault data, preliminarily estimate the fault data distribution model and solve the model, and then verify whether it conforms to the preliminary estimated distribution model. Finally, the failure rate and reliability analysis are carried out to provide a theoretical basis for belt conveyor maintenance.

2. Fault time probability distribution model

In this paper, the XX enterprise belt conveyor is taken as the research object. By collecting the fault data of the equipment, the probability distribution model of the belt conveyor fault interval is established.

2.1. Fault time probability distribution model

Firstly, the belt conveyor time interval data is analysed, and the number of groups is determined according to the empirical formula (1) proposed by sturges [9].

$$K \geq 1 + 3.322 \times lg(n) \tag{1}$$

Where n is the total number of failures.

Through statistics, there are 61 sets of data. After calculation, K=8 can be obtained. The minimum fault interval of the belt conveyor is 0.5h, the maximum fault interval is 666.2h, and the fault interval time range is [0.85, 666.2]. Divided into 8 groups, as shown in Table 1.

Table 1. Fault interval sample.

Number	Upper limit	Lower limit	Mean	Frequency	Rate of Recurrence	Grand total
1	0.85	83.71	42.11	38	0.6393	0.6393
2	83.71	166.92	125.32	12	0.1967	0.836
3	166.92	250.13	208.53	2	0.0328	0.8688
4	250.13	333.34	291.74	3	0.0492	0.918
5	333.34	416.55	374.95	2	0.0328	0.9508
6	416.55	499.76	458.16	1	0.0164	0.9672
7	499.76	582.98	541.37	1	0.0164	0.9836
8	582.98	666.2	624.59	1	0.0164	1

The mean value of each set of fault intervals is the abscissa, the observed value $f(t)$f of the probability density is the ordinate, and $f(t)$ is calculated as equation (2):

$$f(t) = \frac{n_i}{n\Delta t_i} \tag{2}$$

Where $n_i$is the frequency of failures in each group of fault intervals; n is the number of faults 61, Δt_i is the group distance, and the scatter plot of the probability density can be made as shown in Figure 1.

It can be seen from Figure 1 that the distribution of the belt conveyor fault interval does not conform to the normal distribution characteristics, and may obey the exponential distribution or the Weibull distribution.

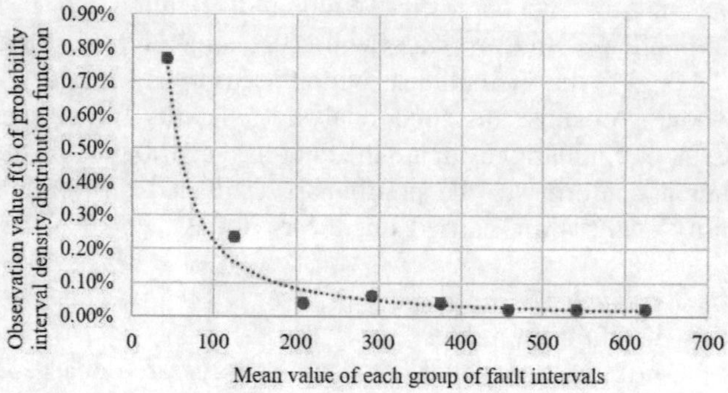

Figure 1. Probability density scatter plot.

3. Determination of the time interval distribution model

According to the distribution law of the empirical model, the shape of the Weibull distribution curve is different depending on the corresponding shape parameters. The exponential distribution can be seen as a special Weibull distribution. In general, the concave curve obeys the three-parameter Weibull model, so it is assumed that the belt conveyor's fault interval is subject to a three-parameter Weibull distribution. The parameter estimation is performed by the least squares method, and the result is checked to determine whether the time interval of the belt conveyor obeys the Weibull distribution.

The probability density function of the Weibull distribution is:

$$f(t) = \frac{\beta}{\eta}\left(\frac{t-\lambda}{\eta}\right)^{\beta-1} exp\left[-\left(\frac{t-\gamma}{\eta}\right)^{\beta}\right] \tag{3}$$

The failure probability function is:

$$\lambda(t) = \frac{\beta}{\eta} \times \left(\frac{t-\gamma}{\eta}\right)^{\beta-1} \tag{4}$$

The reliability function is:

$$R(t) = exp\left[-\left(\frac{t-\gamma}{\eta}\right)^{\beta}\right] \tag{5}$$

Where t is time, $\gamma$is a positional parameter; $\eta$is a scale parameter andβ is a shape parameter.

3.1. Parameter estimation method

The logarithm of the formula (5) can be converted to:

$$ln\{-ln[R(t)]\} = \beta\, ln(t-\gamma) - \beta\, ln(\eta) \tag{6}$$

A set of fault intervals with a capacity of n, arranged in ascending order and defined as t. In order to reduce the error, its reliability is estimated by the median rank method:

$$R(t_i) = \frac{(i-0.3)}{(n+0.4)} \tag{7}$$

Let $y_i = ln\{-ln[R(t_i)]\}$,$x_i = ln\, t_i$. The regression equation that needs to be fitted is:

$$y = ax + b \tag{8}$$

According to the principle of least squares, the regression coefficient a, b and the correlation coefficient R_{xy} are:

$$b = \frac{l_{xy}}{l_{xx}} = \frac{\sum_{i=1}^{n}(x_i-\overline{x})(y_i-\overline{y})}{\sum_{i=1}^{n}(x_i-\overline{x})^2} \tag{9}$$

$$a = \overline{y} - b\overline{x} \tag{10}$$

$$R_{xy} = \frac{l_{xy}}{\sqrt{l_{xx}l_{yy}}} \tag{11}$$

The numerical iterative method is used to determine the estimated value of the position parameter γ, and the positional parameter $\gamma_{01} = min\{t_i\}$ is assumed to be the initial value of γ, and the corresponding correlation coefficient R_{01} is calculated. Similarly, R_{0n} is calculated in turn, and when $\gamma$is estimated correctly, there is a maximum correlation coefficient between x,y, thereby determining the value of the position parameter γ.

The fault interval time data of the belt conveyor can be rearranged and calculated by the parameter estimation method. The results are shown in Table 2.

Table 2. Recalculated fault interval time.

No	t_i	$R_n(t_i)$	x_i	y_i	No	t_i	$R_n(t_i)$	x_i	y_i
1	0.5	0.011401	-0.69315	-4.46836	27	57.12	0.434853	4.066117	-0.56094
2	0.85	0.027687	0.405465	-3.57278	28	58.33	0.45114	4.069027	-0.51097
3	1.5	0.043974	0.774727	-3.10176	29	58.5	0.467427	4.110874	-0.46198
4	2.17	0.060261	0.845868	-2.77816	30	61	0.483713	4.113657	-0.41386
5	2.33	0.076547	1.054312	-2.53029	41	98.92	0.662866	4.405499	-0.04926
6	2.87	0.092834	1.108563	-2.32862	42	104.42	0.679153	4.51086	-0.00497
7	3.03	0.109121	1.991976	-2.15809	43	104.77	0.69544	4.594311	0.039315
8	7.33	0.125407	2.413232	-2.00994	44	109.33	0.711726	4.648421	0.083676
9	11.17	0.141694	2.449279	-1.87866	45	120.8	0.728013	4.651767	0.12821
10	11.58	0.15798	2.70805	-1.76054	46	122.05	0.7443	4.694371	0.173017
11	15	0.174267	2.833213	-1.65295	47	125.75	0.760586	4.794136	0.218208
12	17	0.190554	3.223664	-1.55398	48	132.5	0.776873	4.804431	0.263902
13	25.12	0.20684	3.241811	-1.46218	49	150.67	0.79316	4.834296	0.310238
14	25.58	0.223127	3.330417	-1.37643	50	161.97	0.809446	4.886583	0.357368
15	27.95	0.239414	3.398527	-1.29585	51	162.42	0.825733	5.015092	0.405474
16	29.92	0.2557	3.519869	-1.21972	52	178.83	0.84202	5.087411	0.454768
17	33.78	0.271987	3.541539	-1.14748	53	197.58	0.858306	5.090186	0.505504
18	34.52	0.288274	3.725693	-1.07863	54	259	0.874593	5.186436	0.557995
19	41.5	0.30456	3.769999	-1.01277	55	295	0.890879	5.286144	0.612633
20	43.38	0.320847	3.805106	-0.94957	56	330.28	0.907166	5.556828	0.669923
21	44.93	0.337134	3.824939	-0.88872	57	360.17	0.923453	5.686975	0.730534
22	45.83	0.35342	3.827118	-0.82998	58	412.6	0.939739	5.799941	0.795389
23	45.93	0.369707	3.83579	-0.77312	59	456.5	0.956026	5.886576	0.865815
24	46.33	0.385993	3.839452	-0.71795	60	582.02	0.972313	6.022479	0.943846
25	46.5	0.40228	3.989539	-0.6643	61	666.2	0.988599	6.50169	1.498302
26	54.03	0.418567	4.045154	-0.61201					

From equations (9) and (10), $\beta = b$ =0.81848, a =-3.76548, then $\eta = e^{\left(-\frac{a}{b}\right)}$ =109.508. According to the numerical iteration method, γ = 0.4 can be obtained.

3.2. *d test*

The d test is a difference between the distribution function $F_0(t)$ corresponding to the belt conveyor interval time data and the calculated value $F_n(t)$ corresponding to each fault interval time, and the largest absolute value of the difference is the observed value D_n. Then compare the value of D_n and the critical value $D_{n,\alpha}$. If $D_n \leq D_{n,\alpha}$, the fault data of the belt conveyor conforms to the Weibull three-parameter distribution model. On the contrary, it does not match. After calculation,$D_n = 0.0907902$. In the d test, according to the experience $\mathrm{D_{n,\alpha}} = \frac{1.22}{\sqrt{n}} = 0.156 > \mathrm{D_n}$, the original hypothesis is established, and the fault interval of the belt conveyor conforms to the three-parameter Weibull distribution model.

The probability density function is:

$$f(t) = 0.0075 \times \left(\frac{t\text{-}0.4}{109.508}\right)^{-0.8} exp\left[-\left(\frac{t\text{-}0.4}{109.508}\right)^{0.818}\right] \quad (12)$$

The failure probability function is:

$$\lambda(t) = 0.075 \times \left(\frac{t-0.4}{109.508}\right)^{-0.812} \quad (13)$$

The reliability function is:

$$R(t) = exp\left[-\left(\frac{t-0.4}{109.508}\right)^{0.818}\right] \quad (14)$$

4. Reliability analysis

4.1. Failure probability

The failure rate curve of the belt conveyor can be obtained from the failure probability function. As shown in Figure 2. Analysis of belt conveyor failures found that most of the faults are concentrated on belts, rollers, and motors. Therefore, through quantitative analysis of reliability, it is possible to take preventive maintenance or replace originals, and increase the frequency of regular inspections to reduce the failure rate.

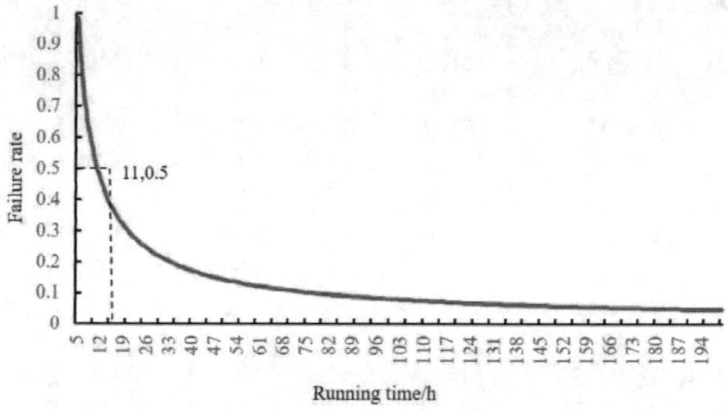

Figure 2. Failure rate curve

4.2. Reliability

Figure 3 shows the reliability distribution curve. It can be seen that as time increases, the reliability gradually decreases, indicating that the longer the time, the less likely the belt conveyor will work properly. Therefore, proper maintenance is required to ensure that the belt conveyor maintains high reliability.

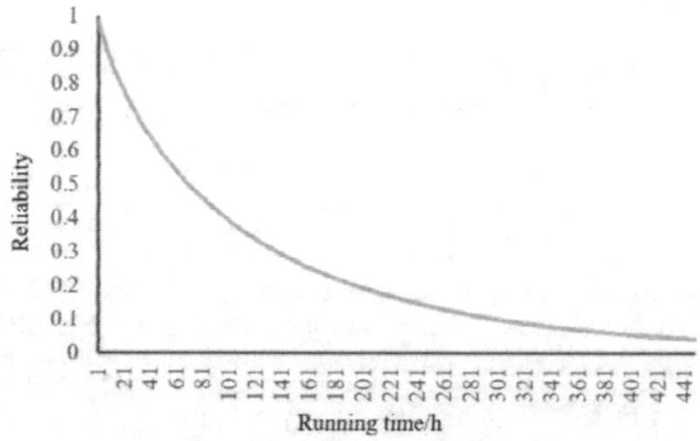

Figure 3. Belt conveyor reliability function

Table 3 shows the operating cycle of the belt conveyor under different reliability. It can be seen that the reliability of the belt conveyor decreases very rapidly with the non-stop operation, and the reliability will be lower than 0.75 in continuous operation for 24 hours. Therefore, if you want to ensure the reliability of the belt conveyor, you need to develop a daily maintenance plan, and carry out equipment maintenance activities every day.

Table 3. Reliability of belt conveyors at different running time

Reliability	0.95	0.9	0.85	0.8	0.75
Running time(hour)	3	7	12	17	24

5. Conclusion

In this paper, for the problem of inaccurate reliability analysis of belt conveyor, the distribution function is fitted based on the belt conveyor fault data. According to the fitting result, the fault interval time may be consistent with the Weibull distribution, and the least square method and correlation coefficient method are used to solve the problem. Finally, the d test is used to prove that the belt conveyor fault interval is in accordance with the Weibull distribution, and the reliability study is carried out. The main conclusions are as follows:

(1) Based on the fault data characteristics, it is concluded that the belt conveyor fault data conforms to the Weibull three-parameter distribution model and is verified. The results show that the belt conveyor fault data conforms to the Weibull three-parameter distribution model and lays a foundation for the reliability analysis of the belt conveyor.

(2) The Weibull three-parameter distribution model is used to analyse the reliability of the belt conveyor. The analysis results show that the belt conveyor will have a reliability of less than 0.75 during continuous operation for 24 hours, which provides a reference for the development of the maintenance cycle.

Acknowledge

This work also financially supported by the Shandong Higher Education Teaching Reform Research Project of China(2015M138), Qingdao Social Science Planning Project of China (QDSKL1801117), Shandong Natural Science Foundation of China (ZR2009HM003) and Ministry of Education, Higher Education Department, Industry, Education and Research Collaborative Education Project of China (201802197019).

References

[1] JING Hui, HAN Gang, Belt Conveyor System Fault Analysis Based on Fuzzy Set. 2015. J. *Coal Mine Machinery*.**36(05)**:305-307

[2] REN Zhongquan, SHEN Xuerong, ZOU Lili,The reliability analysis for the belt conveyor based on FTA. 2009. J. *Mining & Processing Equipment*. **37(17)**:53-55.

[3] Xie Xin, Zhang Liangyou, Reliability analysis for mechanical structure of transmission device of belt conveyer. 2016.J. *Hoisting and Conveying Machinery*. **(09)**:23-26

[4] Weihong Han,Xingshu Li,Yongqing Lv and Haichuan Liang, Application of Fault Tree Analysis to Reliability Analysis of Belt Conveyor. 2011.J. *Colliery Mechanical & Electrical Technology*. **(04)**:79-81+85

[5] Peng Li, *Research on Reliability Analysis and Fault Diagnosis Expert System of Belt Conveyor*. 2016.D(China University of Mining and Technology)

[6] ZHANG Dawei, MA Hongwei, CAO Xian'gang, DU Kang and DONG Ming, Study on fault diagnosis of belt conveyor based on Bayesian network. 2016. J. *Mining & Processing Equipment*. **44(05)**:31-33

[7] Blokus-Roszkowska A, Kolowrocki K, Failure dependency analysis of series system with example of conveyor belt. 2014 (The Slovak Republic :10th International Conference on Digital Technologies) P290

[8] Tomuş, Ovidiu-Bogdan, Andraş and Andrei, Magyari M, Study on the reliability of the underground conveyor belt system installed at Vulcan Mine the Jiu Valley Basin in Romania. 2017(Singapore: MATEC Web of Conferences) P121

[9] Dong Du, Qinghua Pang, *Modern comprehensive evaluation method and case selection*. 2005.M(Beijing: TSINGHUA UNIVERSITY PRESS)

IOP Conf. Series: Materials Science and Engineering **692** (2019) 012010 doi:10.1088/1757-899X/692/1/012010

Manufacturable Gradient-Variable Cellular Structures Design

Gang Li[1] and Yuan Yao[1,2,*]

[1] Rapid Manufacturing Engineering Center, Shanghai University, Shanghai, China
[2] Shanghai Key Laboratory of Intelligent Manufacturing and Robotics & National Demonstration Center for Experimental Engineering Training Education, Shanghai University, Shanghai, China

[*]yaoyuan@shu.edu.cn

Abstract. Topology optimization (TO) has been widely used in the field of additive manufacturing (AM), but the structure designed by the density-based optimization method has obvious material gradient and complex internal structures, which are difficult to be directly manufactured with specific process. This paper proposes a multi-scale design method for designing cellular structure with topology optimization. Firstly, micro units with variable density are designed to support smooth interconnections. Then, we provide a assembling method to generate models with density distributions similar to topology optimization method, while guarantee the manufacturability of the structures. The experimental results show that the generated model does not need support in fabrication, and retains the optimized mechanical properties, which can be used as a TO based fill pattern for 3D printing.

1. Introduction

In the past decade, TO methods for AM technology have become an active research filed. AM is able to create complex functional structures without increasing costs, so it can take advantage of topology-optimized designs to create innovative structures. This combination has an increasingly widespread impact in various industrial fields in recent years. However, applying TO methods in AM can also cause many problems that are difficult to solve.

First, TO methods usually generate internal structures with pores or large holes. Although AM is an efficient approach to fabricate object with complex structures, there are also many limitations. Specifically, when the design object is an internal structure with a complete outer casing. For some AM processes, such as the Fused Deposition Modeling (FDM), Stereolithography (SLA), and Selective Laser Melting (SLM), internal support structures are generated, which is difficult to remove in the post-processing.

Secondly, many applications have different requirements on the microstructures. For example, the pore size and connectivity of the microstructure have a very important influence on the cell attachment and growth in bioscaffold design. Controlling microstructure and macroscopic material distribution at the same time is a challenged task.

By combining the TO and the cellular structures design, this paper proposes a multiscale structural design method TO porous features. We can select and design the right microstructure according to the application needs. At the same time, the global distribution of materials is also considered to ensure that the design of the structure in the macro to meet the requirements of mechanics, lightweight and machine processing.

2. Related works

2.1. Topology optimization

TO method, also referred as layout optimization method, seeks to find the optimal load path for a certain boundary condition, which enables to find the best material distribution in the part that fulfils certain constraints, such as compliance constraints and displacement constraints. Since 1988, Bendse and Kikuchi introduce the homogenization method into the optimization process [1], big progress is have achieved in TO methods. Traditional structural optimization has been extended to solve some new optimization problems such as cellular structure design [2], mass transfer [3], metamaterial design [4], multifunctional material design [5] and optimization of performance and appearance fixed structures [6]. A lot of manufacturing-oriented TO algorithms are presented and the most representative one is density-based algorithm [7]. This method establishes the relationship between the relative density of voxel elements and the physical properties of materials. The relative density of elements is often used as the design variable. However, the most obvious disadvantage of this optimization method is that it produces intermediate density voxel elements that are difficult to manufacture.

The discrete structure-based solid isotropic material penalty function (SIMP) method [8] is a classic density based TO method, which considers that the intermediate density is not manufacturable. The penalty function is used to generate cavity and solid structures. The rational approximation of material properties (RAMP) method [9] is the other commonly used method to solve the problems that do not converge with SIMP [10]. In addition to using the above penalty function to solve the density optimization problem, the bi-directional evolutionary structural optimization (BESO) method [11] is developed to find the optimal value of the density distribution of the voxel material in the structure by using the dichotomy, and gradually discretize the intermediate density to a binary density structure of 0 and 1.

In order to ensure the manufacturability of the design results, more constraints are integrated into the local filter [12] [13]. While the design process is not intuitive and tends to fall into local optimization.

2.2. Cellular structure

Cellular structures such as foam, honeycomb, and lattice structure are used in applications due to their special mechanical properties. There are several existing methods to generate cellular structures, including model boolean operation (MBO), image-based, implicit surfaces, and TO method can be used to design cellular structures. The MBO method assembles new cellular structures by boolean operation on the basic cellular structures [14] [15]. The common used basic structures include rod units, plate units and body units.

Image-based method is to obtain cellular structures by reverse engineering through the micro-CT image. Cellular structures generated by this method are most consistent with the original model, and they are most used in the field of biological implants [16]. Implicit surface modeling method is a method of forming porously connected structures by minimal surfaces such as triply periodic minimal surfaces(TPMS). Using TPMS as the filling structure can improve the robustness of the structure designed by the optimization method [17].

Another direction to design cellular structure units is to use a pure TO method, which uses basic TO methods to design cellular structures that meet specific properties requirements. Although some articles suggest that an optimized self-supporting structure can be designed, it is not a priori and likely to be unstable or even failed [12].

The density-based optimization method can achieve optimal distribution of materials within the structure, but cannot guarantee the manufacturability. In addition, special microstructures are required in many applications. In order to obtain manufactured cellular structures with excellent mechanical properties, this paper proposes a unit assembly method combining density-based TO method and MBO structure design method.

3. Design framework

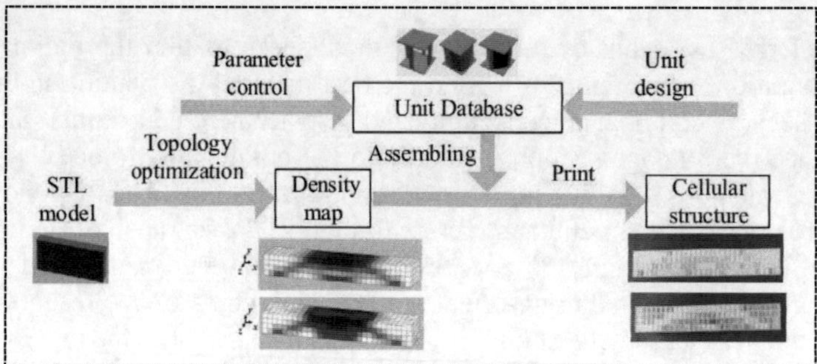

Figure 1. Design framework.

The design framework is shown in figure 1, which consists of three steps. Firstly, the voxel density distribution map of the macrostructure is obtained by the TO. Then, the designed microstructure unit families are selected by control parameters. And finally, the microstructures are assembled into the macrostructure through the mapping function, which generates cellular structure with certain distribution of material density.

3.1. Topology optimization

We take a compliance minimization problem as an example, the objective function is represented as

$$\arg_X \min C(X) = F^T U = U^T K U$$

$$s.t. \begin{cases} F = KU \\ V^* = \sum_{i=1}^{n} X_i \end{cases} \tag{1}$$

In equation (1), F and U represent the load and displacement of the structure, K represents the overall stiffness of the structure, V^* represents the residual material volume of the optimization target, and X_i represents the voxel element density of the macrostructure.

The TO algorithm used in this paper is the SIMP. As mentioned above, in order to make the design structure easier to manufacture the original SIMP algorithm uses penalty strategy. The intermediate density is penalized so that the intermediate density unit approaches both ends to generate the solid and the cavity. In assembly, we use a unit structure with different densities to fit the local density distribution. Therefore, the density distribution of the gradient is more appropriate. In the experiments, different penalty factors are used and tested.

3.2. Unit design

Once we get the density map by SIMP, each element of the map can be replaced with a unit structure with the same density. The key problem is to provide such type of unit, which has the ability to express a wide density range. There are different designs for different application requirements, but some of the must be met. These rules consist of effective structures, stable structures, manufacturable structures, and smooth connectivity. Then these units can be added into a unit database to support fast search and matching.

3.3. Assembling

Assembly is a process that includes one-to-one unit replacement and structure reconstruction. Given an element from the density map, the voxel with relative density less than a certain value X_{min} is determined to be the microscopic structural unit with the smallest material density. The position information of the units with different material densities can also be obtained from the density map. When selecting the same family of microstructural units, it is necessary to consider the problems of connection between different density microstructural units controlled by control parameters. If choose different families of microstructure units, it is still need to consider the connection problems between the units.

4. Unit design

As mentioned in section 3.2, a viable design unit must meet the following requirements: effective structures, stable structures, manufacturable structures, and smooth connectivity. The density range of a structural unit with a cavity is between $[X_{min},1]$. Where X_{min} is a lower limit to its density because the unit must have sufficient material density to ensure its own manufacturing and strength.

4.1. Effective structures

In principle, the material in a certain density can be randomly distributed, but it causes effective structures. Taking a two-dimensional structure as an example, as shown in figure 2. (a) shows the material inside the unit is directly broken off. (b) and (c) show the material inside unit have no contribution to the structural bearing capacity.

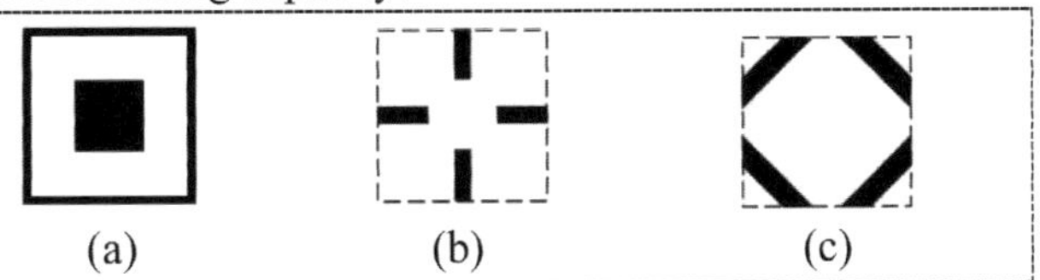

Figure 2. Two-dimensional ineffective unit structures.

There are several three-dimensional ineffective unit structures in figure 3. Figure 3(a) shows hexahedral voxel inner isolated sphere solid model is directly broken off. Figure3 (b) - (d) respectively show that the material in unit does not contribute to the structural bearing capacity.

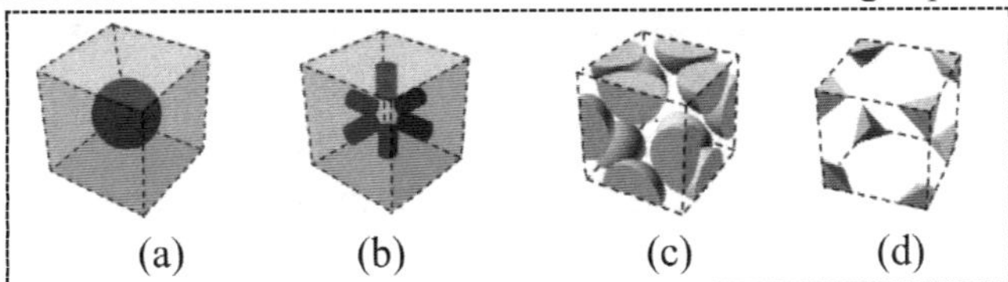

Figure 3. Three-dimensional ineffective unit structures.

4.2. Stable structures

The instability of unit structure includes the minimum size of the structural section and the largest structural stress area of unit. These structures have structural weaknesses as shown by the arrows in figure 4. The structural weakness will not only increase the macrostructural instability, but also reduce structural service life.

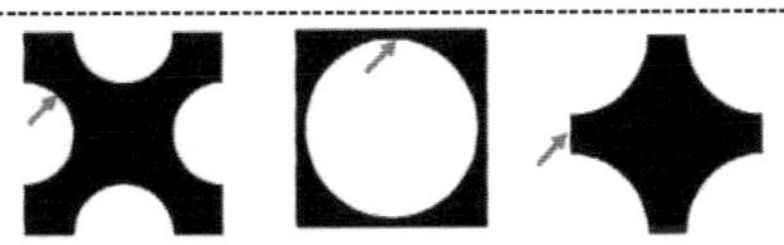

Figure 4. Two-dimensional unit with weakness.

There is also structural weakness in three-dimensional structural unit, as shown in figure 5. The unit consists of a spherical structure that is hollowed out in the cubic unit. Along with the material in the unit decrease, the structure unit becomes more and more unstable.

IOP Conf. Series: Materials Science and Engineering **692** (2019) 012010 doi:10.1088/1757-899X/692/1/012010

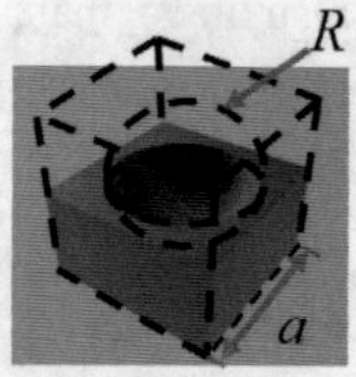

Figure 5. Three-dimensional unit with weakness.

4.3. Manufacturable structures

The manufacturability of the unit is related to the specific processing constraints and the minimum size that can be machined. There is no concept of support structure in the two-dimensional structures. When there is a cantilever structure in the three-dimensional structure, support structures will be required if the angle between the cantilever structure and the printing plane is less than a specific support angle. Take FCC unit in figure 6 as an example. Because of there are suspended beam structures in the upper and lower boundaries, this unit structure needs support structures.

Figure 6. FCC unit.

It is necessary to determine the minimum structural size of the structure and its position. We can judge the minimum size by the intersection of the structure and printing plane, as shown in figure 7. The material distribution of the FCC unit on print plane a and plane b, the minimum size structures on the planes are divided into rectangles, ellipses and quarter circles. Arrows 1, 2 and 3 refer to the minimum size of this structure.

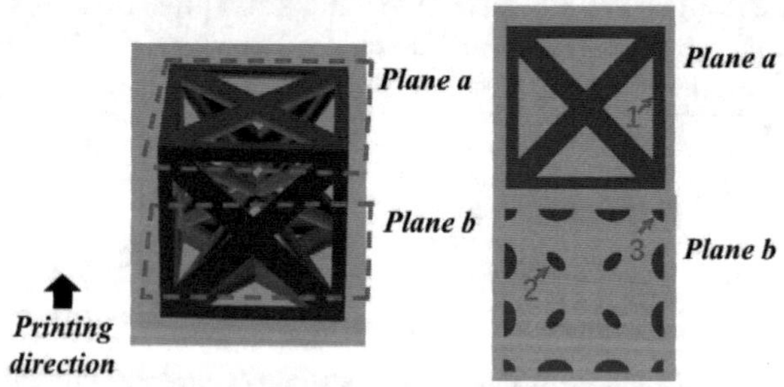

Figure 7. Different material distribution of FCC unit on plane a and plane b.

4.4. Smooth connectivity

The designed units need to be assembled into the macrostructure. Therefore, it is necessary to consider the connection problem between adjacent units. Figure 8(a) shows the mismatch between the same family, and figure 8(b) shows the mismatch situation of different family units. The mismatching phenomenon will not only cause stress concentration even cause structural failure. Similarly, in the three-dimensional structure, there is also a mismatch situation between the adjacent units. In order to avoid the mismatch situation of adjacent units, the units with similar boundary features will be designed or selected.

(a) (b)

Figure 8. Units connection.

4.5. Design strategy

Combining the above-mentioned unit design rules, a two-dimensional unit structure is designed. As shown in figure 9, the idea of the unit design is to design a hollow rectangular structure. This structure not only meets the manufacturability requirements but also ensures that adjacent units can be smoothly connected.

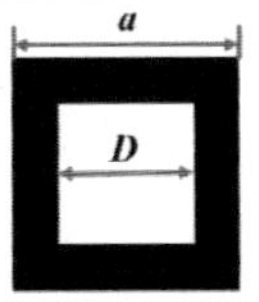

Figure 9. Designed two-dimensional unit.

For two- dimensional situation, the material density of such a unit is controlled by the inner hollow material quadrilateral side length *D*. Assuming that the side length of the quadrilateral unit is *a*, the unit material density can be calculated by equation (2).

$$\rho=(a^2-D^2)/a^2,(D\le a-2d_{\min},2d_{\min}<a) \qquad (2)$$

Where $d_{\min}$ represents the minimum manufacturing size of a manufacturing process.

According to the unit design rules, a three-dimensional structure is designed as shown in figure 10. The basic structural feature is the top and bottom plates in a hexahedral voxel unit, a central cross plate structure, and an inner hexahedral prism structure. The main control parameters of the three-dimensional unit are the top and bottom plate thickness d_1, the cross plate thickness d_2, the cross-sectional square of inner prism side length D. In order to make adjacent units smoothly connected, the dimensions of plate thickness d_1 and d_2 are fixed.

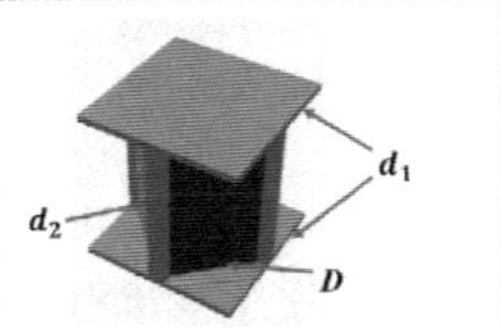

Figure 10. Designed three-dimensional unit.

Assuming the minimum processing size is $d_{\min}=0.5mm$ and parameters $d_1=0.2mm$, $d_2=0.5mm$, the density ρ can be directly calculated by equation(3).

$$\rho=(-D-\sqrt{2}/2)^2\times(a-0.4)+1.4a^2-0.65a+0.1,(0\le D\le\sqrt{2}a/2+\sqrt{2}/4) \qquad (3)$$

The range of the unit density can be obtained by adjusting the structural control parameter *D*. The range is $[(1.4a^2-0.6a+0.1)/a^3,(0.5a^3-0.7a^2-0.325a+0.05)/a^3]$ derived by equation (3), which is too narrow to map voxel relative density. It is necessary to extend the basic structural units of this family. In figure 11, upper row is shown as an overall view of the extend basic unit structures, and lower row is shown as the cross-sectional pattern corresponding to the upper row of structures.

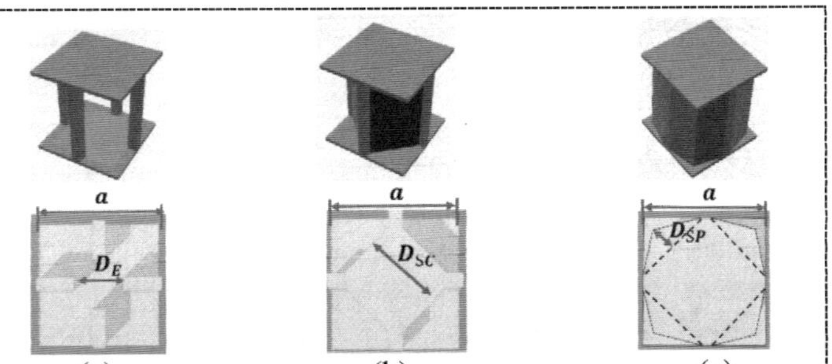

(a) (b) (c)

Figure 11. Basic structures. (a) Low density unit; (b) Intermediate density unit; (c) High density unit.

5. Assembling

Assembling is a process that convert density map from TO to a cellular structure. Given each voxel in the density map, the assembling process finds a microstructure from the unit database with the same material density. The density range of the selected unit family and the scale of unit must be determined, which have a large impact on the accuracy of the finite element analysis (FEA) calculation.

5.1. Determine the scale of unit

Unit assembly is complete based on the material density mapping relationship. Calculating the density range of two-dimensional unit is relatively easy to obtain directly by equation (2). The density range calculation is quite complicated of three-dimensional unit, as shown in equation (4). The density range consists of five subranges of three types of basic units.

$$\rho=\begin{cases}\left[1.4a^2-0.65a+0.1-(a-0.4)D_E^2\right]/a^3,(0\le D_E\le 0.5)\\ \left[1.8a^2-0.4a-(a-0.4)D_E\right]/a^3,(0.5<D_E\le a-0.8)\\ (1.4a^2-0.65a+0.1)/a^3,(0\le D_{SC}\le\sqrt{2}/2)\\ \left[1.4a^2-0.65a+0.1-(a-0.4)(D_{SC}-\sqrt{2}/2)^2\right]/a^3,(\sqrt{2}/2<D_{SC}\le\sqrt{2}/2a+\sqrt{2}/4)\\ 0.5a^3+0.7a^2-0.725a+0.05+\sqrt{2}(a^2-0.9a+0.2)D_{SP},(\sqrt{2}/2a+\sqrt{2}/4\le D_{SP})\end{cases} \quad (4)$$

The structural density control parameters of this family of microstructure units are the inner prism side length D_E, D_{SP}, D_{SC} and the unit scale a, which can provide a wide range of density. For a unit with complex structures, the density is difficult to calculate directly using equation. In this case, the interpolation method is used to calculate the density range.

The parameter a is proportional to the density range, but inversely proportional to the result accuracy of the FEA calculation. Taking a classic three-dimensional MBB structure as an example, the macrostructure size is $150\times 30\times 20mm$, and the smallest manufacturable unit size is set to $0.5mm$. Relationship of the deformation of macrostructure and unit family density range with unit scale are shown in figure 12.

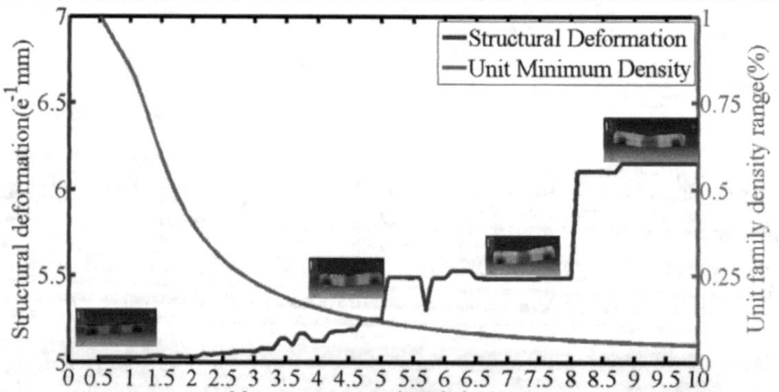

Figure 12. Relationship of deformation of macrostructure and material density with unit scale.

5.2. Structure reconstruction

Taking the microstructures shown in figure 11 as an example, using the limit thought we can know that $\rho_{\min}$ gradually approaches 0 with the increase of the scale of unit. The voxel with relative density between interval 0 and $X_{\min}$ has no corresponding microscopic structural unit. If the voxels in this region are mapped to empty material unit, a large number of support structures generate in macrostructure. This kind of cellular structure is not manufactured, so the strategy to solve above problem is to map voxels with relative density less than $X_{\min}$ to minimum density unit. Although this strategy leads to the material volume fraction of the overall structure larger than the optimized target volume, it can avoid generating support structures.

Given each voxel in the density map, we can get the density, the parameter of the corresponding unit can be calculated from equation (4). Then we can select the unit from the unit database. The replacement process is applied to each voxel of the density map. After these operations, a cellular structure model is constructed. Finally, the cellular model is converted into a mesh model by marching cube algorithm.

6. Experiment and discussion

The experiment is divided into two parts. Firstly, the SIMP algorithm obtains the three-dimensional density map. Then the unit family and the scale of the unit are selected and determined. The cellular structure with optimal material density distribution is constructed by the process mentioned above.

In the experiment, we use a classic MBB beam model with ($150 \times 30 \times 20mm$). The load and boundary conditions are shown in figure 13 (a). In order to compare the stiffness of our method and SIMP, different penalty values are used. The target volume fraction is 20%.

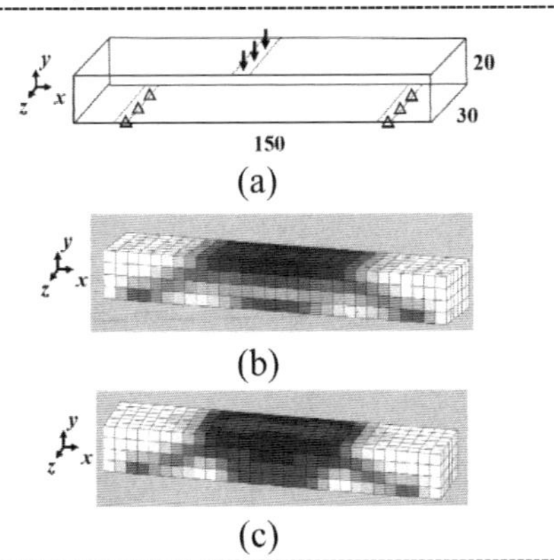

Figure 13. (a) MBB beam structure diagram; (b)Density map with P = 1; (c) density map with P=3.

We use the three-dimensional unit structures shown in figure 11 that can satisfy the requirements of manufacturability. The appropriate unit scale of such unit families has been solved according to figure 12, and it is 4.7mm.

Figure14 gives the fabricated results. Figure 14(a) is designed by SIMP. Its support structure is generated by Cura 3.6.0. Figure14 (b) is designed by our method. The penalty value used in SIMP algorithm is 1. In the assembling process, the voxel density that is less than $X_{\min}$ are mapped to empty material unit. The final model is processed by Cura 3.6.0. Support structures are generated because the cavities cannot be fabricated directly. Figure 14(c) also demonstrates a cellular structure designed by our method. In this model the voxel density that is less than $X_{\min}$ are mapped to the unit with the minimum density.

Figure 14(d) shows a fabricated mode with a shell structure. In this model the cellular structure we designed can be seen as a three-dimensional infill pattern with optimized material distribution. And figure 14(e) shows a shell structure that using a grid support structure generated in Cura 3.6.0. The infill material density is equal to the model shown figure 14(d). Finally, figure 14(f) shows the MBB structure under three-point bending test.

The three-point bending test results are shown in figure 15 and figure 16, respectively. The slopes of the different line graphs show the stiffness of different structures. The middle cliffs in the line graph indicate local failures in these structures. And the end cliffs indicate that the overall structures have failed.

From figure 15, the beam1 has the best overall stiffness. This structure can withstand a maximum load of 757 N and a corresponding maximum displacement of 2.4 mm. The stiffness of the beam3 obtained by our design method has a similar stiffness, which can withstand the maximum load of 733N. The corresponding maximum displacement is 2.87mm. The stiffness of beam2 is the worst among the three structures.

Figure 16 shows the stiffness comparison of beam4 with the structures designed as internal supports and beam5 with grid structures generated by Cura as internal supports. The maximum load that the beam4 can withstand is 196 N, and the maximum displacement is 3.5mm. Beam5 are 435N and 1.97mm. In fact, beam4 is not as good as beam5. The low-density unit structures designed in this paper is not as strong as a continuous grid structures. Because the volume fraction is too small that causes a lot isolated rod structures inside the generated model.

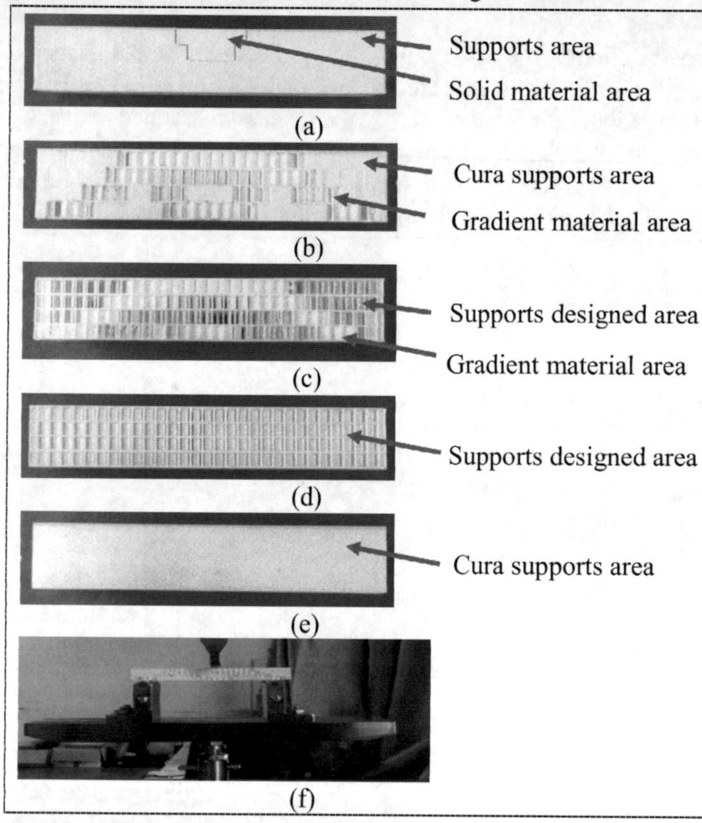

Figure 14. (a) - (e) show the structures of beam1-5, respectively; (f) three-point bending test.

For models generated by SIMP, we regulated with Cura to guarantee that all fabricated models have the same volume fraction. In summary, the stiffness of the structure designed by our method has the similar stiffness with the original SIMP optimized structure, but no additional support structures are needed.

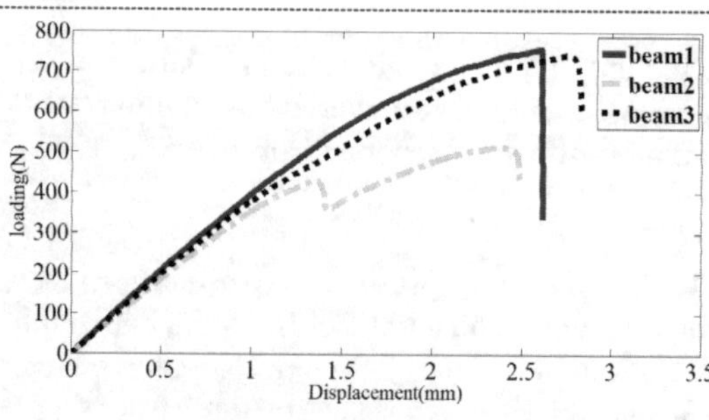

Figure 15. Load-displacement relationship diagram of beam1, beam2 and beam3.

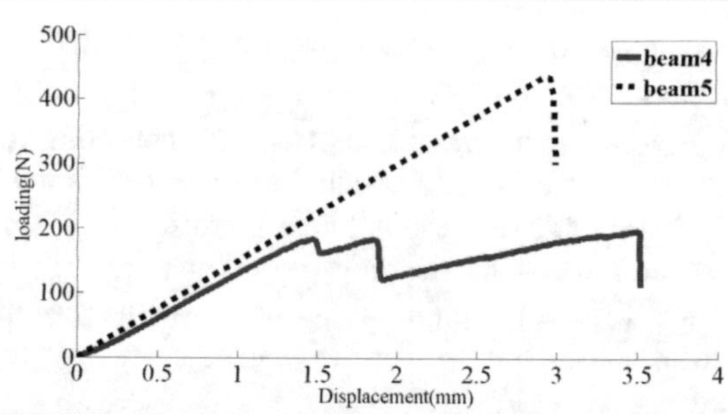

Figure 16. Load-displacement relationship diagram of beam 4 and beam 5.

7. Conclusions

The cellular structure design method proposed in this paper can generate global optimized structures with specified microstructures. The model designed is easier to manufacture. No additional supports are needed. The designed cellular structure can also be seen as a new fill pattern of the 3D printing structure to make the internal structure have the optimal material distribution. The next goals are to design more usable units and extend the internal support filling method to the traditional slicing software.

Acknowledgments

We thank the China-Austria collaborative project "Natural Enhancement for 3D Printing, from Nano to Continuous, for Bio-Inspired Applications" for support.

References

[1] M. P. Bendse, N. Kikuchi, Generating optimal topologies in structural design using a homogenization method, *Computer Methods in Applied Mechanics & Engineering* **71** (1988) 197-224.
[2] P. Zhang, J. Toman, Y. Yu, E. Biyikli, M. Kirca, M. Chmielus, A. C. To, Efficient design-optimization of variable-density hexagonal cellular structure by additive manufacturing: theory and validation, *Journal of Manufacturing Science and Engineering* **137** (2015) 021004.
[3] V. J. Challis, J. K. Guest, Level set topology optimization of fluids in stokes flow, *International Journal for Numerical Methods in Engineering* **79** (2010) 1284-1308.
[4] J. E. Cadman, S. Zhou, Y. Chen, Q. Li, On design of multi-functional microstructural materials, *Journal of Materials Science* **48** (2013) 51-66.
[5] X. Yan, X. Huang, G. Sun, M. X. Yi, Two-scale optimal design of structures with thermal insulation materials, *Composite Structures* **120** (2015) 358-365.
[6] J. Martnez, J. Dumas, S. Lefebvre, L. Y. Wei, Structure and appearance optimization for controllable shape design, *Acm Transactions on Graphics* **34** (2015) 1-11.
[7] H. P. Mlejnek, Some aspects of the genesis of structures, *Structural Optimization* **5** (1992) 64-69.
[8] M. P. Bendsøe, O. Sigmund, Material interpolation schemes in topology optimization, *Archive of applied mechanics* **69** (1999) 635-654.
[9] M. Stolpe, K. Svanberg, On the trajectories of penalization methods for topology optimization, *Structural and Multidisciplinary Optimization* **21** (2001) 128-139.
[10] O. Sigmund, K. Maute, Topology optimization approaches, *Structural and Multidisciplinary Optimization* **48** (2013) 1031-1055.
[11] X. Huang, Y. Xie, Convergent and mesh-independent solutions for the bi-directional evolutionary structural optimization method, *Finite Elements in Analysis and Design* **43** (2007) 1039-1049.
[12] J. Wu, C. C. L. Wang, X. Zhang, R. Westermann, Self-supporting rhombic infill structures for additive manufacturing, *Computer-Aided Design* **80** (2016) 32-42.
[13] J. Wu, Continuous optimization of adaptive quadtree structures. *Computer-Aided Design*, **102** (2018) 72-82.
[14] S. Xu, J. Shen, S. Zhou, X. Huang and Y. M. Xie, Design of lattice structures with controlled anisotropy, *Materials & Design* **93** (2016) 443-447.
[15] T. Tancogne-Dejean, M. Diamantopoulou, M. B. Gorji, C. Bonatti and D. Mohr, 3d plate-lattices: An emerging class of low-density metamaterial exhibiting optimal isotropic stiffness, *Advanced Materials* **30** (2018) 1803334.
[16] W. Sun, B. Starly, A. Darling and C. Gomez, Computer-aided tissue engineering: application to biomimetic modelling and design of tissue scaffolds. *Biotechnology and applied biochemistry* **39** (2004) 49-58.
[17] D. W. Li, W. H. Li, N. Dai, G. Y. Dong, Y. L. Tang and Y. M. Xie, Optimal design and modeling of gyroid-based functionally graded cellular structures for additive manufacturing, *Computer-Aided Design* **104** (2018) 87-99.

IOP Conf. Series: Materials Science and Engineering **692** (2019) 012011 doi:10.1088/1757-899X/692/1/012011

Research on Motion Characteristics of Single Particle Third Body in Braking Process Based on W-M Fractal Feature

Zhihua Sha, Guotao Ru, Jian Yin, Yu Liu*, Fujian Ma, Dapeng Yang, and Shengfang Zhang

School of Mechanical Engineering, Dalian Jiaotong University, Dalian, China 116028

***E-mail:** liuyu_ly12@126.com

Abstract. In order to explore the motion characteristics of the third body particle in the friction of the rough braking interface, based on W-M fractal theory and elastoplastic mechanics theory, the finite element model of the three body micro-extrusion shearing process of rough contact surface is established. The influence of the braking parameters on the motion characteristics of the single-particle third body is analyzed. The analysis results show that the three body contact steady state increases with the increase of brake pressure, and the shear speed affects the three body contact vibration. An increase in the friction factor accelerates the third body discharge friction gap to exacerbate the wear condition.

Key words: third body; rough surface; W-M fractal theory; motion characteristics

1. Introduction

In the process of friction and wear, a series of physical and chemical changes occur on the friction surface through the interaction of heat and force, which causing damage to the friction surface. The abrasive debris generated after the damage of the matrix remains in the friction surface and continues to participate in friction, which is called the third body. The formation of the third body transforms the ideal two-body friction system into a three body friction system. Many scholars have carried out fruitful research on the formation and evolution of the third body.

Godet[1] proposed the concept of a third body to describe the solid particle that existed between the two opposing moving frictional contact surfaces from the matrix. Greenwood[2] et al. proposed an elastic and elastoplastic hybrid contact model between rough surface and smooth surface on the basis of statistics, namely G-W contact model. Edouard[3] shown that the evolution of the vibrational behaviour and of the third-body layer are both gradual. Lin[4] studied plastic deformation zone grains lead to the shear deformation by means of friction stress. Wang[5] used a discrete parallel element method was used to construct a double parallel plate shear model to simulate the infinite shear motion, focusing on the expansion of the particle system under shear. Wang[6] built a caulking and coverage effect model based on the G-W model. Complete the transformation from rough surface contact to three dimensions. Yang J[7] established the 2-D granular matter parallel shearing cell models with different shape of micro-asperity surfaces by the explicit finite element method(FEM). These models were applied to analysing the influence of normal load and factor of friction of contact pairs on the gap. The above work is mainly aimed at the formation of the third body and the influence on the friction performance. However, the research on the existence state of the third body particle in the friction interface is not deep enough. In this paper, the finite element model of elastoplastic meso-three body extrusion shearing process for rough braking interface is established by using ABAQUS. It mainly studies the influence of braking parameters on the deformation and motion of the third body particle, to provide a theoretical basis for understanding the three body friction mechanism.

2. Rough Surface Model Based on W-M Fractal Function

The W-M fractal function is a typical function used to represent random contours[8]. It has self-imitation and can accurately simulate and establish random surfaces with fractal characteristics, continuous and non-differentiable. The mathematical model expression is:

$$Z(x) = G^{(D-1)} \sum_{n=n_1}^{\infty} \gamma^{(D-2)} n \cos(2\pi\gamma^n x) \qquad (1 < D < 2, \gamma > 1) \tag{1}$$

Where, $Z(x)$ is the height of the random surface topography; and x is the position coordinate of the contour; D is the fractal dimension, which describes the irregularity of the function $Z(x)$ at all scales; G is the characteristic scale coefficient, which reflects the magnitude of $Z(x)$ and determines the specific size of $Z(x)$. γ represents the spatial frequency of the contour and for random contours subject to normal distribution, it is suitable for the randomness of high frequency spectral density and phase, generally, γ=1.5; n is a natural sequence number, n = 1, 2, 3,...,100; G is the fixed parameter mesoscale factor; G=0.01. Based on the W-M fractal theory, the simulation of the fractal curve under specific parameters is realized by MATLAB software programming. The fractal dimension of the main parameters can be obtained by measuring the roughness of the friction surface and then converting it. The fractal dimension is determined to be 1.9 and the fractal curve is simulated. Perform spline fitting and finally draw a two-dimensional rough surface profile as shown in Figure 1.

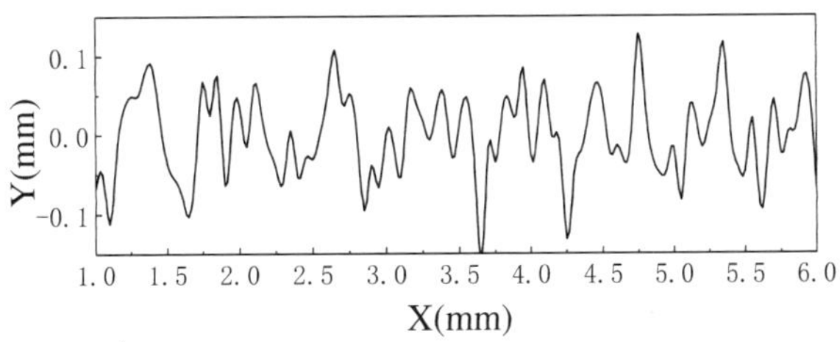

Figure 1. 2D rough surface profile with D=1.9.

3. Establishment of Finite Element Model for Three Body Extrusion Shear of Single Particle

The rough braking surface is damaged by the braking shearing force. After the third body particle fall off the matrix, the third body in the free state exists in the two rough friction surfaces and directly participates in the friction. A single-particle three body extrusion shear finite element model was established using the ABAQUS/Explicit module. The third body particle in the model is considered as mixture particle falling off the matrix, and the size is uniform in the shape of round particle. The upper and lower friction surface topography models are obtained by introducing the Creo fitting with the obtained fractal curve. The model is imported into ABAQUS to complete the assembly as shown in Figure 2.

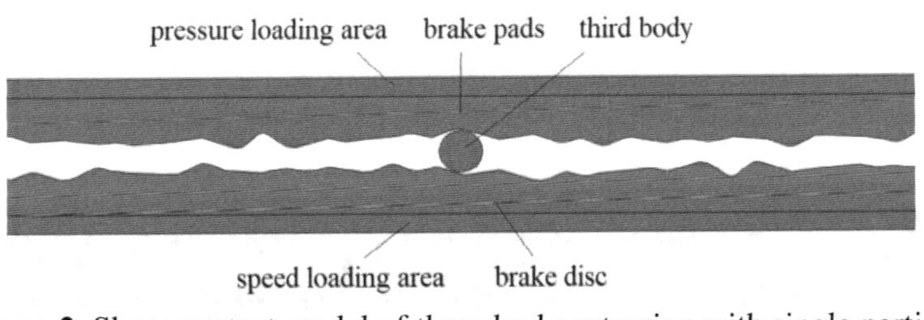

Figure 2. Shear contact model of three body extrusion with single particle.

Considering that the simulation object is a single-particle third body, a large bending moment and deformation may be generated after applying a downward load on the upper surface, which affects the accuracy of the result, and an independent force layer is provided on the upper surface for applying pressure. The material properties of the force layer are set to be extremely dense and rigid, it is not easily deformed, so that the upper plate can be uniformly stressed. Since the force layer is not directly involved in friction, its material properties have little effect on the simulation results.

Downward braking load is applied on the upper surface and the X-direction displacement of the upper surface is limited; the speed is applied to the right on the lower surface and the displacement of the lower boundary is limited in the Y direction. The influence of braking parameters on the stress and strain distribution and motion process of the third body during shearing is studied. The braking parameters are given in Table 1. The loading time is 0.4ms in the first step of simulation. After the steady contact of the three bodies, the speed is applied in the second step of simulation analysis for 2ms. In the shear process of three body friction interface, there is contact friction behavior between particle and surface. The penalty function friction model of elastic slip is adopted. In the model, the particle and upper plate materials are copper-based powder metallurgy materials, and the lower plate materials are Q345B. Specific material parameters are shown in Table 2 and Table 3.

Table 1. Variable table of braking parameters.

Variable name	Variable value					
Brake pressure（p/MPa）	1.5,	3,	4.5,	6		
Motion speed（v/mm·s^{-1}）	400,	800,	1000,	1250,	1600,	2000
Friction factor	0.1,	0.2,	0.3,	0.4,	0.5	

Table 2. Table of material parameter.

Material	Density（kg·m^{-3}）	Elastic Modulus（MPa）	Poisson's ratio
（Upper plate）Copper-based Powder Metallurgy	5250	5.2×10^3	0.3
（Lower plate）Q345B	7850	2.06×10^5	0.28

Table 3. Stress-strain relationship in plastic deformation stage of materials.

（Upper plate）Copper-based Powder Metallurgy	Stress	160	202	238	268	304	344	370
	Strain	0	0.0069	0.0202	0.0358	0.061	0.101	0.15
（Lower plate）Q345B	Stress	345	391	439	519	582	625	
	Strain	0	0.0235	0.0474	0.935	0.1377	0.18	

4. Simulation results and analysis

4.1 Analysis of extrusion shear process

Figure 3 shows the stress distribution of the three body contact model when the braking pressure is 4.5MPa and the speed is 1000mm/s and the braking process is 0.4ms, 0.8ms, 1.2ms and 2.4ms.

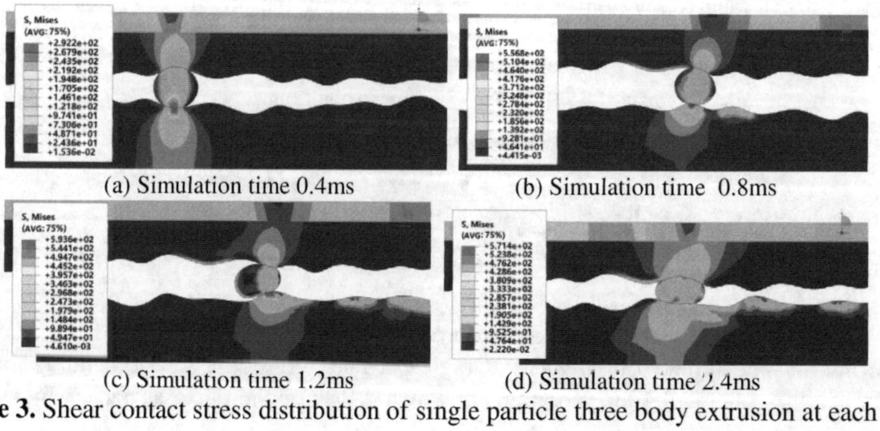

(a) Simulation time 0.4ms (b) Simulation time 0.8ms

(c) Simulation time 1.2ms (d) Simulation time 2.4ms

Figure 3. Shear contact stress distribution of single particle three body extrusion at each time.

The upper plate moves downward with the application of the stress, and the particle are squeezed to the pits of the upper and lower plates. The positions of the particle and the lower plate are constantly adjusted until they reach a stable state. From 0ms to 0.8ms, in the process, the particle remains basically round before reaching the yield limit, and they roll counterclockwise to the right under the action of friction, accompanied by slippage. Because the loading speed from 0.8ms to 2.4ms reaches the maximum value, the particle is affected by the forces of the upper and lower plates. Contact with the lower plate micro-convex body is the cause of stress concentration. During this process, the third body particle are increased in plastic strain region due to the continuous impact of the micro-protrusions, and the particle are elliptical. At this time, the friction force is not enough to cause the elliptical particle to continue to roll, the particle is tightly embedded in the dimples, and a "filling" phenomenon occurs, the plastic deformation occurs in the lower plate due to the scratching action of particles, which results in the ploughing effect. Until the particle slides to encounter a micro-convex body that can produce enough impact force to overcome friction and release it from the pit. When the upper and lower plates are in contact with the particle. Residual stress is generated at the contact position after the two plates are stressed. The larger the residual stress value generated by the contact of the micro-convex with small curvature, the stress is more concentrated.

4.2 Effect of load on the motion characteristics of third body particle

Figure 4 shows the time when the particle reaches the maximum yield stress when the braking pressure is changed. Figure 5 shows that the particle reaches the maximum equivalent plastic deformation when braking pressure is changed. Third body particle can produce plastic deformation, there is only difference in the degree of deformation, so the time to reach the maximum yield stress of the material is the same as the time to start producing plastic deformation. Furthermore, the stress and strain changes of the third body particle in the extrusion shearing process are analyzed.

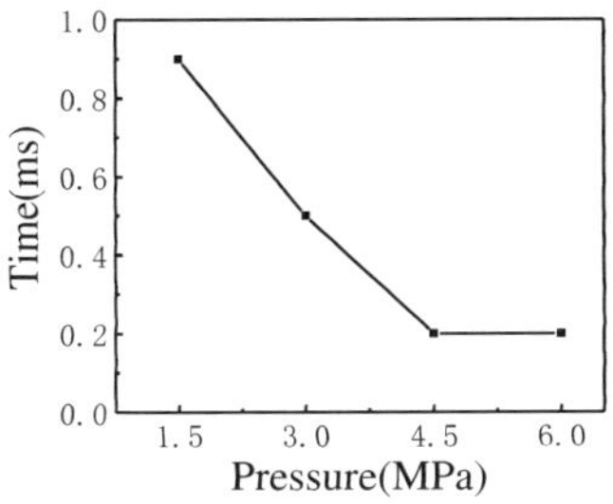

Figure 4. Chart of time with different pressure.

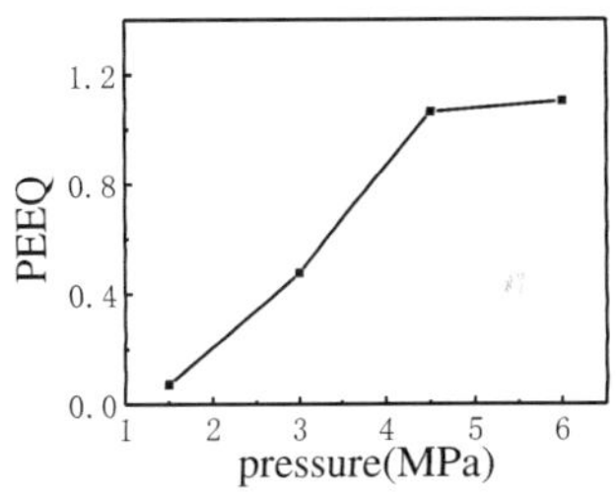

Figure 5. Chart of PEEQ with different pressure.

Brake pressure increased from 1.5MPa to 4.5MPa, the time of maximum stress from 0.9ms gradually reduced to 0.2ms, the maximum stress of the time will not change along with the change of brake pressure as the pressure continues to increase, it remains at 0.2ms, because of the change of the brake pressure directly affect the contact stress, it can make the third body particle quickly reach the yield stress, the stress and the speed of loading takes time, so the brake pressure to add after 4.5MPa, the maximum stress in great changes of time no longer. When the braking pressure is 1.5MPa, it is difficult to reach the yield limit of the material and the plastic shape variable is small. When the braking pressure increases to 4.5MPa, the equivalent plastic strain rapidly increases to 0.4. When the braking pressure increases to 6MPa, the equivalent plastic strain value generated by the third body slowly increases to 1.2, and still shows a trend of increasing with the increase of braking pressure. In addition, the upper plate is lifted due to the contact between the micro-convex body and the particle of the third body in the process, and the plate spacing changes at any time with the contact between the micro-convex body and the third body. In the whole three body extrusion shearing process, when the pressure is 1.5MPa, the fluctuation is the most obvious. The three body contact stability is poor at this time. With the increase of brake pressure, brake pressure makes the three body contact close enough. The stability of three-body contact increases at this time, the fluctuation degree of plate spacing decreases, and the plastic deformation degree of the third body particle increases.

4.3 Effect of motion speed on the motion characteristics of third body particle

Figure 6 shows the time when the particle reaches the maximum yield stress when the shear speed is changed. Figure 7 shows that when the shear speed is changed, the particle reaches the maximum equivalent plastic deformation. It can be seen from Figure 6 that with the speed increases, the plastic deformation time of the third body particle decreases. It means that faster speed will make the third body contact with more micro-convex bodies at the same time. The time for the third body to reach the maximum equivalent plastic deformation is also gradually shortened.

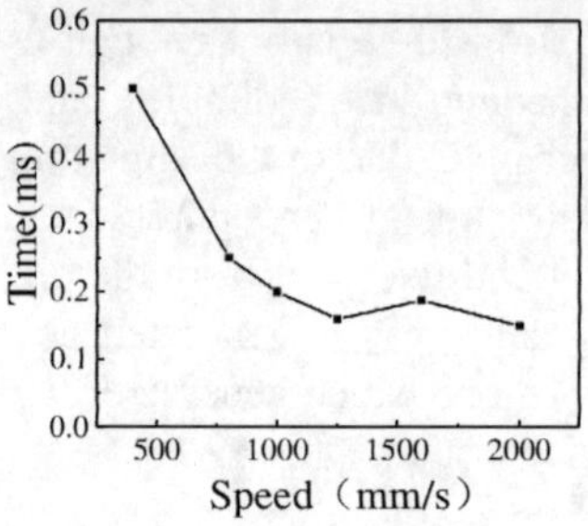

Figure 6. Chart of time with different speed.

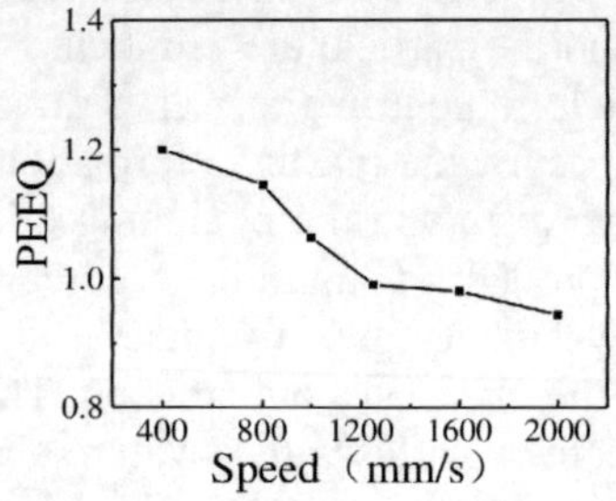

Figure 7. Chart of PEEQ with different speed.

The third body maximum equivalent plastic shape variable showing a downward trend with speed increases from 400mm/s to 1200mm/s. The equivalent plastic shape variable is 1.20 when the speed is 400mm/s, and increases from 1200mm/s to 2000mm/s. The maximum value gradually decreases to around 0.95 and then tends to be stable. Because when the speed is small, the third body particle with two rough contact surface contact state is better, in the process of extrusion shear, the third body particle can be closely contact with each micro-convex body, the contact for a long time, make rough contact surface particle pass more energy to the third body, the third body particle equivalent plastic deformation; the third body particle and two rough contact surface contact state is not stable with speed increase, there is radial runout phenomenon. The contact time between the third-body particle and the micro-convex body becomes shorter in extrusion shear process. At this time, the rough contact surface only transfers limited energy to the third-body particle. The degree of grain deformation in the third body is reduced, and it fluctuates in a smooth range.

4.4 Effect of friction factor on the motion characteristics of third body particle

The force acting on the upper and lower plates acts on the third body particle through the form of extrusion and shear of friction force. In this process, the friction factor has a direct impact on the effect of the third body. Figure 8 shows the time when the particle reaches the maximum yield stress when the friction factor is changed. With the increase of friction factor from 0.1 to 0.5, the time when the third body particle reaches the maximum equivalent plastic deformation fluctuates between 0.2ms and 0.3ms. Removing the influence of random contact surface on simulation results, it can be considered that the friction factor has little effect on the time to reach the maximum stress of the third body.

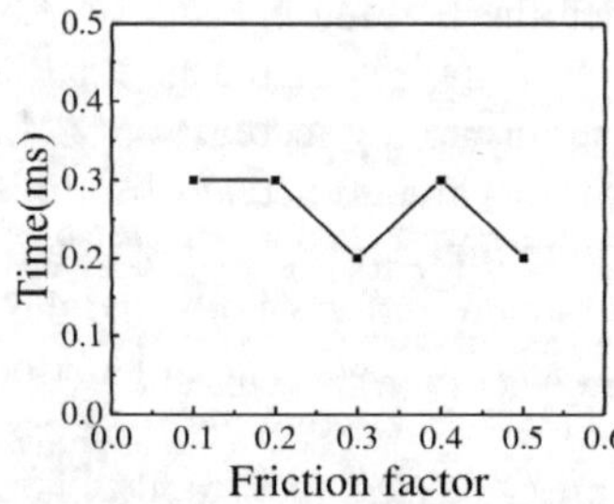

Figure 8. Chart of time with different friction factor.

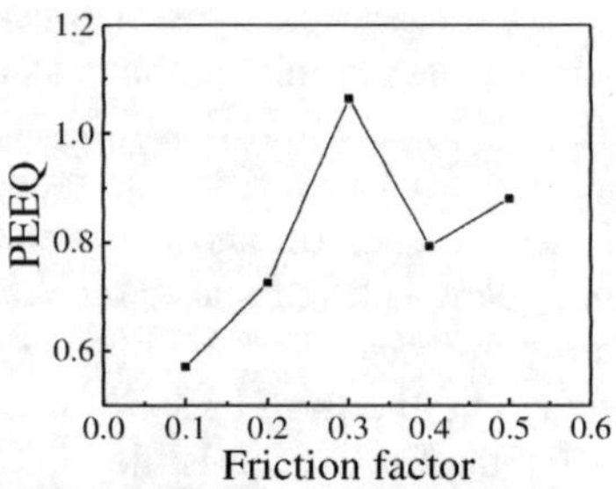

Figure 9. Chart of PEEQ with different friction factor.

Figure 9 shows the maximum equivalent plastic deformation of particle when the friction factor is changed. It can be seen from the fitting curve that the maximum equivalent plastic deformation of the third body particle increases gradually with the increase of friction factor from 0.1 to 0.3, and the minimum equivalent plastic deformation is 0.57 when the friction factor is 0.1. The maximum equivalent plastic deformation is 1.06 when the friction factor is 0.3. The third body particle are rolled to the right under the action of friction force during this process, and with a little slip. When passing through the larger pit on the upper plate, the friction force given by the lower plate is not enough to make it get rid of the pit. The micro-convex body on the lower plate continuously impacts the third body particle that falls into the pit. The friction force of the third body particles increases with the increase of friction coefficient, the force concentration in the contact area between the particle and the lower plate micro-convex body increases, and the deformation degree of the third body increases. In the process of increasing friction factor from 0.3 to 0.5, when the particle falls into the pit on the upper plate, the micro-convex body of the lower plate gives enough force to the particle to get rid of the pit and continue to roll to the right. At this time, the force point of the particle is relatively uniform, and the equivalent plastic deformation of the third body decreases to about 0.85.

5. Conclusions

In this paper, the three body micro-extrusion shear process of brake interface is simulated, the finite element model is established and the dynamic characteristics of single-particle third-body of rough brake interface are analyzed. The results show that the braking conditions have a direct impact on the stress and strain of the third body particle. After the third body is loaded, the particle is embedded in the pit, the lower plate is plastically deformed by particle sliding action, resulting in the "ploughing effect". The increase of brake pressure can effectively improve the stability of three-body contact, and the shear speed mainly affects the vibration of three-body contact. The increase of friction coefficient can shorten the removal time of the third body from the friction surface clearance and increase the wear condition.

Acknowledgments

The authors gratefully acknowledge the financial support of National Natural Science Foundation of China (No.51675075) and Innovative Talents Program of Colleges and Universities in Liaoning Province (No. LR2018048).

References

[1] Godet M 1984 The third-body approach: a mechanical view of wear *j.Wear* **100 (1/2/3)** pp 437-452.
[2] Greenwood J A, Williamson J B P 1966 Contact of Nominally Flat Surfaces *J.Proceedings of the Royal Society of London. Series A* **295(1442)** pp 300-319
[3] Edouard Davin, Anne-Lise Cristol, Jean-François Brunel, Yannick Desplanques. 2019 Wear mechanisms alteration at braking interface through atmoparticle modificatio *J.Wear* pp 426-427
[4] Lin Liu,Yuanyuan Sheng, Ming Liu, Martin Dienwiebel, Zhichen Zhang, Davoud Dastan 2019 Formation of the third bodies of steel sliding against brass under lubricated conditions *J.Tribology Internationa* pp140
[5] Wang W, Shi C, Wang Z , et al. 2015 Effect of multi-parameters on shear expansion process of three-body friction interface *J. Journal of Applied Mechanics* **32(5)** pp 731-737
[6] Wang W. 2015 Three-body contact modeling and mechanical analysis of rough friction interface with granular media *D. Hefei University of Technology*
[7] Yang J, Wang W, Ma S, Liu K. 2013 Effect of surface micro-convex shape on three-body friction interface gap *J. Journal of Hefei University of Technology (Natural Science Edition)* **36(06)** pp 646-649+687
[8] Li C, Dong S. 2000 Characterization trend of three-dimensional surface microtopography *J. China Mechanical Engineering* **11(5)** pp 488-492

Study on Thermal Erosion Process of SiC_p/Al Composite Material in EDM

Yu Liu, Bingchao Wang,Wenchao Zhang, Shengfang Zhang*, Fujian Ma, Dapeng Yang and Zhihua Sha

School of Mechanical Engineering, Dalian Jiaotong University, Dalian, China 116028

*E-mail: zsf@djtu.edu.cn

Abstract. SiC_p/Al composites are composed of two different material properties of aluminium matrix and SiC particles. In order to deeply study the erosion process of SiC_p/Al composite in EDM, a single pulse discharge model of SiC_p/Al composites was established. The melting and solidification model and the volume of fluid (VOF) model in Fluent software were used to simulate the thermal erosion process of EDM. The change of temperature field and the formation of molten pool in the heating process of SiC particles and Al matrix are analysed. The number of SiC particles in molten pool is counted and analysed, and the number of SiC particles in molten pool was counted, the simulation results were verified by the single pulse EDM experiments. The results show that the depth-diameter ratio of the molten pool increases with the increase of the pulse width, and decreases with the increase of the peak current. The number of melted particles in molten pool is basically unchanged.

Keywords: EDM; SiC_p/Al Composite Material; Thermal erosion process; SiC particle; Depth-diameter ratio

1. Introduction

SiC_p/Al composite is a typical particle reinforced metal matrix composites (PRMMCs) with excellent properties such as high thermal conductivity, high elastic modulus and high hardness, it is widely used in military and aerospace fields. Due to its non-contact processing, EDM is not limited by the strength and hardness of the material, and has better advantages in the processing of SiC_p/Al composites. However, EDM machining PRMMCs have serious problems such as severe tool wear and difficult removal of debris, which limits the application of EDM machining PRMMCs. Therefore, it is very necessary to study the erosion mechanism of SiC_p/Al composites in EDM to solve the problems existing in the EDM of PRMMCs. The processing and discharge process is very short, the processing gap is narrow, which is not conducive to direct detection and observation, it is very suitable for simulate the thermal erosion process of EDM by finite element numerical simulation method. At present, in terms of material thermal erosion, Feng Ruiye et al. have simulated the temperature field of TiC/Ni ceramic EDM, established the random particle distribution model, and explored the change rule of discharge parameters on the erosion volume[1]. Gao Yang et al. conducted thermal analysis on the erosion crater of the discharge, and believed that the heat source in the plasma channel was the main driving force for the erosion of the workpiece material. Under different peak current and pulse width, the model could well predict the changes of the depth and radius of the erosion crater[2]. Weingartner et al. studied the temperature field generated by point heat source, dish heat source and time-varying heat source through simulation, found that time-varying heat source was more consistent with the actual processing results[3].

In summary, most of the workpiece materials used in the thermal erosion process of EDM are single component materials or the reinforcement particles and matrix are considered as a whole analysis, the properties of the particles and matrix are rarely distinguished. This paper establishes a model in which SiC particles are uniformly distributed in the aluminium matrix, set different material properties to the SiC particles and the aluminium matrix, and using the melting and solidification model to simulate the single pulse discharge thermal erosion process of SiC_p/Al composites, it is helpful to explore the regularity of erosion change of discharge craters.

2. Establishment of Single Pulse EDM Model

2.1. Physical model

Figure 1 depicts the single pulse discharge process of SiC_p/Al composite material EDM. Under pulse discharge, a plasma channel is formed between tool electrode and workpiece material, which is composed of plasma generated by ionization of dielectric. This process converts the pulsed power into thermal energy and enters the workpiece through heat conduction. At high temperature, the workpiece material is molten and vaporized.

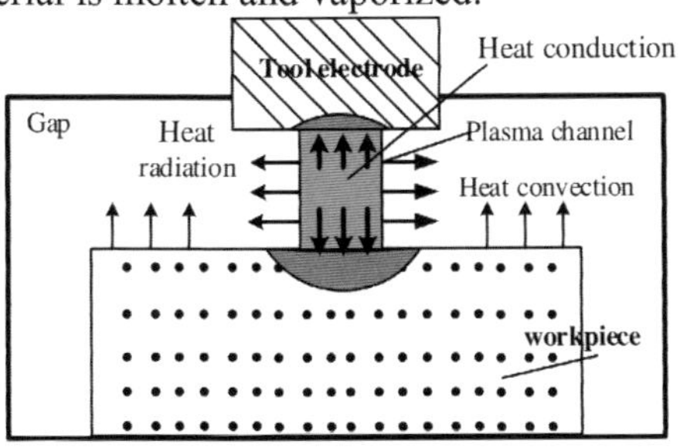

Figure 1. Schematic diagram of the SiC_p/Al composite material EDM.

2.2. Mathematical model

The heat source distribution of EDM is close to the Gaussian distribution, and the heat flux density in the plasma channel at the point *r* away from the center of discharge point can be expressed as[4]:

$$q(r) = \frac{k \cdot \eta \cdot U(t)I(t)}{\pi R^2} \cdot \exp(-k\frac{r^2}{R^2(t)}) \tag{1}$$

where, $U(t)$(V) is the voltage, $I(t)$ (A) is the current; η for the energy distribution coefficient of anode workpiece, which is 0.3; $R(t)$ (m)is t moment plasma channel radius; k is the concentration factor of the heat source, which is 3.

2.3. Model Schematics and Boundary Conditions

Figure 2 shows the composite material model of thermal erosion phase. The matrix and the particles are simplified into a two-dimensional model. To reduce the amount of calculation, a symmetric model is established, and only the influence of the heat source on the workpiece is considered. The depth-to-diameter ratio of the discharge crater is generally less than 1, so the length of the model along the radial direction is 250μm, and the length along the depth direction is 100μm. According to Yu Xiaolu et al[5], the particle size of SiC is distributed in 3-8μm, the model has a particle size of 8μm. The uniform distribution model of the particles, the circular area is the particles, and the rest is the aluminium matrix, which respectively gives their respective material properties. The interface boundary is set as the connection between particle and matrix, which can realize data exchange. Table 1 shows material selection and parameter settings.

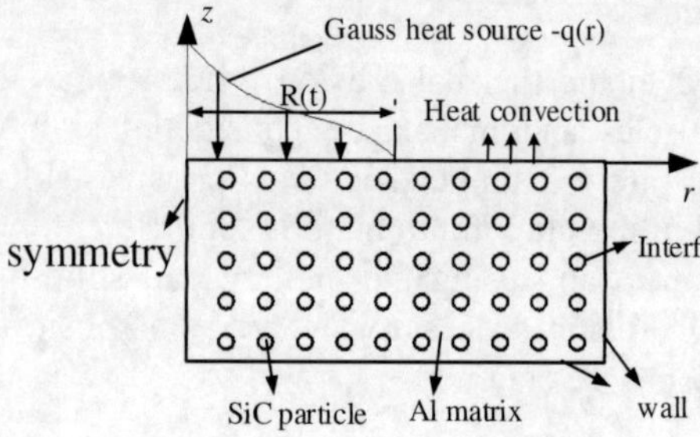

Figure 2. Model Schematics and Boundary Conditions.

Table 1. Simulation parameters settings.

Workpiece material	Solver	Voltage (V)	Peak current(A)	Pulse Width (μs)
SiC_p/Al composite materials	Pressure-Based	45	6,8,10,15,20	50
		45	10	10,20,30,40,50

3. Simulation results and analysis

3.1. Changes of molten pool morphology at different discharge times

Figure 3 shows the formation of the molten pool on the surface of the workpiece during discharge and the presence of the particles. In red, the liquid fraction is 1, representing the completely melted material. In blue, the liquid fraction is 0, indicating that the material is solid. As shown in Figure 3 (a), the matrix material begins to melt at 1μs, and partial particles begin to melt because the melting point of the SiC particle phase is higher than the Al matrix. As the discharge progresses, at 50μs, molten pool tends to be flat, the shape is basically no change, and some particles exist in the molten matrix because they have not reach melting point, as shown in Figure 3 (b).

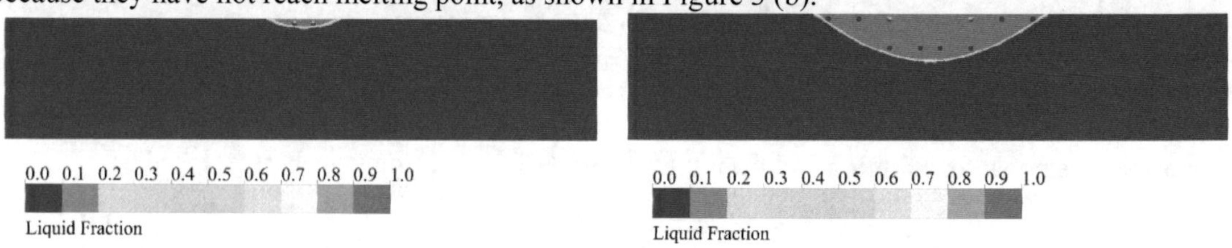

(a) 1μs. (b) 50μs.

Figure 3. Changes of molten pool morphology at different discharge times.

3.2. Effect of pulse width on thermal erosion of materials

Figure 4 and Figure 5 are the curves of temperature along the radial direction and depth direction at different pulse widths when the peak current is 15A. The smaller the pulse width is, the greater change of temperature is. As the pulse width increases, the highest temperature on the surface of the workpiece first decreases and then rises when the pulse width is 40μs. Because as the radius of the discharge channel expands, the energy gradually decreases, and the temperature decreases. After the optimal pulse width is reached, the discharge channel no longer changes, the heat in the molten pool accumulates, and the temperature begins to rise. The temperature does not change substantially in the radial direction at 15μs, but the temperature in the depth direction drops sharply. This is attributed to the fact that the EDM heat source is mainly a surface heat source, and is approximately Gaussian distribution, the heat source is directly loaded in the radial direction. Due to the large temperature gradient in the depth direction, the temperature changes rapidly in a small range.

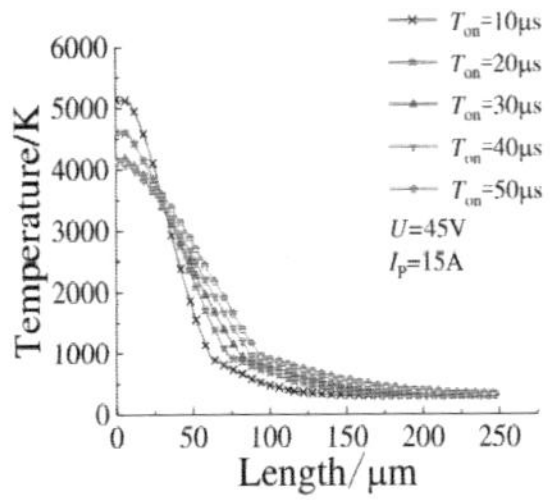

Figure 4. Temperature change in radial.

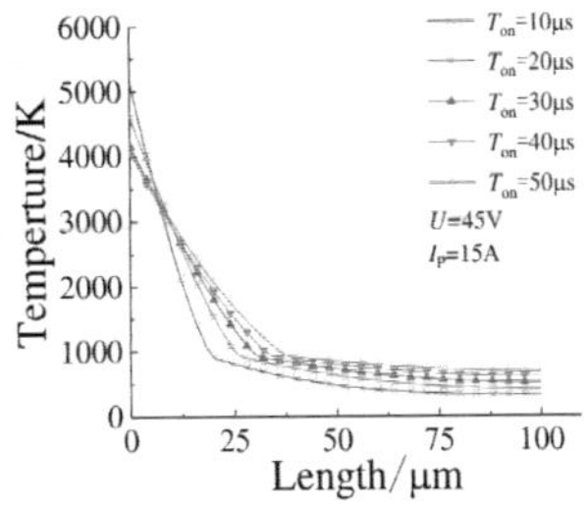

Figure 5. Temperature change in depth.

Figure 6 shows the radius and depth of the molten pool increase with pulse width increase, and the change of radius length is large at first and then low with pulse width increased. Because there is thermal convection in the radial direction and heat dissipation of the dielectric, while the depth direction is only thermal conduction. In addition, when the optimal pulse width is reached, the radius of plasma channel remains unchanged, because the temperature of the heat source center is higher than the edge, so the depth direction changes faster than the radius. From Figure 7, the depth-to-diameter ratio of the molten pool increases with pulse width increases, and basically unchanged from 30us to 40μs pulse width.

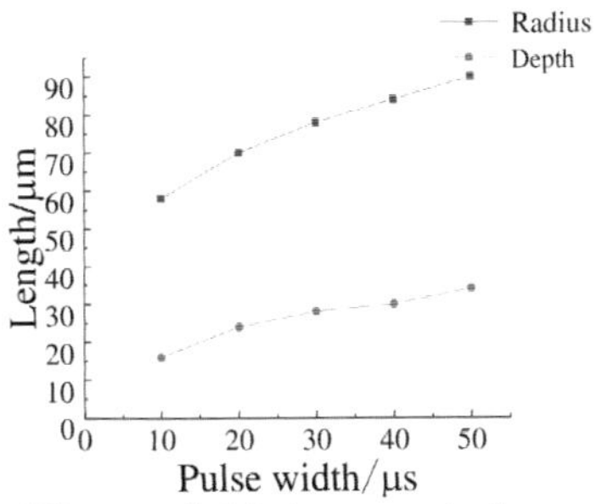

Figure 6. Molten pool size.

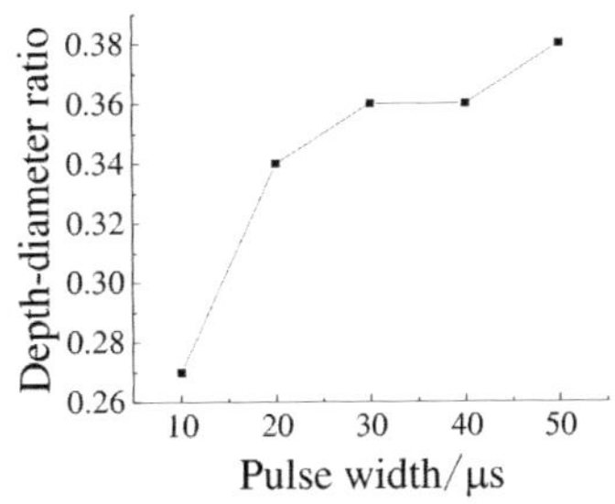

Figure 7. Molten pool depth-diameter ratio.

3.3. *Effect of peak current on thermal erosion of materials*

Figures 8 and 9 are the curves of the temperature in the radial direction and the depth direction at different peak currents with a pulse width of 50μs. It can be seen that the temperature in the radial direction and the depth direction increase with the peak current increases. Because discharge current directly affects the value of the heat flux density, and the increase of discharge energy will lead to the increase of overall temperature, but the change of depth direction is small.

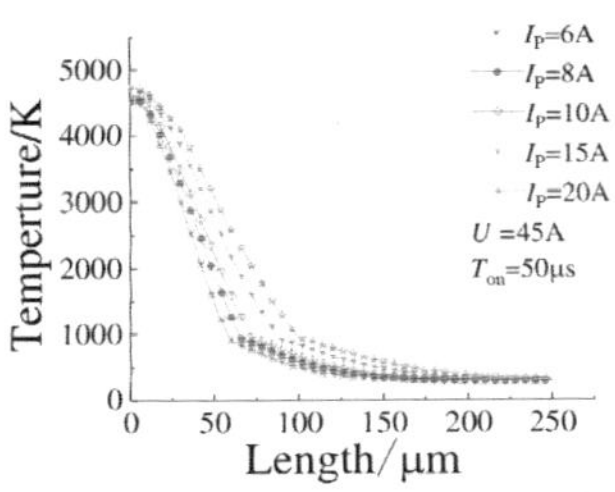

Figure 8. Temperature change in radius.

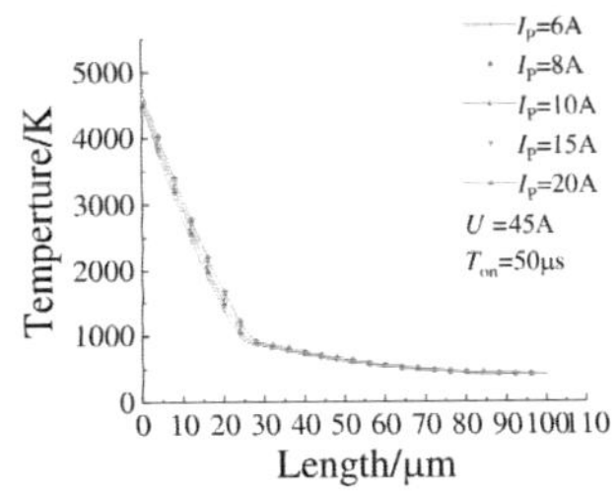

Figure 9. Temperature change in depth.

Figure 10 shows both the radius and the depth increase with current increases, but the radius changes significantly, while the depth does not change after 15A. This is because when the current increases, and the surface temperature of the workpiece rises and material melts, the coefficient of thermal conductivity gradually decreases, while the radial direction not only transfers heat from the molten matrix, heat source acts directly on the surface of the material, so the temperature along the radial direction keeps increasing, but the depth is not changed when it increases to a certain value. The curve of the depth-to-diameter ratio of the surface molten pool under different peak current conditions is shown in Figure 11. The peak current increases, the depth-to-diameter ratio decreases continuously.

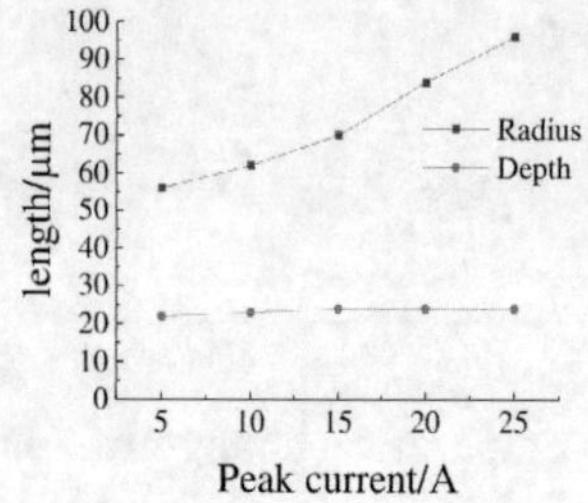

Figure 10. Molten pool size.

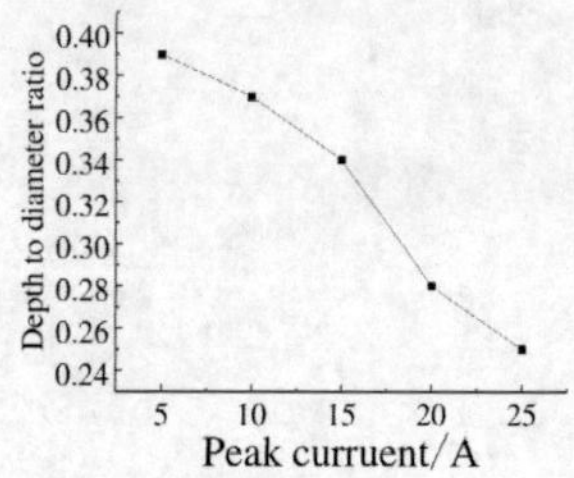

Figure 11. Molten pool depth-diameter ratio.

3.4. Changes and statistics of particles in molten pool

Figure 12 shows the variation of the particles inside the molten pool at different discharge moments. The results show that at 0.6μs the two particles near the discharge center begin to melt, and becomes an irregular circular shape. At 40μs, it is found that the third and fourth particles also began to melt and is melted completely at 70μs. Comparing the edge of the crater at 40μs and 70μs, it can be seen that the bottom of the crater is not flat at 40μs. This is because the particles on the edge of the crater did not reach the melting point, and the particles were between the molten pool and the solid matrix, making the bottom of the crater not flat.

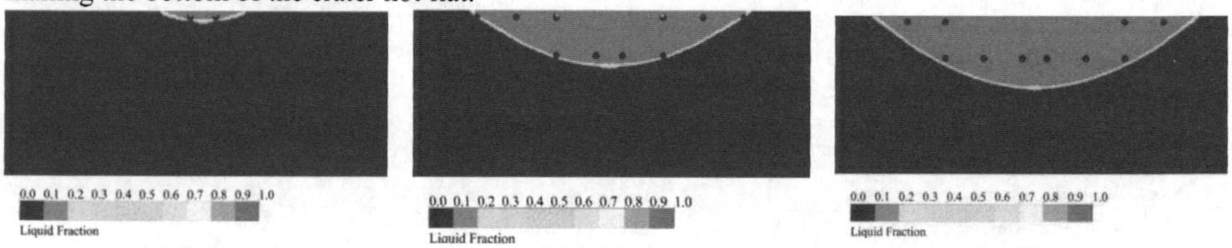

(a) 0.6μs. (b) 40μs. (c) 70μs.

Figure 12. Particles state in molten pool.

Figure 13 shows the change curve of the number of particles in the molten pool under different peak currents when the pulse width is 50μs. As the peak current increases, the number of particles in the molten pool also gradually increases, because the area of the molten pool becomes larger with current increasing. However, there are two particles that are melt completely when the current is 5A. And there are still only two particles when the current is 25A, which means that the amount of molten particles do not change with the increase of the peak current. In actual processing, if the number of particles increase in the molten pool, it will hinder or even stop the next discharge, which is not conducive to the discharge. Therefore, proper current reduction is beneficial to the erosion of the material.

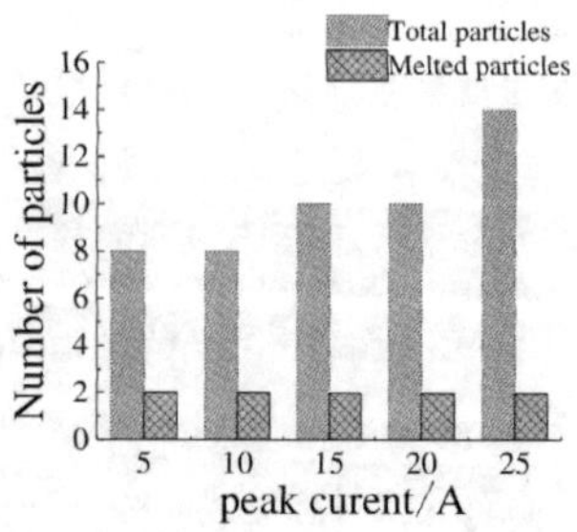

Figure 13. The number of particles in the molten pool.

4. Experimental results and analysis

The experiments using single pulse EDM power supply, table 2 shows the selection of machining parameters.

Table 2. Experimental machining parameter.

Wokepiece	Tool	Pulse Width	Peak current	Voltag	Dielectric
SiC_p/Al composite	Brass	50μs	25A	45V	Kerosene

The experimental results are shown in Figure14. The maximum depth of the crater is 17μm and the width is 196μm. Comparing the simulation results obtained by the same simulation parameters, the crater diameter is basically the same, and the depth of the experimental processing is smaller than that of simulation. Because the debris exists in the molten pool during the actual processing, making the processing depth affected, but these factors are not considered in the simulation, and the simulation results are basically reasonable within the allowable error range.

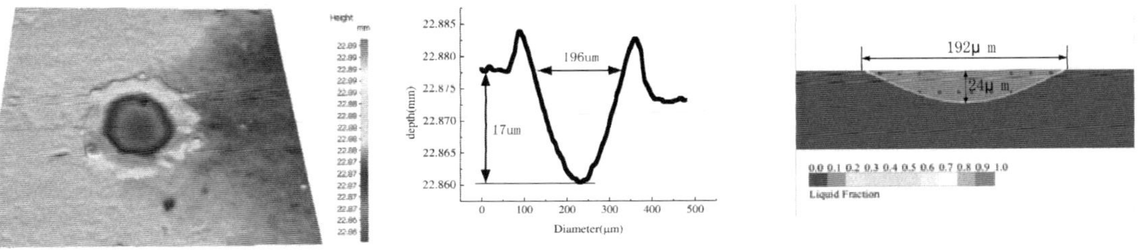

(a) The crater morphology of the experimental results. (b) Experimental measurement results. (c) Simulation results.

Figure 14. Comparison of EDM experiments and simulation results.

5. Conclusion

In this paper, considering the composite properties of SiC_p/Al materials, the model of SiC_p/Al composites with uniform distribution of SiC particles was established. The material thermal erosion process of single pulse EDM was simulated and the simulation was verified by experiments. The results show that the temperature in the workpiece decreases first and then increases with the increase of pulse width, increases with the increase of peak current. The depth-to-diameter ratio increases with the increase of pulse width, and decreases with the increase of peak current. The number of melted particles are basically unchanged and reducing the current appropriately will benefit the discharge in the processing.

Acknowledgments

The financial support from the National Natural Science Foundation of China under grant no. 51875074, Scientific Research Platform Foundation of Liaoning Province under grant no. JDL2016006, and General Program of Natural Science Foundation of Liaoning Province under grant no.20180550425 is acknowledged.

References

[1] Feng Ruiye. Study on Temperature Filed Simulation During Electro-Discharge Machining of TiC/Ni Cermet *J. Electric Machining &Mould*, 2016(2):1-5

[2] Gao Yang, Liu Lin, *et al.* Finite Element Thermal Analysis of EDM Crater *J. Electric Machining &Mould*, 2008(2): 8-11.

[3] E. Weingärtner, F. Kuster, *et al.* Modeling and simulation of electrical discharge machining *J. Procedia Cirp*, 2012, 2(7): 74-78

[4] Zhang Qingfen. Modeling and simulation of SiCp/Al composite EDM *D. Harbin Institute of Technology*, 2011.

[5] Yu Xiaolu, Zhang Haitao, *et al.* Quantitative Research of Statistical Characters of Particle in SiC Particle Reinforced Aluminum Matrix Composit *J. China Materials Science Technology & Equipment,* 2011(4):30-32.

IOP Conf. Series: Materials Science and Engineering **692** (2019) 012013 doi:10.1088/1757-899X/692/1/012013

Effect of Alum Sludge and Metakaolin as a Partial Replacement Cement Adding Superplastizer

Gunalaan Vasudevan

Department of Construction Management, Faculty of Built Environment, Tunku Abdul Rahman University College Malaysia

gunalaanv@tarc.edu.my

Abstract. The most important activities that improve the economy of Malaysia is the construction industry. Hence it will cause the usage of the concrete increase. When the concrete usage increase, it also increase the cement use because cement is the primary material that forms the concrete and during the production of the cement, it will cause the emission of carbon dioxide which will harm the environment as well as people in Malaysia. After that, alum sludge also caused a lot environment problem because it is lack of potential to be reused in other industry and it needs to be treated before it disposed of, which will cause the financial problem to the government. After the treatment, the economical ways to dispose of it is by using landfill method. But due to a limited landfill site in Malaysia and it is the temporary solution for the disposal of the waste, it becomes a problem to Malaysia and the number of the alum sludge keeping increase year by year because of the population increase year by year too. The purpose of the study is to find out the effectiveness of this combination. The research method that use in this study is by taking out some experiment such as slump test also known as workability test, compressive strength test, flexural strength test, ultrasonic pulse velocity (UPV) test (direct and semi-direct method) and scanning electron microscope (SEM) with the concrete day-age 7,14 and 28 days. The result gets these test is the 10% alum sludge as a partial replacement for metakaolin in concrete adding superplasticizer (Sika Viscocrete 208) is the best replacement level in term of the workability ,compressive strength , Flexural strength, speed of the UPV and in scanning electron microscope. After 10% replacement level, the reading all suffer a downturn. Additionally, when the day-age of the concrete increase, the reading will be increase too. This happens to all the replacement level and the 10% replacement level with 28 day-age achieve the highest reading in this study. The contribution and implications of the findings are the using the 10% of the alum sludge as a partial replacement for metakaolin in concrete adding superplasticizer (Sika Viscocrete 208) is the best replacement level compare to others.

1. Introduction

Metakaolin is the pozzolanic material that is very effective and it will enhance the strength of concrete [1]. It helps to deduct the use of cement in concrete and in the same time it also enhance the strength of concrete when the metakaolin use in concrete carried the experimental test that related to the replacement of the cement with metakaolin, [2] the results showed that the compressive strength of mortar increase about 50% when the cement by metakaolin, the result showed that the compressive strength started to decrease. Alum sludge is a waste that generated during the process of purification of domestic and industrial waste water. As the population increases every year, the generation of alum sludge waste in Malaysia has been rapidly increasing 3.2 million cubic meters of domestic sludge every year that produces by Malaysia and increases up to 4.3 Million cubic metres in 2005. A lot of researchers have been cooperation to create the new and suitable solutions in order to fix the problem that causes by the alum sludge. They also cooperate with the conventional treatments to show health and environmental hazards[3].In current years, the alum sludge only handle through filling and application in land, disposal to the ocean and use for agriculture. The most economical method of the alum Sludge disposal is land disposal. By using land disposal method, the crops will be available to grow on the poor land. According to [4] they say that the sludge is the biggest scale poor land. According to [4], they say that the sludge is the biggest scale compared to other by-products that produced in the process of sewage treatment and they also found that the sludge contains of heavy metals.

2. Problem Statement

Nowadays, the construction industry is one of the most important activities that improve the economy of Malaysia. With the increase of the construction activity, the use of concrete will increase and this will cause the cement needs also increase. In order to replace the cement used, metakaolin takes an important role in this replacement. With this replacement, the emission of the carbon dioxide will decrease because it is replaced the cement used in concrete. Other than that. It also will Increase the service life of the building. This is because when metakaolin replaced the cement, it will increase the durability and strength properties of the concrete [5] then, the alum sludge caused a lot of environmental problem and it is the largest waste produce per years in Malaysia. The reason that the alum sludge is increasing every tear is that the population of the Malaysia always increase and this causes the alum sludge increase too. According to [6], they say that Malaysia generated about 4.5 million cubic meters of domestic sludge since 2005 and it is keeping raising annually. Even there have some methods to stabilize and dewater the alum sludge, but the landfill is the method that most of the sludge is ended up to be disposed of even after treated. This is because it is the most economical ways to dispose of the waste. However, the landfill is only a temporary solution for the disposal of alum sludge waste because there is limited space for the sludge waste to be disposal of. Due to this problems, some researches had been carried out to determine the potential use of alum sludge as partial cement replacement materials, in bricks, clay and ceramic materials in order to increase the use of alum sludge. The increased importance of the reuse of alum sludge becomes very useful to reduce the environmental impact. The alum sludge can be used in cement manufacture by reuse the sludge as cement partial [7]. But alum sludge is a waste material that is no stable. Considering the pozzolanic properties of metakoalin and the possibility of fixing heavy metals in the pozzolanic hydrates, the use of metakoalin should help to improve the properties of concrete containing alum sludge [8]

3. Literature Review

3.1 Pozzoloric Materials

3.1.1 Metakaolin. Metakaolin is one of the most widely used mineral admixtures these days for high-performance concrete mixes. It helps concrete to gain both higher performance and economy. However, metakaolin is different from other mineral admixtures, it is not an industrial by-product as the others. Kaolin is one of the most abundant natural minerals and when it is heat-treating, it will become metakaolin. Instead, it is specially produced by high-quality kaolin using thermal treatment technology yielding a highly reactive amorphous material with pazzolonic and latent hydraulic reactivity that is very suitable for cementing applications especially for the high-performance concrete. Metakaolin typically contains 50-55% silicon dioxide (SiO2) and 40-45% aluminium oxide ($A1_2O_3$) [9]. When the cement is replaced by metakaolin, the result showed better performance than control mortar, particularly in terms of mechanical properties. Saving up to 40% of clinker reduces energy consumption and the amount of carbon dioxide emissions to the atmosphere.

3.1.2 Alum sludge. Alum sludge is a by-product and it is one of the final products of wastewater treatment at Sewage treatment plants. polyaluminum chloride (PACI) and large amount inorganic matters. Alum sludge acted as a chemical conditioner and physical conditioner improved the alum sludge dewatering. It indicated that the addition of alum sludge reduced the dosage of polyacrylamide (PAM) and decreased the moisture content of alum sludge.Investigated the possibility of the alum sludge to reuse in the cement industry. They discovered that the alum sludge has a similar particle size and chemical composition with Portland cement. 10 to 30% of alum sludge are used to replace the cement use in mortars and the mechanical strength of the mortars is investigated by them. From the results, the mixes are lower than the control cement and the slump is decline. With the result, they found that the chemical compositions are the most important factor that affect the alum sludge to be reuse.

4. Result and Discussion

4.1 Workability

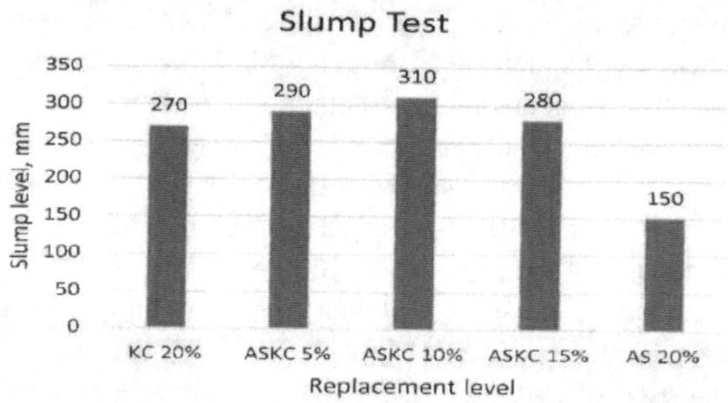

Figure 1 Slump Test

For the slump test, the concrete that using alum sludge act as the partial replacement for metakaolin in the concrete by using water content ratio 0.48 with adding superplasticizer into the concrete. The fresh concrete result for all the mix is already recorded and showed in figure 1. From the table below, the slump level for the concrete that contains 20% metakaolin is 270mm. After that, the alum sludge is added into the concrete to replace the metakaolin content, the slump level increase by 20mm by replaced 5% of the alum sludge to the metakaolin and keep increasing up to 310mm in 10% of the replacement. But after the replacement of the 10%, The slump level is continuously dropping and in the 20% of the replacement of the alum sludge to the metakolin, it suffered a very big decline up to 150mm, which is the lowest slump level compare to other the result find that by using 15% of the alum sludge replaced the metakaolin and combined with the superplasticizer in the concrete. The slump level is achieved 80mm and this is the highest level that they reached. But in this test, the highest slump level achieved is 310 mm and only by using 10% replacement, the slump level is dropping. In conclusion, the 10% of the alum sludge mix with 10% of the metakaolin will reach highest slump level.

4.2 Compressive Strength Test

For the compressive strength test 7, 14 and 28 day-age cubes form in the 150mm × 150mm × 150mm cube mould. The figure 2 below shows the result of the 20% of the metakaolin. 5%, 10%, 15% and 20% of the alum sludge that replace the metakaolin contest is the concrete cube. In the day-ago for each of the control mix, there are different result Compare to the other day-age of the cube. They age cube is 16.35 MPa, it is keeping increasing about 12.00 MPa when the they-age increase and in 28 day-age, it obtains 42.30 MPa compressive strength. Then for the 10% of the replacement in 7 days to 14 days, the compressive strength of it increase about 22.00 MPa, which is the highest increasing compared with another control mix. After the 14 days, it is increased mostly compared to another control mix that after 14 days. The replacement for the metakaolin in 7 day-ago concrete. The Compressive strength of the cube is 16.35 MPa. After replaced by 5% of the alum sludge to the metakaolin it becomes increase up to 18.34 MPa and keeping increasing up to 23.85 MPa when replacing 10% of the alum sludge. However, after 10% of the replacement it drops to 15.98 MPa and when the alum sludge is fully replaced the metakaolin content in the concrete, the strength of it is 13.15 MPa, which is lower than 20% of the metakaolin concrete.

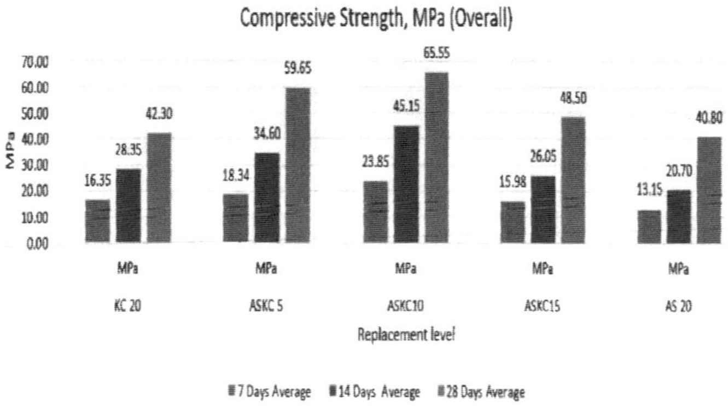

Figure 2 Compressive Strength

4.3 Splitting Tensile Strength Test

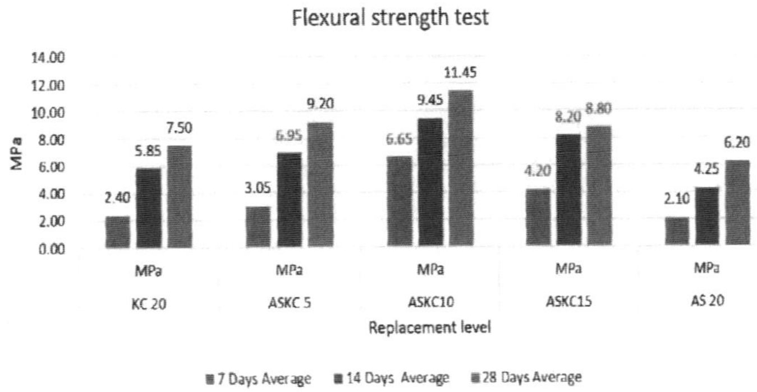

Figure 3 Flexural strength Test

In flexural strength test 7, 14 and 28 days cubes are carried out in order to obtain the result. Figure 3 shows the result of the 5%, 10%, 15% and 20% of the alum sludge that replace the metakaolim content and last 20% of the metakolin in concrete. It is same as the resuly of compressive strength test, which is the result of the different day- age for each of the control mix are different. In the 7 day -age if the 20% of the metakolin concrete, the flexura strength of it is 2.40 MPa. It is keeping increase more than 2MPa when the day-age when the day-age increase. In 28 day-age, it obtains 7.50 MPa flexural strength. For 10% alum sludge that replaces the metakaolin content in concrete, the flexural strength of the 7 day-age concrete is 6.65 MPa and in 14 day-age the strength is 9.45 MPa. Futhermore, in 28 day-age concrete, the strength if it is 11.45 MPa and the difference compared to 14 day-age is about 2.00 MPa.Next, for the comparison of the control mix with the same day-age, before the replacement for metakaolin in 7 day-age concrete, the flexural strength of the concrete is 2.40 MPa. After the 5% of the alum sladge the metakaolin, it becomes increase up to 3.05 MPa. Flexural strength is keeping increase when the alum sludge content is increased, which is 10%of the alum sludge replacement obtain 6.65 MPa. However, after 10% of the replacement, the strength is starting to drop up to 4.20 MPa and when the alum sludge is fully replaced the metakaolin content in the concrete, the strength of it is 2.10 MPa. Compare with 20% of the metakaolin, 20% of alum sludge is better than it and the difference strength between them is about 0.30 MPa. For 28 day-age concrete, 10% of the alum sludge achieve the highest strength, which is 11.45 MPa and the difference between 5% of the replacement is about 2.25 MPa

4.4 Ultrasonic pulse velocity (UPV)

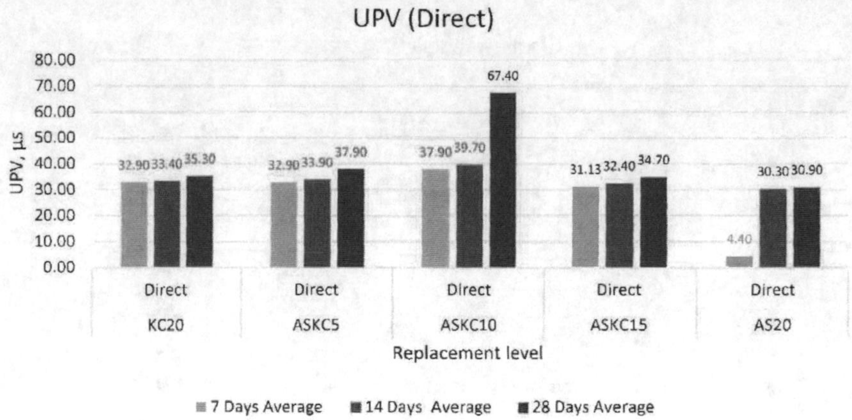

Figure 4: Direct method

This method is the most sensitive compared to the indirect method and it is used to detect the void that contains the concrete block. When the transmitter placed opposite of the receiver, it will provide maximum sensitivity, give a clear path length and more accurate. In this test, the material using is same as the previously conducted test. Figure 4 showing the comparison of the day-age for 20% metakaolin concrete, the reading of it is 32.90 us. It is a continuous increase but not so much when the day-age increase and in 28 day-age, the reading of it is 35.30 us. This situation is same for 5% and 15% of the alum sludge that replace the metakaolin in the concrete. For the 10% of the replacement, at 7 day-age the reading of it is 37.90 us. When the concrete block is up to 28 day-age, the reading is increased about 28 us. Which is up to 67.40 us and this is the highest increment compared to others control mix. Next is the comparison of the same day-age but different control mix. For 7 ay-age of 20% of metakaolin concrete, the reading of it is 32.90 us. With the replacement up to 10% of the alum sludge, the reading is keeping increase but not so much and it is up to 37.90. However, in 15% of the replacement, the reading is dropping about 6 us. When the metakaolin is fully replaced by the alum sludge in concrete, the reading for the ultrasonic pulse velocity drops a lot, which is up to 4.40 us. This is the reading that suffers most decrements compare to others control mix in the same day-age. However, in 28 day-age 10% of the replacement, the reading is 67.40 us, which have about 30 us difference compared to 5% of the replacement, which is the highest increment and it is the highest reading compare to others control mix.

5. Conclusion and Recommendation

The performance of alum sludge as a partial replacement for metakaolin adding superplasticizer in concrete is still new in the construction industry. This is because there only has one test that done relating to this topic. The engineering properties, physical & mechanical properties and the optimum value of in this replacement will be drawn out in this chapter. The engineering, physical and mechanical properties test like workability test, compressive strength test, tensile strength test, ultrasonic pulse velocity (UPV) test and scanning electronic microscope (SEM) test are carried out to achieve the objective of the study. The laboratory test was carried out according to the method that discovers in literature review and result will answer the issue that state in the objectives.

1. To investigate the engineering properties of the combination of concrete.
2. To determine and assess the physical and mechanical properties of alum sludge when cooperating with metakaolin in concrete by replacement 5%, 10%, 15%, 20%

3. To identify the optimum value 5%, 10%, 15% and 20% of alum sludge to replace for metakaolin.

Through the laboratory, the objectives of the study were determined and discovered from the test result that obtains in the test. Therefore, conclusion and recommendation on the performance of alum sludge as a partial replacement for metakaolin adding superplasticizer in concrete were extracted. The finding of this study will be summarized and concludes in this chapter. Recommendations to improve the performance of alum sludge as a partial replacement for metakaolin adding superplasticizer in concrete.Here has some recommendation that can be suggested after done this study. This recommendation can further improve the performance of the control mix concrete. There are: Strength properties of concrete with partial replacement of metakaolin with alum sludge need investigation for a longer period (56, 90, 180 days).More studies have to be carried on the effect of using superplasticizers on the percentage reduced of cement content on the concrete mixes. In the present study, 20 percent replacement of cement has been considered. The other percentages like 30, 40, and 50 percent need to investigate. It should be investigated whether alum sludge is possible to mix with others by- products. Further studies need to be carried out that by using 0.60 water content for the concrete mix.

Reference

[1] Aiswarya S, Prince Arulraj G and Dilip C, 2013.A review on use of metakaolin in concrete.

[2] Haider M. O., Hamid, R., Abudullah, S.R.S., Tan Kofli, N. and Mohd R. T., 2013.Physical and Mechanical Properties of High Performance Concrete with Alum Sludge as Partial Cement Replacement.

[3] Kim Guan, T, 2015. The use of sewage sludge ash (SSA) as partial replacement of cement in concrete.

[4] Fontes,C.M.A.,Barbose,M.C.,Toledo Filho,R.D.and Goncalves,J.P.,2004.Potentiality of sewage sludge as as mineral additive in cement mortar and high performance concrete.

[5] Prasada Rao,D.V and Venkata Maruthi,S.,2016.Effect of Nano-silica on concrete containing Metakaolin

[6] Brooks,J.J.and Megat Johari,M.A.,2017.Effect of metakalolin on creep and Shrinkage of concrete.

[7] Courard, L., Darimont, A., Schouterden, M., Ferauche, F., Willem, X. and Degeimbre, R.2003. Durability of mortars modified with metakaolin.

[8] Cyr, M., Rachida, I. and Escadeillas, G., 2012. Use of metakaolin to stabilize sewage sludge ash and municipal solid waste incineration fly ash in cement-based materials.

[9] Poon,C.S.,Kou,S.C.and Lam,L.,2017.Pore sixe distribution of high performance metakaolin in concrete.

IOP Conf. Series: Materials Science and Engineering **692** (2019) 012014 doi:10.1088/1757-899X/692/1/012014

Simulation Study on Cutting Force and Cutting Temperature in Turning of Titanium Alloy

Dapeng YANG, Yu LIU, Fujian MA, Zhihua SHA, Shengfang ZHANG*

School of Mechanical Engineering, Dalian Jiaotong University, Dalian, China

*zsf@djtu.edu.cn

Abstract. The objective of this paper is to study the cutting force and cutting temperature in turning of titanium alloy. The simulation of cutting force and cutting temperature by finite element simulation software ABAQUS is conducted. The orthogonal experiment is designed. The influence of various factors on cutting force and cutting temperature is obtained. A prediction model of cutting temperature and cutting force is obtained by performing multiple linear regression analysis on the data obtained from orthogonal experiments.

1. Introduction

Titanium alloys represent a significant metal portion of the aircraft structural and engine components due to their excellent properties such as high strength-to-weight ratio, toughness, superb corrosion resistance, bio-compatibility and high fatigue strength[1]. ASTM grade 5 (Ti6Al4V) is the most common Ti-alloy used and accounts for more than 50% of all Ti-alloy used[2]. Unfortunately, titanium alloys are generally classified as difficult-to-machine materials because of certain inherent properties[3]. Low thermal conductivity causes heat dissipation difficulty within the machining zone which creates high thermal gradients leading to rapid tool wear. Lower elastic modulus causes excessive workpiece deflection and moves away from the cutting tool, which corresponds to badly machined surface integrity. The great deflection of thin-walled structures results in tool vibration, chatter, and poor surface finish. Due to the chemical affinity between the workpiece and tool in high temperature, chemical wear occurs in the cutting tool. All these problems are related to the severe high cutting force and temperature-induced during the cutting process.

Significant researches on machining titanium alloys have been performed to enhance the tool life and improve the machined surface integrity. The machined surface stress distribution has a direct impact on the fatigue strength of the machined part. Stress, temperature, and vibration in the machining zone are the three main factors that enhance the cutting tool damage[4]. During the machining process, the high temperatures are generated by strength and hardness of the enhancement of the material which accelerates tool wear[5]. Titanium alloys retain their strength even at high temperatures, up to about 550 °C[6]. Titanium is highly reactive with the tool materials at high temperatures. For getting higher productivity, cutting speeds are required higher. The tool life is very short and results in a poor surface finish at high cutting speed. For instance, machining at speeds over 90 m/min had a tool life less than 10 min[7]. By optimizing the process parameters, abundant studies have been conducted to obtain better cutting conditions[8-10]. The tool geometry combination may significantly reduce cutting force and temperature which will improve the machinability of titanium alloys. Using advanced lubrication and cooling methods such as minimum quantity lubrication (MQL) and cryogenic cooling techniques has some benefits to improve the machining process with environmental friendliness.

In this research, a numerical analysis of cylindrical external turning on Ti6Al4V alloy is presented. A Johnson-Cook constitutive equation was implemented as a flow stress model and adapted to the Ti6Al4V alloy. Through the 2D finite element simulation procedure, the cutting forces and temperatures were gotten with several sets of cutting conditions such as different combinations of cutting parameters and tool geometries. Orthogonal experiments have been designed for optimizing the cutting and tool parameters with five factors and four levels. Finally, a prediction model has been obtained by performing multiple linear regression procedure.

2. Finite Element Simulation

In this paper, the numerical simulation procedure is aimed at testing the feasibility of a material constitutive model for the Ti6Al4V when implemented in a 2D FE model of a turning operation under dry cutting conditions. 2D FE model for orthogonal cutting has been simulated by utilizing the commercial FEA code ABAQUS/Explicit™. Workpiece material of Ti6Al4 was modeled as plastic. The cutting tool was modeled as a rigid body. Figure 1 shows the system of relative motion between workpiece and tool.

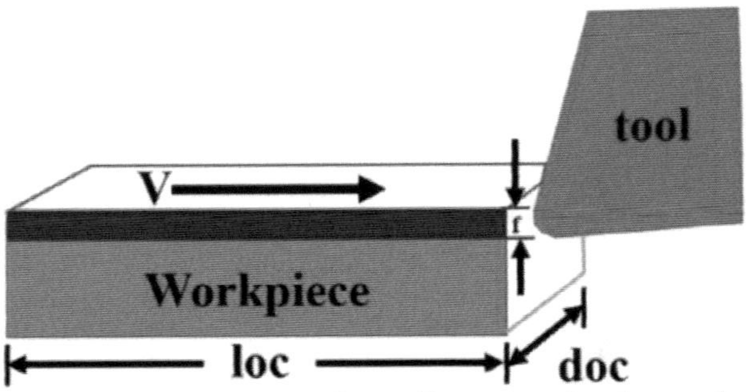

Figure 1. Schematic diagram of two-dimensional orthogonal cutting

To avoid serious distortion of the mesh in simulation, an improved Lagrange mesh generation method is adopted. The mesh generation is not uniform, but relatively fine near the cutting part for simplifying the simulation procedure. The FE simulation model is shown in Figure 2.The number of elements used in simulations was about 9000 (8000 for upper 40 layers and 1000 for lower 5 layers) for workpiece. The density deviation method was used for tool meshing. In the simulations, X-axis represented the cutting direction and Y-axis represented the feed direction. Boundary conditions of the workpiece were determined with respect to various cutting speeds.

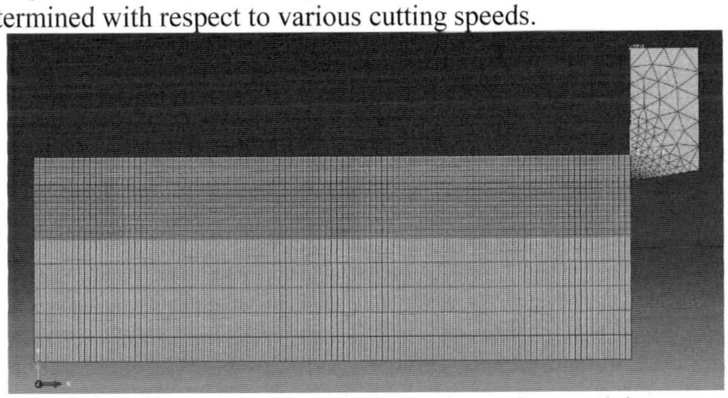

Figure 2. The finite element simulation model

The material and tool properties are given in Table 1[11]. The room temperature T_0 was set equal to 20°C for dry cooling. The fracture criterion for chip separation based on equivalent plastic strain was used in this paper.

Table 1. Properties of workpiece and tool

	Elastic modulus (GPa)	Poisson ratio	Thermal conductivity (W/m°C)	Specific heat (J/kg°C)
Workpiece (Ti6Al4V)	113	0.342	7	546
Tool (carbide)	800	0.2	46	203

Johnson-Cook (J-C) material model was used to determine the strain rate sensitivity in the simulations. Table 2 shows different coefficient values suggested for the J-C Model of Ti6Al4V alloy from previous studies [12].

$$\sigma=(A+B\varepsilon^{n})\left(1+C\ln\left(\frac{\dot{\varepsilon}}{\dot{\varepsilon}_0}\right)\right)\left(1-\left(\frac{T-T_0}{T_m-T_0}\right)^{m}\right) \quad (1)$$

Table 2. Johnson-Cook material model parameters for Ti6Al4V

A (MPa)	*B* (MPa)	*C*	*n*	*m*
1098 1	1092	0.93	0.014	1.1

The cutting forces and temperatures are partially generated by the frictions between the chip and rake face. The sliding zone and bonding zone describe the two different contact states. The shear stress at each point in the bond zone is substantially the same, and the friction in the slide zone decreases along the rake angle. Due to the complexity of friction configuration, the penalty function method was adopted to deal with the sliding friction in this paper.

3. Orthogonal experiments design

The study aims to investigate experimentally effects of cutting parameters and cutting tool geometry on the cutting main force and temperature in the turning of Ti-6Al-4V alloy. From single factor simulation experiment, one can find the changing tendency affected by a single cutting parameter or tool geometry. The cutting process is a complicated procedure and the cutting parameters and tool geometry have strong dependent relation acting on the cutting forces and temperature. Among these factors, the effect degrees for cutting force and temperature were checked by designing several orthogonal experiments.

The design of experiments has a major effect on the number of experiments needed. Therefore, it is essential to have a proper design of experiments. A factorial design was selected in this work so that all the interactions between the independent variables can be investigated. In this study the five parameters namely rake angle, relief angle, tool nose radius, cutting speed and cutting depth of the cutting tool were selected for the experimentation. The five selected factors with four levels are shown in Table 3.

Table 3. Cutting parameters and tool geometry of titanium alloy turning

Level	Rake angle (°)	Relief angle (°)	Tool nose radius (mm)	Cutting speed (m/min)	Cutting depth (mm)
1	0	6	0. 1	30	0.1
2	5	8	0. 2	50	0.125
3	10	10	0. 3	60	0.15
4	15	12	0. 4	80	0.175

The Taguchi method is executed for finding the optimum number of experiments with the combinations of parameters. An L16(4^5) orthogonal array was formed based on the five factors-four level design. A total of sixteen experiments have been completed using uncoated carbide tools for machining at different cutting conditions selected through L16 orthogonal array. After sixteen runs of simulation, the cutting forces and temperatures have been gotten. The factor effect degrees are determined by equation (2) where R_j (j = 1,..., 5 corresponding 5 factors) is the average (n = 4 for 4 levels) range of K_{ij}. K_{ij} denotes the summation of n sets of simulated cutting force (temperature) with the ith level fixed for the jth factor.

$$R_j = \frac{(K_{ij})_{max} - (K_{ij})_{min}}{n} \quad (2)$$

The average range R_j have been derived from sixteen runs of simulation experiment shown in Figure 3 where (a) indicates the cutting force range changement and (b) for the cutting temperature one. By observing the range curve, one can find that the effect degree for cutting force is depth of cut a_p > rake angle γ_0 > tool nose radius r > cutting speed V_c > relief angle α_0, and the effect degree for cutting temperature is cutting speed V_c > depth of cut a_p > rake angle γ_0 > relief angle α_0 > tool nose radius r.

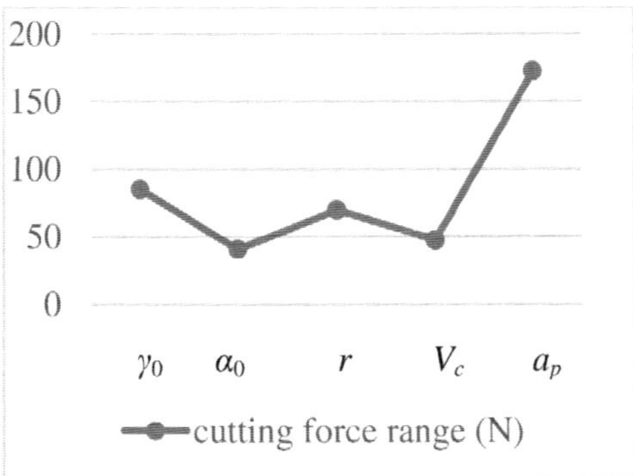

(a)

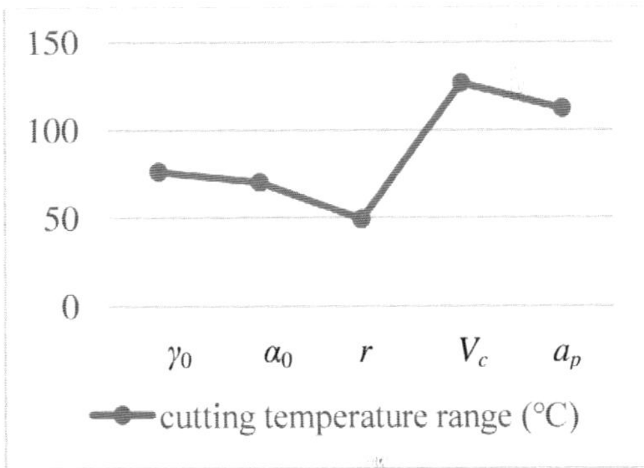

(b)

Figure 3. The average range R_j of five factors for cutting force (a) and temperature (b)

4. Prediction model establishment

In the section, linear regression analysis methods were used to analyse the data derived from previous experiments to find out mathematical relation between cutting force (temperature) and cutting/tool geometric parameters. Regression equations were finally given for all experimental trials to represent the relation between dependent and independent variables. The model of predicted cutting force, F_R can be expressed as equation (3).

$$F_R = C\gamma_0^{b_1}\alpha_0^{b_2}r^{b_3}V_C^{b_4}a_p^{b5} \quad (3)$$

where C, b_1, b_2, b_3, b_4, and b_5 are model parameters. By logarithmic transformation equation (3) can be written as

$$Y = c + b_1R + b_2A_0 + b_3R + b_4v + b_5A_P \quad (4)$$

where $Y = \ln F_R$, $R = \ln \gamma_0$, $A_0 = \ln \alpha_0$, $R = \ln r$, $v = \ln V_c$, and $A_p = \ln a_p$. Residual analysis was executed and the abnormal experimental data has been eliminated. The regression model was solved on MATLAB™. Figure 4 is an example of residual analysis for deleting the abnormality of trials results where the dashed line indicates the abnormal data. After the multivariable linear regression procedure, the prediction model between cutting force and cutting parameters was built as equation (5) with solved unknown equation coefficients.

$$F_R = 1436.12\gamma_0^{-0.0724}\alpha_0^{0.1175}r^{0.1512}V_C^{0.0961}a_p^{0.8964} \quad (5)$$

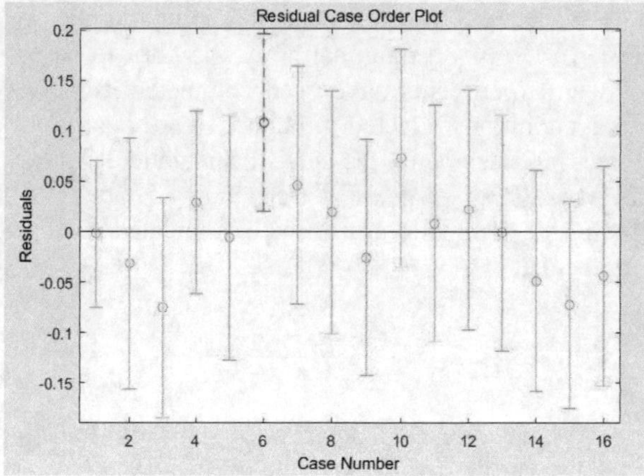

Figure 4. Residual diagram for cutting force regression procedure

By utilizing a similar procedure, the prediction model between cutting temperature and cutting parameters has been solved as shown in equation (6).

$$T = 635.49\gamma_0^{0.0575}\alpha_0^{-0.0171}r^{-0.0355}V_C^{0.0474}a_p^{0.1677} \tag{6}$$

For examining the accuracy of the prediction models, three trials have been randomly selected. The predicted and simulated values of cutting force are represented in Figure 5.

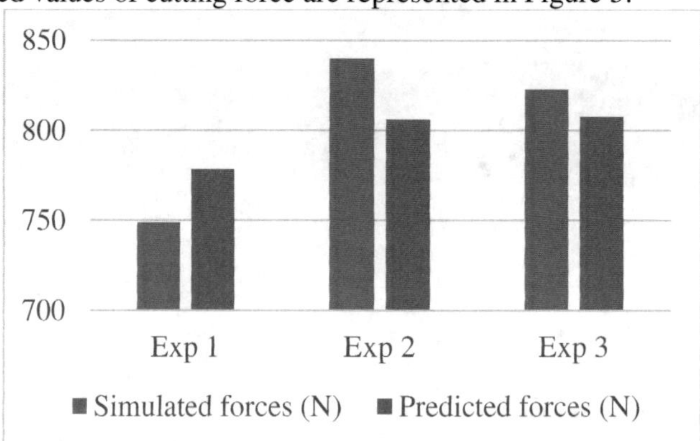

Figure 5. Comparison between simulated and predicted main cutting forces during dry machining

5. Conclusion

This paper presents an FE model of turning on Ti6Al4V under dry cooling conditions. Johnson-Cook model was implemented and coupled with a hybrid bounding – sliding friction condition in the cutting contact zone. The influence on cutting force and temperature has been checked by designing orthogonal experiments using the Taguchi method with five cutting factors with four levels. The effect degree for cutting force is $a_p > \gamma_0 > r > V_c > \alpha_0$, and the effect degree for cutting temperature is $V_c > a_p > \gamma_0 > \alpha_0 > r$. Predictive models for determining the cutting forces and temperature were developed in the turning operation of Ti6Al4V.

Acknowledgments

The authors gratefully acknowledge the support by the Educational Department of Liaoning Province (grant number JDL2016014) and Department of Science & Technology of Liaoning Province (grant number 20170540146).

References

[1] OZEL T, ULUTAN D. Prediction of machining induced residual stresses in turning of titanium and nickel based alloys with experiments and finite element simulations [J]. CIRP Annals - Manufacturing Technology, 2012, 61(1): 547-50.
[2] GUPTA K, LAUBSCHER R F. Sustainable machining of titanium alloys: A critical review [J]. Proceedings of the Institution of Mechanical Engineers, Part B: Journal of Engineering Manufacture, 2017, 231(14): 2543-60.
[3] PRAMANIK A. Problems and solutions in machining of titanium alloys [J]. The International Journal of Advanced Manufacturing Technology, 2014, 70(5): 919-28.
[4] ZHAO H, BARBER G C, ZOU Q. A study of flank wear in orthogonal cutting with internal cooling [J]. Wear, 2002, 253(9): 957-62.
[5] SHOKRANI A, DHOKIA V, NEWMAN S T. Environmentally conscious machining of difficult-to-machine materials with regard to cutting fluids [J]. International Journal of Machine Tools & Manufacture, 2012, 57(none): 83-101.
[6] REVURU R S, POSINASETTI N R, VSN V R, et al. Application of cutting fluids in machining of titanium alloys—a review [J]. The International Journal of Advanced Manufacturing Technology, 2017, 91(5): 2477-98.
[7] ARRAZOLA P J, GARAY A, IRIARTE L M, et al. Machinability of titanium alloys (Ti6Al4V and Ti555.3) [J]. Journal of Materials Processing Tech, 2009, 209(5): 2223-30.
[8] KHANNA N, DAVIM J P. Design-of-experiments application in machining titanium alloys for aerospace structural components [J]. Measurement, 2015, 61(280-90.
[9] CHAUHAN S, DASS K. Optimization of Machining Parameters in Turning of Titanium (Grade-5) Alloy Using Response Surface Methodology [J]. Advanced Manufacturing Processes, 2012, 27(5): 531-7.
[10] WANG Z, NAKASHIMA S, LARSON M. Energy Efficient Machining of Titanium Alloys by Controlling Cutting Temperature and Vibration ☆ [J]. Procedia Cirp, 2014, 17(523-8.
[11] BORDIN A, IMBROGNO S, ROTELLA G, et al. Finite Element Simulation of Semi-finishing Turning of Electron Beam Melted Ti6Al4V Under Dry and Cryogenic Cooling [J]. Procedia CIRP, 2015, 31(551-6.
[12] CHEN L, EL-WARDANY T I, HARRIS W C. Modelling the Effects of Flank Wear Land and Chip Formation on Residual Stresses [J]. CIRP Annals, 2004, 53(1): 95-8.

Calculation of Parameters of the Cryogenic Rotor-Blade Engine for the Drive of the Refrigeration Unit for Truck

D Uglanov*, O Tereshchenko, S Zaika and I Neverov and R Panshin.

Department of Heat Engineering and Heat Engines, Samara University, 443086, Moskovskoye shosse, 34, Samara, Russia

*E-mail: dmitry.uglanov@mail.ru

ABSTRACT. Cars with internal combustion engines using hydrocarbon fuels occupy the main market share. The cost of these products is increasing, and this negatively affects the prosperity of the population and the cost of freight transportation. The growth in energy consumption and the rapid growth of cars deplete non-renewable natural oil and gas resources, leading to intense combustion of atmospheric oxygen.

One of the promising types of environmentally friendly engine is a cryogenic fuel engine. Such engines have higher energy performance and lower cost in comparison with electric vehicles, and their operating costs are comparable with traditional cars, taking into account the ecological aspect.

The study presents the development of methods for calculating the rotor-blade engine and its design. Since refrigeration plants consume 3% of the internal combustion engine power, it is advisable to replace them with autonomous units using a cryogenic working fluid, thereby reducing emissions of harmful substances into the atmosphere.

1. INTRODUCTION

Every year the number of trucks and cars in the world increases. Daily a large number of freight traffic around the world, logistics between countries is developing rapidly, requiring for this a large number of trucks.

The increase in energy consumption and the rapid growth of cars with internal combustion engines deplete the non-renewable natural resources of oil and gas, which leads to intense combustion of atmospheric oxygen. The processes that occur during fuel combustion create serious problems in the field of ecology and emissions of harmful substances into the atmosphere. A large concentration of carbon dioxide in the state is a barrier to the thermal radiation of the Earth's Sun, which leads to a greenhouse effect.

2. MODERN CONDITION OF ENGINES FOR CRYOGENIC FUEL

One of the promising types of environmentally friendly engine is a cryogenic fuel engine [1,2]. Such engines have higher energy characteristics and lower cost in comparison with electric vehicles, and their operating costs are comparable to traditional cars with engines, taking into account the ecological aspect [3].

3. MODERN CONDITION OF ENGINES FOR CRYOGENIC FUEL

In England the production company Dear has already begun serial production of the cryogenic engine

[1].

The Dearman engine works as follows:

1) the hot heat carrier (ethylene glycol) is fed into the cylinder;

2) then, liquid nitrogen is introduced into the cylinder, which enters into interaction with the high-temperature liquid, and due to this it begins to expand;

3) heat from the heat carrier is absorbed by the expanding gas, as a result of which isothermal expansion occurs;

4) the piston is urgently lowered, the exhaust valve is opened, and the mixture of gas and hot coolant is removed from the engine;

5) Then, the hot heat carrier is separated from the mixture in the separator and returned to the tank, while nitrogen or air is discharged into the atmosphere. In this research [3] the power plant operating on liquid nitrogen (LN2) is presented, it is intended for transformation of thermal energy of environment at gasification of liquid nitrogen into mechanical energy, which can be used, in particular, for working a cryogenic car.

The power plant 'figure 1', operating on liquid nitrogen, includes a special Dewar vessel 1, a heat exchanger 5 with shut-off regulating and measuring equipment that ensure the conversion of liquid nitrogen to high-pressure gaseous nitrogen with a temperature close to the environment, and an air motor.

The working process of this engine is carried out for one revolution of the crankshaft in three cycles: input under excess pressure of nitrogen gas, working stroke and exhaust gases. The nitrogen gas is introduced into the inlet cavity 20 by supplying liquid nitrogen from the Dewar 1 vessel under excess pressure with the valve 2 open through the gasifier 5 in which it passes from the liquid state to the gaseous due to heat inflow from the environment.

From the gasifier 5, through the pipeline 6, gas enters the common cavity 17 for the two cylinders of the air motor, which acts as a receiver from which the gas inlet to the cavity 20 occurs when the crankshaft 13 rotates. In this case, the push rod 15, screwed into the piston 11, moved by the connecting rod 14, The valve 18 opens and the inlet cavity 20 is filled with a high-pressure gas. In position B, the valve 18 is closed. With further rotation of the crankshaft from position B to position B, expansion of the gas takes place with the accomplishment of useful work (working stroke). When the crankshaft rotates to position B, openings 16 in the cylinder wall are opened, which allows release of not fully expanded gas (exhaust) into the atmosphere. The energy of the working stroke is perceived by the flywheel 9 and ensures the further rotation of the crankshaft and the cycle is repeated.

The schematic representation of the power plant is shown below in figure 1.

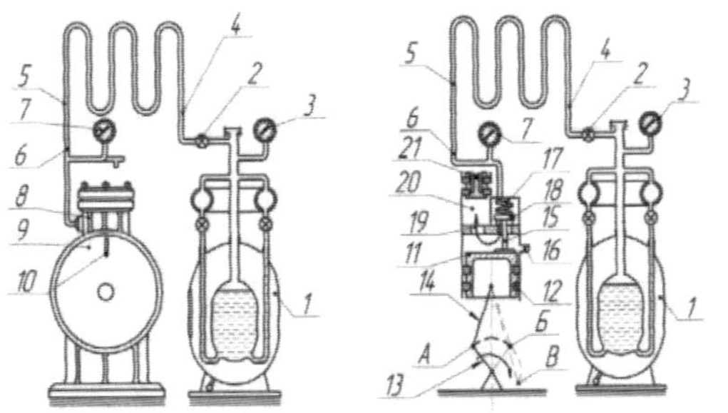

Figure 1. Schematic representation of the power plant.
1 - Dewar vessel, 2 - valve, 3.7 - manometers, 4, 6 - pipeline, 5 - gasifier, 8 - inlet cavity, 9 - flywheel, 10 - crankshaft position sensor, 11 - piston, 12 - O-ring, 13 - crankshaft, 14 - connecting rod, 15 - pusher, 16 - outlet, 17 - receiver, 18 - valve, 19 - bypass opening 20 - inlet cavity, 21- piston volume adjustment.

This installation has significant advantages:

A) complete ecological cleanliness;

IOP Conf. Series: Materials Science and Engineering **692** (2019) 012015 doi:10.1088/1757-899X/692/1/012015

B) inexpensive technology for the production and maintenance of CS;

C) the effective efficiency can reach 50 ... 60%, which significantly exceeds the maximum efficiency of internal combustion engines;

D) fire safety;

E) availability and relatively low cost of the working fluid (liquid nitrogen);

The main disadvantage of such installations is insufficiently effective use of low-temperature cryoproduct potential.

4. METHOD OF CALCULATING ROTARY-LOBE CRYOGENIC EXPANDER

Calculation procedure and a description of an installation contributing to reducing emissions of harmful substances by road transport are presented in this research.

As a result, the capacity for the refrigeration unit is calculated by the following formula:

$$Q_п = Q_{ст} + Q_{пр} + Q_т \quad (1)$$

The plant scheme with a cryogenic cylinder and a rotor-blade expander is shown below in figure 2.

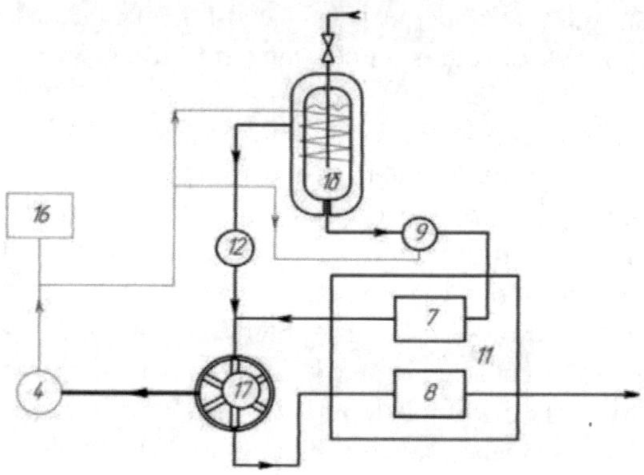

Figure 2. Diagram of the plant with a cryogenic cylinder and a rotary-blade expander. 1b - universal gas cylinder; 4 - electric generator; 7 - heat exchanger of the cryogenic working fluid; 6 - a thermal chamber; 7 - cryogenic pump; 8 - refrigerant evaporator; 11 - a thermal chamber; 12 - gas reducer; 16 - battery pack.

The cryogenic liquid from the balloon 1b, by means of a pump 9, enters the heat exchanger 7, where boiling occurs. Specific work (heat of vaporization, heating of single-phase flow) and fluid flow determine the cooling capacity of the proposed power plant. The refrigerant in the form of steam is supplied by the vtoronno-lopa expander17, the work received through the drive goes to the power generator 4, the flow, expanding in the inter-port channel, is stalled, which allows reusing steam in the heat exchanger 5.

The procedure for calculating the rotor-blade expander consists of the following steps:

1) Evaluation of the influence of the number of plates on the parameters of the expander.
2) Estimation of the influence of eccentricity on the speed of the expander.
3) Estimation of the influence of the width of the plate on the stability of work.
4) Estimation of the influence of the length of the rotor on the speed of the expander.
5) Selection of the angles of the location of the edges of the suction and discharge windows.
6) Choice of plate material. For the production of plates in expanders with lubrication, various materials are used: asbestos-textile, textolite, glass-textolite, steel 85 [4,5].
7) Determination of the main dimensions of the expander [6].
8) Finding the power of the expander [7].

The effective power on the expander shaft (W).

$$N_в = N_i + N_{fr} \quad (2)$$

where N_{fr}— power used to overcome frictional forces.

Using the presented technique, we make the determination of the parameters of the rotor-blade expander.

1. Theoretical action speed :

$$S_{\Gamma} = \frac{S}{\lambda} \tag{3}$$

2.The optimal number of plates:

$$Z_{opt} = \pi \left(\frac{R(\bar{\lambda}+1)}{3\delta} \right)^{1/3} \tag{4}$$

3. Angle between plates:

$$\beta = \frac{2\pi}{Z_{opt}} \tag{5}$$

4. Stator radius:

$$R = \left(\frac{2\pi S_{\Gamma}}{C\bar{\lambda}K_L u} \right)^{1/3} \tag{6}$$

5. Rotor length:

$$L = K_L R \tag{7}$$

6. Rotor eccentricity:

$$e = \bar{\lambda} r \tag{8}$$

7. Plate thickness:

$$\delta = \bar{\delta} R \tag{9}$$

8. Plate height:

$$h = 0.4R \tag{10}$$

9. Rotor radius:

$$r = R - e \tag{11}$$

10. Number of turns:

$$n = \frac{U}{2\pi R} \tag{12}$$

11. Angular velocity:

$$\omega = 2\pi n = 52 \tag{13}$$

12. Maximum angle of expansion:

$$\alpha_{\mathrm{p}} = \arccos \left[2 \left(\frac{p_0}{p_k} \right)^{\frac{1}{n}} - 1 \right] \tag{14}$$

13. Current gas pressure in the cell:

$$p_{\phi} = p_0 \left(\frac{\beta + 2\sin\frac{\beta}{2} + \frac{\bar{\lambda}}{2}\sin\beta - \frac{\bar{\lambda}}{2}\beta}{\beta + 2\sin\frac{\beta}{2}\cos\varphi + \frac{\bar{\lambda}}{2}\sin\beta\cos 2\varphi - \frac{\bar{\lambda}}{2}\beta} \right)^n \tag{15}$$

14. The pressure drop between two adjacent cells, one of which is connected to the suction inlet:

$$\Delta p_{\phi} = p \left\{ \left[\frac{\beta + 2\sin\frac{\beta}{2} + \frac{\bar{\lambda}}{2}\sin\beta - \frac{\bar{\lambda}}{2}\beta}{\beta + 2\sin\frac{\beta}{2}\cos(\varphi - \frac{\beta}{2}) + \frac{\bar{\lambda}}{2}\sin\beta\cos 2(\varphi - \frac{\beta}{2}) - \frac{\bar{\lambda}}{2}\beta} \right]^n - 1 \right\} \tag{16}$$

15. The pressure drop between two adjacent cells, one of which is connected to the discharge pipe:

$$\Delta p_{\phi} = p_i - p \left[\frac{\beta + 2\sin\frac{\beta}{2} + \frac{\bar{\lambda}}{2}\sin\beta - \frac{\bar{\lambda}}{2}\beta}{\beta + 2\sin\frac{\beta}{2}\cos\left(\varphi - \frac{\beta}{2}\right) + \frac{\bar{\lambda}}{2}\sin\beta\cos 2\left(\varphi - \frac{\beta}{2}\right) - \frac{\bar{\lambda}}{2}\beta} \right]^n \tag{17}$$

16. Pressure drop between two adjacent cells:

$$\Delta p = p \left\{ \begin{array}{l} \left[\frac{\beta + 2\sin\frac{\beta}{2} + \frac{\bar{\lambda}}{2}\sin\beta - \frac{\bar{\lambda}}{2}\beta}{\beta + 2\sin\frac{\beta}{2}\cos\left(\varphi + \frac{\beta}{2}\right) + \frac{\bar{\lambda}}{2}\sin\beta\cos 2\left(\varphi + \frac{\beta}{2}\right) - \frac{\bar{\lambda}}{2}\beta} \right]^n \\ - \left[\frac{\beta + 2\sin\frac{\beta}{2} + \frac{\bar{\lambda}}{2}\sin\beta - \frac{\bar{\lambda}}{2}\beta}{\beta + 2\sin\frac{\beta}{2}\cos\left(\varphi - \frac{\beta}{2}\right) + \frac{\bar{\lambda}}{2}\sin\beta\cos 2\left(\varphi - \frac{\beta}{2}\right) - \frac{\bar{\lambda}}{2}\beta} \right]^n \end{array} \right\} \tag{18}$$

IOP Conf. Series: Materials Science and Engineering **692** (2019) 012015 doi:10.1088/1757-899X/692/1/012015

17. Coefficient $\zeta_{сж}$:

$$\zeta_{сж} = 1 + 0{,}12\frac{\tau}{U} \quad (19)$$

18. Indicated horsepower:

$$N_i = N_T\zeta_{сж}\zeta \quad (20)$$

19. Power expended on overcoming the friction forces of the plates in the rotor slots from inertia forces:

$$N_i = \frac{4}{\pi}m\omega^3R^2\bar{\lambda}Z\mu_1\mu_2\frac{1+\bar{\lambda}}{1-\bar{\lambda}}\left[1 - \frac{h}{2R} + 2\bar{\lambda}^{2}\right] \quad (21)$$

20. Power expended to overcome the forces of the pressure difference in the neighboring cells:

$$N_r = \frac{1}{2\pi}L\bar{\lambda}^{2}\mu_1R^2\omega p \times 10^{1.175(e/h)+1.325}k^{0.291} \times lg\frac{p_i}{p} \quad (22)$$

21. Power expended on overcoming the friction forces of the plates along the cylinder from inertia forces:

$$N_{ic} = m\omega^3R^2Z\mu_2\left[1 - \frac{h}{2R} + 2\bar{\lambda}^{2}\right] \quad (23)$$

22. Power expended to overcome the friction forces of the plates on the cylinder from the pressure difference in the neighboring cells:

$$N = \frac{1}{2\pi}L\bar{\lambda}^{2}\mu_1\mu_2R^2\omega p \times 10^{1.24(e/h)(\frac{1+[e/(2h)\sin\psi]}{\cos\psi})+1.4}k^{0.528} \times lg\frac{p_i}{p} \quad (24)$$

23. Total power:

$$N_e = N_i + N \quad (25)$$

24. Mechanical efficiency

$$\eta_m = \frac{N_i}{N_e} \quad (26)$$

The results of calculation of parameters of the rotor-blade expander are shown below in the table 1.

Table 1. The results of calculation of the main parameters

S_r, m^3/s	Z_{opt}	β	C	R (m)	L(m)	e	δ (m)	h (m)	h_I (m)
0.0625	10	45°	12.44	0.22	0.88	0.022	0.0044	0.088	0.089
r, m	F_h (m^2)	m (kg)	n (s^{-1})	Ω	Φ	γ_0	δ_1	δ_2	δ_3
0.198	0.006	2.6	5	34	112	4	21.6	134	18
δ_4	α_p	p_Φ (Pa)	Δp_Φ (Pa)	Δp (Pa)	q_{12} (Nm^{-1})	q_{22}, (Nm^{-1})	q_{23} (Nm^{-1})	q_1 (Nm^{-1})	N_T (Watt)
18	128	125080	66670	11240	112.8	573.4	3400	33.1	102
$\zeta_{сж}$	N_i (Watt)	P_i (MPa)	T_0 (kH)	N_3 (kWatt)	μ_2	Ni (Watt)	Nr (kWatt)	N_{ic} (kWatt)	N, (kWatt)
1.07	136000	1.74	600.491	77	0.05	419.1	9.7	4.5	1.9

5. CONCLUSION

In the course of the research, a technique was developed for calculating the rotor-blade expander. The efficiency is 44%, so there is no need for variation in parameters to ensure a higher efficiency. The need to change a number of specified parameters arises, only with an efficiency of less than 40%.

It was found that the main source of pollution are cars running on hydrocarbon fuels, the proportion of which falls on cars with a refrigerated chamber.

Since refrigeration plants consume 3% of the ICE power, it is advisable to replace them with autonomous units using a cryogenic working fluid, thereby reducing emissions of harmful substances into the atmosphere.

REFERENCES

[1] Novotelnov, V.N. Cryogenic machines [Text]: textbook for high schools / VN. Novotelnov, A.D. Suslov, V.B. Poltaraus - St. Petersburg: Politechnica, 1991. 335 p.
[2] Bogdanova, SN / Refrigeration equipment. Air conditioning. Properties of substances [Text]: reference book / Bogdanova SN - St. Petersburg: SPbGAKHPT, 199. - 320 p http://dearman.co.uk/dearman-technologies/dearman-engine/
[3] http://www.holodcatalog.ru/entsiklopedii/promyshlennoe-oborudovanie/dvigatel- dearman-na-zhidkom-azote/
[4] S.I. Bondarenko, I.N. Kudryavtsev, A.Ya.Levin, N.M. Levchenko, B.N.Murinets-Markevich, A.I.Pyatak, Development of a cryogenic power plant for an environmentally friendly car, ISSUES OF ATOMIC SCIENCE AND TECHNIQUES, Series: Vacuum, pure materials, superconductors No. 14, 2004, P.152-157.
[5] Chumak, I.G. Refrigerating plants. Designing [Text]: a tutorial for universities / IG. Chumak, D.G. Nikulshina. - Moscow: Higher School, 1988. - 280 p.
[6] Blagin E.V., Dovgyallo A.I., Nekrasova S.O. и др. Estimation of the energy efficiency of cryogenic filled tank use in different systems and devices // Applied Thermal Engineering. 2016. Vol. 101. Pp. 537-544
[7] Blagin E.V., Uglanov D.A., Dovgyallo A.I. About LNG Energy Utilization Efficiency Estimation // Procedia Engineering, 2016. Vol. 152. Pp. 209-218)

Automobile Connecting Rod Forging Optimization Based on DEFORM and Orthogonal Experiment

XU Hong-wei[1], JING Yan-bo[1], QI Xiao-ming[1], LI Shou-dong[1] and LIU Dou-wei[2]

[1] Shandong Hua Cheng China-Germany Transmission Equipment co., LTD, Zibo 255000, China
[2] School of Mechanical Engineering and Automation, Beijing University of Aeronautics and Astronautics, Beijing 100191, China

E-mail: 13051160303@163.com.

Abstract. Taking blank temperature, mould temperature and friction coefficient as test factors, the orthogonal experiment of three factors and three levels of automobile connecting rod forging was designed. Using Deform-3d, the forging experiment under different factors is simulated, and the forging results under different technological parameters were obtained. The process parameters of connecting rod forging are optimized by using the comprehensive measurement method and taking deformation uniformity and maximum forming load as indexes. The results show that the optimized process parameters can reduce the forming load and improve the deformation uniformity.

1. Foreword

Connecting rod is an important part of automobile engine, which plays an important role in converting reciprocating motion of engine piston into rotation motion of crankshaft, and bears higher impact load and stress [1]. At present, the main processing method of automobile connecting rod is forging. With the improvement of engine performance requirements, connecting rod production technology is also continuously improved, and the hot forging technology, quality and efficiency of connecting rod are required to be improved [2].

In order to improve the production quality of connecting rod forging, this paper takes the forging processing of a certain automobile connecting rod as the research objective, according to the orthogonal experiment method, designed the orthogonal experiment with three factors and three levels, including blank temperature, mold temperature and friction coefficient, and adopted Deform-3d software to conduct the forging simulation experiment simulation.

2. Establish simulation model of connecting rod forging

UG is adopted to establish the three-dimensional model of connecting rod forging blank and upper and lower moulds, which is imported into Deform-3d, and position interference adjustment is carried out to establish the simulation model of connecting rod forging, as shown in the Figure 1 simulation model of connecting rod.

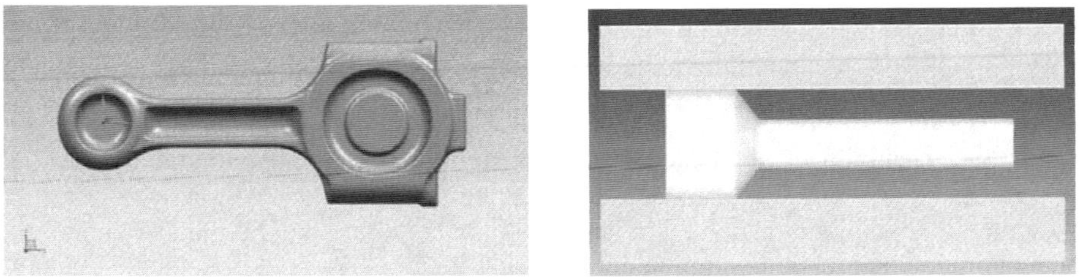

Figure 1. Simulation model of connecting rod forging.

Blank material for 1045 # steel, can be directly obtained from the Deform, the blank according to the tetrahedron grid, grid nodes is established for 12699, 11850 polyhedron unit, up and down mold movement by mechanical pressure, set up each step by step into the distance and step length, according to the experimental parameters to the blank temperature, mold temperature, the coefficient of friction gives corresponding numerical values.

3. Orthogonal experimental design

Orthogonal test design method is a method to arrange and analyse multi-factor experiments scientifically by using normalized orthogonal table. It can select representative experimental schemes from many experimental schemes, and analyse these experimental schemes to obtain the optimal production process [3]. The design process of orthogonal experiment generally includes four steps: selecting table and designing table head, defining and completing experimental program, calculating range to determine the order of primary and secondary factors, and determining the optimal program. In this paper, the orthogonal experimental design of automobile connecting rod forging is designed by taking automobile connecting rod forging process parameters as experimental factors and connecting rod forging quality optimization as experimental objective.

3.1 Design experimental factors and levels

Preheating temperature of blank is an important parameter of hot forging process. Preheating treatment of blank can improve the atomic fluidity inside metal, which is conducive to forging, but too high temperature will easily cause metal surface sintering, affecting the quality of forgings [4]. Forging process, the die and blank exposure will cause the decline of blank temperature, makes the metal blank resistance increases, is not conducive to the forging forming, at the same time die contact surface and the contact surface temperature difference, thermal stress, can lead to mould deformation, and affects the quality of the forging forming, therefore also need to analyse die preheating temperature [5]. The friction between the die and the blank will reduce the metal fluidity and the friction coefficient will directly affect the deformation uniformity. Therefore, this paper selected blank temperature, mould temperature, friction coefficient three experimental factors, and according to the actual forging process parameters, the blank temperature is set at 1100°C, 1150°C, 1200°C, mould temperature is set at 200°C, 250°C, 300°C, friction coefficient is set at 0.3, 0.35, 0.4.

The experiment of three factors and three levels was designed according to the orthogonal experiment method, and the orthogonal experiment table of L_9（3^4） was selected. The horizontal design of experimental factors was shown in Table 1.

Table 1. Factors and levels of orthogonal experiment.

	A Blank Temperature /°C	B Mould Temperature /°C	C Friction Coefficient
1	1100	200	0.40
2	1150	250	0.35
3	1200	300	0.30

IOP Conf. Series: Materials Science and Engineering **692** (2019) 012016 doi:10.1088/1757-899X/692/1/012016

Orthogonal experiment has three factors and three levels. As shown in the Table 2, the orthogonal table of L_9 (3^4) is adopted. Meanwhile, in order to analyze the variance of orthogonal experiment, an empty column is added to the head design [6].

Table 2. Design of gauge head for connecting rod forging experiment.

Factor	A	B	C	Vacant Column
Number	1	2	3	4

3.2 Determine the experimental indexes

In the process of connecting rod forging, the better the deformation uniformity of the forgings, the lower the internal stress of the forgings, and the higher the performance and quality of the forgings. The equivalent strain is an important basis for measuring the deformation uniformity. In this paper, the maximum equivalent strain and minimum equivalent strain difference are selected as the indexes to measure the uniformity of forging deformation. Considering the influence of forming load on the service life of die in forging process, this paper takes the maximum forming load in forging process as the index to measure the service life of die.

3.3 Experimental scheme

According to the orthogonal table of L_9 (3^4), 9 groups of experiments with different factor levels were set up. Forging simulation is carried out according to different factor levels of each group, and the simulation results of one group of experiments are shown in Figure 2. According to the experimental indexes, the forming load and equivalent strain obtained by each group of experiments are derived, and the data are analysed.

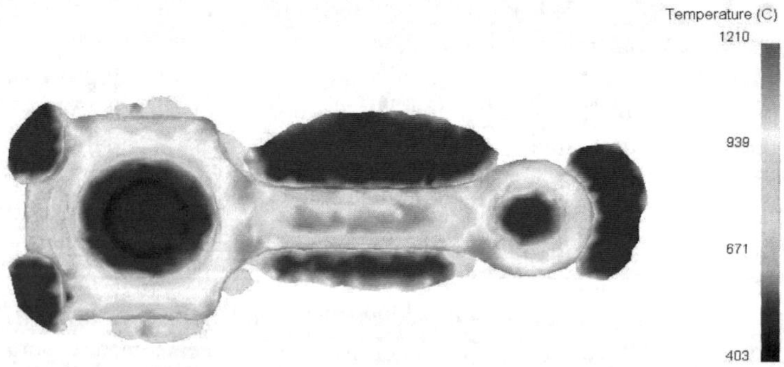

Figure 2. Simulation results.

4. Results analysis and optimization

4.1 Orthogonal experimental results

Data required by each group of experiments were extracted, and Table 3 orthogonal experimental data table was obtained.

Table 3. Orthogonal experimental data sheets.

Number	Blank Temperature /°C	Mould Temperature /°C	Coefficient of Friction	Vacant Column	Forming Load /× $10^6 N$	Equivalent Strain Difference
	1	2	3	4		
1	1200	200	0.40	1	10.3	4.216
2	1200	250	0.35	2	9.8	4.320
3	1200	300	0.30	3	9.3	4.490
4	1150	200	0.35	3	10.3	4.194
5	1150	250	0.30	1	10.9	4.481
6	1150	300	0.40	2	10.2	4.511
7	1100	200	0.30	2	11.6	4.407
8	1100	250	0.40	3	12.3	4.394
9	1100	300	0.35	1	11.3	4.290

4.2 Forging parameter analysis based on forming load

Taking forming load as optimization parameter, the range analysis of orthogonal experimental table is carried out, as shown in Table 4 Analysis of forming load range.

Table 4. Analysis of forming load range.

Sum	Blank Temperature	Mould Temperature	Coefficient of Friction	Vacant Column
T_{1j}	29. 4	32. 2	32. 8	32. 5
T_{2j}	31. 4	33. 0	31. 4	31. 6
T_{3j}	35. 2	30. 8	31. 8	31. 9
R_j	5. 8	2. 2	1. 4	0. 9

The range analysis shows that the blank temperature has a great influence on the forming load. In this experiment, the selected temperature is 1200°C, followed by mould temperature and friction coefficient, respectively 300°C and 0.35 °C. The combination of these parameters did not appear in the orthogonal experiment. The experimental design was carried out for these parameters, and the experimental results were compared with the third experiment with the smallest forming load. As shown in FIG. 3-1 load formation curve, the experimental results show that when the blank temperature, mould temperature and friction coefficients are 1200°C, 300°C and 0.35 respectively, the forming load in the forging process decreases somewhat, and the maximum forming load is 8. 9× 10^6N.

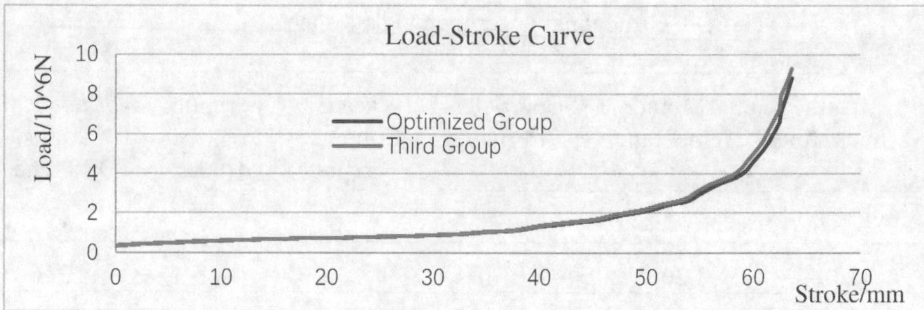

Figure 3. Load stroke curve

4.3 Forging parameter analysis based on equivalent strain difference

Taking the equivalent strain difference as the optimization target, the range analysis of forging parameters is carried out, and the range analysis of the equivalent strain difference in table 5 is obtained. From range analysis, it can be known that the minimum equivalent strain difference can be obtained. The blank temperature should be 1200°C, the mould temperature 200°C, and the friction coefficient 0.35.

Table 5. Equivalent strain difference

Sum	Blank Temperature	Mould Temperature	Coefficient of Friction	Vacant Column
T_{1j}	13. 025	13. 113	13. 121	12. 988
T_{2j}	13. 398	13. 195	12. 804	13. 237
T_{3j}	13. 092	13. 290	13. 398	13. 078
R_j	0. 306	0. 177	0. 594	0. 249

Experimental results are shown in Figure 4 strain optimization results, with the maximum strain of 4.22, the minimum strain of 0.217 and the strain difference of 4.003. By comparing the strain difference of 4.194 in the fourth group of the optimal results in the orthogonal experimental table, the feasibility of the optimization results is proved.

Figure 4. Strain optimization results

4.4 Multi-index analysis based on comprehensive balance method

In the orthogonal experiment with multiple indicators, the combination of factors and parameters corresponding to each indicator may be inconsistent. In the comprehensive analysis, the comprehensive balance method needs to be considered. According to the orthogonal experimental results, the trend chart of experimental factors and indicators is drawn, and the results are shown in Figure 5. Trend chart of factor indicators.

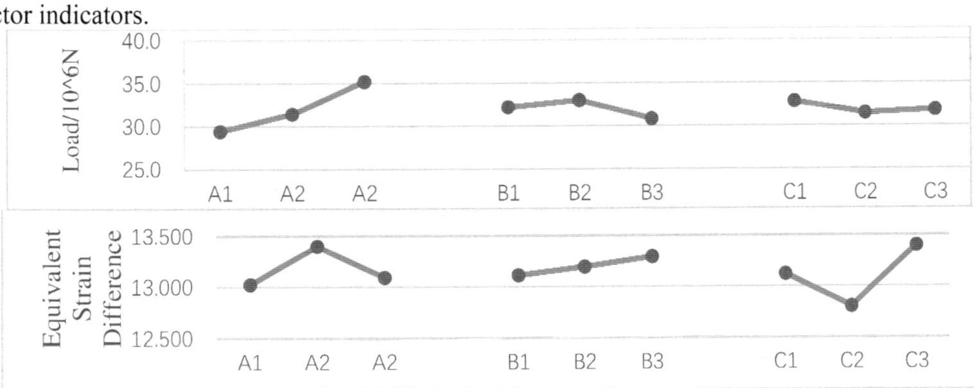

Figure 5. Trend chart of factor indicators

As can be seen from the trend chart of factor indicators, when the forming load and equivalent strain difference are the minimum values, the corresponding blank temperature and friction coefficient are 1200°C and 0.35 °C, and the corresponding mold temperature is 300°C and 200°C. It can be seen from range analysis that, among the three factors, mold temperature has the smallest influence on the equivalent strain difference, and the influence on the maximum forming load is between blank temperature and friction coefficient, so mold temperature should be based on the maximum forming load and choose 300°C.

5. Conclusion

The blank temperature, mould temperature and friction coefficient were optimized through the three factors and three levels orthogonal experiment. The results show that the blank temperature has the greatest influence on the maximum forming load. The friction coefficient had the greatest influence on forging deformation uniformity. When the blank temperature, mold temperature and friction coefficient were taken as 1200°C, 300°C and 0.35 °C respectively, the maximum forming load in the forging process was the smallest and the deformation uniformity was better.

References

[1] Bo, L., Liu, F., Wang, S., Ji, F., Sun, J., Chen, L., Wang, F. (2016) Microstructure and mechanical properties of Fe-C-Cu powder forged automobile engine connecting rod. Journal of Metal., 52:41-50.
[2] Wu, X., Sun, Y., Li, Y., Xia, X. (2014) Research on the influence of hot forging process parameters on die wear of connecting rod. Hot processing technology.,43:99-101.
[3] He, Y., (2017) Influence analysis of hot forging process parameters on connecting rod forming based on DEFORM 3D. Hot processing technology., 46:182-185.
[4] Wu, X. (2014) Numerical simulation of forging of connecting rod and optimization of process parameters. Nanjing university of science and technology Publishing, Nanjing.
[5] Haiyan, S. (2017) Mathematical Statistics. Bei Hang University Publishing, Beijing.
[6] Zhao, X., Zhao, G., Wang, G. (2000) Shape optimization design method of forging die based on deformation uniformity. Chinese journal of mechanical engineering., 64:67-85.

Design of Magnetic Circuit and Finite Element Analysis of Stepped Ferrofluid Seal with Small Sealing Gap

Ruibo Zhang[1], Xiaolong Yang[1, *] and Guohong Wang[1]

[1]School of Mechanical Engineering, Guangxi University of Science and Technology, Liuzhou 545006, China

[*]Corresponding author e-mail: 09116324@bjtu.edu.cn

Abstract. For the sake of study the pressure capability of ferrofluid seals with small sealing gap, a diverging stepped ferrofluid structure with small sealing gap was designed according to the theory of stepped ferrofluid and magnetic path design, and the length and cross-sectional area of the permanent magnet (NeFeB) are obtained. On the basis of these parameters, the ferrofluid distribution of seal gap is calculated by using the magnetic ferrofluid element method, and the pressure capability is obtained, then the reliability of the magnetic path design is obtained according to the capability of withstand voltage. There are some findings as following after analysis and comparison. It is the magnetic flux leakage at the junction of the pole boot and NeFeB that cause pressure capability calculated by the magnetic path method is lower than that calculated by the finite element method.

1. Introduction

Ferrofluid seals is a kind of technology that ferrofluid seals generates sealing ring to the response characteristics of external magnetic field to achieve the function of resisting pressure difference on both sides of the sealing structure[1]. It is widely used in fields of aerospace, defense, machinery, chemical, petroleum, environment, instruments due to its advantages of zero leakage, long life and low friction[2-3]. At present, there are many researches on common ferrofluid seals in small gap. Research on vacuum, dustproof and small diameter magnetic liquid static sealing with low pressure capability and dynamic sealing technology with small sealing gap from 0.1mm to 0.3mm has been relatively mature, but the research on high pressure, high temperature, low temperature and high vacuum is few, which limits its application[4]. Wang Ruijin[5] et al summarized the general law of ferrofluid seals, and focused on the pressure capability of seal gap under different speeds and parts processing quality, then concluded that the seal gap is too small, which can not improve the pressure capability, but will decline. Gu Jianming[6] et al drew a conclusion that the sealing capability of the ferrofluid increases with the increase of seals, and decreases with rise the seal gap through experiments. Besides, the effect is better when the seal gap is 0. 05-0.20 mm, at the same time, the sealing series has an optimum value. Li Decai[7] et al numerically studied the pressure capability of small gap and concluded that sealing pressure capability is greatly decreases in rise sealing gap, but the sealing gap should not exceed 0.3 mm. These studies were only completed on the base of a non-stepped, but the study of stepped ferrofluid seals with small gap has not been carried out. In 2014, Yang Xiaolong and Li Decai et al began to carry out numerical analysis and experimental research on stepped ferrofluid seals with large sealing gap, and compared it with the pressure capability of ordinary ferrofluid seal, then came to a conclusion that the stepped ferrofluid seal is an effective method to improve the sealing performance of large gap ferrofluids. Yang Xiaolong and Chen Fan[8] demonstrated that the pressure capability of the diverging stepped ferrofluid seal with large sealing gap decreases with the increase of the radial seal gap width based on the theory and experiment of sealing. Yang Xiaolong et al drew a conclusion that the pressure capability of the diverging stepped ferrofluid seal decreases with the increase of the radial seal gap using the magnetic path design method and the finite element analysis method. However, there are no systematic studies about the sealing property of stepped ferrofluid seals with small sealing gap.

In order to improve the pressure capability of the stepped ferrofluid seals with small sealing gap, this paper designs a structure of the diverging stepped magnetic path with small sealing gap based on the theory of magnetic path design, and studies the effects of parameters such as the axial clearance, radial clearance, the number of axial teeth, the number of radial tooth on its sealing capability through finite element method. The research results provide important theoretical guidance for designing a high-reliability stepped ferrofluid seal structure with small gap.

2. Design of magnetic path

Figure 1 is a diverging stepped ferrofluid seal structure, and its equivalent magnetic circuit is shown in Figure 2. It can be seen from the Figure 1 that the magnetic circuit is mainly composed of permanent magnet, pole piece, magnetic fluid and stepped shaft. The permanent magnet generates magnetic field which binds the magnetic fluid in the radial and axial sealing gap formed between the stepped shaft and the pole piece. The magnetic force on the magnetic fluid is to resist the pressure difference between the two sides of the seal structure and to achieve the purpose of blocking the leakage channel.

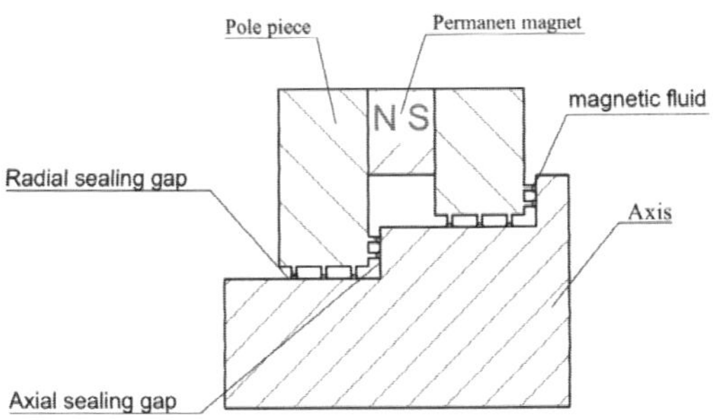

Figure 1. Structural Chart of Stepped ferrofluid Seal

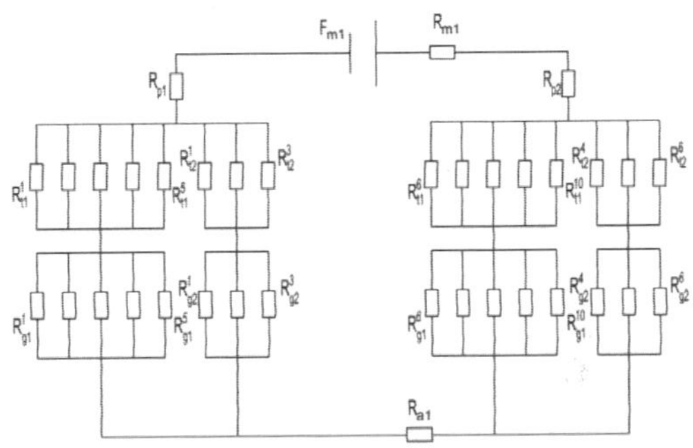

Figure 2. Diagram of equivalent magnetic path

It is assumed that the gravity of the ferrofluid is much smaller than the magnetic force, and the ferrofluid line can be replaced by an arc; the isomagnet line is considered to coincide with the ferrofluid line, and the surface tension of the ferrofluid can be ignored. Then the pressure capability of ferrofluid static seals can be approximately expressed by following formula

$$\Delta P = \mu_0 M_s \sum_{i=1}^{N} (H_{max}^{i} - H_{min}^{i}) = M_s \sum_{i=1}^{N} (B_{max}^{i} - B_{min}^{i}) \quad (1)$$

where H_{max}^i and H_{min}^i and B_{max}^i and B_{min}^i are, respectively, the maximum and minimum magnetic field strengths and the maximum and minimum magnetic flux densities under the ith is pole tooth, while N is the total number of sealed poles.

The total sealing capability of the converging stepped ferrofluid as:

$$\Delta P = \sum_{i}^{N} (P_{ia} + \lambda P_{ir}) \quad (2)$$

In(2), P_{ia} and P_{ir} are the pressure capabilities of the ferrofluid seal in the axial and in the radial sealing gaps formed by the ith pole piece and the stepped shaft. If $P_{ia}<P_{ir}$, then $\lambda = 1$, otherwise

$\lambda = 0$, $P_{ia}$and P_{ir} can be calculated by formula (1).

Two assumptions are contained in magnetic path design of parallel ferrofluid seals in this paper: first, ignore the magnetic flux leakage; second, ignore the edge effect. According to Kirchhoff's first law,

$$\sum \Phi_i = 0 \tag{3}$$

That is, at any point in the magnetic path, the sum of the magnetic flux algebra entering the region equals to the sum of the magnetic flux algebras leaving the region. According to the symmetry of the magnetic path structure,

$$\Phi_2 = \Phi_1 \tag{4}$$

Φ_1 and $\Phi_2$respectively represent the magnetic fluxes in the higher pressure side pole shoe and the atmosphere side pole shoe, among them

$$\Phi_1 = 5B^1_{g1}S^1_{g1} + 3B^1_{g2}S^1_{g2} \tag{5}$$

B^1_{g1} and B^1_{g2} respectively represents the first radial gap under the first pole teeth on the first pole shoe on the high voltage side and the magnetic flux density in the first axial gap under first pole teeth in first pole shoe, respectively; S^1_{g1} and S^1_{g2} denotes the annular area of the first radial gap under the first pole tooth and the first axial seal gap under the first pole teeth on the high pole side first pole shoe, respectively.

$$\Phi_2 = 5B^6_{g1}S^6_{g1} + 3B^4_{g2}S^4_{g2} \tag{6}$$

B^6_{g1} and B^4_{g2} respectively represents the sixth radial gap of sixth pole tooth on the atmospheric side pole shoe and the magnetic flux density in the fourth axial gap of the fourth pole teeth on the middle pole shoe; S^6_{g1} and S^4_{g2} denotes the annular area of the sixth radial seal gap under the sixth pole teeth on the atmospheric side pole shoe and the fourth axial seal gap under the fourth pole tooth on the atmosphere side, respectively.

$$\Phi_{m1} = B_{m1}S_{m1} = \Phi_2 = 5B^6_{g1}S^6_{g1} + 3B^4_{g2}S^4_{g2} \tag{7}$$

Where Φ_{m1} B_{m1} and S_{m1} represent the magnetic flux, magnetic flux density and annular area of NeFeB on the atmospheric pressure side respectively.
According to Kirchhoff's second law, $\sum H_i l_i = \sum N_i i$

$$F_1 = H_{m1}L_{m1} = \left(5B^6_{g1}S^6_{g1} + 3B^4_{g2}S^4_{g2}\right)\left[R_{p2} + \frac{\left(R^6_{t1}+R^6_{g1}\right)\left(R^4_{t2}+R^4_{g2}\right)}{3\left(R^6_{t1}+R^6_{g1}\right)+5\left(R^4_{t2}+R^4_{g2}\right)}\right] + \left(5B^1_{g1}S^1_{g1} + 3B^1_{g2}S^1_{g2}\right)\left[R_{p1} + R_{a1} + R_{m1} \frac{\left(R^6_{t1}+R^6_{g1}\right)\left(R^1_{t2}+R^1_{g2}\right)}{3\left(R^1_{t1}+R^1_{g1}\right)+5\left(R^1_{t2}+R^1_{g2}\right)}\right] \tag{8}$$

Wherein F_1 represents the magnetomotive force of the NeFeB; H_{m1}, L_{m1} represent the ferrofluid strength and length of the NeFeB; R_{p1}, R_{a1} and R_{m1} represent the first pole piece, the stepped shaft, and the NeFeB, respectively. R^6_{t1}, R^6_{g1}, R^4_{t2}, R^4_{g2} respectively represent the magnetic drag of the sixth radial pole tooth on the atmospheric side pole shoe and the sixth radial seal gap of the sixth radial pole tooth, and the atmospheric side pole shoe ,The magnetic drag of the fourth radial seal gap of the upper fourth axial pole teeth and the fourth radial pole tooth.
Equation (5) multiply (6)

$$V_{m1} = L_{m1}S_{m1} = \frac{F_1\Phi_{m1}}{B_{m1}H_{m1}} \tag{9}$$

In order to reduce the volume and weight of the NeFeB and improve the utilization rate of the NeFeB, the NeFeB should be operated at its maximum magnetic energy product, and therefore,

$$V_{ml} = L_{ml}S_{ml} = \frac{F_l \triangleleft_{ml}}{(BH)max} \tag{10}$$

Divide equation (5) by equation (6)

$$\frac{S_{m1}}{L_{m1}} = \frac{\left(5B^1_{g1}S^1_{g1} + 3B^1_{g2}S^1_{g2}\right)H_{m1}}{F_1 B_{m1}} \tag{11}$$

Equation (8) Multiply (9).

$$S_{m1} = \left(5B^1_{g1}S^1_{g1} + 3B^1_{g2}S^1_{g2}\right)\sqrt{\frac{H_{m1}}{(BH)_{max} B_{m1}}} \tag{12}$$

Divide equation (8) by equation (9)

$$L_{m1} = F_1\sqrt{\frac{B_{m1}}{(BH)_{max} H_{m1}}} \tag{13}$$

Since the magnetic permeability of the material of the pole shoe and rotating shaft is much larger than that of the air and the NeFeB, the obstruction of the pole shoe and the rotating shaft can be neglected when calculating the length of the NeFeB.

3. Finite element analysis of the ferrofluid

The dimensions of each part of magnetic path are shown in table 1. If the sealing pressure capability is required to be not less than 2×10^5 Pa , we know that the total static magnetic induction difference in the sealing gap is 6.5T according to the general static sealing withstand voltage formula. The length of NeFeB is 5.6 mm, the cross-sectional area is 243 mm^2 according to the symmetry of magnetic path and the design of the parallel ferrofluid seals.

Table 1. Parameter of converging stepped ferrofluid seals.

Item	Value	Item	Value
Inner radius of the 1/2 pole piece (mm)	17/21	Number of radial teeth under the 1/2 pole piece	2
Outer radius of the 1/2 pole piece (mm)	30		
Length of the 1/2 pole piece (mm)	5.3	Width of pole teeth (mm)	0.3
Number of axial teeth under the 1/2 pole piece	3	Axial sealing gap width (mm)	0.1
Slot width(mm)	0.7	Radial sealing gap height (mm)	0.1
Slot depth (mm)	0.8		

Create a physical environment of ferrofluid-tight in the preprocessor of ANSYS finite element analysis software. Due to the symmetry of the structure, the three-dimensional axisymmetric problem of the ferrofluid seals can be simplified to a two-dimensional one. NdFeB magnet is chosen as the NeFeB of diverging stepped ferrofluid seals structure with small gap, the coercive force H_c is 1.356×106 A/m, the magnetic permeability is 1.05. The material of the pole shoe and the stepped shaft are 2Cr13, and oil-based ferrofluid, with saturation magnetization of 30.7 KA/m is selected as ferrofluid. Since the ferrofluid strength generated by the NeFeB in the gap under each pole piece is greater than the saturation magnetization of the ferrofluid, the ferrofluid is saturated and magnetized. The saturation magnetization of the ferrofluid is almost the same as that of the air, so the ferrofluid can be treated as air. Each part is endowed with corresponding material properties, and Smart Meshing, with the accuracy of 4, is selected; A boundary condition that parallels with lines of force is applied to the model boundary.

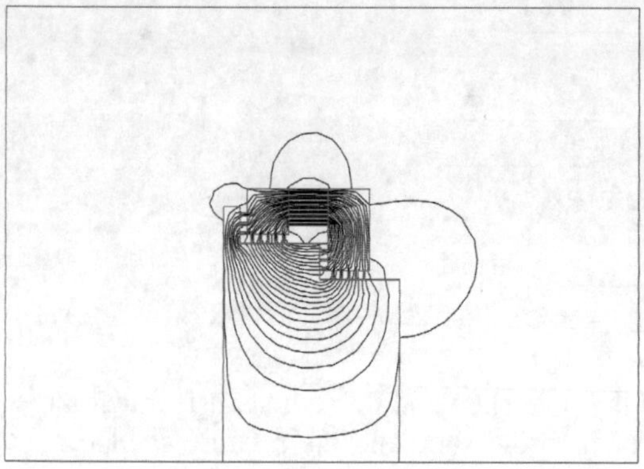

Figure 3 Distribution of magnetic line of force

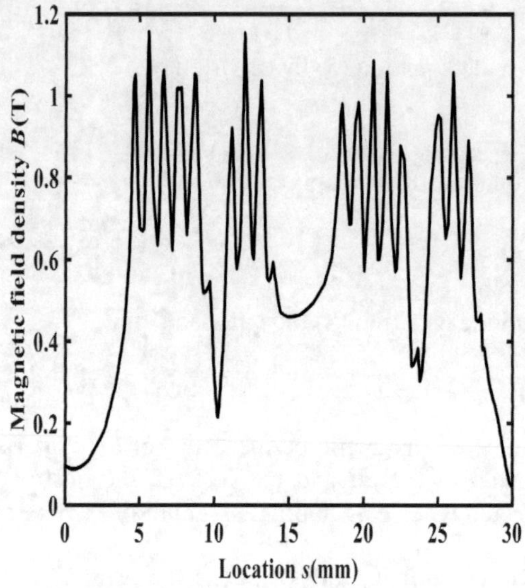

Figure 4 Axial path BSUM curves

Figure 3 and **Figure** 4 of magnetic flux density curve are obtained by finite element analysis., verify the accuracy of the magnetic path design according to the withstand voltage formula of stepped ferrofluid seals, the data of the Table 1 and **Figure** 4, the NeFeB data calculated by magnetic path design, So sealing pressure capability Δp is

$$\Delta P = \mu_0 M_s \sum_{i=1}^{N}(H^i_{max} - H^i_{min}) = M_s \sum_{i=1}^{N}(B^i_{max} - B^i_{min}) = 0.307 \times 5.8631 = 1.8 \times 10^5 \text{pa} < 2 \times 10^5 \text{ Pa}$$

Therefore, it is calculated that the pressure capability of the diverging ferrofluid seals structure is less than that required by the magnetic path design, 2×10^5 Pa, through the finite element method. Magnetic flux leakage，which is caused by the direct contact of magnet, pole piece and shell, is the main reason for this phenomenon. In order to improve the accuracy of the magnetic path design and the pressure capability of the ferrofluid seals, it is possible to select a non-magnetically conductive housing and increase the axial length of the intermediate pole piece to reduce the leakage magnetic energy.

4. Conclusion

- The finite element analysis results verify the reliability of the magnetic path design method.
- The pressure resistance of divergent stepped ferrofluid seal calculated by magnetic path design method is slightly larger than that of finite element analysis, mainly due to magnetic leakage at the direct contact between NeFeB and pole shoe and shell.
- We can reduce the leakage of magnetic energy by choosing non-permeable shells.

Acknowledgments

The authors gratefully acknowledge the support of the National Nature Science Foundation of China (Grant No. 51905114), the support of the Science and Technology Project of Guangxi Province (Grant No. 2016GXNSFBA380213), the support of the Innovation Project of GuangXi University of Science and Technology Graduate Education (Grant No. GKYC201901) and the support of the Science and Technology Project of Liuzhou (Grant No. 2017BC20204).

Reference

[1] Rosensweig R E 1997 Ferrohydrodynamics(New York: Dover Publications)

[2] Urreta H, Aguirre G, Kuzhir P, Lopez de Lacalle LN. 2018 Seals based on ferrofluids for high precision spindles of machine tools. *Int J Precis Eng Manuf* (Volume 19(4)) PP.495-503.
[3] Y.B Chen,D.C Li and Y.J Zhang Numerical Analysis and Experimental Study on Ferrofluid Reciprocating Seals *IEEE Trczns Magn* , (Volume 55(1)) PP.495-503.
[4] Yan C and Decai L 2011 Design and Experiments for Ferrofluid Seal of Tank Panoramic Mirror *Acta Armamentarii* (Volume 32(11)) PP 1428-1432.
[5] Ruijin W, Yunhong L and Ming Z 2001 Effect on pressure tight capacity for gap of ferrofluid seals *Vacuum* (2) PP 42-44
[6] Jianming G and Yongxing X 1999 Effect of the Gap of Ferrofluid Seal on Seal apacity *JOURNAL OF SHANGHAI JIAOTONG UNIVERSITY* (Volume 33(3)) PP 136-138
[7] Mingli K and Decai L Numerical Simulation for Ferrofluid Sealing Device *Journal of Beijing Jiaotong University* (Volume 32(1)) PP 116-118
[8] Xiaolong Y, Peng S, Fan C, Fuxiang H, Decai L and Peter J T 2019 Numerical and Experimental Studies of a Novel Converging Stepped Ferrofluid Seal *IEEE TRANSACTIONS ON MAGNETICS*(Volume 55(3))

IOP Conf. Series: Materials Science and Engineering **692** (2019) 012018 doi:10.1088/1757-899X/692/1/012018

Experimental and Numerical Studies of Ice Accretion on Atmospheric Icing Laboratory at Different Icing Pattern and Time

Chen Zhao[1]*, Panfeng Liu[2] and Haijiang Wang[1]

[1] Southern University of science and technology, Shenzhen 518055, China.
[2] China Aero-Polytechnology Establishment, Beijing 100191, China.

*zhaochen913@163.net

Abstract: The purpose of this paper was to explore a suitable icing test method which can achieve the requirement of the experimental conditions better and faster in atmospheric Icing Laboratory. So, the effects of varying the spray method and spray time of different conditions were studied in experiment. The ice shape and profile of ice accretion obtained in atmospheric icing laboratory were investigated with variable air temperature. The icing object was exposed to the one aerosol cloud——freezing rain. In addition, the heat balance model of dynamic process of single droplet would be set up, the experimental condition of distance between nozzles and icing object was analysed by simulation calculation for the model. At the same time, appears the effects of different distances on the mass, ice shape and ice accreted of profile under different conditions in experiment.

1 Introduction

The atmospheric icing was a complex natural phenomenon, while it will present major hazard to many areas of modern equipment such as aircraft、power lines and warship. For aircraft, it would destroy the normal flow of air, reduce the aircraft's performance and lift, increase resistance. For power line, it could lead failure of power line conductors [1]. For warship, it could cause serious influence on stability and floating.

Over the past years, research mostly focus on the impact and freezing processes of a water droplet on different smooth cold solid surfaces[2],[3]. The Hayashi et al. divided the frost formation process into three different periods: namely, the crystal growth, the frost layer growth, and the frost layer full growth periods [4]. Ice accretions at mixed conditions were test and it shows ice accretion shapes on a cylinder model deduced from NRC studies at RATFac [5],[6]. Extensive ice particle impact studies have been conducted by Hauk et al [7]. Wu et al. illustrated the icing of droplets on the surface [8]. Hoke observed the frost on a variety of plates and found that the water distribution on the plate at the end of the condensation period was strongly dependent on plate temperature, humidity [9]. The author also carried out wind tunnel tests and numerical simulations on the blade icing both for HAWT and VAWT [10]. Homola researched the turbine size and temperature dependence of icing on wind turbine blades [11].

Most of the existing experimental data corresponds to the air temperature and air humidity. But in most practical situations, especially in the icing climatic& environmental test based on the MIL-STD-810G[12], the experimental conditions not only involve those factors which were the air temperature, air humidity and droplet diameter, but there were other important influence factors which were spray method, spray time and the distance between nozzle and icing object. Because those factors will influence the test process and test results. But those experimental data for a continuous variation of distance or spray method appear to be lacking in the literature, and the validation of theoretical models for distance remains difficult in the absence of such experimental.

The spray method in experience may be inclined in three different spray ways relative to the droplet trajectories as sketched in (i) vertical; (ii) oblique; (iii) parallel. As well as two different sprays time which was (i) continuous and (ii) interrupt. So, in this paper, the main goal of the present studied was to study the effects of the different spray methods and spray time defined with the ice accreted on variable icing conditions. A subsidiary goal was to study the effect of different distance for ice accreted under different icing conditions. Finally, those findings can give certain guidelines for implementation process in testing technology.

2 The experimental setup

Experiments were performed in the atmospheric icing Laboratory (5m×2.5m×3m) at the Aero-Polytechnology Establishment in Beijing. The temperature could be cooled down to -25℃. To vary spry ways, the nozzles were fixed to some adjustable on the frame. To different spray time, the program was supported with different interval time for obtaining different spray time.

Different spay methods and spray time were applied to model different air temperature for different ice shape. In order to try to achieve a uniform distribution of impinging water along the test specimen. The nozzles were supplied with tap water at 0℃~10℃. A constant wind speed of 0.2m/s was measured near the nozzle. The wind and droplet supply were two independent systems, so, the wind speed was deemed steady. The air temperature was performed under -15℃, -10℃, -7℃ and -3℃, and the precipitation rate was 25mm/h that was same as MIL-STD-810G [12]. Through the accumulation of a large number of engineering tests, it was known that the mass of test specimen was less than 1.5kg, the interval of reaching stability state of temperature was about 60min. Therefore, when the temperature of atmospheric icing laboratory reached the setting values, the icing experiments was started after 60min.

During the icing period, all inputs were held constant. At the end, the detail of ice accretion was measured ice length. And it would also take digital photographs of ice accretion from different perspectives.

3 Numerical calculation and theoretical analysis

As we known, the distance from nozzle to icing object was a very important factor in icing experimental. Because this distance could make the droplet become super-cooled. And this distance was also need to be long for droplet to be mixed in the air flow. In this section, single droplet model would be established, and thermodynamic analysis and calculation would be carried out.

3.1 The droplet model

A brief description of droplet model was presented as follows. In this model, the left side of droplet was set to the fan in order to simulation the uniform air speed, the opposite side was set to the outflow and the other sides were set to the adiabatic boundary.

3.2 The computation method

The temperature field and the boundary condition will be dispersed by the finite difference method and the cable element method.

For the anyone control volume *V*, the surface area was *S*, the energy equation without inner thermal source was:

$$\frac{\partial}{\partial t}\int_V \rho e dV = \int_S k \cdot \nabla T \cdot \vec{n} dS \tag{1}$$

Where: *e*—the internal energy of per mass, *k*—the thermal conductivity, *T*—the temperature of control volume, *n*—the normal vector outside the surface.

Due to $\rho e = \rho h - PV$,the pressure had nothing to do with the time, so:

$$\frac{\partial}{\partial t}\int_V \rho dV = 0 \tag{2}$$

The Equ.1 could be solved by:

$$\frac{\partial}{\partial t}\int_V \rho h dV = \int_S k \cdot \nabla T \cdot \vec{n} dS \tag{3}$$

The enthalpy and temperature of equation as a stay function, it was suitable for all regions, then the Equ.3 become

$$\rho \cdot \frac{\partial h}{\partial t} = k \cdot \nabla^2 T \tag{4}$$

Where $\nabla^2 = \frac{\partial^2}{\partial x^2} + \frac{\partial^2}{\partial y^2} + \frac{\partial^2}{\partial z^2}$;

The relationship between the enthalpy and temperature was $\frac{dh}{dT} = c$,*c* was specific heat. So the Equ.4 could become

$$\rho c \cdot \frac{\partial h}{\partial t} = k \cdot \nabla^2 h \tag{5}$$

Then, the relationship between enthalpy and temperature was:

$$T = \begin{cases} h / c_{p_s} & h \le c_{p_s} T_m \\ T_m & c_{p_s} T_m \le h \le c_{p_l} T_m + r \\ [h - r - (c_{p_s} - c_{p_l})] / c_{p_l} & h \ge c_{p_l} T_m + r \end{cases} \tag{6}$$

Where T_m was phase change temperature of control volume, *r* was latent heat of phase change.

$$H(T) = \begin{cases} c_{p_s} .T & T < T_m \\ c_{p_l} T_m + r & T = T_m \\ r + (cp_s - cp_l) T_m + c_{p_l} T & T > T_m \end{cases} \tag{7}$$

The effective thermal in phase change process:

$$cp_{eff} = \frac{r}{T_{m2} - T_{m1}} + cpl \tag{8}$$

The relationship of unit of heat capacity and temperature was:

$$c_p(T) = \begin{cases} c_{p_s} & T < T_{m1} \\ \frac{r}{T_{m2} - T_{m1}} + c_{p_l} & T_{m1} \le T \le T_{m2} \\ c_{p_l} & T > T_{m2} \end{cases} \tag{9}$$

3.3 The results and analysis

In this model, the diameter of droplet was 100μm, the initial droplet temperature was 3℃,the air temperature was -5℃ and the air speed was 1m/s.

Through a series of computation based on the above computation method, the temperature area of droplet presents in Figure.1.

Figure.1 presents that the temperature of droplet decreases very quickly before 0.03s, and the temperature would keep a long time around 0℃. When t=0.65s, the release of the phase change latent heat of droplet would basically complete. In order to analysis temperature changes of a single water droplet better, the temperature of several typical points of water droplets was detected, such as the front end, the last end, the center of the droplet and the side edge of the droplet.

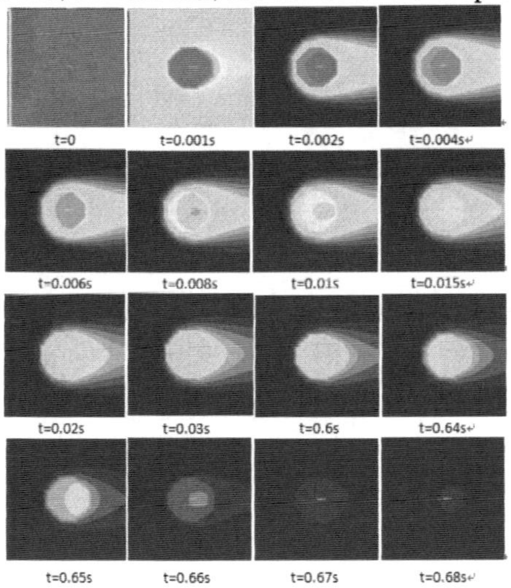

Figure 1. The process of single droplet

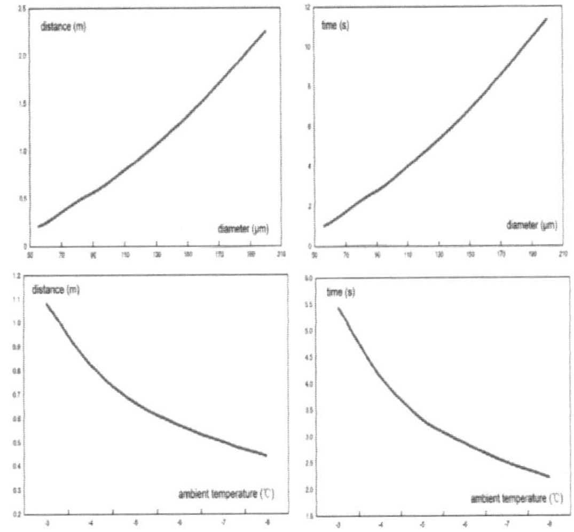

Figure2. The analysis diagram of different particle size, ambient temperature

In order to determine the effect of some factors further, the computation would be repeat for different droplet diameter and air temperature.

Figure.2 shows the influence of droplet diameter on the distance (the droplet temperature dropped to air temperature) and time (the droplet temperature dropped to air temperature) of the ice accretion. Other conditions remain unchanged. From Figure.2, it could see that all computation results increase with increase droplet diameter. It means that the distance between nozzle and icing object would be longer with the increase of droplet diameter in icing experimental. But actually, the freezing rain droplet diameter was almost lower than 100μm, otherwise, the length or height of the test equipment would be limited. Therefore, the suitable droplet diameter must be chosen to achieve the icing experiment. In other words, in the actual ice test process, different distances according to different test equipment and test conditions by the calculation results would be chosen. This influence law had important guiding significance for better determining the process of test implementation.

Figure.2 also shows the influence of air temperature on the distance (the droplet temperature dropped to air temperature) and time (the droplet temperature dropped to air temperature) of the ice accretion. From Figure.2, it could see that all computation results decrease with decreased air temperature. That means that the lower air temperature, the shorter distance between nozzle and icing object and time in icing experimental.

4 Experiment and discussion

In above section, through numerical calculation and theoretical analysis, it was known that the distance, the air temperature had very important effect on icing test. So, in this section, the effect of different spray method, two spray time, four different air temperature and three different distances on the ice shape of ice accretion for freezing rain conditions with numerical calculation would be discussed by experiment. And the range of boundary conditions as given in Table 1.

IOP Conf. Series: Materials Science and Engineering **692** (2019) 012018 doi:10.1088/1757-899X/692/1/012018

Table.1 Test boundary conditions with variation steps and range

Boundary condition	Variation range
Fluid flow parameter	
Spray method	Vertical, Oblique, Parallel
Spray time	Continuous, Interrupt
Droplet diameter	50um~100um
Air temperature	-3℃,-7℃,-11℃,-15℃
Distance	0.5m,1m,1.5m
Wind speed	0.2m/s
Exposure time	10min,20min,30min

4.1 Spray methods

Figure.3 shows the time of the shape of ice deposit obtained by spray methods of vertical, oblique (45°) and parallel under freezing rain condition. From the phenomenon of overall icing thickness, under the same test time, air temperature and distance, the icing thickness of vertical jet was the thickest, followed by oblique 45 degrees, and finally parallel. The specific values were shown in Table2.

Table.2 The results of different spray method and time

Method	Oblique(45°)		Parallel		Vertical	
Time	Continuous	Interrupt (spray 2min stop 1min)	Continuous	Interrupt (spray 2min stop 1min)	Continuous	Interrupt (spray 2min stop 1min)
T	-7℃					
Distance	50cm					
10min	11mm	6mm	8mm	4mm	10mm	6mm
20min	17mm	11mm	13mm	9mm	18mm	17mm
30min	20mm	16mm	23mm	13mm	32mm	22mm

For the ice thickness, when the test sample body was injected vertically, it was placed directly on the ground of the test box. When the droplet was ejected from the nozzle, it will be affected by both gravity and nozzle pressure at the same time. The droplet reaches the surface of the test sample body, the latent heat of droplet had not been fully released, so the droplet also appears liquid, and the droplet would collide with the test sample body and the ground, which would make the droplet become a droplet with a smaller diameter, and then the smaller droplet would backsplash surface and surrounding of the test sample body. Otherwise, because the adhesion of droplets to the surface of the test piece will increase under low temperature environment, the droplets with smaller particle size would mostly adhere to the surface of the test piece under these two kinds of impact conditions. The number of droplets on the test piece would be more than that of oblique and parallel spraying, which will eventually lead to the thickness of the vertical ice layer greater than that of oblique and parallel spraying.

From the point of view of ice type, there were all hard ice. This was consistent with meteorological phenomena. And this also proves the correctness and effectiveness of experiment.

In terms of uniformity, there was little difference between oblique and parallel directions, and the vertical direction was the worst. Comparing oblique injection with parallel injection, it could be seen from Figure. 3 that the ice thickness of the lower half of the test piece decreases significantly in oblique, while the ice thickness of the whole test piece was relatively uniform in parallel, showing symmetry of axis. However, there was still much room for improvement of the surface of the whole test piece, and the study of improving the ice uniformity would be done in the future.

IOP Conf. Series: Materials Science and Engineering **692** (2019) 012018 doi:10.1088/1757-899X/692/1/012018

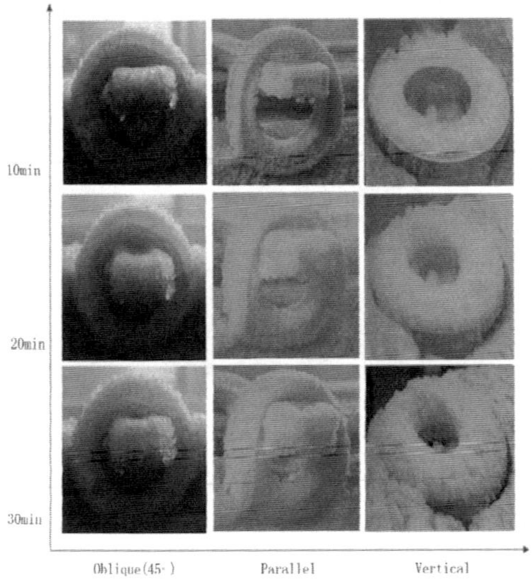

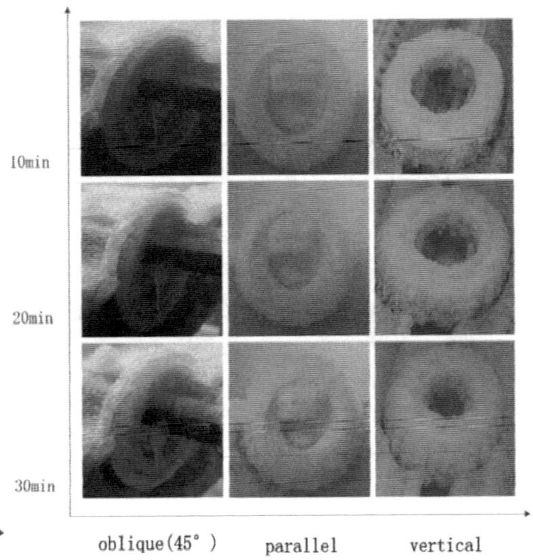

Figure.3 Map of different spray method **Figure.4** Icing results map of interrupt spray time

4.2 Spray times

Since there was a process of releasing latent heat inside the droplet after the droplet was ejected from the nozzle, it could be seen from the previous simulation calculation that certain distance and time were needed. In the actual ice test process, this distance could be predicted by calculation, but there were still some deviations. So, in order to make the droplets fully release latent heat after contacting with the test piece, changing the injection time sequence, change the continuous injection to 2 minutes and stop for 1 minute, and the experimental results were shown in Figure. 4 and table2. There were three injection modes.

It could be seen from Figure. 4 that the regularity of icing phenomenon was consistent with the performance of Figure. 3. The thickness of vertical ice layer was still the largest. When oblique jet was applied, the lower half of the ice layer of the test piece was missing a lot. Compared with vertical and oblique, parallel injection show good performance in terms of ice shape, thickness and uniformity. Therefore, for the icing test in laboratory, the parallel injection method would be adopted. The determination of this method had important guiding significance for the implementation of the icing test and was conducive to the repetition of the icing test. In addition, compared with figure. 3, and table2, it could be seen that the ice thickness of continuous spraying was obviously better than that of interrupt spraying, and the ice shape had not changed, and the uniformity of the ice layer on the test pieces was basically the same. Therefore, through the experimental study in this paper, it was suggested that the continuous spraying method should be used in the freezing test of atmospheric icing laboratory, and the test quality could be guaranteed in high efficiency.

4.3 Different air temperature

In order to verify the influence of different air temperatures on the freezing test, four different temperatures (-3℃, -7℃, -11℃and -15℃) would be tested by three different spraying methods. The specific test results were shown in Figure. 5.

It could be clearly seen that the ice shape were rain ice when the air temperature was -3℃ for the three spraying modes, and the ice shape were hard ice when the air temperature was – 7℃, and when the air temperature were – 11℃ and – 15℃ respectively, the ice type produced by the three spraying methods were all soft frost, which were all the same as the actual meteorological conditions. For the ice thickness, shows that the ice thickness would be bigger in the lower temperature for three spray method at the same test time. In other words, the time to complete the icing test would decrease as the temperature decrease.

Therefore, according to the test results, it could effectively guide the requirements of different ice types in the subsequent freezing test of the test chamber.

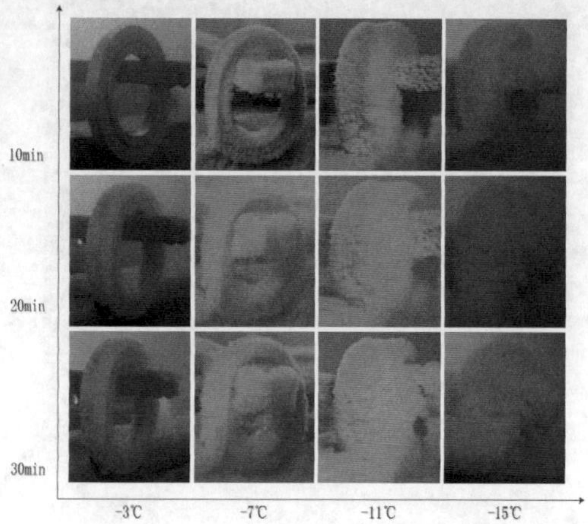

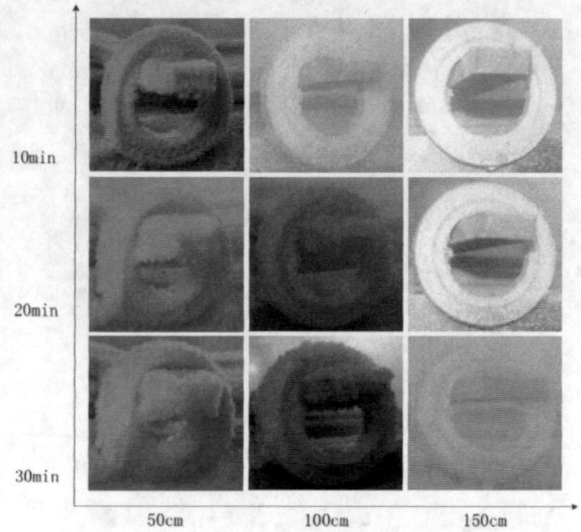

Figure.5 Icing results map of different air temperature **Figure.6** Icing results map of different distance between nozzle and body

4.4 Different distance between nozzle and icing object

In order to verify the influence of different nozzle-to-specimen distance on freezing test, two typical different injection methods would be used to test the different parameters of distance 0.5m, 1m and 1.5m.

Under the same droplet flow rate, the most effective freezing process for freezing test was that the droplets could collides with the test piece immediately when the latent heat of the droplet was releases. This could make the droplet have liquid property before contacting the test piece. When contacting the test piece instantly, the final latent heat was released immediately and adhered to the test piece to form an ice layer.

Figure.6 presents a series of pictures on the profiles of ice deposits obtained under freezing rain condition of distance between jet and object with different spray method. And shows the results of ice thickness for different distance. It could be seen from the figure that for the injection modes, the farther the distance was, the smaller the ice thickness was. The reason was that the longer the distance was, the longer the latent heat release time. Therefore, when the freezing test was carried out in this laboratory, the larger the distance was, the latent heat release of the droplets were completed before reaching the test piece, and the ice layer would be formed and cannot adhere to the test piece. In addition, there was another factor was the number of droplets. When the distance was 50 cm, it shows the characteristics of maximum ice thickness. According to the observation of the test process, when the droplets reach the test piece, the droplets are not completely frozen. The reason for the thickness was that the closer the distance was, the larger the number of droplets contacted with the test piece. So, the droplet volume was obviously larger than the other two test distances, the ice thickness was larger than the other two test data. In conclusion, the distance between nozzle and test piece had obvious influence on freezing test. Specifically, the distance should be adjusted before each test, according to different test pieces and test conditions. At the same time, from the view of the ice type, it could be seen that the ice type was hard ice for the parallel and oblique in different distance, but there was a slight change for the oblique in the ice type.

5 Conclusions

The dependence of the thickness and uniformity of ice deposits accreted on different test object in different spray method and time were investigated in atmospheric icing Laboratory. Three different spray method and two different spray time were applied to different air temperature and distance between nozzles to object. The effects of the different factors as they were varied may be summarized as follows:

a) From the ice thickness, the three-icing spray method of vertical, parallel and oblique could meet the test requirements, but from the uniformity and consistency, this result means that the parallel spray method and oblique spray for arbitrary air temperature could be validated, but the parallel was the best.

b) The icing effect of continuous injection was better than that of discontinuous injection for atmospheric icing Laboratory.

c) Experiments with different temperature appear the ice shape would be changed with air temperature. Rain ice was sawed when the air temperature was greater than -5℃, and rime ice would be observed when the air temperature was less than -10℃,and hard frost would appear when the air temperature was between -5℃~-10℃. The important of this result was emphasized by the fact that the same phenomena were achieved at different ice conditions and geometry in atmospheric icing Laboratory.

d) The distance between the nozzle and the test piece had a great influence on the freezing test, we could use the single droplet model to predict the icing condition of distance between nozzle and object, by using values from those calculated, it might have been possible to get better icing affect in atmospheric icing Laboratory.

References

[1] Parent O, Ilnica A. Anti-icing and de-icing techniques for wind turbines: Critical review. Cold Reg Sci Technol 2011;65:88–96;

[2] G. Yang, K. Guo, N. Li, Freezing mechanism of supercooled water droplet impinging on metal surfaces, Int. J. Refrig. 34 (2011) 2007–2017;

[3] Z. Jin, Z. Wang, D. Sui, Z. Yang, The impact and freezing processes of a water droplet on different inclined cold surfaces, Int. J. Heat. Mass Transf. 97 (2016);

[4] Y. Hayashi, A. Aoki, S. Adachi, K. Hori, Study of frost properties correlating with frost formation types, J. Heat Transfer 99 (1977) 239–245;

[5] Myers T G, Charpin J.P.F. A mathematical model for atmospheric ice accretion and water flow on a cold surface. International Journal of Heat and Mass Transfer, 2004, 47: 5483-5500;

[6] P. Struk, T. Bencic, J.-C. Tsao, D. Fuleki, D. KKnezevici, Preparation for scaling studies of ice-crystal icing at the NRC research altitude test facility, in: 5th AIAA Atmospheric and Space Environments Conference, AIAA-2013-2675, 2013;

[7] T. Hauk, E. Bonaccurso, I. Roisman, C. Tropea, Ice crystal impact onto a dry solid wall. Particle fragmentation, Proc. R. Soc. A 471 (2015) 20150399;

[8] X. Wu, W. Dai, W. Xu, L. Tang, Mesoscale investigation of frost formation on a cold surface, Exp. Therm. Fluid Sci. 31 (2007) 1043–1048;

[9] T. Currie, D. Fuleki, A. Mahallati, Experimental studies of mixed-phase sticking efficiency for ice crystal accretion in jet engines, in: 6th AIAA Atmospheric and Space Environments Conference, AIAA-2014-3049, 2014;

[10] Li Yan, Li Shengmao, Tagawa Kotaro. Numerical simulation on icing effects on the pressure field around blade airfoil for straight-bladed VAWT. Int J Agric Biol Eng 2011; 4 (3):41–7;

[11] Homola Matthew, Wallenius Tomas, Makkonen Lasse, Nicklasson Per, Sundsb Per. Turbine size and temperature dependence of icing on wind turbine blades. Wind Eng 2010; 34 (6):615–27;

[12] MIL-STD-810G, Department of defense test method standard, 521.3-1-521.3-7

IOP Conf. Series: Materials Science and Engineering **692** (2019) 012019 doi:10.1088/1757-899X/692/1/012019

A Study on the Design Approach and Theory for the Hydrostatic Propulsion Drive System of a Tamping Machine

Wenlin Wang[1*] **and Xiong Cai**[2]

[1]School of Mechanical Engineering, Dongguan University of Technology, Dongguan 523808, China.
[2]Dongfeng Honda Automobile Co., Ltd, Wuhan 430056, China

[*]Corresponding Author: E-mail: pianowwl@163.com

Abstract. With the fast developing of modern high-speed railways, it is urgent and meaningful to introduce effective design approach and theory for advanced railcar development. A design approach and theory for the hydrostatic propulsion drive system of a tamping machine is proposed in this study, the design approach and theory includes the overall design method, specification of the hydraulic pump and motor, the optimally matching of the hydraulic pump and diesel engine and the force equilibrium verification. A design case study is finally conducted and a series of good performance validates the effectiveness of the proposed design approach and theory: the designed closed-type hydrostatic propulsion drive system is smaller and more integrated in configuration, and is with high efficiency; the designed system is stable and fast in dynamic response with acceptable overshoot and errors, the displacement performance of the tamping machine is satisfactory for its cyclic tamping operations, and the tamping machine has also got a wide scope for high-speed mobility with satisfactory acceleration performance. Thus, the proposed design approach and theory in this work will be instructive for further tamping machine product development and design optimizations.

1. Introduction

A hydraulic tamping machines is a fluid-power-driven railway machinery which is crucial and efficient in modern railway construction and maintenance [1-3]. The hydrostatic propulsion drive system [4-6] of a tamping machine usually works in two modes: the low-speed mode for cyclic operation and another high-speed mode for mobility. Thus, it is meaningful to understand and study the design approach and theory for hydrostatic propulsion drive system of a tamping machine.

Kache [4] proposed an all-hydraulic hybrid system for railway machineries, Wang investigated the hydrostatic propulsion drive system of the 08-32 tamping machine [5], which has different acceleration modes by using a hydraulic proportional pump and a hydraulic variable motor; Zhou [6] studied the whole fluid power drive system of a continuous tamping machine. Yan [7] designed a hydrostatic propulsion drive system capable of running both on high speed for mobility and on low speed for tamping, Hu et al. improved the pressure stability [8] of the hydraulic clamping system of a tamping machine by using a resistance orifice, Zhai et al. [9] investigated the adjustment and maintenance technology for the hydraulic system of the 08-32 tamping machine. In addition, Dasgupta [10] and Ho et al. [11] introduced novel key hydraulic component technologies for use in the hydrostatic transmission system to improve the dynamic performance of the machine.

In this study, a design approach and theory for the hydrostatic propulsion drive system of a tamping machine is proposed. The design approach and theory includes the overall design method, specification of the hydraulic pump and motor, the optimally matching of the hydraulic pump and diesel engine and the force equilibrium verification process. A design case study is finally conducted and the simulation rcsults show that the tamping machine has got a series of good performance. The proposed design approach and theory will be instructive for further tamping machine product development and design optimizations.

2. Design approach and theory

2.1. Overall design

There are usually two types of hydrostatic propulsion schemes for a tamping machine, as shown by Figures 1 and 2, the open-type and the closed-type.

The hydrostatic propulsion drive system using the open-type hydraulic power circuit (Figure 1) is popularly used in current tamping machines. The biggest advantage of open-type is flexible and easy to maintain, but the drawback is low efficient.

The closed-type hydrostatic propulsion drive system lets the oil circulate between the pump and motor, so it is smaller and more integrated in configuration, and is with high efficiency. In recent years, several more advanced tamping machines have used the closed-type hydrostatic propulsion drive system, and practices have validated the feasibility and efficiency of the closed-type system.

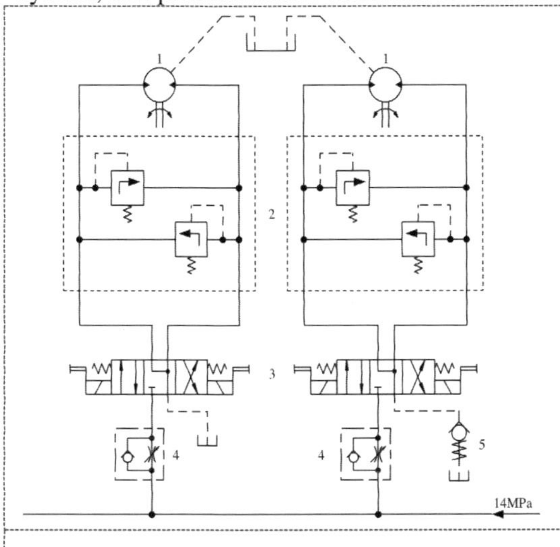

Figure 1. Open-type hydraulic power circuit of the hydrostatic propulsion drive system (1. hydraulic motor, 2. safety relief valve, 3. electro-hydraulic directional control valve, 4. throttle and check valve, 5. back-pressure valve).

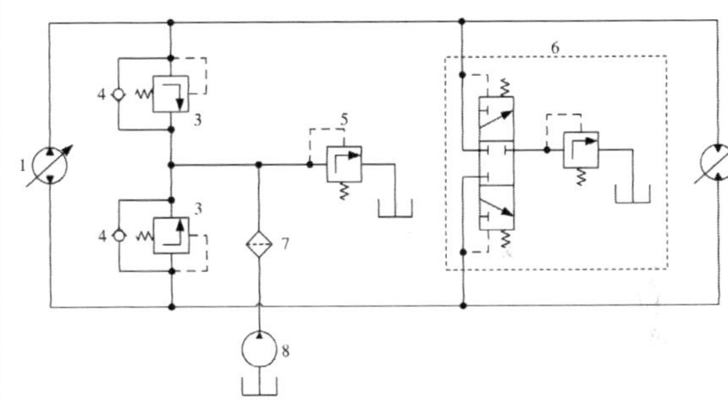

Figure 2. Closed-type hydraulic power circuit of the hydrostatic propulsion drive system (1. variable pump, 2. variable hydraulic motor, 3. high-pressure relief valve, 4. check valve, 5. fluid compensation relief valve, 6. flush valve, 7. filter, 8. fluid compensation pump).

2.2. Specification of the hydraulic pump and motor

The hydraulic pump and motor are the most crucial devices in the hydrostatic propulsion drive system, they should be firstly specified.The required maximum rotation speed of the hydraulic motor n_{max} in a tamping machine is described as

$$n_{max} = \frac{iv_{max}}{60\pi D} \tag{1}$$

where i is the gear ratio, v_{max} is the maximum speed of the tamping machine, and D is the wheel diameter of the tamping machine. Thus, the displacement of the hydraulic pump q_b can be formulated by

$$q_b = \frac{Q_b}{n_b} = \frac{Q_m}{n_b \eta_v} = \frac{n_{max} q_{min}}{n_b \eta_v} \tag{2}$$

where Q_b, n_b, η_v are the flow, rotation speed and volumetric efficiency of the pump, q_{min} is the minimum displacement of the hydraulic motor.

2.3. Optimally matching of the hydraulic pump and the diesel

In mobile machinery applications, the swash plate type variable pump is more suitable for the working conditions, Figure 3 illustrates a type of variable pump with proportional electro-hydraulic control. Once the hydraulic pump is specified, it should be optimally matched with the diesel engine.

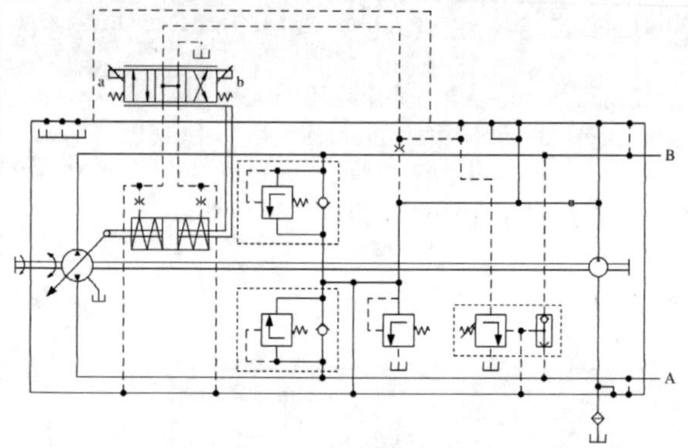

Figure 3. Swash plate type variable pump with proportional electro-hydraulic control.

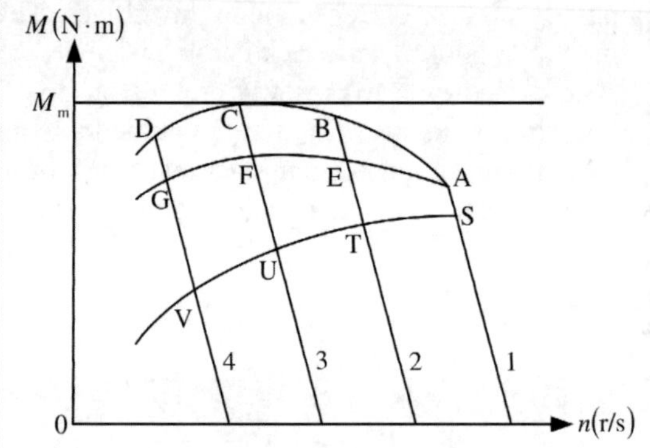

Figure 4. Optimally matching of the hydraulic pump and the torque-speed characteristics of the diesel engine.

The output power of the diesel engine N_e is given by

$$N_e = \frac{\pi}{3000} M_e n_e \tag{3}$$

where M_e and n_e are the torque and rotation speed of the engine. The output power of the hydraulic pump N_b is given by

$$N_b = \frac{P_b Q_b}{60} = \frac{P_b q_b n_e}{60000} = \frac{\pi M_b n_e}{3000} \tag{4}$$

where P_b is the output pressure of the pump, M_b is the input torque of the pump, and M_b can be described as

$$M_b = \frac{P_b q_b}{2\pi} \tag{5}$$

Because there is a relationship between the output powers of the hydraulic pump and the diesel engine

$$N_b = N_e \eta_1 \eta_2 \tag{6}$$

where η_1 is the mechanical transmission efficiency between the diesel engine and the hydraulic pump, η_1 could be 1 when the hydraulic pump is rigidly connected with the diesel engine; η_2 is the efficiency of the hydraulic pump and usually be regarded as 0.95. Thus, substituting Equations (3)-(5) to Equation (6) to obtain

$$M_{\mathrm{b}} = \frac{P_{\mathrm{b}} q_{\mathrm{b}}}{2\pi} = 0.95 M_{\mathrm{e}} \tag{7}$$

As demonstrated by Figure 4, although when the engine works in ABCD, the maximum power can be obtained, but the engine is easily to be overloaded and stopped, so the best torque-speed performance of the engine is AEFG. When the engine works in its best points, it is usually assumed that the output torque of the engine remains constant, therefore, the optimally matching of the hydraulic pump and the diesel engine can be described as

$$M_{\mathrm{b}} = 0.95 M_{\mathrm{e}} = const \tag{8}$$

2.4. Verification

When the overall design, specification of the hydraulic pump and motor, optimally matching of the hydraulic pump and the diesel engine are completed, the verification of force equilibrium should be carried out.

The total resistance force $\sum F$ of a moving tamping machine can be formulated by

$$\sum F = mg\left(f_0 + kv\right)\cos\beta + mg\tan\beta + \frac{C_{\mathrm{D}} A v^2}{21.15} + m\delta\frac{\mathrm{d}v}{\mathrm{d}t} \tag{9}$$

where m, A and C_{D} are the total resistance force, the mass, the windward area and the drag coefficient of wind of the tamping machine, (f_0+kv) is the coefficient of rolling friction of the wheels, β is the slope angle, δ is an effective coefficient of the rotation mass of the tamping machine, and δ is given [12] by

$$\delta = 1 + \frac{\sum I_{\mathrm{w}}}{mr^2} + \frac{I_{\mathrm{f}} i_0^2 i_{\mathrm{C}}^2 \eta_t}{mr^2} \tag{10}$$

where I_{w} and I_{f} are the moments of inertia of the wheel and the fly wheel of the diesel.

The driving force of the hydrostatic propulsion F_{C} is formulated by

$$F_{\mathrm{C}} = \frac{M_{\mathrm{C}}}{r} = \frac{\Delta p n q_{\mathrm{m}} i \eta_{\mathrm{mv}} \eta_{\mathrm{m}}}{2\pi r} \tag{11}$$

where M_{C} is the output torque of hydraulic motor, r is the wheel radius, Δp, q_{m} and η_{mv} are the pressure difference, displacement and volumetric efficiency of the hydraulic motor, n is the number of driving axles, i and η_{m} are the gear ratio and mechanical efficiency of the gear box.

3. A case study

Using the design approach and theory proposed in Section 2, a case study of the hydrostatic propulsion drive system design for a tamping machine is conducted. The open-type hydraulic power circuit is used, the pump and motor specification, the optimally matching of pump and diesel and the force equilibrium verification are performed, the designed parameters of the hydrostatic propulsion drive system are summarized in Table 1.

Table 1. The designed parameters of the hydrostatic propulsion drive system of a tamping machine.

Parameter	Value	Parameter	Value
Rotation speed of engine (r/min)	2300	Empty laden mass of tamping machine (Kg)	52000
Total maximum displacement of variable pump (mL/r)	305	Full laden mass of tamping machine (Kg)	60000
Displacement of fluid compensation pump (mL/r)	68.1	Wheel radius (m)	0.42
Set pressure of main relief valve (MPa)	40	Maximum friction force (N)	9172.8
Set pressure of fluid compensation relief valve (MPa)	2.5	Rolling resistance force (N)	3057.6

Relief pressure of flush valve (MPa)	2	Drag coefficient of wind (N/(m/s)^2)	5.55
Total maximum displacement of hydraulic motor (mL/r)	480	Slope angle (°)	0
Minimum displacement of hydraulic motor (mL/r)	165.6	Volumetric and mechanical efficiencies of hydraulic pump and motor	0.95
Gear ratio of gear box	5.944	Mechanical efficiency of wheel gear box	0.92

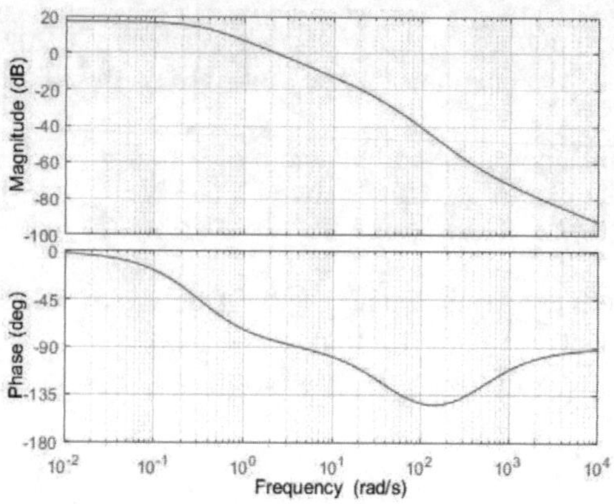

Figure 5. Bode plot of the hydrostatic propulsion drive system.

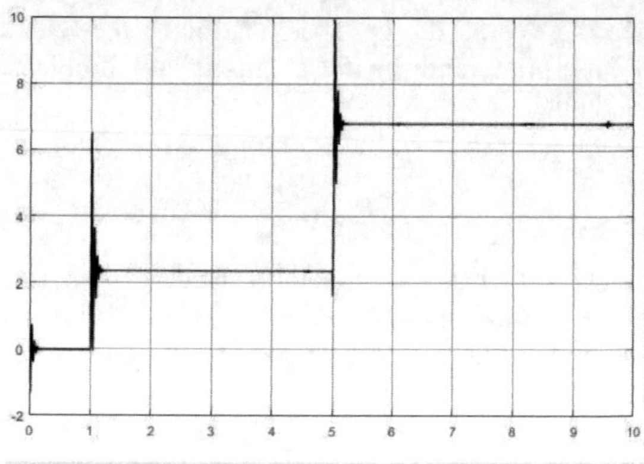

Figure 6. Step response of the hydrostatic propulsion drive system.

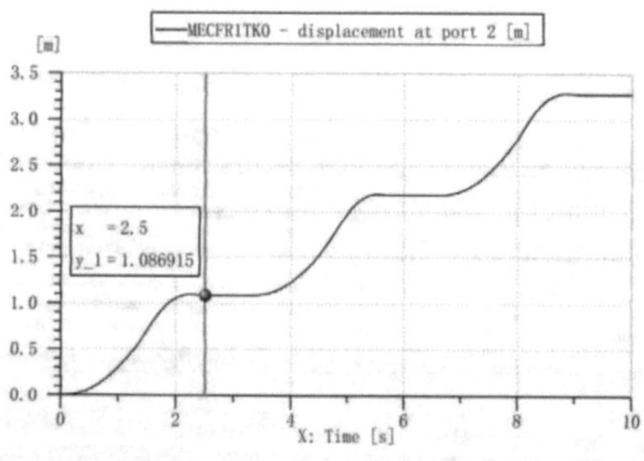

Figure 7. Displacement performance of the tamping machine during its cyclic tamping operation.

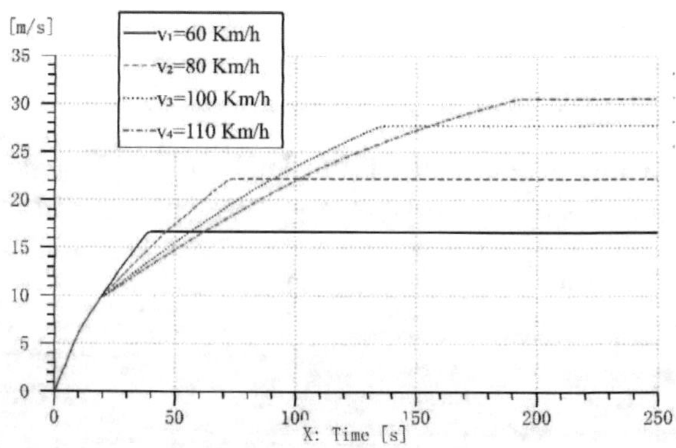

Figure 8. Acceleration performance of the tamping machine during its fast moving.

A simulation model of the designed hydrostatic propulsion drive system is built in the AMEsim software environment, and both the system dynamic response and operation performance of the tamping machine are evaluated, the simulation results are illustrated by Figures 5-8.

As shown by Figure 5, the Bode plot of the hydrostatic propulsion drive system illustrates that the hydraulic system has a large magnitude margin with a cross-over frequency of 2.28 rad/s, the corresponding phase margin is 95.31°, thus, the system is very stable.

The step response of the hydrostatic propulsion drive system, Figure 6, demonstrates that the system performs rapidly with acceptable overshoot, and the system becomes stable with small errors.

Figure 7 shows that with the designed hydrostatic propulsion drive system, the step displacement is 1.087 m, so the displacement performance is satisfactory for the tamping machine during its cyclic tamping operations.

Figure 8 shows that the acceleration periods are 42 seconds, 74 seconds, 137 seconds and 197 seconds or so when the vehicle is accelerated to 60 km/h, 80 km/h, 100 km/h and 110 km/h, respectively. Thus, the tamping machine with the designed hydrostatic propulsion drive system has got a wide scope for high-speed mobility with satisfactory acceleration performance.

4. Concluding remarks

- The closed-type hydrostatic propulsion drive system is smaller and more integrated in configuration, and is with high efficiency, so it is a trend in the future tamping machine applications.
- A design case study validates the effectiveness of the proposed design approach and theory for the hydrostatic propulsion drive system of a tamping machine, including the hydraulic pump and motor specification, the optimally matching of the hydraulic pump and diesel engine and the force equilibrium verification: The designed system is stable and fast in dynamic response with acceptable overshoot and errors, the displacement performance of the tamping machine is satisfactory for its cyclic tamping operations, and the tamping machine has also got a wide scope for high-speed mobility with satisfactory acceleration performance.
- The proposed design approach and theory for the hydrostatic propulsion drive system of a tamping machine in this work will be instructive for further tamping machine product development and design optimizations.

Acknowledgements

The authors are grateful for financial support from the National Natural Science Foundation of China (NSFC) under Grant No. 11572123 and the Research Fund for High-level Talent of Dongguan University of Technology under Project No. GC200906-30.

References

[1] Famurewa S M, Xin T, Rantatalo M and Kumar U 2015 Optimisation of maintenance track possession time: A tamping case study. *Proc. IMechE, Part F: J. Rail & Rapid Transit* **229** 12–22

[2] Wen M, Li R and Salling K B 2016 Optimisation of preventive condition-based tamping for railway tracks. *European J. Operational Research* **252** 455–465

[3] Aursudkij B 2007 *A laboratory study of railway ballast behaviour under traffic loading and tamping maintenance* (Ph.D Thesis, Nottingham: University of Nottingham, UK)

[4] Kache M 2014 Investigation an all-hydraulic hybrid system for diesel-hydraulic rail cars *European Transport Research Review* **6** 181–189

[5] Wang X 2008 *Research on the hydrostatic transmission running system of 08-32 tamping machine* (Master's Degree Thesis, Changsha: Central South University, China)

[6] Zhou Y Z 2011 *Research on the fluid power drive system of a continuous tamping machine* (Master's Degree Thesis, Changsha: Central South University, China)

[7] Yan C G 2010 The analysis of 08-16 tamping machine's hydraulic drive system in the high-speed running *Equipment Manufacturing Technology* **10** 28–29

[8] Hu Y, Hu J K and Fang J K 2014 Research on pressure stability of clamping hydraulic system in railway tamping device *J. Railway Science and Engineering* **11** 146–150

IOP Conf. Series: Materials Science and Engineering **692** (2019) 012019 doi:10.1088/1757-899X/692/1/012019

[9] Zhai S C and Huang Z J 2008 Tamping equipment's hydraulic system of 08-32 tamping machine and it's use and service *Hydraulics Pneumatics & Seals* **28** 55–57
[10] Dasgupta K 2000 Analysis of a hydrostatic transmission system using low speed high torque motor *Mechanism and Machine Theory* **35** 1481–1499
[11] Ho T H and Ahn K K 2010 Modeling and simulation of hydrostatic transmission system with energy regeneration using hydraulic accumulator *J. Mech. Sci. Tech.* **24** 1163–1175
[12] David A C 2009 *Automotive Engineering* (Oxford: Butterworth-Heinemann)

IOP Conf. Series: Materials Science and Engineering **692** (2019) 012020 doi:10.1088/1757-899X/692/1/012020

Structural Topology Optimization of Cantilever Crane Boom Based on Equivalent Moving Load Method

Jian Yin, Shengfang Zhang, Yu Liu*, Zhihua Sha, Fujian Ma, Dapeng Yang and Dongfang Yuan

School of Mechanical Engineering, Dalian Jiaotong University, Dalian, 116028, China

*E-mail: liuyu_ly12@126.com

Abstract. As a key structure in the process of lifting heavy objects by cantilever cranes, the boom is of great significance. When the cantilever crane carries heavy objects, the load on the boom is a typical moving load. The traditional structural optimization method cannot fully consider the influence of the moving load on the structure. Based on the variable density structure topology optimization method, the equivalent moving load method is proposed to transform the motion of heavy objects on the boom into multiple equivalent working conditions. And according to the structural characteristics and motion form of the boom, the weight coefficient of every equivalent working conditions is assigned. The number of equivalent moving loads is determined in the form of multiple growth, and finally the lightweight cantilever crane boom structure is obtained, which is 32.08% lighter than the original structure, thus verifying the structural topology based on the equivalent moving load method proposed in this paper. The feasibility and effectiveness of the optimization method are also provided for other structural optimization that are subject to moving loads.

1. Introduction

The crane has a complex structure and a large number of parts, which is of great significance for its related lightweight design. At present, the related research work on crane light weighting mainly focuses on the optimization of the structure of the whole machine or parts of the crane by using related design methods such as topology, shape and size optimization. Wang[1] et al. studied the topology optimization of the crane telescopic boom section, and used the topology optimization method to design the boom section. Li[2] et al. used the continuum topology optimization method to design a new type of main beam of the bridge crane, which greatly reduced the weight of the main beam under the condition of ensuring the strength and rigidity of the main beam. However, for structures with special types of loads such as cantilever cranes, traditional structural optimization methods are often difficult to obtain better optimization results. If it can combine the characteristics of its mobile load and consider the influence of different position forces on structural optimization, and establish a structural topology optimization model including structural stiffness and satisfying the manufacturing process, it will greatly help the lightweight research of cantilever crane structure.

2. Moving load and EML method

2.1. Load on the boom

In the process of carrying heavy objects, the cantilever crane moves in a straight line on the track formed by its boom. During this process, the gravity of the weight is applied as a load on the boom of the cantilever crane. The direction and magnitude of the gravity of the object do not change with time, but as the object moves, the position of the load on the boom changes with time. Therefore, during the lifting process, the weight applied to the boom is a kind of moving loads[3,4].

The equivalent moving load(EML) method is designed for special types of loads, such as moving loads. It divides the process of continuous action into many parts. Similar to the data acquisition for the entire motion process, N points with static load information at a certain moment are obtained and passed. These points are weighted through the trajectory feature, and finally N static loads are respectively established into N equivalent working conditions, thereby providing services for subsequent related work, such as force analysis and structural optimization.

2.2. Equivalent moving load method

2.2.1. Physical model. The equivalent moving load method is applied in the optimization of the boom structure of the cantilever crane. Considering the actual movement as a cycle in the whole movement process of the moving load on the boom. At this time, the movement track of the load on the boom is regarded as a straight line is shown in Figure 1(a); the moving load formed by the gravity of the weight carried by the cantilever crane on this straight line can be equivalent to the point in Figure 1(b); the equivalent process is completed by fitting the trajectory line at the load point equivalent to the load method, as shown in Figure 1(c).

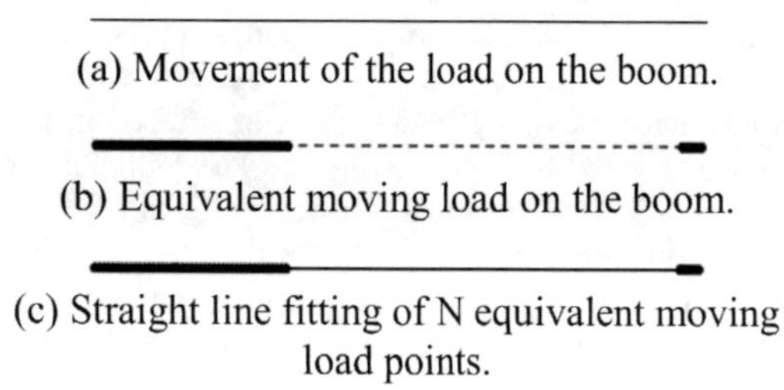

Figure 1. Principle of the EML method.

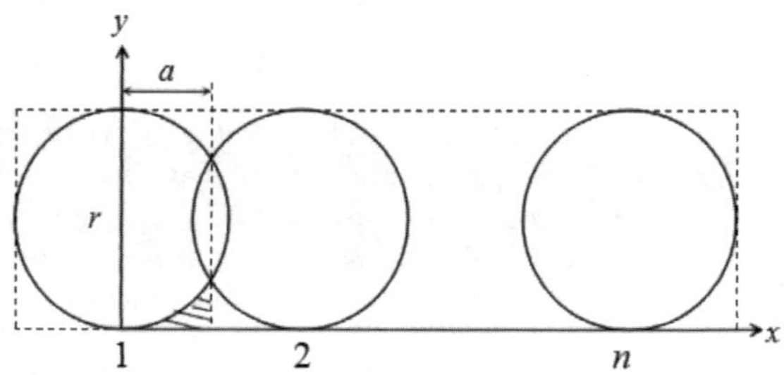

Figure 2. Calculation of equivalent accuracy of equivalent moving load method.

2.2.2. Mathematical model. After the equivalent, the position points of each working condition are sequentially connected to form the total length L of the moving load moving track, and the distance between the adjacent two points is:

$$a = \frac{L}{N-1} \tag{1}$$

Let the radius of the circle connected to the total length L of the moving load be r, as shown in Figure 2, according to the geometric relationship, the curve formula for calculating the shaded portion of the circle which is intercepted by the straight line $x=a$ in the first quadrant is:

$$y = r - \sqrt{r^2 - x^2} \tag{2}$$

Let the area of the shaded part be s, and the total area of the inscribed circle and the plane of the boom is S, then:

$$\begin{aligned} s &= \int_0^a \left(r - \sqrt{r^2 - x^2}\right) dx \\ &= ar - \frac{a}{2}\sqrt{r^2 - a^2} - \frac{r^2}{2}\arcsin\frac{a}{r} \end{aligned} \tag{3}$$

$$S = 4s(N-1) \tag{4}$$

Bring equation (3) into equation (4),

$$S=4(N-1)\left(ar-\frac{a}{2}\sqrt{r^2-a^2}-\frac{r^2}{2}\arcsin\frac{a}{r}\right) \tag{5}$$

When $N\to+\infty$ and $a\to 0$, $S\to 0$, which means the more the number of circles, the closer it is to the shape of the plane of the boom; similarly, in the EML method, the larger the number of equivalent working conditions is, the closer it is to the real moving load after equivalent.

2.2.3. Calculation of weight coefficient of equivalent working condition. Taking the cantilever crane boom as an example, the weight calculation of each working condition is deduced. As shown in Figure 3, the cantilever crane boom is simplified as a cantilever beam, and N equivalent working conditions are equally divided on it, and when the total number of the equivalent working conditions is N, the weight of the N-th equivalent working condition is a_N.

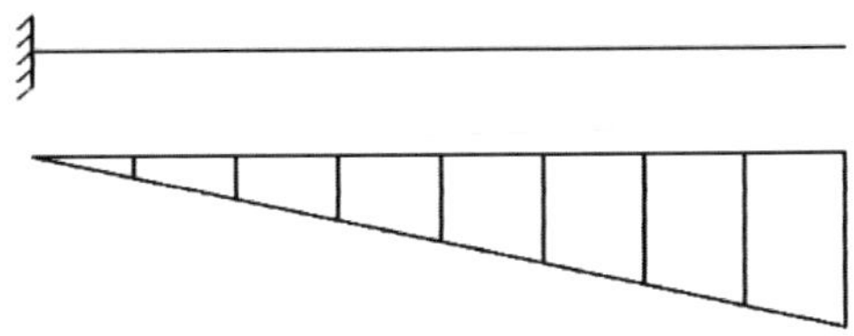

Figure 3. Weight calculation of the EML method.

If the sum of the weight of N equivalent working conditions is 1, and the weight of every equivalent working condition is proportional to the its located position, then:

$$\begin{cases} a_1+a_2+\cdots+a_N=1 \\ \dfrac{a_1}{1}=\dfrac{a_2}{2}=\cdots=\dfrac{a_N}{N} \end{cases} \tag{6}$$

where, N is the total number of the equivalent working conditions, a_N is the weight of the N-th equivalent working condition. By solving equation (6), the weight of the N-th equivalent working condition when the total number of operating conditions is N can be obtained, that is,

$$a_N=\frac{2}{N+1} \tag{7}$$

3. Mathematical model of topology optimization based on EML method

In order to obtain the optimal performance of the optimization result, a comprehensive analysis of the determination of the deletion or retention of the element under multiple working conditions is required [5,6]. In the variable density method, each working condition is usually used as a single objective optimization function. By using the multi-objective optimization method, all working conditions are weighted to establish a comprehensive objective function. If and only if the relative density of a certain element can be 0 under every working condition, the element is deleted from the structure. The established objective function is as follows:

$$\begin{cases} \min\ F(x)=\sqrt{\sum_{i=1}^{N}\left[a_iC_i(x)\right]^2} \\ \text{s.t.}\ \begin{cases} V\le fV_0 \\ F_i=K_iU_i\left(i=1,2,\cdots,N\right) \\ 0<x_{\min}<x\le 1 \end{cases} \end{cases} \tag{8}$$

where, x is the design variable, which is also the relative density of each element, $x_{\min}$ is the minimum of x, to avoid singularity, it is usually set to 0.001. $F(x)$ is the integrated objective function, i is the serial number of the working condition, $C_i(x)$ is the structural compliance value under the i-th working condition, $C_{\min}$ and $C_{\max}$ are the minimum and maximum structural compliance value respectively. V is the volume of the optimized structure, and V_0 is the original volume. f is the volume percentage. F_i, K_i, and U_i are the external force matrix, stiffness matrix and displacement matrix of the structure in the i-th condition respectively.

4. Topology optimization design method and process

Flow chart of the topology optimization block diagram based on the EML method is shown in Figure 4. The process of the object moving along the cantilever crane boom is equivalent to the uniform multiple working conditions through the EML method[7]. During the moving process, the size and direction of the load are unchanged, only the position of the load point changes. Therefore, the effect of the moving load on the structure is basically the same, and the optimization objective function can be optimized by the method of equation (8).

5. Structural topology optimization finite element model

Based on the above method, the geometric model of the cantilever crane boom is established by Pro/Engineer 5.0 software, as shown in Figure 5. Based on the established geometric model, the finite element optimization model of the cantilever crane boom is established by using the OptiStruct module in the HyperWorks 11.0 software, as shown in Figure 6. The material used for the boom is S45C, which is meshed into 200337 elements. The fixed constraint is applied to the leftmost position where the cantilever crane column is connected, and the gravity of the weight is applied to the boom as a load. The lower edge rises in different positions. Set the riser around the boom to a non-optimized area with the inner material of the boom set to the optimized area. The number of equivalent operating conditions is set according to the condition of 2 times increase, and can be stopped when the optimization result tends to be stable, which is 1, 2, 4, 8, 16, ... According to the number of working conditions, the weighting coefficients of each working condition can be calculated separately by combining equation (1) and (2).

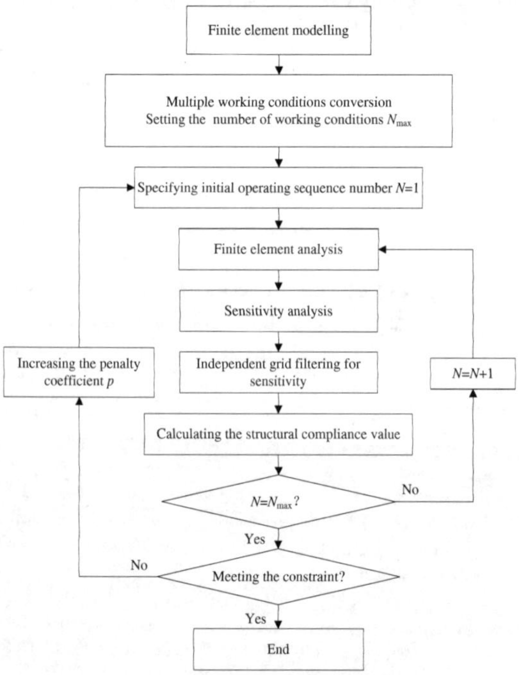

Figure 4. Flow chart of structure topology optimization of cantilever crane boom based on equivalent moving load method.

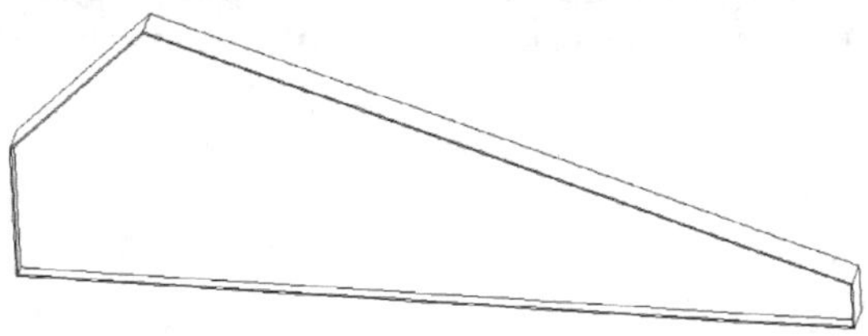

Figure 5. Geometry model of cantilever crane boom.

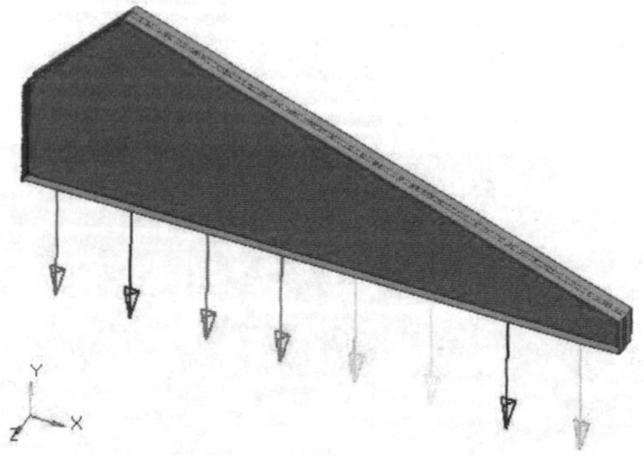

Figure 6. Finite element optimization model of cantilever crane boom.

6. Optimization results

According to the above structure topology optimization model and optimization process, the topology of the boom structure with the different number of equivalent working conditions is optimized. The topology optimization results are shown in Figure 7. Area with color filled in the figure shows the material left, and the white area indicates that the material is removed here.

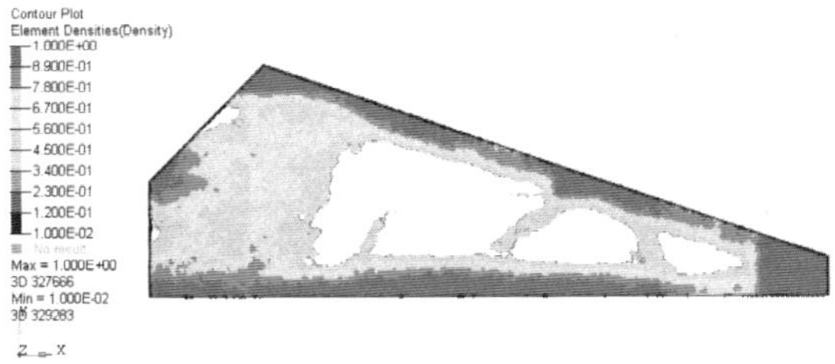

(a) N_{max}=1

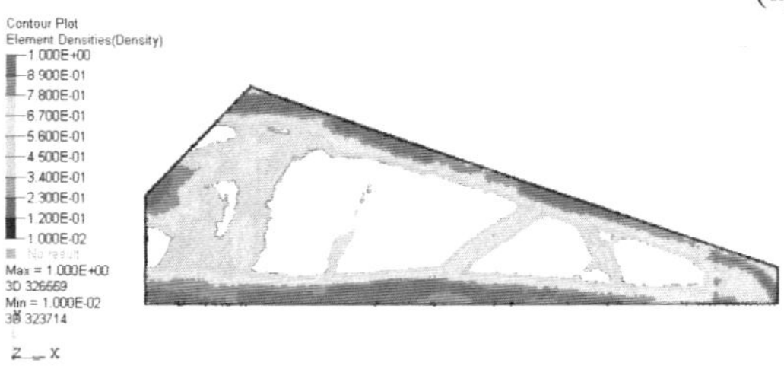

(b) N_{max} =2

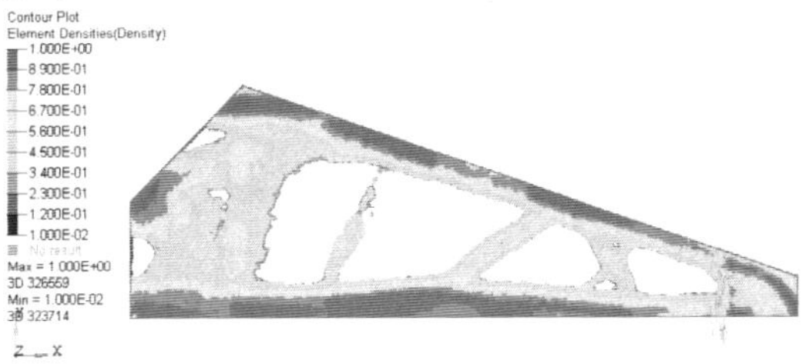

(c) N_{max} =4

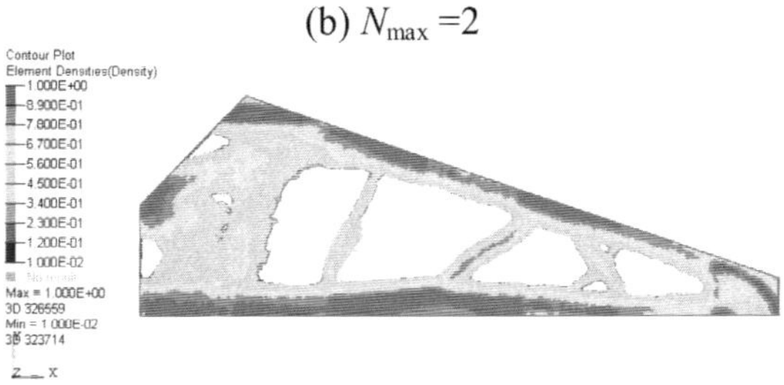

(d) N_{max} =8

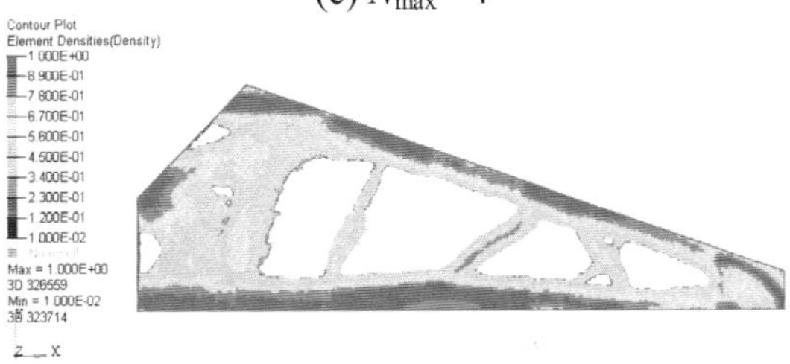

(e) N_{max} =16

Figure 7. Optimization results with the different number of equivalent working conditions.

When the number of the equivalent working conditions is small, especially in Figure 7(a), the color boundary of most of the material areas on the left side is turbid, and the main support structure has not yet formed. Therefore, when the number of the equivalent working conditions is small, the optimization process can obtain feasible solutions, but the numerical instability in the topological result is serious, and the key features have not been fully formed. The number of the equivalent moving load conditions increases by a factor of 2, as shown in Figure 7(b), 7(c), and 7(d). The numerical instability of the large material area in Figure 7(a) has been controlled, and the key support structure in the boom structure is gradually clear.

As the number of the equivalent working conditions increases exponentially, the optimization results gradually become stable. In the optimization result, the contact position between the optimized area and the non-optimized area is all close to red, and the relative density of the element is close to 1, indicating that the material needs to be greatly retained, so that the cantilever crane boom structure has greater rigidity. The boom will not be severely deformed during heavy loads; the material near the contact position of the boom and the column (left side of Figure 7(d), 7(e)) is sufficient, and the root structure of the cantilever beam is subjected to the load process. In the case of large stress concentration, sufficient material is required to release the deformation energy; the rest of the boom is relatively removed from the material, and the structure is mostly in the form of a truss, which ensures the rigidity of the cantilever crane boom is sufficient. Underneath, the weight of the boom is greatly reduced, and the purpose of weight reduction is achieved.

According to the stable topology optimization result, the cantilever crane boom structure is reconstructed, and the lightweight cantilever crane boom shown in Figure 8 is obtained. The optimized cantilever crane boom structure is better than the original structure. The quality is reduced by 32.08%.

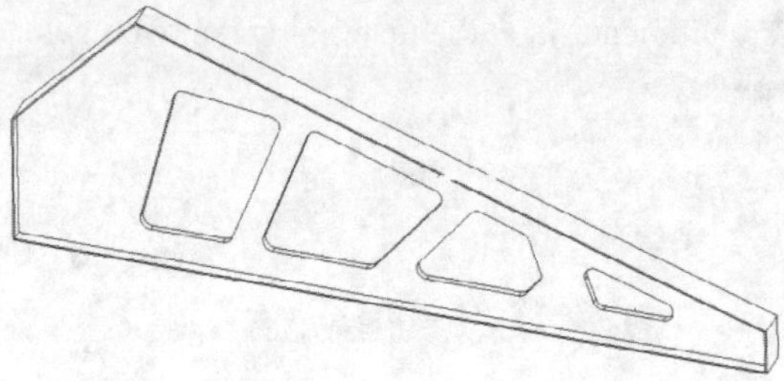

Figure 8. Lightweight cantilever crane boom.

7. Conclusions

In this paper, based on the variable density structure topology optimization method, aiming at the moving load of the cantilever crane boom, a structural topology optimization method based on the EML method is proposed to study the lightweight structure of the cantilever crane boom structure. By transforming the EML of the heavy object lifting process, the number of the equivalent working conditions is set by the multiple growth mode. When the number of the equivalent working conditions is 8, the optimization result is stable; the boom is light according to the topological result. After lightweight, the quality of the structure is reduced by 32.08% compared with the original structure. The optimization process shows that the structural topology optimization method based on the EML method proposed in this paper is effective and feasible, and provides design ideas for other lightweight design of structures subjected to moving loads.

Acknowledgments

The authors gratefully acknowledge the financial support of National Natural Science Foundation of China (No. 51675075) and Innovative Talents Program of Colleges and Universities in Liaoning Province(No. LR2018048).

References

[1] WANG Xin, HUANG Lin, GAO Yuan, et al. Topological optimization for crane telescopic boom-section[J]. *Journal of Dalian University of Technology*, 2009, 49(3):p 374-379.

[2] LI Zuobin, CAO Xuyang, WANG Peng, et al. The Design for a New Type of Girder of Overhead Crane Based on Topology Optimization[J]. *Machinery Design & Manufacture*, 2017(10):p 175-178.

[3] ALFOUNEH M, TONG L. Maximizing modal damping in layered structures via multi-objective topology optimization[J]. *Engineering Structures*, 2017, 132:p 637-647.

[4] OUYANG H, MOTTERSHEAD J E, LI W. A Moving-Load Model for Disc-Brake Stability Analysis[J]. *Journal of Vibration & Acoustics*, 2003, 125(1):p 53-58.

[5] SHA Z, YIN J, ZHANG S, et al. Structure Topology Optimization Design of Brake Pad in Large Megawatt Wind Turbine Disc Brake[C]. *International Conference on Information Sciences, Machinery, Materials and Energy*, Chongqing, 2015.

[6] ZHANG Zhifei, XU Wei, XU Zhongming, et al. Double SIMP Method for Gray scale Elements Suppression in Topology Optimization[J]. *Transactions of the Chinese Society for Agricultural Machinery*, 2015,46(11):p 405-410.

[7] ZHANG S F, YIN J, LIU Y, et al. Multiobjective Structure Topology Optimization of Wind Turbine Brake Pads Considering Thermal-Structural Coupling and Brake Vibration[J]. *Mathematical Problems in Engineering*. 2018.

IOP Conf. Series: Materials Science and Engineering **692** (2019) 012021 doi:10.1088/1757-899X/692/1/012021

Research on Braking Vibration Characteristics of Grooved Brake Interface of Disc Brake

Zhihua Sha, Panli Kang, Jian Yin, Yu Liu*, Shengfang Zhang, Fujian Ma and Dapeng Yang

School of Mechanical Engineering, Dalian Jiaotong University, Dalian 116028, China

*E-mail: liuyu_ly12@126.com

Abstract. With the good performance and wide application of surface texture in various aspects, more and more scholars are devoted to investigating the effect of texture interface on tribological properties. Disc brake friction pair is a typical friction brake interface. In order to consider the influence of grooved brake interface on vibration characteristics during braking, a simplified finite element model of brake interface is established in this paper. By processing parameterized grooves with different lengths and angles on the surface of the brake disc, the effects of "blind groove", "through groove" and "angle" on the vibration characteristics of the brake disc during braking process are simulated and studied, which can provide reference for the structural design and optimization of the brake disc texture.

1. Introduction

The brake has the function of decelerating, stopping or maintaining the stop state of the moving machinery, and it is an important part to ensure the safety of the device[1]. At present, surface texture technology shows good application potential in tribology. It is of great significance to apply texture technology to the surface of brake pairs to study the influence of texture parameters on braking performance during braking process. Andersson et al. studied the effect of texture on surface lubrication behavior on steel surface. The results showed that texture greatly reduced surface friction and wear[2]. Saw et al. studied the influence of groove arrangement on brake system instability. The results show that when the groove inclines at a certain angle, the degree of unstable vibration of the brake system decreases[3,4]. Wang et al. proposed that the impact of the grinding ball on the groove would produce low-frequency friction fluctuation. As a discontinuous excitation, it can effectively suppress the formation of high-frequency components of interface friction and achieve the effect of reducing friction noise[5]. Based on the common operating conditions of high-power wind turbine disc brake, this paper studies the influence of brake interface groove parameters on the vibration characteristics of brake disc by establishing parametric grooves on the surface of brake disc, which can provide reference for the structural design and optimization of brake disc.

2. Establishment of brake rigid body model

2.1. Establishment of three-dimensional brake model

The brake mainly includes upper and lower calipers, compensation mechanism and friction parts components. This paper focuses on the influence of groove on braking process, without considering secondary braking, wear and other conditions. Therefore, the compensating mechanism and brake

IOP Conf. Series: Materials Science and Engineering **692** (2019) 012021 doi:10.1088/1757-899X/692/1/012021

calipers can be subtracted, but the inertia provided by them should be equivalent attached to the central axis connected with the brake disc. The upper and lower friction pairs in the brake are symmetrical, so the center plane of the brake disc is used as the symmetry plane, and the upper half of the disc is combined with the upper brake to model. The model is shown in Figure 1.

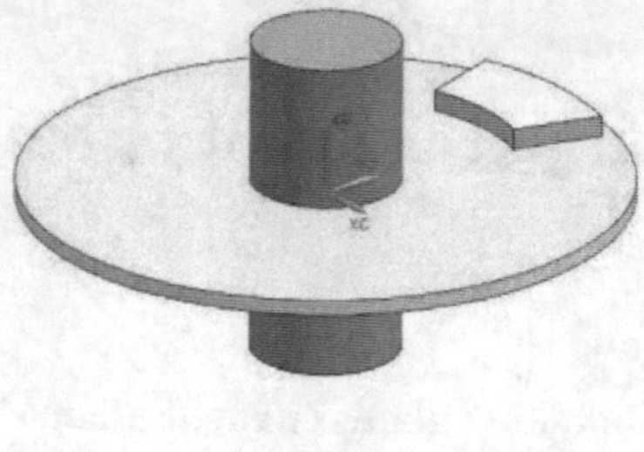

Figure 1. Brake rigid body three-dimensional model.

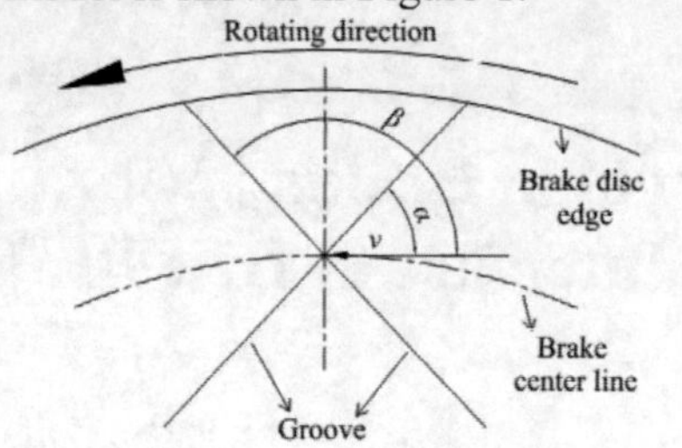

Figure 2. Definition of groove angle.

2.2. Groove type brake interface parameter establishment

The model in Figure 1 is based on the material and size of friction pairs components in wind turbine disc brakes. The brake disc is of arc structure, and the material is copper-based powder metallurgy. Its size is 520mm in Inner diameter, 760mm in outer diameter, 26mm in thickness, and 36° in center angle. The material of the brake disc is Q345b, with an outer diameter of 800mm, an inner diameter of 200mm and a thickness of 20mm. The material parameters are shown in Table 1.

Table 1. Material parameters

Brake component	Material	E (GPa)	μ	ρ (kg/m^3)
Brake disc	Q345b	210	0.31	7850
Brake pad	Copper-based powder metallurgy	5.2	0.3	5250

On the upper surface of the brake disc, four uniformly distributed rectangular grooves are formed according to the angle, and the length is l mm, the cross-sectional width is a=2mm, and the height is b=1mm. The groove angle of the brake disc is defined as shown in Figure 2. Taking the brake center line of the matching position as the boundary, the angle between the direction of the brake disc rotation speed v and the groove edge of the brake disc periphery is defined as α[6] (0°<α<90°). For the convenience of the following description, it is called a small angle, and its complementary groove angle β=180-α(α>90°), which is called a large angle.

In Figure 3, a simulation angle groove is set up every 15° between 0° and 180°. In Figure 3(a), the groove length l of the surface of the brake disc is limited according to the radius $R1$, $R2$ and a certain angle, and at least one side of the length direction of the groove is outside the radius of two curvatures of the brake pad. In this case, the groove on the brake disc surface is called “through groove”. In Figure 3(b), the groove length l of the surface of the brake disc is a fixed value of 120mm, and the length thereof is included in the two curvature radius of the brake pad. In this case, the groove on the surface of the brake disc is called “blind groove”.

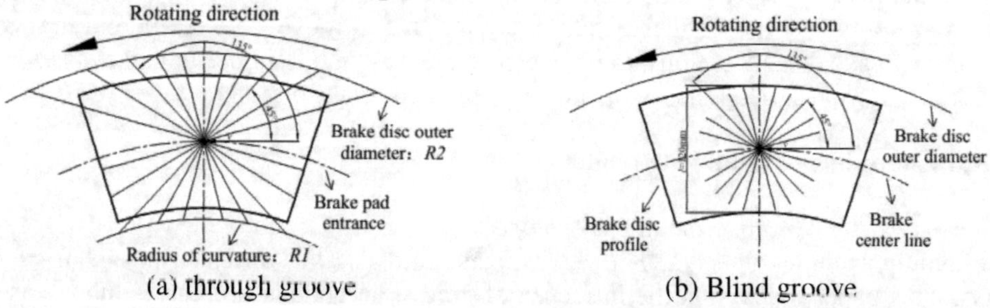

(a) through groove (b) Blind groove

Figure 3. Groove parameters of the brake disc surface in the mating view.

3. Establishment of brake finite element model

The modeling problem of contact collision has always been a very important and difficult problem to solve in multi-body dynamics systems. So far there is no universal model. The models of such problems are mostly based on experience and related experiments[6]. If the deformation of the object is not considered, the ideal solution result can be obtained by the multi-rigid method and the rigid body contact theory[7]. However, during the braking process, the brake disc and the brake pad in the brake belong to the easily deformable component, and the research focuses on the instantaneous deformation of the component in the contact collision and the result of the deformation reaction on the contact collision. Therefore, object deformation can not be ignored, it can not be limited to rigid body and classical collision model assumptions[8]. In this paper, the deformable brake disc and brake pad are flexibly processed in the HyperMesh software, then import both into the ADAMS software to replace the rigid body parts. Finally, the simulation is carried out after the constraints of each component and the initial conditions are set. The constraints are shown in Table 2.

Table 2. Model constraints.

Constraint component	Constraint type	Constraint component	Constraint type
Brake disc and central shaft	Fixed pair	Central shaft and ground	Rotating pair
Upper brake pad and ground	Mobile pair	Upper brake pad and brake disc	Contact pair

The simulation conditions adopt the common working conditions of the wind turbine brake, that is, the initial speed of the brake disc is n=1000r/min, the braking force applied to the brake pad is F=17000N, and the friction coefficient between the brake disc and the brake pad is μ=0.3. The completed dynamic model diagram is shown in Figure 4.

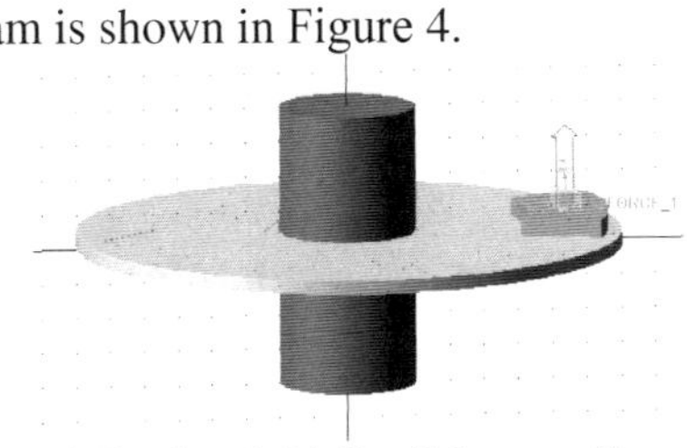

Figure 4. Brake rigid-flexible coupling model.

4. Vibration characteristics analysis of grooved brake disc

This paper does not consider wear and tear, and investigates the influence of the groove interface on the braking performance based on braking time and braking stability. Braking time can be expressed by the angular velocity or kinetic energy curve of the brake disc. This paper uses the kinetic energy curve. Because the square of the velocity or angular velocity is included in the kinetic energy calculation formula, the braking rate can be more intuitively distinguished.

4.1. Simulation results and analysis of through-groove brake disc

As mentioned above, α smaller than 90° is called a small angle, and β larger than 90° is called a large angle, where β=180-α is a complementary angle of α. In Figure 5(a), the vibration amplitude curves of the through groove brake discs of 15°-30° and their corresponding complementary angle sections are almost all in a straight line. That is to say, when α is small or β is large, the change of angle has no obvious influence on the vibration of brake disc during braking. In the 30°-150° angle range, the vibration amplitude curve rises sharply first, and the upward trend is moderated at 60°. The amplitude reaches the maximum at 75°, and then the curve drops sharply and the downward trend becomes slower at 105°. At the limit of 90°, the complementary angle is α-β, and the vibration amplitude of the small angle in the through groove brake disc is larger than the large angle. In Figure 5(b), the strain energy amplitude curve of the brake disc has the same regularity as the vibration amplitude curve in Figure 5(a), and it is considered that the vibration and strain of the brake disc mutually promote each other. That is, the larger the axial impact, the greater the amount of strain change, and the better the strain recovery performance of the impact, which in turn exacerbates the vibration amplitude.

From Figure 5(a) and (b), the two figures are normal distributions with 90° as the center line and offset to the left. One reason for the offset is that when the through groove angle is close to 90°, the span of the groove length becomes smaller, and the shock oscillation time is shortened when the brake pad passes the groove, and the vibration and strain do not form a chain interaction effect. Another reason is that the vibration and impact of the small-angle grooved brake disc during braking are greater than or equal to the large angle.

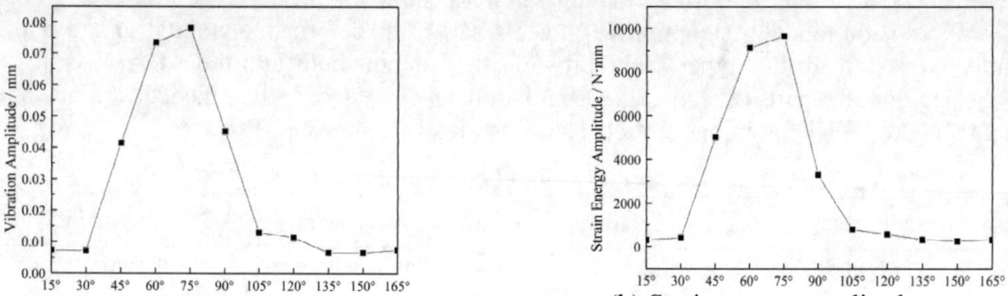

(a) Vibration amplitude curve (b) Strain energy amplitude curve

Figure 5. Through-groove brake disc amplitude curve.

As shown in Figure 6, the kinetic energy curve of the 45°-90° through groove brake disc is clearly separated from other angle curves, and the kinetic energy declining trend is $E_{75°}>E_{60°}>E_{45°}>E_{90°}>$other. Referring to Figure 5, when the groove angle α is small, the amplitude of the vibration/strain energy of the brake disc is close to the corresponding complementary angle β amplitude. This phenomenon causes the two curves on the kinetic energy curve to coincide. When the groove angle α is gradually increased and gradually approaches 90°, the vibration/strain energy amplitude of the brake disc is significantly different from the amplitude of the corresponding complementary angle. At the same time, the kinetic energy curves of the complementary angle α-β of the through-groove brake discs is obviously separated, and the kinetic energy of the through groove brake disc with larger amplitude of vibration/strain energy decreases faster. It is concluded that the magnitude of the vibration/strain energy of the brake disc has a one-to-one correspondence with the speed of the brake. That is, the greater the axial vibration/strain energy amplitude of the brake disc, the faster the kinetic energy of the corresponding angle through-groove brake disc drops. Moreover, with 90° as the boundary, in the complementary angle pair α-β, the small angle groove brake is not less than the large angle groove.

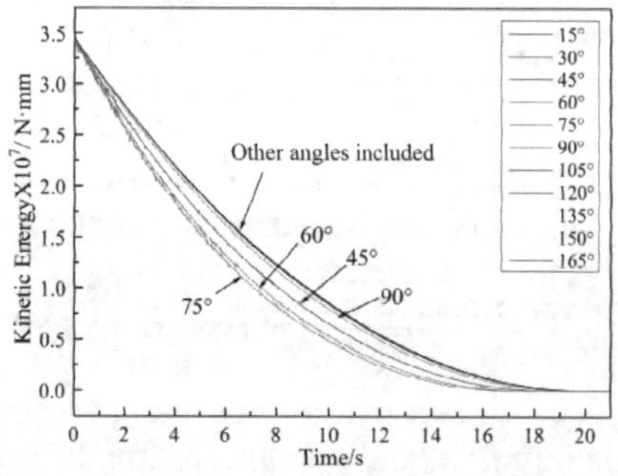

Figure 6. Kinetic energy drop curves of through-groove brake discs.

The above analysis reflects the influence of the through-groove brake disc on the braking performance, and the change in the groove angle has a significant effect on the braking process. In short, thc change in the through-groove angle significantly affects the fluctuation amplitude of the vibration/strain energy, which in turn affects the braking kinetic energy drop curve. The amplitude of the through grooved brake disc varies greatly, the vibration is more severe, and the energy loss of the system is also faster, which is the main factor affecting the braking rate. Therefore, the through-groove brake disc is suitable for working conditions where the system vibration requirements are not critical but require rapid braking.

4.2. Simulation results and analysis of blind groove brake disc

In Figure 7(a), the vibration amplitude curve first oscillates increase with the angle change; at 90°, the curve rises sharply and the amplitude reaches the maximum at 105°; then the curve decreases to a large extent. The strain energy amplitude curve in Figure 7(b) is consistent with the change trend of Figure 7(a), where them also have a mutually reinforcing relationship. However, with 90° as the boundary, the vibration amplitude of the large-angle blind slot brake disc is larger than the small angle. Because under the premise that the blind groove brake disk has a small degree of fluctuation, when the groove is turned to the brake pad with the brake disc, in the complementary angle pair α-β, the large angle β blind groove is in contact with the brake pad first on the side away from the central axis, and the small angle α blind groove is opposite, as shown in Figure 8. However, the closer to the outside along the radial direction of the brake disc, the greater the linear velocity, and the shock vibration caused by the brake pad passing through the groove is relatively large.

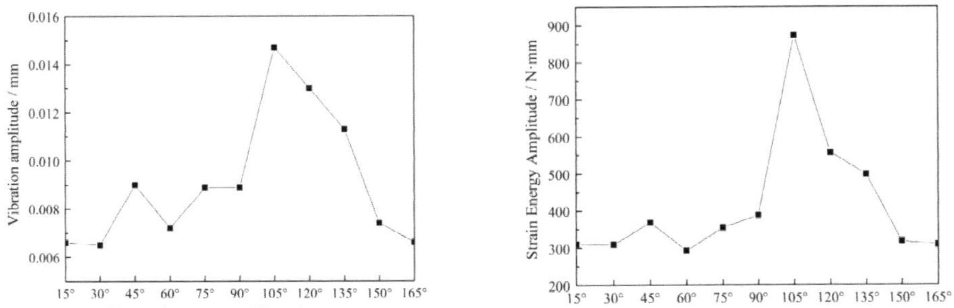

(a) Vibration amplitude curve (b) Strain energy amplitude curve

Figure 7. Blind groove brake disc amplitude curve.

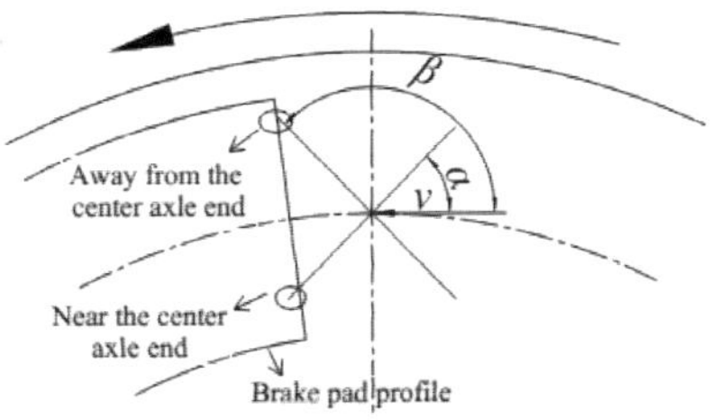

Figure 8. Contact condition of the brake pad when passing through the blind slot.

In Figure 9, the kinetic energy curves of the blind grooved discs at different angles overlap. Because the blind groove is completely contained within the two curvature radii of the brake pad, and the arc surface of the two radii of the brake pad does not directly participate in the impact when passing through the groove. Therefore, the braking process is gentler than the through groove, and the vibration/strain energy amplitude is also much smaller than the through groove, which is not enough to cause the kinetic energy to drop significantly. In this case, the change in the angle of the blind groove has little effect on the braking process.

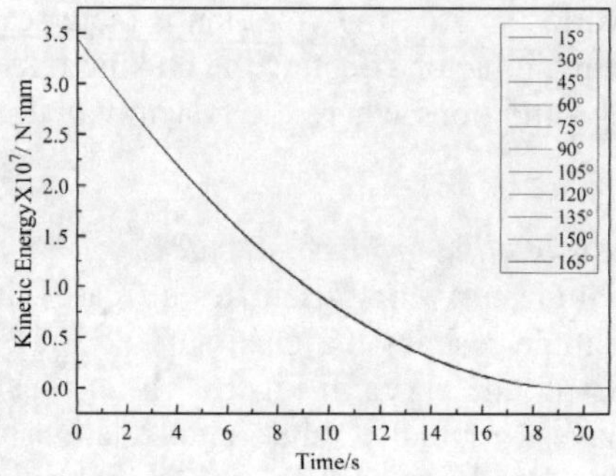

Figure 9. Kinetic energy drop curves of blind groove brake discs.

The above analysis reflects the effect of the blind groove brake disc on the braking performance during braking. The change of the blind groove angle has no obvious influence on the kinetic energy drop of the brake disc. The influence on the braking performance is mainly reflected in the vibration/strain energy amplitude fluctuation (small compared with the through-groove amplitude fluctuation). This paper focuses on the influence of the groove parameters on the braking process, and does not consider the advantages of heat dissipation and chip resistance of the groove on the surface of the brake disc under actual working conditions. Therefore, the blind groove brake disc can be used in a working condition that is dedicated to heat dissipation, chipping, and requires less system vibration, but has no obvious requirements on braking time.

5. Conclusions

In this paper, by establishing parameterized grooves on the surface of the brake disc, the related laws of vibration and kinetic energy curves of the series of grooves in braking are obtained, which provides reference for the design and optimization of the texture structure of the brake disc.

- The change of the through groove angle of the brake disc affects the vibration/strain energy amplitude change of the brake disc, further affecting the braking rate; the vibration/strain energy amplitude of the small-angle groove is greater than or equal to the large-angle groove, and the larger the vibration/strain energy amplitude of the brake disc is, the faster the brake is.
- The change of the blind groove angle of the brake disc has no obvious influence on the braking rate. The influence on the braking performance is mainly reflected in the fluctuation of the vibration/strain energy amplitude of the brake disc; the large-angle blind groove amplitude fluctuation is larger than the small-angle blind groove, but much smaller than the through groove.
- The strain energy amplitude and vibration amplitude curve of the grooved brake disc have a good consistency, and the two have mutually reinforcing relationship.

Acknowledgments

The authors gratefully acknowledge the financial support of National Natural Science Foundation of China (No.51675075) and Innovative Talents Program of Colleges and Universities in Liaoning Province (No. LR2018048).

References

[1] Zhang Y N 2016 Traction elevator brake and its technical inspection *J. China Equipment Engineering* (10) pp 47-48

[2] Andersson P et al. 2007 Microlubrication effect by laser-textured steel surfaces *J.* Wear 262 pp 369-379

[3] C.L. Saw, C.G Choong et al. 2011 Disc brake squeal suppression through chamfered and slotted pad *J. International Journal of Vehicle Structures and Systems* 3 pp 28-35

[4] C.L. Saw, A.R. AbuBakar et al. 2009 Supressing disc brake squeal through structural modifications *J. Jurnal Mekanikal* 29 pp 67-83
[5] WANG Z G, MO J L et al. 2013 Effect of grooved surface texture on interface friction vibration and noise characteristics *J. Vibration and Shock* 32(23) pp 175-179
[6] Zhang Y B 2007 Research on multi-body dynamics contact and collision modeling *D. Huazhong University of Science and Technology*
[7] Xiong X J, Zhang X X et al. 2009 Nonlinear Finite Element Analysis of Rolling Bearing Contact *J. Journal of Test and Measurement Technology* 23(1) pp 23-27
[8] Hao J 2016 Research on dynamic characteristics of braking and compensation mechanism of wind turbine disc brakes *D. Dalian Jiaotong University*

IOP Conf. Series: Materials Science and Engineering **692** (2019) 012022 doi:10.1088/1757-899X/692/1/012022

Fatigue Life Calculation of High-Power Disc Brake under Thermal-Mechanical Coupling

Shengfang Zhang, Zhiyu Han, Qiang Hao, Yu Liu*, Zhihua Sha, Fujian Ma and Dapeng Yang

School of Mechanical Engineering, Dalian Jiaotong University, Dalian, China 116028

*E-mail: liuyu_ly12@126.com

Abstract. The fatigue failure mode of the brake disc is the initiation and expansion of the fatigue crack on the surface of the brake disc until the brake disc brakes. Therefore, considering the fatigue failure mode of the brake disc, the fatigue life of the brake disc is calculated. Firstly, the thermo-mechanical coupling model of the brake disc and pad is established by Abaqus software, so that temperature and stress of the brake disc could be obtained. Then, the surface crack initiation life of the brake disc is calculated by fe-safe software Finally, the fatigue crack growth life of the brake disc is calculated by nCode software. The calculation and analysis results show that the fatigue crack of the brake disc appears on the outer side of the brake disc friction surface under the working condition of this paper. The crack initiation life is that the number of braking is 12830 and the crack propagation life is 695000for the number of braking.

1. Introduction

Disc brake has many advantages, such as simple structure, light weight, fast heat dissipation and convenient adjustment. It is widely used in large-scale equipment, such as railway vehicles, mining machinery, and metallurgical machinery, as a deceleration device for shaft rotating parts[1]. With the increasing load, speed and power of the machine, the braking condition becomes more and more severe. Especially, under the emergency braking conditions, the braking friction generates a lot of heat energy. Most of heat is absorbed by brake discs and pads and diffused through heat conduction. The heat that is too late to diffuse causes the temperature of the brake disc to rise sharply, and a temperature gradient is generated inside. In addition, structural constraints cause thermal stress inside the brake disc. Under the action of cyclic thermal stress, the brake disc appears the phenomenon of thermal fatigue cracking[2-3]. In recent years, a large number of scholars have studied the fatigue failure of brake discs. Based on the direct coupling analysis results of friction pair thermal force and the Manson-Coffin fatigue prediction model, considering the cyclic stress-strain curves of the brake disc, Yang Wei used the fe-safe software to calculate the number of braking of the brake disc during the whole life cycle. , [4]. During the non-destructive testing process, F. Bagnoli et al. found that a large number of fire truck brake discs used to support aviation activities cracked, and the crack length was 1-7.5cm. The analysis concluded that the repeated cycles of thermal stress were the causes of cracks[5].In this paper, Abaqus is used to conduct thermo-mechanical coupled finite element analysis of the brake disc. The temperature and stress results are imported into fe-safe software for surface fatigue failure analysis, so as to obtain the location, life and crack length of fatigue crack initiation on the brake disc surface. Finally, the nCode software is used to analyze and calculate the crack life of the brake disc, and the complete fatigue life prediction method of the brake disc is obtained.

2. Thermo-mechanical coupling simulation of brake disc

2.1. Modeling of brake disc

In order to simulate the thermo-mechanical coupling of brake disc, it is necessary to establish the model of the brake disc and pad. The dimension parameters of a certain type of brake disc and pad are shown in table 1.

Table 1. Dimensional parameters of brake disc and pad.

	Inner radius r_0	Outer radius r	Thickness h	Wrap angle θ
Brake disc	100mm	400mm	40mm	360°
Brake pad	260mm	380mm	24mm	60°

The brake disc material is Q345b, and the brake pad material is copper-based powder metallurgy. The thermo-mechanical coupling parameters of the material are shown in table 2 and table 3.

Table 2. Material parameters of brake disc and pad.

Material parameters	Density (kg/m^{-3})	Thermal conductivity (w/m·°C)	Specific heat capacity (J/kg·°C)	Modulus of elasticity (Pa)	Linear expansion (/°C)	Poisson ratio
Brake disc	7850	48	462	2.08×10^{11}	1.28×10^{-5}	0.31
Brake pad	5500	30	550	5.2×10^{9}	1.5×10^{-5}	0.3

Table 3. Material parameters of brake disc varying with temperature[6].

Temperature (°C)	Thermal conductivity (w/m·°C)	Specific heat capacity (J/kg·°C)	Modulus of elasticity (Pa)	Linear expansion (/°C)
100	48	462	2.08×10^{11}	1.28×10^{-5}
200	40.9	507	2.12×10^{11}	1.34×10^{-5}
300	39.6	548	1.94×10^{11}	1.39×10^{-5}

Since the brake disc topography and force are symmetrical about the middle section, half of the brake disc and one brake disc are taken as simulation models for simulation.

2.2. Setting of thermo-mechanical coupling simulation conditions

Thermo-mechanical coupling simulation of the braking process was performed using Abaqus. The unsteady heat conduction differential equation of the three-dimensional temperature field is described as follows.

$$\frac{\partial T}{\partial t}=\frac{\lambda}{\rho c}\left(\frac{\partial^2 T}{\partial x^2}+\frac{\partial^2 T}{\partial y^2}+\frac{\partial^2 T}{\partial z^2}\right)+\frac{q}{\rho c} \tag{1}$$

where, T is the surface temperature of the brake disc; t is the braking time, s; λ is the thermal conductivity; ρ is the density; c is the specific heat capacity; q is the heat flux.

The braking load of 17000N was applied to the upper surface of the brake pad. The initial braking speed is 1000r/min and the braking time is 19.16s.

IOP Conf. Series: Materials Science and Engineering **692** (2019) 012022 doi:10.1088/1757-899X/692/1/012022

2.3. Simulation results and analysis

The braking time of the maximum stress in the braking process is selected. It can be seen from figure 1-3 that the stress, temperature and axial strain on the brake disc surface are basically the same, and are distributed in an annular shape, and the maximum value is in the outer region of the brake disc friction region.

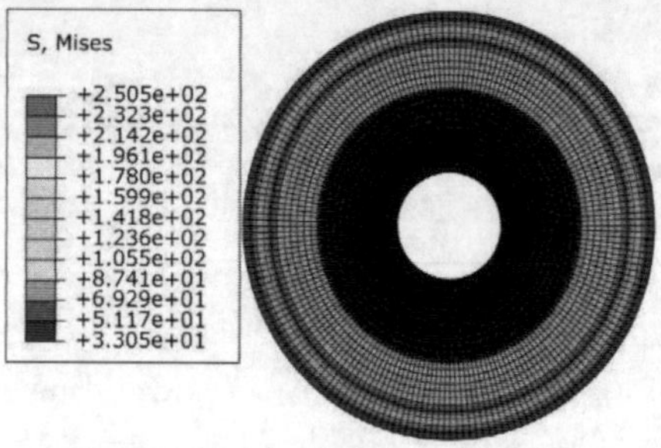

Figure1. The cloud diagram of the brake disc stress at 14.06 seconds.

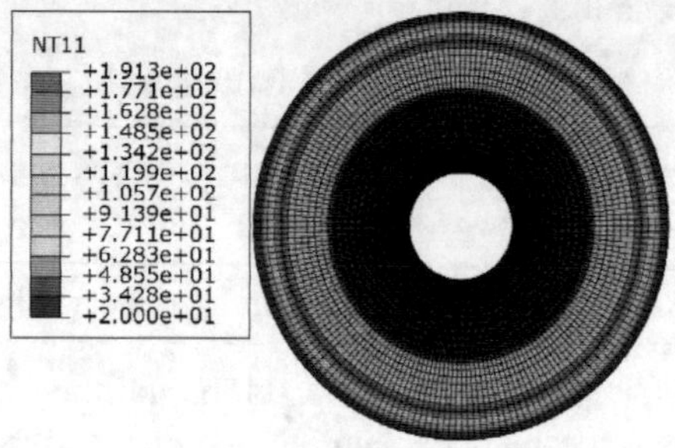

Figure 2. The cloud diagram of the brake disc temperature at 14.06 seconds.

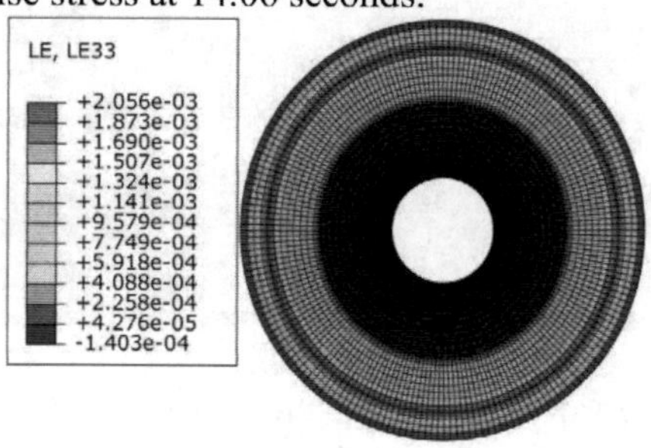

Figure 3. The cloud diagram of the axial strain of the brake disc at 14.06 seconds.

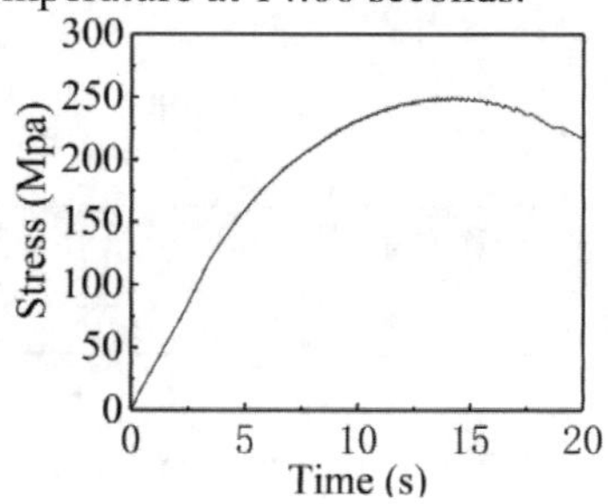

Figure 4. Stress curve of the maximum stress point on the surface of the brake disc.

It can be seen from figure 4-6 that the curve trend of stress, temperature and axial strain at the stress concentration of brake disc is basically the same.

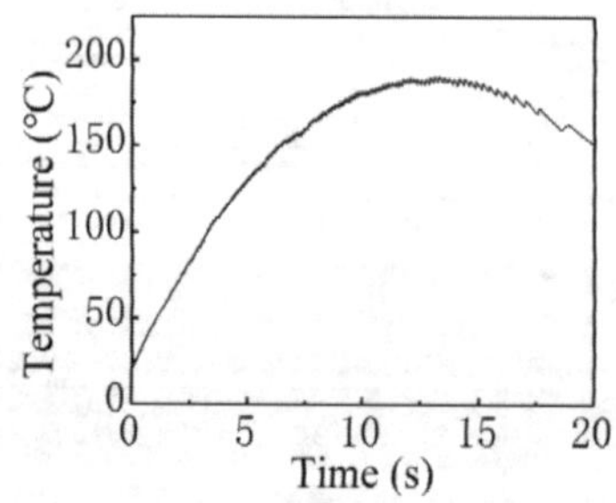

Figure 5. Temperature curve of the maximum stress point on the surface of the brake disc.

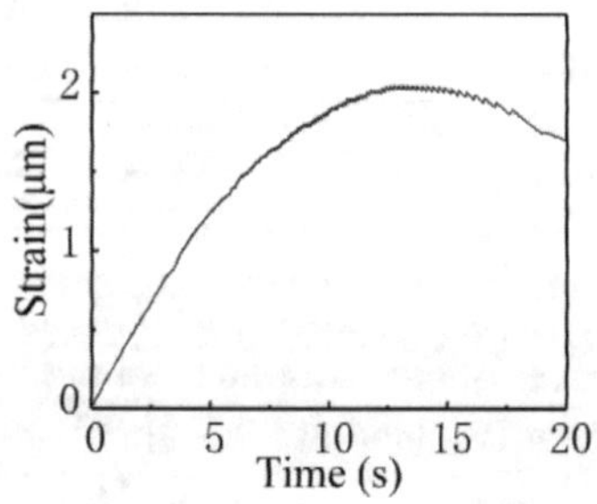

Figure 6. Axial strain curve at the maximum point of stress on the surface of the brake disc.

The axial strain curves of the contact surfaces of the brake disc and pad are as shown in figure 7 and figure 8. It can be seen that the axial strain of the brake disc and pad in the external friction region is larger at 14.06 s. Therefore, when the brake pad is in contact with the surface of the brake disc, the contact pressure of the larger strain portion is greater than the contact pressure of the remaining portion.

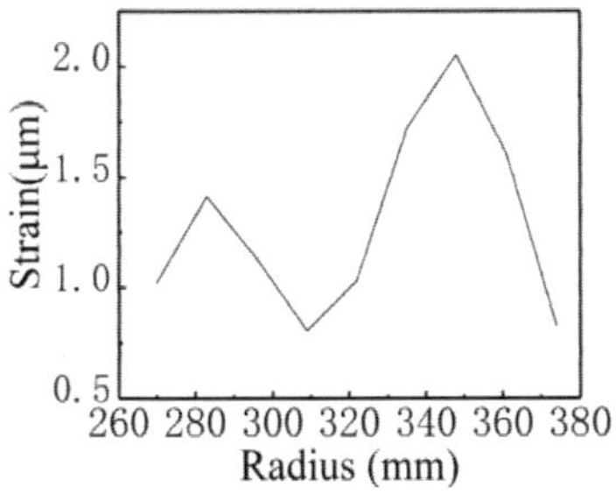

Figure 7. Axial strain curve of the contact surface of the brake disc at 14.06 seconds.

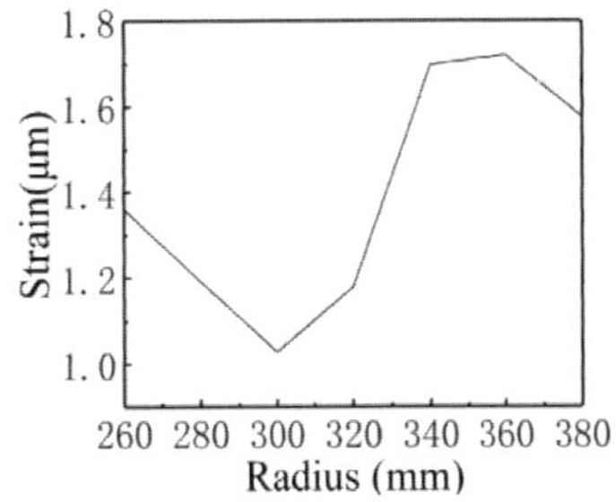

Figure 8. Axial strain curve of the contact surface of the brake pad at 14.06 seconds.

By analysing the above results, it can be concluded that the maximum temperature stress occurs in the external friction region between the brake disc and pad, which is due to the larger external linear velocity of the brake disc, resulting in more heat production and higher temperature in the external friction region of the brake disc, so the thermal deformation caused by temperature rise is more obviously than other contact regions. Thermal deformation will lead to uneven distribution of contact pressure between the brake disc and pad, which will increase the brake pressure in the external friction region of brake disc, and further affect the heat production and temperature there. Under the combined action of temperature and thermal deformation, the stress concentration of brake disc in the external friction region is caused.

The crack direction of brake disc is generally radial and axial propagation, and this fatigue crack direction is generally caused by surface circumferential stress. It can be seen from figure 9 and figure 10 that the distribution of circumferential stress is basically the same as that of stress. The stress curve of the maximum stress point on the brake disc surface is basically the same as that of the circumferential stress curve, which proves that the circumferential stress is the main stress on the brake disc surface.

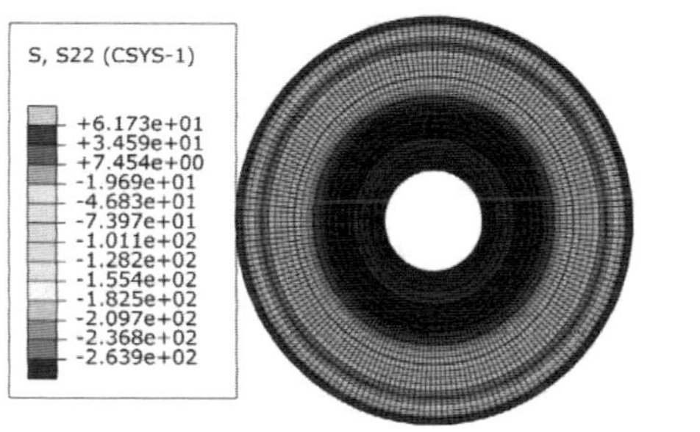

Figure 9. The cloud diagram of the brake disc circumferential stress at 14.06 seconds.

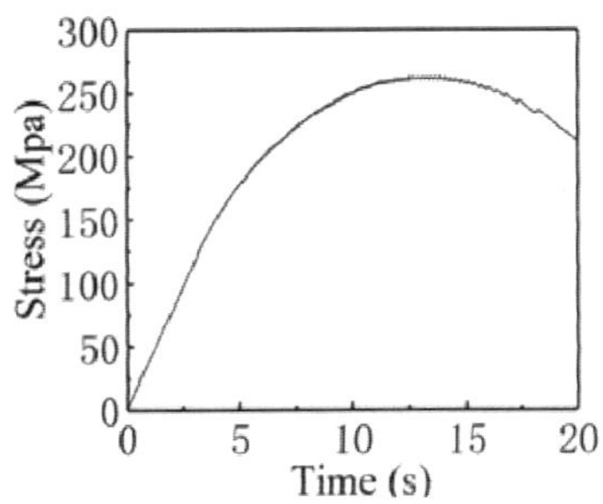

Figure 10. Circumferential stress curve of the maximum stress point on the brake disc surface.

3. Brake disc fatigue failure analysis

3.1. Fatigue analysis method of brake

On the basis of thermo-mechanical coupling analysis, the crack initiation life and the crack initiation location of the brake disc are obtained by fe-safe surface fatigue analysis. Then, with the circumferential stress of the location as input, the crack propagation life of the brake disc is analysed by nCode software, and the complete fatigue life of the brake disc is obtained.

3.2. Crack initiation life of brake disc

The surface fatigue of the brake disc is analysed by fe-safe stress-life analysis method. Because the temperature of the brake disc varies greatly during braking, the S-N curve of the brake disc material Q345b at different temperatures is needed. The safety factor is 2.0, and the fitting result of the stress-life curve of the material is as shown in table 4.

Table 4. Fitting results of stress life of Q345b at different temperatures[7].

Temperature(°C)	Stress-life curve	Stress ratio
20	$S_a = 148.25 / N_f^{0.1065} + 109.25$	
150	$S_a = 178.32 / N_f^{0.1237} + 112.08$	
250	$S_a = 224.60 / N_f^{0.1432} + 113.85$	-1
300	$S_a = 637.67 / N_f^{0.2268} + 116.32$	

Fe-safe software uses Miner's linear fatigue cumulative damage criterion to calculate the fatigue damage caused by combined loading under various working conditions. The formula is as follows:

$$D = \sum_{j=i}^{k} D_i = \sum_{j=i}^{k} \frac{n_i}{N_i} \quad i = 1, 2, 3, \ldots \tag{2}$$

where, D is the fatigue damage value of the brake disc, when $D = 1$, it indicates that the surface of the brake disc has fatigue failure; k is the total number of loads obtained by rain flow counting method; N_{i} is the number of cycles required for the fatigue failure of the brake disc under i load; n_{i} is the number of cycles of the brake disc under i load, so D_{i} is the fatigue damage value of the brake disc under i load. When the sum of damage value of all loads is 1, fatigue failure occurs on the surface of the brake disc.

It can be seen from figure 11 that the location of the fatigue failure on the surface of the brake disc is the location of the maximum stress on the brake disc surface, that is, the external friction region of the brake disc. The crack initiation life is that the number of braking is 12830.

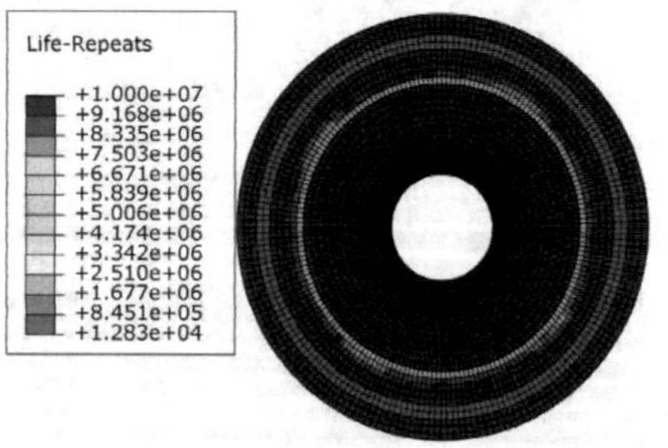

Figure 11. Fatigue life nephogram of brake disc surface.

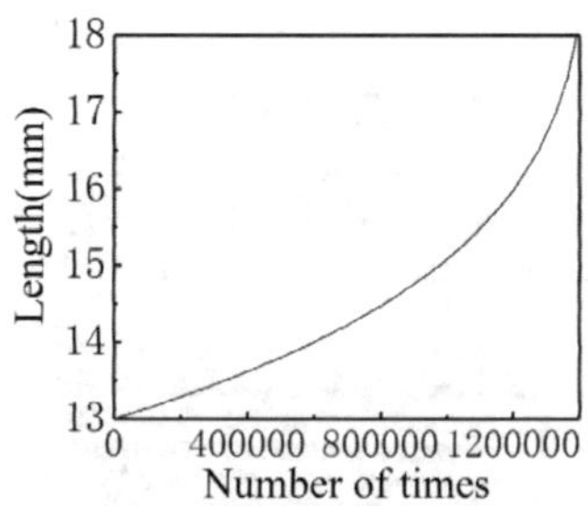

Figure 12. Crack depth-life curve.

3.3. Crack propagation life of brake disc

The location of the crack initiation node can be obtained by fe-safe, and the circumferential stress of this node is taken as the load of the crack propagation model. Data processing is carried out by rainflow technique, and then Paris formula is used for calculation. Paris formula is as follows:

$$\frac{\mathrm{d}a}{\mathrm{d}N} = C\left(\Delta K\right)^m \tag{3}$$

where, C and m are material constants, determined by experiments, C is 5.5131e^{-12}, m is 3.3231[7]; a, N and delta ΔK are crack length, load cycle number and stress intensity factor amplitude.

The main parameters of the crack are crack length $2c$ and crack depth a. The surface crack is taken as the model for software calculation. The width of the model is 80mm and the thickness of the model is half 20mm. The tensile force of the model is the circumferential stress amplitude of the crack initiation node braking once. As can be seen from figure 11, fatigue failure occurs in two elements in the radial direction of the brake disc. Therefore, it can be inferred that the length of the crack is two elements and the length of the element is 13 mm, so the length of the crack is 26 mm. The selected crack shape is semi-circular, so the crack depth a of the brake disc is 13 mm.

The fatigue failure is assumed to extend the crack depth to the complete fracture of the model. The calculation results of crack growth are shown in figure 12. The crack growth life is that the number of braking is 1390000. Taking the safety factor as 2.0, the crack propagation life of the brake disc is that the number of braking is 695000.

4. Conclusions

Taking Q345b brake disc as the object of analysis, this paper established a thermo-mechanical direct coupling finite element analysis model. The temperature and stress of brake disc during braking process were calculated, and the crack initiation life and crack propagation life of brake disc under this braking condition were analysed. In this paper, the following conclusions are obtained by calculation:

- The maximum stress of the brake disc appears in the friction region outside the brake disc. The reason is that the linear velocity of the brake disc outside leads to a larger heat production, a faster temperature rise, and a larger thermal stress than the rest of the surface, and obvious thermal deformation, resulting in uneven distribution of contact pressure, which increases the heat production of the friction region outside the brake disc and makes the stress rise.
- By fe-safe stress-life analysis method, the crack initiation life of brake disc is 12830 times. The location of crack initiation is the same as that of stress concentration during braking process, which is the external friction area of brake disc.
- The crack propagation model is established, and the crack propagation life of brake disc is that the number of braking is 695000. The total life of brake disc is that the number of braking is 707830.
- This paper provides a complete fatigue life analysis and calculation method for brake discs, which makes the life analysis of brake discs more accurately.

Acknowledgments

The authors gratefully acknowledge the financial support of National Natural Science Foundation of China (No.51675075) and Innovative Talents Program of Colleges and Universities in Liaoning Province (No. LR2018048).

References

[1] Yang X 2015 Research on fatigue resistance of biomimetic coupling surface applied to brake disc *D. Jilin University*
[2] Jin W W, Wang C C, Zhang H F, etc. 2017 Research on Fatigue Characteristics of Brake Disc *J. Rail Transportation Equipment and Technology* **(2)** pp 18-20 (in Chinses)
[3] Yang Y 2008 Research on thermal evaluation method of SiCp/A356 composite brake disc for high-speed train *D. Beijing Jiaotong University*
[4] Yang W 2016 The thermos-mechanical coupling and model credibility assessment for friction pair of wind turbine spindle brake *D. Nanchang University*
[5] Bagnoli, F F, Dolce F, and Bernabei M 2009 Thermal fatigue cracks of fire fighting vehicles gray iron brake discs *J. Engineering Failure Analysis* **16(1)** pp 152-163
[6] Hu Y Y 2014 The thermos-mechanical coupling finite element analysis for friction pair of spindle brake of wind turbine *D. Nanchang University*
[7] Qu X G 2016 Research on fatigue damage of steel Q345b used in casting crane at elevated temperature D. Lanzhou University of Technology

IOP Conf. Series: Materials Science and Engineering **692** (2019) 012023 doi:10.1088/1757-899X/692/1/012023

Design Method of Engine Room Equipment Layout Based on Improved Fuzzy Comprehensive Evaluation

Huang Zhigang, Zhang Xinyu and Liu Jingyu

Harbin Engineering University, Harbin 150001, China

384972014@qq.com

Abstract. Engine room is the place where marine power equipments, pipelines and other equipments are arranged centrally. In order to optimize the design process of engine room equipment layout, this paper proposes a method for the optimization design of engine room layout baesd on improved fuzzy comprehensive evaluation. By constructing the personnel activity space model and equipment space model, formulating the engine room environment constraint conditions dominated by personnel experience factors and obtaining the feasible scheme of engine room equipment layout. By designing the evaluation model of equipment layout plan in the engine room, formulating the evaluation method dominated by fuzzy comprehensive evaluation algorithm and obtaining the optimal layout scheme. Based on fuzzy comprehensive evaluation method, the improved method directly retains the influence of the design experience factors of personnel, which can make engine room equipment layout and pipelines more in line with engineering requirements, as well as provide better equipment accessibility and personnel accessibility.

1. Introduction

Nowadays, the research on ship engine room layout is developing towards intellectualization [1]. Lee K H adopted artificial intelligence technology to optimize the layout of ship engine room [2]. Jin Hyung developed the unit generation method to automatically process the piping design of ship engine room [3]. Zhang used the genetic algorithm to optimize the layout of key equipments and carried out the 3D layout of engine room [4]. Various literatures mainly focus on the intelligent ship layout algorithm. They would like to use the algorithm model and 3D modeling to optimize the layout of the equipments in the ship's living room or the pipelines in engine room. However, excessive reliance on algorithms will weaken the designer's experience factor. At the same time, some valuable intelligent layout models only consider the living room, whose environment is greatly different from engine room. Therefore, it is necessary to redesign the layout model for the engine room environment.

In order to overcome these shortcomings, this paper designs a fuzzy comprehensive evaluation method. Compared with the algorithm-oriented design idea, this method takes the engine room environment constraint condition as the evaluation index before the algorithm evaluation, and directly exerts the influence of design experience factors on the scheme, so that the engine room equipment layout scheme is more in line with the actual needs. Compared with other evaluation index models, the evaluation index model developed by the invention focuses on the special environment of the engine room, making the algorithm more fit the requirements of engine room equipment layout, and has certain pertinence.

2. Design of engine room environmental constraint condition

Engine room environmental constraint condition refers to the state conditions that must or should be satisfied during equipment layout, including equipment level, location space, space model type and restrict indexes. On the basis of analyzing personnel activity space model, building equipment space model by the design experience of personnel, and determining the constraint conditions of engine room environment to obtain feasible layout schemes.

2.1. Personnel activity space model

Considering the personnel channel requirements and setting personnel attributes. Setting the standard figure of male is 1.78m, the shoulder width is about 440mm, the side width is about 210mm, and the arm length plus hand length are 680mm.

In terms of height, the main equipment layout area in engine room is between the steel plate and the upper deck. It is set that the minimum height for personnel to keep walking upright is 1.8m and the range of 0 ~1.8m from the steel plate is regarded as the personnel activity space. If the height of a certain area is lower than 1.8m, people need to bend down. If the height is lower than 1.6m, people need to squat down.

In terms of width, we take a shoulder width of 660mm as the minimum width of main evacuation passageway, and take a side width plus arm length of 680mm as the minimum width of maintainable space.

In addition, defining the main pipeline space area as the space above the height of 1.8m from the steel plate, which only affects the equipment layout. The restricted pipeline space is defined as the range of 1.6~1.8m height from the steel plate, which has an impact on equipment layouts and personnel activities.

The dimension table of personnel activity space model is shown in table 1:

Table 1. Dimension table of personnel activity space model.

Personnel activity space model area		Height\Width
High space	Minimum height of person walks upright	1.8m
	Stoop walking height area	1.6~1.8m
	Crouch walking height area	Less than1.6m
Width space	Minimum width of main evacuation passageway	660mm
	Minimum maintainable space width	680mm
Pipeline space	Main pipeline space area	Higher than 1.8m
	Restricted pipeline space area	Higher than 1.8m

2.2. Equipment space model

(1) Unit space:

Unit space [5] is an envelope of equipment model, which is created according to the inherent properties of the device model.

(2) Pipe coverage space:

Defining the unit space of pipeline and its vertical downward space area as pipeline coverage space, where is not recommended to place equipments in order to avoid putting influence on personnel passage and maintenance work.

(3) Free space:

Equipment accessibility refers to the state in which the maintenance personnel can see the equipment and carry out maintenance operation during maintenance [6]. The space which is set aside for equipment maintenance in the initial design stage is called free space, which expresses the maintenance operation space.

The modeling process of free space is as follows: Analyzing the accessibility requirements of equipment, determing the size of space allowance, and assigning scope "base" to the equipment unit space. The "base" represents the size of the surrounding operating space, which is called free space allowance. There will be superimposed areas in the free space. The superimposed area is also regarded as a free space that satisfies the conditions and does not constitute interference with each other.

(4) Channel space

The space which is set aside for personnel activities in the initial design stage is called channel space, which expresses the main activity area and evacuation area. The channel space in engine room is mainly arranged around the main engines, gear boxs, pedals, generator sets and main switchboards, as well as in front of escape doors and escape straight ladder passages.

The process of channel space modeling is the same as free space modeling. Defining minimum width of main evacuation passageway of 660mm as channel space allowance. The overlapping channel space also satisfies the conditions and does not constitute interference with each other.

2.3. Engine room environment constraint conditions

Using personnel activity space model and equipment space model. Then according to three principles of "First master, then auxiliary. First large, then small. First heavy, then light" to grade engine room equipments [7]. Building the engine room environment constraint conditions table as shown in table 2:

Table 2. Table of equipment constraint conditions.

Equipment level	Equipment name	location space	space model	restrict indexes
Level-1 equipment	Main engine, main engine exhaust silencer	Bottom structure layer	Free space model	No interference Balance of center of gravity Spare room for gear boxes and pedals
	Door, straight escape	Personnel activity space area	Free space model	Near the main personnel corridor Follow the engine room environment
Level-2 equipemnt	Seawater filter, Sea chest, seawater pipeline	Bottom structure layer	Pipeline space model	Sea chest should avoid midship station Pipe coverage space is minimal Equipments must not be layouted above sea water piping
	Engine inlet and exhaust ducts	Pipeline coverage space area	Pipeline space model	No interference Pipe coverage space is minimal Try not to layout equipments under pipelines
Level-3 equipment	Main engine and auxiliary engine start the battery	Personnel activity space area	Unit space model	No interference layout around the main engine As far as possible layout in the free space, the main channel and pipe space outside

3. Design of fuzzy comprehensive evaluation method for engine room equipment layout

Using fuzzy mathematics theory and statistics principle to evaluate the comprehensive of engine room layout. Under the expert evaluation foundation, using the linear weighted method to assign rating labels for each objective function of multi-objective optimization problem. Then, multipling rating labels and weight coefficient of this objective function to get fuzzy comprehensive evaluation value so as to transform the multi-objective problem into a single-objective problem. Now the optimal solution of single objective problem is an effective solution of multi-objective optimization problem.

3.1. Analysis of engine room layout evaluation index

The evaluation of the layout of engine room is based on the evaluation value of each factor. Starting from the equipment constraint conditions and setting the evaluation model of engine room layout scheme, as shown in figure 1:

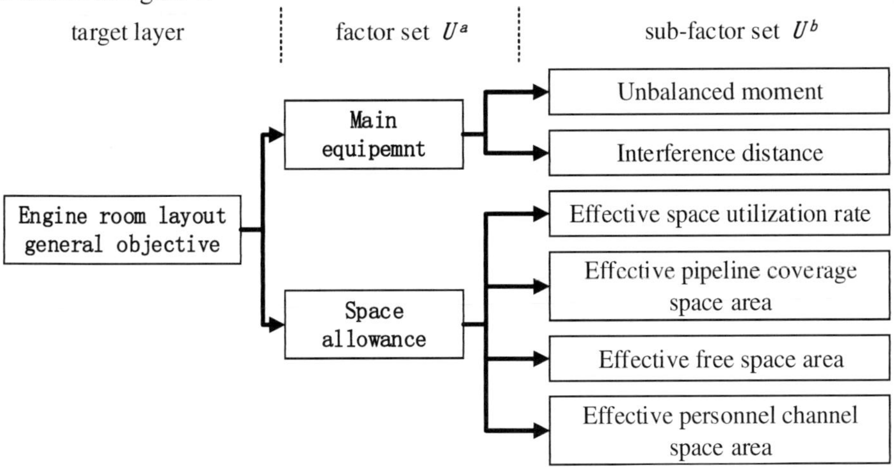

Figure 1. Evaluation model of engine room layout scheme.

The evaluation indexes of sub-factor set are shown in the following:

(1) The unbalanced moment M is the minimum

In order to avoid the ship's lateral inclination and to ensure the strength of each deck platform, equipments should be arranged in such a way that the absolute value M of the algebraic sum of moments in the middle distance section is as small as possible.

$$M = \left| \sum\nolimits_1^n m_i y_i \right| \quad (1)$$

Where m_i is the weight of equipment i. y_i is the horizontal coordinate of center of mass of equipment i, and n is the total number of equipments.

(2) The interference distance f_1 is the largest

For the sake of operational safety, a certain distance should be kept between equipments in engine room. The larger the interference distance f_1 between equipment, the better.

$$f_1 = \sum\nolimits_{i=1,\, j=1}^{q} \left| X_i - X_j \right| \quad (2)$$

Where q is the total number of equipments requiring interference distance. X_i and X_j are the location parameters of equipments requiring interference distance.

(3) The effective space utilization rate E is the largest

Increasing the effective space utilization rate E of engine room can facilitate the personnel maintenance activities in engine room. Since the equipments cannot be reduced, the larger the effective space utilization rate E is, the better it will be for the work.

$$E = 1 - \frac{\sum_{i=1}^{n} S_i^1 + \sum_{j=1}^{m} S_i^2}{A} \quad (3)$$

Where S_i^1 is the space area of engine room occupied by equipments. S_j^2 is the area of engine room corners (invalid space) that cannot be utilized. i is the equipment number. The value of E is between 0 and 1, and the larger the value is, the higher the effective space utilization is [8].

(4) The effective pipeline coverage space area S_1 is the minimum

$$S_1 = \sum_{i=1}^{n} S_i^1 - \sum_{j=1}^{m} S_j^2 \tag{4}$$

Where S_i^1 is the pipeline coverage space of pipeline i. S_j^2 is the pipeline coverage space overlap area of pipelines j.

(5) The effective free space area S_2 is the largest

$$S_2 = \sum_{i=1}^{n} S_i^1 - \sum_{j=1}^{m} S_j^2 \tag{5}$$

Where S_i^1 is the free space area of equipment i. S_j^2 is the free space overlap area of equipments j.

(6) The effective channel space area S_3 is the largest

$$S_3 = \sum_{i=1}^{n} S_i^1 - \sum_{j=1}^{m} S_j^2 \tag{6}$$

Where S_i^1 is the channel space area of equipment i. S_j^2 is the channel space overlapping area of equipments j.

3.2. Establish weight coefficient evaluation index

Constructing pairwise comparison matrix of factors u_1, u_2..., u_n from factor set U^a and sub-factor set U^b to obtaine the importance index weight matrix [9]. Then, solving the characteristic roots respectively, and obtained the relative weight ω_i of each index under a single criterion after normalization.

Assuming the weight of factor i in U^a be ω_i^a, then the U^a weighted set is [10] :

$$A = \{\omega_1^a, \omega_2^a, ..., \omega_i^a, ..., \omega_m^a\} \tag{7}$$

Where m is the total number of evaluation indexes of factor set.

Assuming the weight of factor j in U^b is ω_{ij}^b, then the U^b weighted set is:

$$B_i = \{\omega_{i1}^b, \omega_{i2}^b, ..., \omega_{ij}^b, ..., \omega_{im}^b\} \tag{8}$$

Where i represents the sub-factor set under the i U^a, and m is the total number of evaluation indexes of sub-factor set.

3.3. Design fuzzy comprehensive evaluation set

Due to the nature of evaluation indexes and the units of measurements are different, there is often a lack of correlation between indexes. In order to avoid evaluation failure, establishing a expert team and providing effective evaluation set. Each evaluator makes an evaluation conclusion for each factor (that is, for a certain element in the evaluation set), and then calculates the percentage of various evaluation conclusions to obtain the membership degree S_i. Membership degree S_i constitute a single factor evaluation set r:

$$r = [S_1, S_2, S_3, ..., S_m]^T \tag{9}$$

Where m represents the total number of evaluation results.

single factor evaluation set r_i constitute a single factor evaluation matrix R_i:

$$R_i = [r_1, r_2, r_3, ..., r_n]^T \tag{10}$$

Where n is the total number of evaluation indexes of a sub-factor set.

The sub-factor weighted set B_i is introduced to calculate the first-level fuzzy comprehensive evaluation set C_i of factors i:

$$C_i = B_i \times R_i \tag{11}$$

Where C_i is the fuzzy comprehensive evaluation of factor set i [11][12]. The larger the C_i value is, the better the effect of factor set i is.

C_i only reflects the unilateral evaluation value, so the factor set weighted set A is introduced to calculate the second-level fuzzy comprehensive evaluation set:

$$D = A \times \left[C_i\right]^T \tag{12}$$

Where $\left[C_i\right]$ is the second-level single factor evaluation matrix. D represents the evaluation value of target layer, namely the scheme evaluation value. The larger the value of D, the better the effect of the scheme.

4. Method of equipment layout in engine room

Combined with personnel design experience factors and fuzzy comprehensive evaluation algorithm to guide the layout of equipment in engine room, constructing a method flow of equipment layout in engine room as is shown in figure 2:

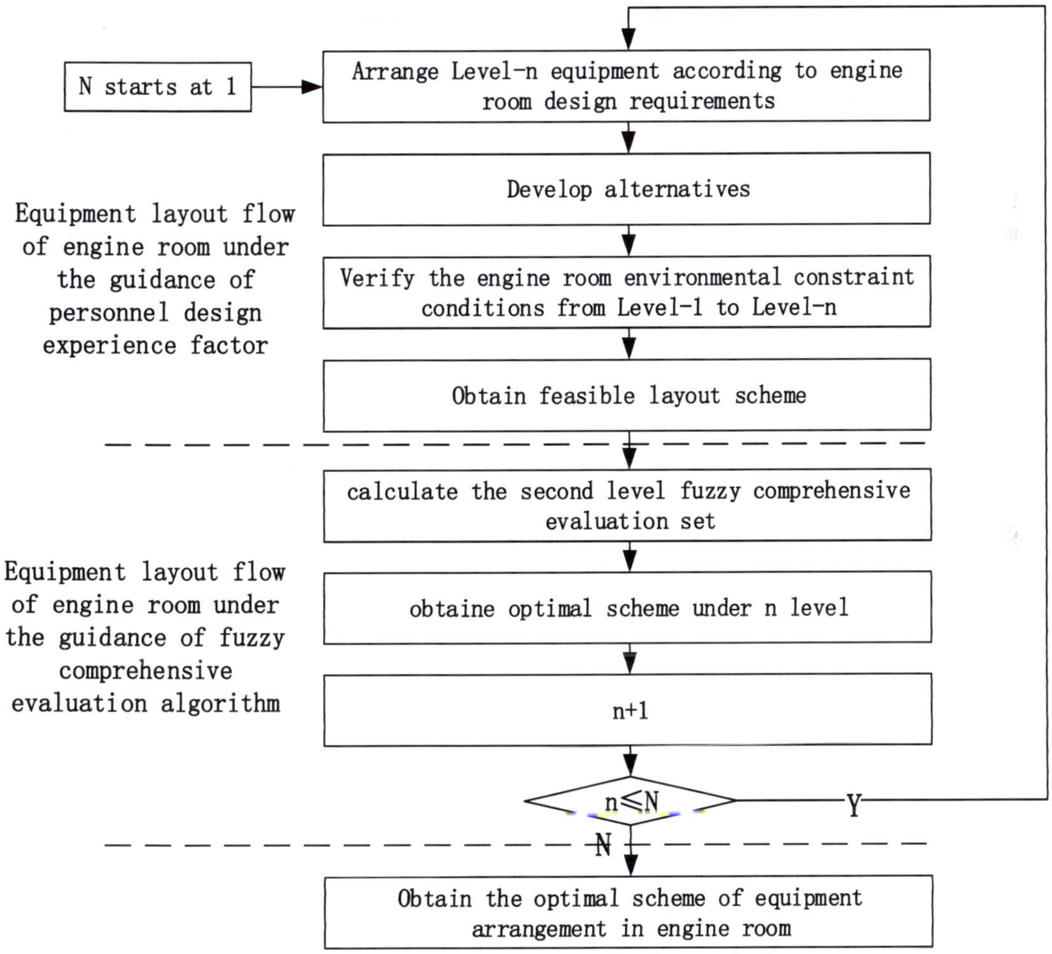

Figure 2. Improved fuzzy evaluation method.

The method includes the following steps:

(1) Arranging the level-n equipments (from the level-1 equipment) according to the design requirements of engine room. Adjusting the position of equipment based on the design experience of personnel, and form 3 to 4 alternative schemes;

(2) Verifing the engine room environmental constraint conditions of each scheme from grade 1 to n, and obtaining the feasible layout scheme which meets the conditions. If n=1, only verifing the engine room environmental constraint conditions of current level.

(3) Using the fuzzy comprehensive evaluation algorithm to evaluate the feasible layout scheme, and obtaining the second-level comprehensive evaluation set D. Retaining the best scheme, namely the scheme with the maximum D.

(4) Cycling (1) and (3) until n= N (the toatal number of quipement levels) , and end the cycle. At this time, all equipments are arranged, and the scheme with the maximum D value is retained, that is, the best scheme.

For example, when level-1 equipments satisfy constraint conditions, we design multiple feasible schemes. Checking the space constraint conditions of a scheme, the effect as shown in figure 3 can be obtained:

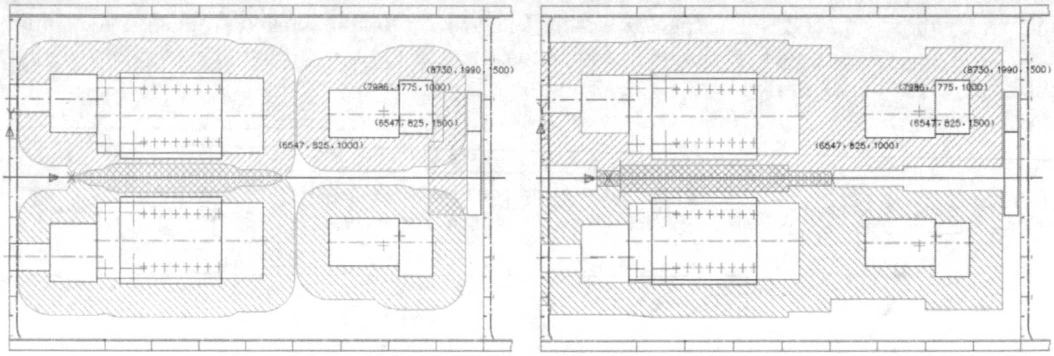

(a) check graph of free space (b) check graph of personnel space

Figure 3. Equipment free space diagram.

Optimizing engine room equipment layout of a certain ship. The level-2 feasible scheme layouts are shown in figure 4. Where, compared with scheme A, scheme B moves 200cm towards the bow of the generator set:

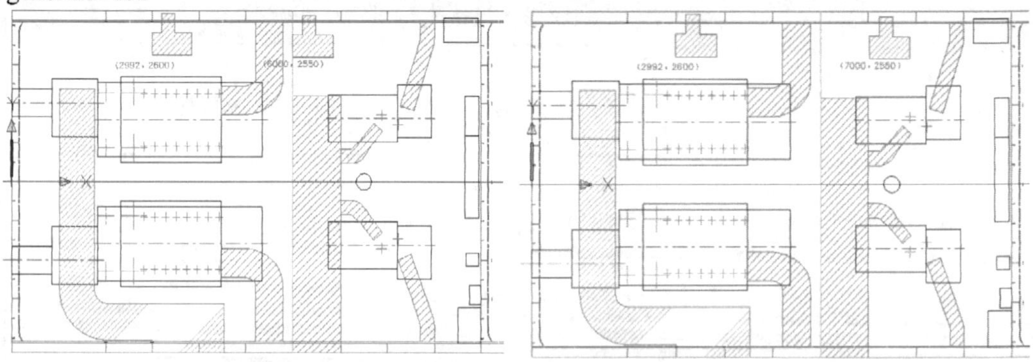

(a) scheme layout of scheme A (b) scheme layout of scheme B

Figure 4. Schemes of three-level equipment layout.

Obtaining the optimal target layer evaluation value D under scheme A is 0.79. Compared with the optimal target layer evaluation value D under scheme B is 0.81. Selecting the evaluation value of 0.81 to retain the optimal scheme B.

5. Conclusion

Aiming at the problem of engine room equipment layout, this paper puts forward a method of engine room equipment layout optimization design with more engineering practice value. This method considers the influence factors of personnel experience and puts forward the preliminary equipment layout scheme according to the engine room environmental constraint conditions. Then through the fuzzy comprehensive evaluation indexes to accomplish the algorithm level of the pros and cons of scheme comparison to obtain the optimal scheme. This method increases the influence of human decision on the basis of algorithm evaluation, conforms to the design specification and avoids subjective speculation, as well as ensures the scientific analysis and comprehensive measurement of the equipment layout in engine room.

References

[1] Lee D and Lee K H. An approach to case-based system for conceptual ship design assistant[J]. Expert Systems with Applications, 1999, 16(2): 97-104
[2] Lee K H, Lee J K and Park N S. Intelligent approach to a CAD system for the layout design of a ship engine room[J]. Computers & industrial engineering,1998,34(3): 599-608
[3] Park Jin Hyung and Storch R L. Pipe-routing algorithmdevelopment:case study of a ship engine room design[J]. Expert Systems with applications, 2002, 23:299-309
[4] Zhang Guozhong and Liu Yujun. 3D optimal layout design of engine room based on genetic algorithm [J]. Shipbuilding Technology,2005,6:25-31
[5] Su Chengguan. The Format Criterion of Maste's Degree Paper of DMU[D].Dalian Maritime University,2009:21-31
[6] Hu Xiaotang, Zhang Wenyao and Li Zhibin. Analysis of ship maintainability design [J]. China Shiprepair, 2007(06):46-48
[7] He Wang, Liu Cungen and Wang Xuefeng. Study on intelligent layout of ship engine room [J].Ship Engineering, 2014,36(06):85-88+109
[8] Wang Yunlong, Wang Chen, Ji Zhuoshang and Zhao Xueguo. Study on intelligent layout optimization design method of ship living quarters [J]. Shipbuilding of China, 2013, 54(03):139-146
[9] Li J R, Khoo L P and Tor S B. Desktop virtual reality for maintenance training: an object orientedprototype system(V-REALISM)[J].Computers in Industry,2003,52(9):109-125
[10] Wang Chao, Shi Hongquan and Liang Yizhi. Research on software quality evaluation of warship equipment based on AHP [J]. Ship Electronic Engineering, 2004,(03):117-119
[11] Maria Lu's cia Leite Ribeiro Okimoto and Eliana Remor Teixeira. Proposed procedures for measuringthe lifting task variables required by the Revised NIOSH Lifting Equation-A case study[J]. InternationalJournal of Industrial Ergonomics,2009,39(1) :15-22
[12] Zhang Leilei. Application research on marine engine room planning[D]. Jiangsu University of Science and Technology, 2017: 85-87

IOP Conf. Series: Materials Science and Engineering **692** (2019) 012024 doi:10.1088/1757-899X/692/1/012024

Numerical Simulation of Wave Landing Loads Characteristics of Twin-Float Seaplane

HU Qi[1,2], WU Bin [1,2], WANG Ming-Zhen[1,2], HUANG Miao[1,2] and JIANG Rong[1,2]

[1] China Special Vehicle Research Institute, Hubei Jingmen 448035
[2] Key Laboratory of High-Speed Hydrodynamic Aviation Science and Technology, HuBei Jingmen 448035

E-mail: 1252847359@qq.com

Abstract. The coupled Euler-Lagrangian method is used to simulate the load characteristics of a Twin-Float seaplane when it is landing in waves. The fluid domain is described by Euler, the structure is described by Lagrangian, and the coupling interaction between the Euler material and the Lagrangian element is simulated by the enhanced immersion boundary method. Firstly, a three-dimensional numerical wave tank is established, and the accuracy of numerical wave-making is verified by linear wave theory. Then, the numerical calculation of seaplane wave landing load at different wave heights with the same wave length is carried out, Lastly the numerical results compared with the model experiment of towing tank. The numerical simulation results are in good agreement with the model experiment results. The results show that there will be obvious bouncing phenomena when the seaplane landed in the wave, and the higher the wave height is, the greater the load will be generated when the wave length is the same.

1. Introduction

With its advantages of large loading capacity, wide flight range and ultra-low altitude aircraft, seaplane has incomparable advantages over roadbed aircraft in search and rescue at sea, monitoring and protection of marine environment, protection of marine rights and interests, and low altitude flight experience. The landing of a seaplane in the ocean environment is the most important working condition during its mission. The impact process of landing on the water surface is short and the load is large. The impact load is the most important problem affecting the safety of the seaplane. Especially in the sea area where the wave height is relatively high, the landing load of the aircraft is very large. Therefore, it is of great significance to study the wave landing loads of seaplane [1].

In recent years, domestic and foreign scholars have carried out some research work on the dynamic response of seaplane, amphibious aircraft and civil aircraft under water in calm water. Luo [2] used ALE coupling algorithm to analyze the influence of attitude angle and sinking speed on load characteristics of amphibious aircraft in calm water. Ma [3] established a numerical regular wave model based on linear wave theory, and studied the dynamic response of amphibious aircraft landing in water at different sinking velocities with different wave strike positions. Chu [4] studied the landing load of ground effect aircraft in calm water surface by theoretical and experimental means. Sun[5] used Dytran general coupling algorithm to simulate the surface landing of amphibious aircraft in calm water, and compared it with the model test results to verify the correctness of the simulation calculation. Yao[6] analyzed the dynamic characteristics of amphibious aircraft landing on water based on CEL algorithm, and compared the dynamic response of aircraft structure under different attitude angles and vertical velocities. Wang [7] simulated the evolution and development of water-entry voids of spheres based on ABAQUS explicit CEL method, and compared it with ballistic data of water-entry experiments, which showed the effectiveness of CEL method in solving fluid-solid coupling interaction.

In summary, the horizontal velocity, sinking velocity and attitude are the most important factors affecting the landing load of seaplane and amphibious aircraft. Some scholars have studied the landing loads and motion responses of amphibious aircraft at different wave encounters, but they have not been compared with the experiments.

In this paper, the ABAQUS explicit dynamic solver Explicit is used to establish a three-dimensional fluid-structure interaction coupling model using CEL algorithm. Firstly, the pusher wave-making method is used to simulate the wave model, and the accuracy of the numerical wave is verified by the wave height recognition technology. Then, the landing cases of the seaplane numerical model with different wave heights with the same wave length are calculated. The curve of the vertical displacement and the vertical velocity changing with time in the process of landing is obtained. Compared with the simulation results, the motion response and load characteristics under different wave heights are analyzed.

2. Numerical Method

2.1. CEL Method

CEL is usually called coupled Eulerian-Lagrangian method, which can simulate the coupling interaction between Eulerian materials and Lagrangian elements. The Euler-Lagrangian contact formula is based on the enhanced immersion boundary method. In this method, the Lagrangian structure occupies the empty area in the Euler mesh. The contact algorithm automatically calculates and traces the interface between the Lagrangian structure and the Euler material. The remarkable advantage of this method is that it does not need to generate a coordinated grid for Euler domain. A simple and regular Euler element usually produces better accuracy.

CEL is based on the fluid volume method in simulating hydrodynamic implementation. When material flows through the grid, it traces material by calculating its Euler volume fraction in each element. By definition: if a material is fully filled with a unit, its volume fraction is 1.0; if there is no material in a unit, its volume fraction is 0. In this paper, the method is fully utilized to track the volume fraction of the element near the free surface, and a wave elevation recognition technology based on CEL is developed. The accuracy of numerical wave generation is verified by this technology.

2.2. Fluid Model

The CEL in Explicit, an ABAQUS dynamic solver, approximates incompressible viscous laminar flow of water controlled by N-S equation using Mie-Grüneisen equation of state in linear form [8].

The volume response of water is described by Mie Grüneisen equation. Its most commonly used form is:

$$p = K\varepsilon_{vol} \tag{1}$$

Where K is the bulk modulus of the material. The linear equation is as follows:

$$p = \frac{\rho_0 c_0^2 \eta}{(1-s\eta)^2}(1-\frac{\Gamma_0 \eta}{2}) + \Gamma_0 \rho_0 E_m \tag{2}$$

Where c_0 and s is the two parameters of the linear shock wave, $U_s = c_0 + sU_p$, U_s is the wave velocity and U_p is the particle velocity. $\eta = 1-\rho_0/\rho$ is equal to the nominal volume compression strain ε_{vol}. If $s=0$ and $\Gamma_0 = 0$ is defined, the volume response is equal to the bulk modulus. Tab. 1 shows the material properties of water.

IOP Conf. Series: Materials Science and Engineering **692** (2019) 012024 doi:10.1088/1757-899X/692/1/012024

Table 1. Material Properties of Water

Parameter	value
ρ	1000kg/m^3
γ	0.0089Ns/m^2
c_0	1450m/s
s	0
Γ_0	0

3. Numerical Model

3.1. Structure and Fluid Model

Two-dimensional view of a Twin-Float seaplane is shown in Fig 1 [9].

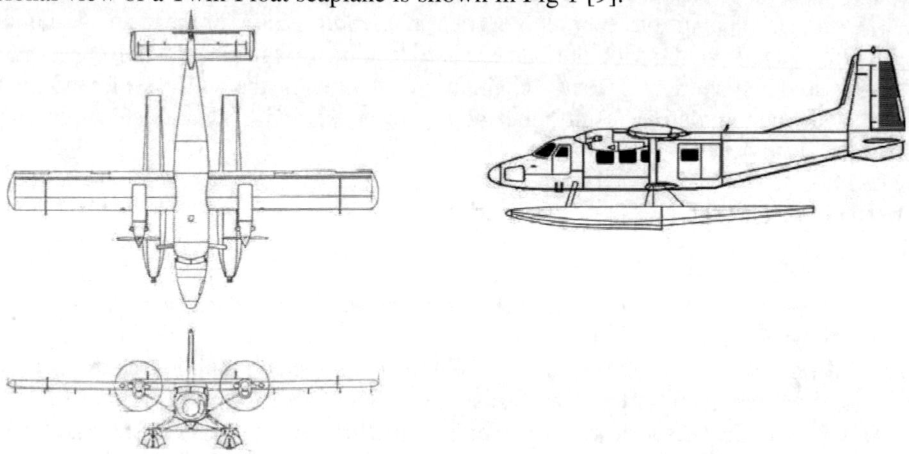

Figure 1. Two-dimensional view of a Twin-Float seaplane

In order to reduce the calculation cost and shorten the calculation time, it is simplified to floating wave landing. According to the calculation of wing aerodynamic lift, the lift, drag and pitch moments at the center of gravity at the initial horizontal velocity and attitude angle are obtained, and the aerodynamic loads are treated equivalently in the process of landing numerical simulation. In the numerical simulation of landing, the concentrated load curve of reference point is defined, and the equivalent aerodynamic force is applied to the centre of gravity of the seaplane. The validity of this aerodynamic equivalent loading has been verified by numerical calculation [3].

The Twin-Float is defined as a rigid body by reference point, and a 1/2 float model is adopted. The float structure adopts Lagrangian shell element S4R and Euler element EC3D8R for water fluid. The geometric model is shown in Fig 2. The size of float mesh is 10mm and the number of units is 7761. The minimum mesh size of the water is 10 mm and the number of elements is 1936550. The density of float is 2100 kg/m^3, the modulus of elasticity is 69 Gpa and Poisson's ratio is 0.33.

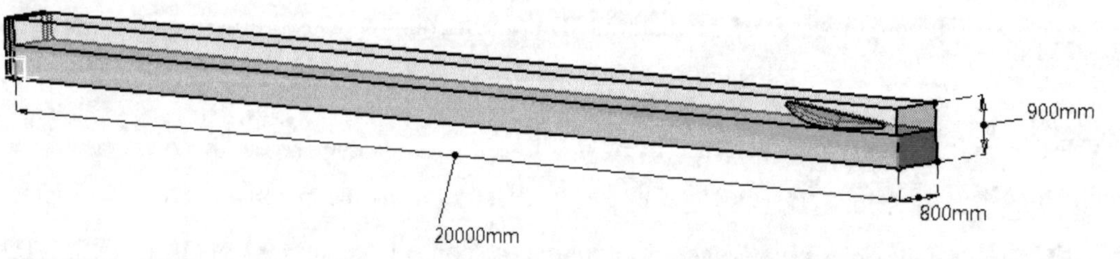

Figure 2. Water landing of seaplane geometric model

3.2. Fluid-Structure coupling Model

IOP Conf. Series: Materials Science and Engineering **692** (2019) 012024 doi:10.1088/1757-899X/692/1/012024

According to the actual parameters of the Twin-Float seaplane [9], the mass and pitching inertia of the reference point are set. Symmetrical boundary conditions are defined on the symmetrical surfaces of fluid and float, and Euler non-reflective boundary conditions are defined on other boundary surfaces of fluid. The reference point of the buoy is constrained, and only the displacement in X and Z directions and rotation around Y axis are allowed. The vertical displacement and vertical velocity of the reference point are recorded with time. The volume faction tool of CEL is used to calculate the percentage of material in the water near the free surface aiming at verifying the wave accuracy.

4. Numerical Wave Tank

4.1. Push-plate Wave-Making Theory

The numerical push-plate wave-making is the basis of studying the interaction between structures and waves. The push-plate wave-making method is used to simulate the regular waves. The initial position of the push plate is $S/2$, ω is the push frequency. Based on linear shallow water theory, the relationship between motion range of push plate and wave height is given [10, 11]:

$$\frac{H}{S}=\frac{2(\cosh(2k_p h)-1)}{\sinh(2k_p h)+2k_p h} \tag{3}$$

Where k_p is wave number, h is water depth and H is wave height.

4.2. Wave Elevation Recognition Technology

Based on the volume fraction of Euler material, the material volume percentage of Euler element at free liquid level is recorded, and the material percentage of all Euler elements near free liquid level is added up. By arranging the numerical wave altimeter on the free surface with Euler element, but the Euler element on the free surface must cover the maximum wave height range, as shown in Fig 3.

Figure 3. Layout of Wave Altimeter

4.3. Regular Wave Simulation

If the depth H of water is 0.5m, the movement period T is 1.3724s and the push distance S is 33.168mm. The theoretical wave height is obtained from the above formula (3), and wave height is 0.4m, the wave number is 2.5133, the wave length is 2.5m. Because value of wave height, depth and steepness, it can be judged that the wave belongs to weak non-linear wave with finite water depth. A numerical wave elevation monitor is set up at 2.5m and 7.5m behind the push board. Fig. 4 shows the wave elevation time-history curves at 2.5m and 7.5m. From the graph, it can be seen that the numerical wave-making technology based on CEL meets the requirement of wave landing of seaplane.

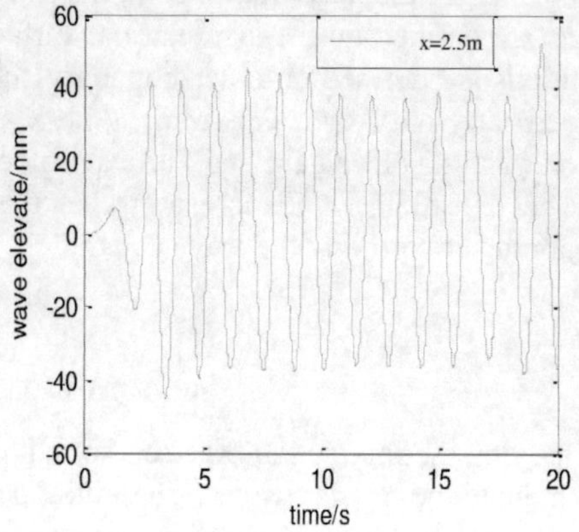

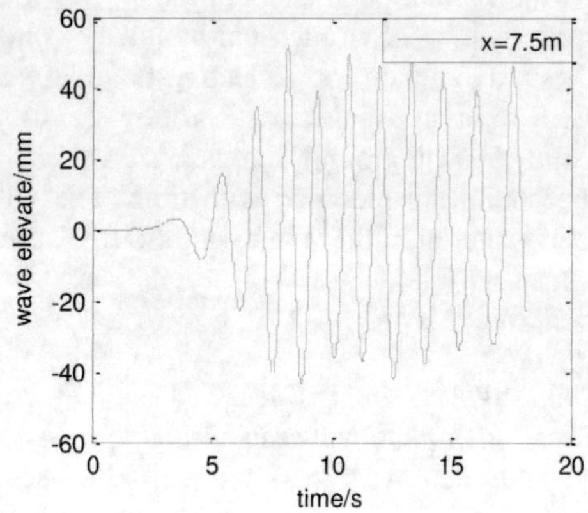

Figure 4. Wave Elevation Time-History Curves at Positions x=2.5m and x=7.5m

5. Model Experiment

According to Furude similarity criterion, the model of Twin-Float seaplane is designed. The seaplane is suspended on the motion control system by wire rope, and the initial attitude angle and roll angle are adjusted by front, back, left and right support rods, just because of this the model can kept stable during the acceleration process. After the model is installed, the initial height of the model from the water surface is debugged through the vertical motion mechanism. During the test, the trailer is started, the model is accelerated to a certain horizontal speed, and then the wire rope is disconnected by using the electromagnetic throwing device to make the model impact freely on the wave surface. The model landing wave experiment of a Twin-Float seaplane is shown in Fig. 5.

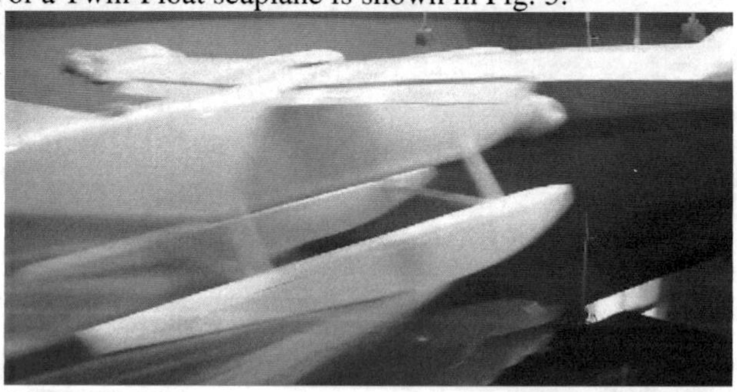

Figure 5. Wave landing test model of Twin-Float seaplane

6. Results and Discussion

6.1. Calculation Cases

In order to study the effect of wave height on impact loads, the parameters in Tab. 2 are the same except wave height. Horizontal velocity corresponds to stall velocity when the surface of the real aircraft glides down 0.88g under overload, and vertical velocity corresponds to the sinking velocity of the real aircraft 1.0m/s. Height wave ratio is less than 1/20, which accords with the theory of wavelet wave under linear assumption. The bottom step of the float is sliding surface, and the broken step is near the centre of gravity. Therefore, it is safer for the seaplane to occur the step and impact water than for the bow and stern of the float then the attitude of the seaplane is 5 ~7 degree. So the initial attitude of seaplane is 6 degree in this paper.

Table 2. Initial state of wave landing of seaplane

cases	Wave Height/m	Wave Length/m	Attitude /Degree	Horizontal velocity/m/s	Vertical velocity/m/s
1	0.04	5.0	6	15.19	0.236
2	0.06				
3	0.08				
4	0.10				

6.2. Motion characteristics

According to the velocity and displacement curves of reference points, the landing motion characteristics of seaplane under different wave heights are analyzed. Velocity and displacement curves reflect the motion characteristics of the aircraft in underwater waves. As shown in Fig. 5-6, the higher the wave height, the greater the response of landing motion. After the float encounters waves for the first time, it bounces, reaches the maximum vertical displacement, glides for a certain distance under the action of aerodynamic lift and gravity, and then encounters waves for the second time. Obviously, the motion response of the first encounter wave is larger than that of the second encounter wave. During the second encounter wave, the motion response of different wave heights is not much different.

Fig. 6 also shows that the higher the wave height, the larger the bouncing motion of the float after the first wave encounter, but the later the second wave impact.

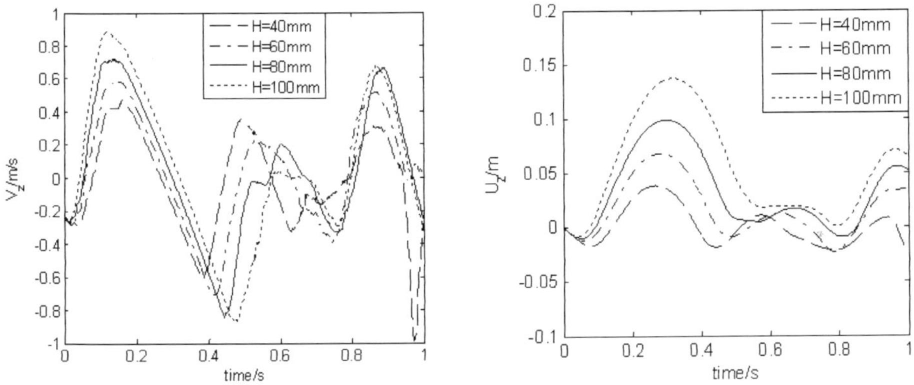

Figure 6. Vertical displacement and velocity curves of seaplane at different wave heights

6.3. Load characteristics

The displacement curve can be double differentiated to obtain the vertical acceleration curve of the seaplane with time, as shown in Fig. 7. The figure also shows the landing experiment results of a Twin-Float seaplane model with a wave height of 40mm and a wave length of 5 m. Through comparative analysis, the vertical acceleration curve obtained by numerical simulation is in good agreement with the data collected by the acceleration sensor of the model experiment. It can be seen in Fig. 7 that the first wave landing process of the seaplane model has been completed in about 0.15 seconds. So the water process is a dynamic process of instantaneous impact. During the landing experiment of seaplane, the impact point between the model and the wave surface has a high randomness. When numerical simulation is carried out, the initial state of the model is just in contact with the water surface and hull baseline close to the wave surface. So the acceleration by the landing of the numerical simulation increases rapidly. In the wave landing experiment, the seaplane does not necessarily impact the wave surface when it touches the water. So the acceleration growth rate by model experiment is slower than that in numerical simulation.

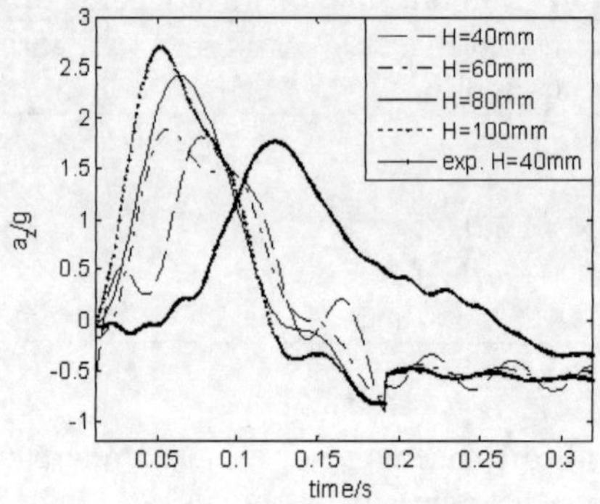

Figure 7. Comparison curves of vertical acceleration for different wave heights and model experiment

7. Conclusions

In this paper, the CEL method in ABAQUS/Explicit solver is used to simulate the push-plate wave-making and the interaction between float and water. The accuracy of numerical wave-making is validated by the wave elevation recognition technique developed by free surface volume fraction. The model experiment results are compared with the simulation results of wave landing, which verifies the accuracy of the simulation results of wave landing.

The landing results of the seaplane at different wave heights and the same wave length are analyzed. The seaplane's wave landing causes obvious bouncing phenomena, and the higher the wave heights, the greater the motion response. Generally speaking, the motion response of the first encounter wave is larger than that of the second encounter wave, and the higher the wave height, the later the second encounter wave is.

References

[1] HUANG Ling-cai, YONG Ming-pei. Key technologies and industrial application prospects of amphibian aircraft [J]. Acta Aeronautica ET Astronautica Sinica , 2019,40(1):18-34 (in Chinese)

[2] LUO Lin-yin, YANG Shi-fu, LV Ji-hang. Analysis and numeral simulation of water landing response model for amphibian [J]. Journal of Machine Design, 2013, 30(08):86-89. (in Chinese)

[3] MA Zeng-hui, LIU Li-sheng, YUE Zhen, et al. Numerical investigation on seakeeping performance of amphibious aircraft landing on waves [J]. Journal of Computational Mechanics, 2018, 35(3):380-386. (in Chinese)

[4] CHU Lin-tang, WU Bing, WANG Min-Zhen, etal. Theoretical calculation and experiment on impact loads of landing of winging round aircraft on water surface [J]. Acta Aeronautica ET Astronautica Sinica, 2016, 37(12):3698-3705. (in Chinese)

[5] SUN Pei-cheng, GAO Xiao-peng, DONG Zu-shun, Simulation analysis of amphibious aircraft water landing [J]. Ship & ocean engineering, 2014, 43(06):171-174. (in Chinese)

[6] Yao Xiao-hu, Huang Yu-tai, Ou Zhi-cheng, et al ,CEL algorithm-based analysis of dynamic characteristics of amphibious aircraft landing on water [J]. Journal of South China University of Technology (Natural Science Edition) ,2015,43(06):110-115. (in Chinese)

[7] WANG Xiao-hui, CHU Xue-sen, FENG Guang. Numerical simulation of sphere's water-entry based on Coupled Eulerian-Lagrangian method [J]. Journal of Ship Mechanics, 2018,22(07):838-844. (in Chinese)

[8] M. Ahmadzadeh,B. Saranjam,A. Hoseini Fard,A.R. Binesh. Numerical simulation of sphere water entry problem using Eulerian–Lagrangian method [J]. Applied Mathematical Modeling, 2014, 38(5-6).

[9] Liu Yong-jun. The Design of Y-12 Floatplane variant [D].Nanjing University of Aeronautics and Astronautics,2008 (in Chinese)

[10] SHANG GUAN Zi-ning, HAN Duan-feng, MA Qing-wei. Application of SPH-ALE method in numerical wave making problem [J]. Journal of Ship Mechanics, 2015, 19(Z1):43-51. (in Chinese)

[11] CHENG Shen-Tao. Wave numerical simulation based on linear wave theory[C]. China Ocean Engineering Society. Papers of the 17th China Ocean (Coastal) Engineering Symposium (Part II) China Ocean Engineering Society: Ocean Engineering Branch of China Ocean Society, 2015:5. (in Chinese)

IOP Conf. Series: Materials Science and Engineering **692** (2019) 012025 doi:10.1088/1757-899X/692/1/012025

Research on Machining Fixture Layout Optimization for Near-Net-Shaped Jet Engine Blade

Kaiyao Zhang [1, 2], Dongbo Wu [2 *] and Jiao Wang [1]

[1] Yantai University. Yantai, Shandong 264005, P. R.China

[2] State Key Laboratory of Tribology, Beijing Key Lab of Precision/Ultra-Precision Manufacturing Equipment and Control, Department of Mechanical Engineering, Tsinghua University, Beijing, China 100084, P.R. China.

[*]Corresponding author's e-mail: wdb4451253@126.com

Abstract. The design of machining fixture layout optimization for near-net-shaped jet engine blade is a major problem in the blade processing and manufacturing process. In this paper, the mathematical model of machining fixture layout optimization is established, and the influence of the clamping point position on the machining deformation of the blade-fixture system is analysed. Firstly, the mathematical model of machining fixture layout optimization is established according to the process requirements and clamping characteristics. Secondly, the positioning clamp points of machining fixture is optimized by the combination of finite element analysis (FEA) and genetic algorithm under the optimization target of the minimum value of the maximum deformation of the blade. Finally, cutting experiments are carried out to verify the reliability of the optimal layout fixtures, and the feasibility of the algorithm is verified by comparing with the minimum value obtained by the traditional Monte Carlo statistical method.

1. Introduction

The manufacturing of aero-engines precision forging blade is transforming towards to intelligent manufacturing. In the process of precision forging blade, the traditional process is that the low-melting alloy casting process is used as the positioning clamping tool for machining processing, which is not only seriously pollution, but also unable to meet efficiency requirements. Therefore, the clamping scheme of multi-point directly positioning is a good choice for the current blade precision process [1, 2]. Fixture design and optimization will be a major problem.

Many scholars have conducted research on the optimization design of thin-walled clamping fixture. Gene et al. proposed a dynamic model and analysis method for workpiece-fixture system, and analysed the influence of three parameters of fixture size, position and sequence on fixture [3]. HAMEDI et al. studied a nonlinear finite element analysis system combining artificial neural network and genetic algorithm, and analysed the optimal clamping force of fixture [4]. Wang et al. used FEA to optimize the clamping deformation of thin-walled curved parts [5]. Qin established a mathematical model based on the historical dependence of friction caused by friction, solved the influence of clamping force and clamping sequence on deformation by FEM [6]. Zhou et al. used genetic algorithm to optimize the fixture layout and clamping force of fixture [7, 8].

Based on the above analysis, it can be seen that the current research has focused on the optimization of the fixture for the simple aluminum alloy shapes (arc or bottomless). There is no

IOP Conf. Series: Materials Science and Engineering **692** (2019) 012025 doi:10.1088/1757-899X/692/1/012025

optimized analysis method for the precision-forged blade fixture of titanium alloy materials. In fact, there are still many problems in the low-melting alloy casting fixture for replacing the precision forging blade. Although the new fixture solves the problems of pollution and processing efficiency, it also reduces the rigidity of the workpiece- fixture system, and introduces large blade deformation during the process.

2. Fixture layout optimization model

In the process of precision forging blade Tenon root, the new fixture mainly adopts multi-point positioning and direct clamping to clamp the blade body. Among them, the precision forged blade is a thin-walled material whose surface is curved, which is titanium alloy (elastic modulus 110GPa), and the clamping element uses PEEK-GF30 (elastic modulus 7GPa).

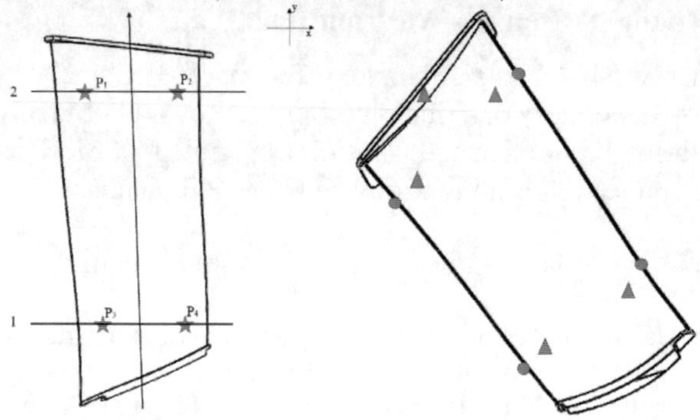

(a) (b)

Figure 1 Positioning clamp point layout of fixture

In actual machining, the blade types are varied and varied, so a new type of fixture is required for each blade. In this paper, a three-stage stator compressor blade is taken as the example, and its containment boundary is 200x100x5mm. The clamping of the new fixture consists of 4 clamping points, 5 positioning points and 4 auxiliary positioning points, as shown in Figure 1. During the processing of the blade Tenon root, the fixture mainly clamps the blade body by four clamping points at the back of the blade. In the selection of the initial four clamping points, the blade body reference point in the blade precision forging coordinate system is empirically selected, but in the actual blade boring process, the position layout of the four clamping points is determined for the blade- fixture system stiffness. Therefore, rationally arranging the position layout of the four clamping points of the new fixture can effectively improve the stiffness of the blade-tooling system, thereby improving the processing quality of the blade Tenon root.

2.1 Model assumptions

To facilitate the final problem solving, make the following assumptions:

(1) During the processing of the blade Tenon root, the rigidity of the system is insufficient. The blade is subjected to the cutting load under the condition of clamping and will be elastically deformed in the longitudinal direction. Therefore, the model can be simplified and simplified into 8 cylinders to clamp the blade.

(2) The contact areas of the four clamping points are the same, and the clamping force is the same.

(3) The blade Tenon root is subjected to a time-varying cutting load during the actual machining process, but the blade deformation is greatest when the load acts on the blade Tenon root. Therefore, it is assumed that a static load is applied at the Tenon root of the blade [9].

2.2 Finite element model

A finite element model is established and solved by genetic algorithm.

(1) Meshing the simplified blade- fixture system. There are 21,700 nodes and 10070 units.

(2) The boundary conditions of the finite element model are:

- There are 8 elastically constrained contact faces between the simplified blade and the fixture component, wherein $\boldsymbol{P_i}$ (i=1~4) is in contact with the blade as frictional contact, friction coefficient μ=0.48, $\boldsymbol{Q_j}$ (j =1~4) Contact with the blade is defined as a binding contact.
- Defining $\boldsymbol{Q_j}$ (j=1~4), the bottom end of the four cylinders does not produce displacement. The four cylinders at $\boldsymbol{P_i}$ (i=1~4) are clamped by 2200N pressure, and the 8 N force acts perpendicularly at the end of the blade Tenon root. Simulate external loads.

(3) Using finite element software to solve the blade deformation.

2.3 Mathematical modelling

The blade Tenon root is affected by the time-varying cutting load during the actual machining process. Therefore, the dynamic equation in the classical mechanic's theory can be used to describe the deformation during the machining process caused by the cutting force, as shown in equation (1).

$$[M](\ddot{d}) + [C](\dot{d}) + [K]\{d\} = \{F_{(t)}\} \tag{1}$$

In order to facilitate the solution of the model, the dynamic process is transforming into a static process under the under the assumption of the model. That is, the influence of inertia and damping on the blade- fixture system is not considered, and the stiffness matrix is continuous. Therefore, equation (1) can be converted to linear static structure analysis, as shown in equation (2).

$$[K]\{d\} = \{F_{(t)}\} \tag{2}$$

In order to determine the optimal position of the fixture layout, the layout of each clamping position within the blade containment boundary is optimized (the initial position of the blade clamping point is shown in $\mathbf{P_1}$, $\mathbf{P_2}$, $\mathbf{P_3}$ and $\mathbf{P_4}$ in Figure 1(a), which is the precision forging coordinate. The reference point of the blade body under the system. The position of the clamping point is assumed in Table 1,

Table 1 Blade body clamping point position

Clamping point	P_1	P_2	P_3	P_4
Position	x_1, y_1	x_2, y_2	x_3, y_3	x_4, y_4

The minimum value of the maximum deformation amount of the blade Tenon root is selected as the objective function in all cases, as shown as Equation 3

$$d = \min(\max(|d_1|, |d_2|, \ldots, |d_k|, \ldots, |d_n|)) \tag{3}$$

where, d represents the optimal deformation amount, d_i represents the deformation amount of the blade during machining process, and n represents the number of selected samples, and the larger n indicates the closer to the actual working condition.

The position change range of the position point is as shown in Equation 4.

$$\begin{gathered} 24.2mm \le x_1 \le 29.7mm, \quad 2.5mm \le y_1 \le 3.1mm \\ 26mm \le x_2 \le 31.8mm, \quad 1.3mm \le y_2 \le 1.6mm \\ 21.3mm \le x_3 \le 26mm, \ 103.7mm \le y_3 \le 126.7mm \\ 21.3mm \le x_3 \le 26mm, \ 100.1mm \le y_4 \le 122.3mm \end{gathered} \tag{4}$$

The cutting force load is shown in Equation 5.

$$\mu|F_{ti}| \ge \sqrt{F_{xi}^2 + F_{yi}^2} \quad i = 1, 2, 3, 4 \tag{5}$$

F_{ti} is the clamping force required for to clamping force, F_{xi}, F_{yi} are the tangential force between the clamping element and the blade body, and μ is the friction factor.

IOP Conf. Series: Materials Science and Engineering **692** (2019) 012025 doi:10.1088/1757-899X/692/1/012025

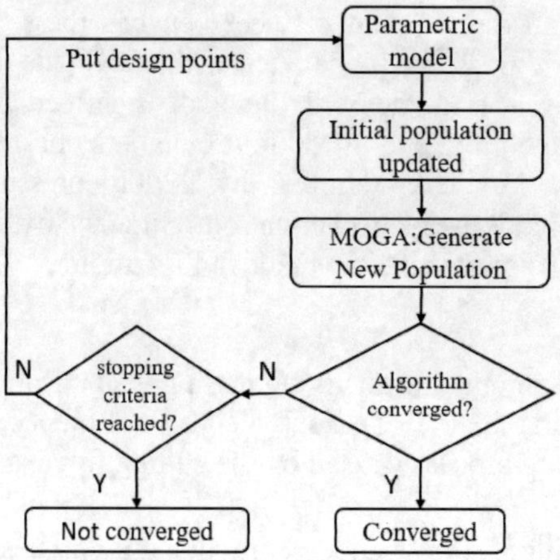

Figure 2 Flow chart of the best solution for optimizing blade tooling

2.4 Model solving

The genetic algorithm is simply a randomized search method that simulates the survival of the fittest and the survival of the fittest. The genetic algorithm is used to determine the clamping component layout of the blade fixture. The position of the clamping component needs to be randomly reorganized within the containment boundary, and combined with the finite element method to solve multiple sets of solutions, and finally the best blade fixture layout is obtained. Solve the flow chart of the optimal layout scheme of the blade tooling, as shown in Figure 2.

In traditional genetic algorithms, the choice of roulette is used to select the appropriate chromosome. However, the operator's ability to select is not high, and it is prone to premature convergence. Pareto proposed the concept of Pareto Optimal in 1896, and now ANSYS replaces the choice of roulette with the Pareto advantage. When goals and constraints conflict with each other and there is no ideal group, the Pareto advantage provides a set of solutions. Always sacrifice at least one target or constraint group while improving at least one other target or constraint. This paper uses the Paste advantage to solve the model.

3. Experimental verification

Table 2 shows the best position and the fixture layout obtained by combining the genetic algorithm and the finite element solution. Figure 3 shows the convergence trend of the optimization algorithm.

Table 2 The optimized clamping point position

Parameter	Candadate point 1	Candadate point2	Candadate point3
X_1	29.634	29.583	29.582
Y_1	2.6693	2.7404	2.7402
X_2	31.174	31.092	30.829
Y_2	1.4384	1.3685	1.5778
X_3	21.84	21.625	21.994
Y_3	110.79	105.31	104.46
X_4	22.643	21.362	21.949
Y_4	115.35	101	110.75
Deformation	0.10438	0.10618	0.10646

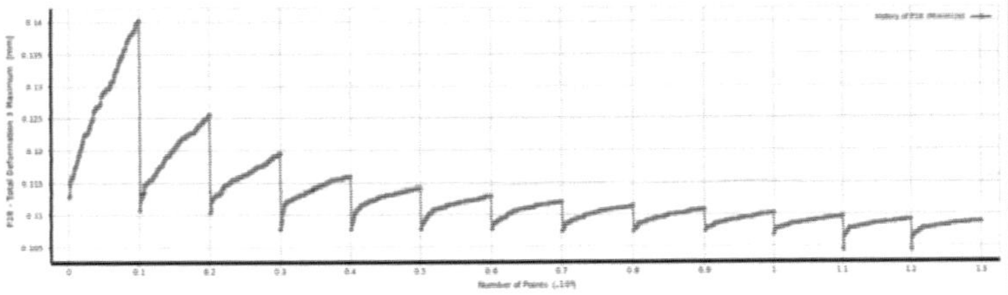

Figure 3 Convergence trend

In order to verify that the optimized fixture has good dynamic behavior, the cutting experiment analysis of the fixture is performed. Figure 4 shows the fixture structure at different positioning positions. Four sets of typical positions are taken for the cutting dynamics comparison experiment. Fixture C is designed according to the optimized positioning point.

Figure 4 Four different sets of fixture layouts

The experimental site is shown in Figure 5. The laser displacement sensor is used to measure the dynamic response signal of the fixture during the cutting process. Under the same cutting process parameters, the same cutting force is obtained. By comparing the changes of the displacement response, the fixture stiffness is analysed.

Figure 6 shows the change of the displacement response signal during the cutting process. Comparing the four fixtures, it can be analyzed that the displacement response of the fixture C is the smallest, so the stiffness of fixture C is significantly better than other fixtures. This also shows that the proposed fixture optimization design analysis method is reliable.

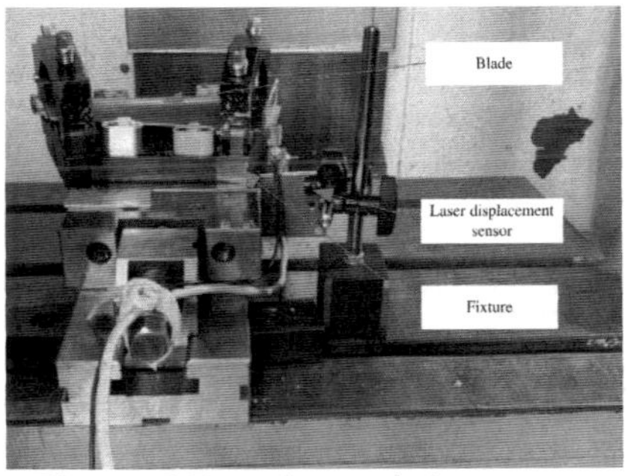

Figure 5 Experimental site

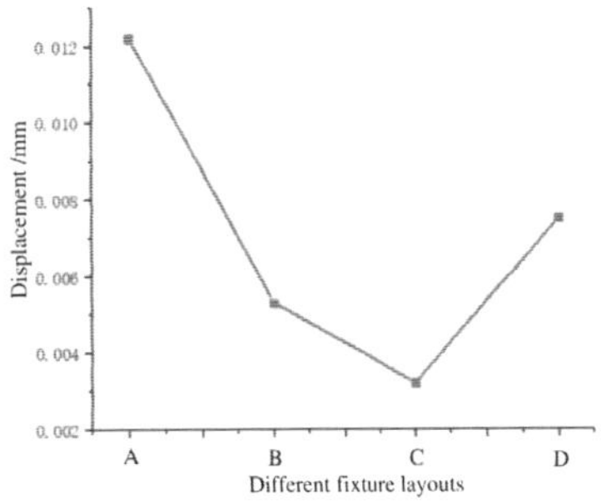

Figure 6 Displacement response signal

4. Conclusion

In this work, the design of machining fixture layout optimization is analysed by mathematical model, FEA and genetic algorithm. It is verified by experiments that the obtained optimal fixture is reliable, the rigidity of the fixture is obviously improved, and the proposed analysis method is reliable.

Acknowledgements

The author would like to acknowledge the support and contributions of our colleagues in Xi'an Aero-Engine (Group) Ltd. National Natural Science Foundation of China (Grant 51575310). The

Project was supported by The Tribology Science Fund of State Key Laboratory of Tribology (SKLTKF18B12).

References

[1] Wang H, Wu B H, Li X Q. Advanced Machining Technology of New Generation Commercial Aeroengine Blade[J]. Aeronautical Manufacturing Technology, 2014,20:29-30. (in Chinese)
[2] Wu D B, Wang H, Zhang K Y, et al. Research on Formation Mechanism and Optimization Method of Surface Waviness of TC4 blisk blade[J]. Journal of Manufacturing Processes, 2019, 35:305-326.
[3] Y.J. Gene Liao, S. Jack Hu. Flexible multibody dynamics based fixture-workpiece analysis model for fixturing stability[J]. International Journal of Machine Tools & Manufacture, 2000,40: 343-362.
[4] MOHSEN HAMEDI. Intelligent Fixture Design through a Hybrid System of Artificial Neural Network and Genetic Algorithm[J]. Artificial Intelligence Review, 2005, 23:295-311.
[5] Wang Y Q, Mei Z Y, Fan Y Q. Finite Element Optimization of Machining Fixture Layout of Thin-walled Arc Workpiece[J]. Chinese Journal of Mechanical Engineering, 2005, 41(6): 214-217.
[6] Qin G H, Wu Z X, Zhang W H. Analysis and Control Technique of Fixturing Deformation Mechanism of Thin-walled Workpiece[J]. Chinese Journal of Mechanical Engineering, 2007, 43(4): 211-216.
[7] Zhou X L, Zhang W H, Qing G H, et al. On Optimizing Fixture Layout and Clamping Force Simultaneously Using Genetic Algorithm[J]. Mechanical Science and Technolgy, 2005,24(3): 339-342.
[8] Chen W F, Ni L J, Wang N S. Investigation on Optimization Method for Fixture Layout and Clamping Forces[J]. China Mechanical Engineering, 2007, 18(12):1413-1417.
[9] Zhang K Y, Wu D B, Guo X F, et al. High-resolution tenon distortion test system for blade processing[J]. Mechanical Science and Technolgy, 2019, 04.

IOP Conf. Series: Materials Science and Engineering **692** (2019) 012026 doi:10.1088/1757-899X/692/1/012026

Comparative Analysis of Power Plants Using Low Potential Heat of Liquefied Natural Gas (LNG)

Uglanov Dmitry, Blagin Evgeny, Panshin Roman, Tereshchenko Olga and Biryuk Vladimir

Samara University, Department of Heat Engineering and Heat Engines, 443086, Moskovskoyeshosse, 34, Samara, Russia

*E-mail: dmitry.uglanov@mail.ru

Abstract. The idea of using low-potential energy of cryogenic liquids has recently become more widespread. This is primarily due to the increasing production of cryogenic substances, as well as the general trend of energy saving and the energy potential existing in the cryoproduct in the form of energy previously used for its liquefaction.

Low-potential energy is the energy of a colder environment, having the same pressure as the consumer's energy converter. Currently, one of the most frequently used cryogenic substances is liquefied natural gas (LNG), and methods of regasification and use continue to improve.

This paper proposes a method for a comparative analysis of the use of liquefied natural gas as a working fluid, a study of the efficiency of the regasification process. The main parameters of energy installations were calculated and the main advantages and disadvantages were identified on the basis of the results, and the most optimal one was selected.

1. Introduction

Currently, natural gas is becoming one of the most important sources of energy. Natural gas in a compact form for transportation and storage may be contained in compressed (gaseous) and liquefied states. The most relevant is the use of liquefied natural gas (LNG) as an energy carrier for the same purposes as ordinary natural gas, since it allows reducing the weight and size characteristics of storage and transportation tanks.

LNG makes it possible to re-gasify objects remote from main pipelines by creating a reserve of LNG directly at the consumer, avoiding the construction of expensive pipeline systems.

In a liquefied state, natural gas at a pressure of 1 atmosphere is 600 times denser than under normal conditions. LNG makes it possible to gasify facilities remote from main pipelines by creating a reserve of LNG directly at the consumer, avoiding the construction of expensive pipeline systems.

2. Regasification process

Usually, the regasification of liquefied natural gas (LNG) occurs due to the heat of the environment. It is considered that the use of environmental heat is not an energy-consuming process. However, it should be borne in mind that considerable energy was used to liquefy LNG (about 1 kWh of electricity per 1 kg of LNG), which is simply discharged into the environment with this method of gasification. Thus, liquefied natural gas (like any other cryoproduct) contains the energy potential that could be used when it returns to its original gaseous state, and, therefore, the LNG regasification process (figure

1) itself has significant energy saving potential.

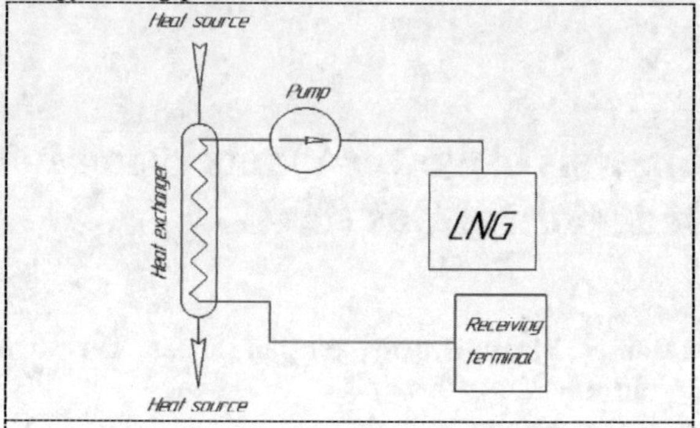

Figure 1. Regasification process.

For maximum compensation of the energy spent on liquefaction, the process of regasification of the cryoproduct can be carried out using any thermodynamic (figure 2) cycle (most often, the open Rankin cycle).

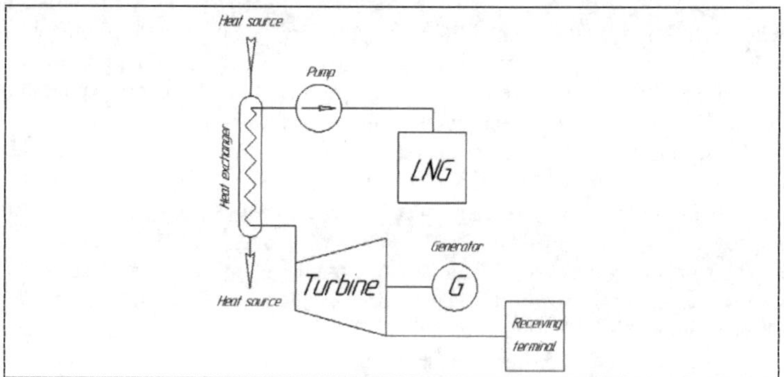

Figure 2. Regasification process using open thermodynamic cycle.

3. Comparative analysis of power plants using low-potential LNG heat

To assess the effectiveness of using low-potential heat of cryogenic substances, the following parameters are used:

- Carnot Efficiency - a parameter showing the maximum efficiency:

$$\eta_{Carnot} = 1 - \frac{T_1}{T_2}; \quad (1)$$

where T_1 и T_2 – are the lowest and highest temperatures.

- Thermal efficiency - shows the efficiency of heat conversion into the work of a cycle:

$$\eta_{Term} = \frac{N_\Sigma}{Q_\Sigma}; \quad (2)$$

where N_Σ – the total capacity of the entire plant; Q_Σ – total heat.

- Exergic Efficiency - a parameter showing energy efficiency:

$$\eta_{Ex} = \frac{N_\Sigma}{E_{\Sigma\, ex}}; \quad (3)$$

where $E_{\Sigma\, ex}$ – total exergy installation of power plant.

- Specific operation of LNG is a parameter indicating the amount of work of a cryogenic substance (LNG) from the entire installation operation:

$$L_{LNG} = \frac{N_\Sigma}{G_{LNG}} * \left(\frac{E_{LNG}}{E_{\Sigma\, ex}}\right); \quad (4)$$

where G_{LNG} – LNG consumption in power plant; E_{LNG} – LNG exergy.

- The energy recovery coefficient of the cryoproduct - is a parameter indicating the fraction of the return of the energy of the cryogenic substance used for gasification.

$$k_B = \frac{W}{W_{exp} * G_{LNG}}. \quad (5)$$

where W - amount of energy obtained from gasification of LNG; W_{exp} – energy spent to liquefy a cryogenic substance (0,6-0,8 $\frac{kWh}{kg}$).

The objects of comparison will be the schemes of power plants that use low-grade heat LNG, described in an array of articles [1-15]. Each of the schemes uses a different thermodynamic cycle or a combination of several cycles. Also, in the schemes apply various methods of regasification of the cryoproduct. The analysis of the design features of each scheme is presented in Table 1. The estimated parameters are presented in Table 2.

Table 1. Constructive analysis analysis of power plants.

№ power plant (PP)	*heat sources*	working fluids	thermodynamic cycles	advantages	disadvantages
[1]	Sea water	LNG, propane	Closed Rankine cycle, direct extension cycle	High rates of LNG specific work, high power, a supply of heat without burning fuel	The complicated structure, low efficiency, low energy conversion efficiency, the need for proximity to sea-water source
[2]	Warm environment	LNG	Direct extension cycle	High rates of LNG specific work, heat supply without burning fuel	Low efficiency, low power, low energy conversion efficiency
[3]	Warm environment	LNG, nitrogen	Open the Rankine cycle, closed Brayton cycle	High rates of LNG specific work, a supply of heat without burning fuel	Low efficiency, low energy conversion efficiency
[4]	Secondary heat (exhaust gas)	LNG, water	Open and closed Rankine cycle	High rates of LNG specific work, high efficiency	Low rates of energy conversion efficiency
[5]	Secondary heat (exhaust gas)	LNG, R116, R227ea	Closed Rankine cycle	High rates of LNG specific work, high rates of energy conversion efficiency	The complicated structure, low efficiency
[6]	Secondary heat (exhaust gas)	LNG, water-ammonia mixture	Closed Rankine cycle	Simple design, proven technology	Low efficiency, the low energy conversion and a low share of LNG operation.
[7]	Burned fuel, secondary heat (exhaust gas)	LNG, air, nitrogen	Brayton open cycle, open and closed Rankine cycle	The high efficiency, high energy conversion indicators and LNG specific work	Complicated construction
[8]	Burned fuel	LNG, water vapor	Closed Rankine and Brayton cycle	High efficiency, high power	The complicated structure, low rates of LNG specific work
[9]	Solar radiation	LNG, CO_2, water vapor	Open Rankine cycle	Heat supply without burning fuel	Complicated construction, low efficiency, low energy conversion rates, low LNG specific work rates
[10]	Secondary heat	LNG, air, water vapor	Brayton cycle using SOFC, Rankin closed cycle	The highest rates of energy conversion efficiency, high efficiency, high rates LNG specific work	Complex construction, unprocessed production technology, part of LNG is used as fuel

[11]	Warm environment	LNG	Closed Brayton cycle	LNG is a working fluid	Low efficiency, low energy conversion efficiency
[12]	Air	LNG, water-ammonia mixture	Open and closed Rankine cycle	High rates of LNG specific work	Low efficiency, low energy conversion efficiency, low power
[13]	Water vapor	LNG, CO_2	Closed Rankine cycle	High rates of LNG specific work, high efficiency	Low rates of energy conversion efficiency
[14]	Burned fuel	LNG, air	Open Brayton cycle	High rates of LNG specific work, high power	Complicated construction
[15]	Secondary heat (exhaust gas)	LNG, air	Open Brayton cycle	High power	Low efficiency, low energy conversion efficiency, part of LNG is burned as fuel

Table 2. The estimated parameters of power plants.

Power plant	T_{min} (K)	T_{max} (K)	G_{LNG} ($\frac{kg}{s}$)	N_{Σ} (kWh)	η_{Carnot} (%)	η_{Term} (%)	η_{Ex}(%)	L_{LNG} ($\frac{kJ}{kg}$)	k_B(%)
PP-1 [1]	111	300	2.175	418.4	63	14.4	13.6	150.6	8.9
PP-2 [2]	112	300	0.06	36.42	62.7	31.7	40.5	433.7	10.65
PP-3 [3]	111.67	537.15	1	311.41	81.2	23.8	22.62	268.65	14.42
PP-4 [4]	111.67	440.15	16.36	7488.7	74.6	59.2	34.8	296.51	12.23
PP-5 [5]	171.15	393.15	8.3	2821.1	56.5	45.5	65.1	226.33	15.73
PP-6 [6]	111.5	473.13	2.463	433.83	76.4	17.2	23.8	296.31	8.15
PP-7 [7]	113	1563	65.03	217.1	92.8	82.6	74.9	316.21	13.7
PP-8 [8]	108	1573	0.028	2765.9	93.1	59.4	73.8	8.403	4.39
PP-9 [9]	111.52	359.4	0.131	9.1273	69	9.16	10.1	5.776	10.72
PP-10 [10]	113.67	1312.6	4.68	4310.4	91.3	67.2	97.3	725	16.9
PP-11 [11]	111.5	288	1	164	61.2	16.7	24.1	164	4.7
PP-12 [12]	111.5	473	0.12	46.86	76.4	34.4	28.7	305.5	9.3
PP-13 [13]	111	923	1	293.15	88	79	42.1	230	13.6
PP-14 [14]	111.5	1573	267.5	191.67	92.9	60.2	45.5	208.96	1.7
PP-15 [15]	111.5	1523	100	148.23	92.7	31.8	20.6	219.9	9.4

4. Regasification process

Comparative analysis means by a comparison of the energy characteristics of different plants (figure 3). However, it is obvious that for an installation in which heat is supplied at a higher temperature, the thermal efficiency, other things being equal, should be higher than that of an installation in which heat is supplied at a lower temperature. The criterion that shows the ratio of the level of temperatures for the supply and removal of heat is the limiting efficiency of the Carnot cycle. From the obtained values, we plotted and analyzed the point diagrams of dependences of the thermal (figure 4 a) and exergic

IOP Conf. Series: Materials Science and Engineering **692** (2019) 012026 doi:10.1088/1757-899X/692/1/012026

efficiency (figure 4 b) and the specific work of LNG (figure 4c) on the Carnot efficiency. The efficiency of the Carnot Cycle shows the ultimate efficiency of the cycle depending on the temperature level. Obviously, with a larger value of the marginal efficiency, the value of the thermal efficiency should be greater. However, the specific work of LNG does not depend on the upper temperature level, but on the parameters of the LNG state used in each specific cycle. Accordingly, it is expected that the total data array $\eta_{Term} = f(\eta_{Carnot})$ and $\eta_{Ex} = f(\eta_{Carnot})$ will be located in an ellipsoidal region, the greater semi-axis of which will have a slope corresponding to $0 < \frac{\eta_{Term}}{\eta_{Carnot}} < 1$, and $L_{LNG} = f(\eta_{Carnot})$ - will be located in the ellipsoidal region, the large semi-axis of which is a horizontal line.

■ PP-1	● PP-2	● PP-3	× PP-4	✱ PP-5
● PP-6	+ PP-7	□ PP-8	◇ PP-9	△ PP-10
× PP-11	✱ PP-12	o PP-13	+ PP-14	■ PP-15

Figure 3. Color designation of the various proposed schemes installations using LNG.

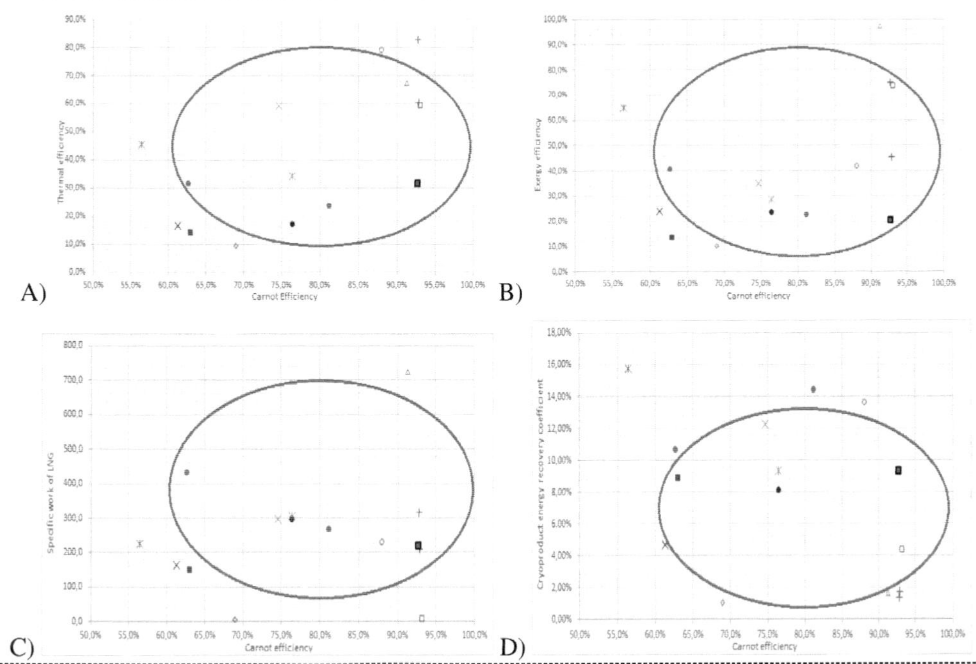

Figure 4. calculation results: (a) - the dependence of the thermal efficiency and efficiency of Carnot; (b) - exergy efficiency; (c) - specific work of LNG; (d) - energy return coefficient..

Based on the results obtained, it can be noted that the dual-circuit scheme using a solid oxide fuel cell stands out from the others with the highest rates of LNG specific work, while having high efficiency. This scheme can be taken as optimal for detailed inspection.

5. Conclusion

The compiled method of comparative analysis allows to evaluate the efficiency of utilization of low-grade heat of liquefied natural gas, taking into account the peculiarities of the working process. Different installations were analyzed, general tendencies of indicators dependencies were found. The features of the working process of these plants were analyzed, their advantages and disadvantages were identified. The installation with the use of a solid oxide fuel cell turned out to be the most optimal.

IOP Conf. Series: Materials Science and Engineering **692** (2019) 012026 doi:10.1088/1757-899X/692/1/012026

References

[1] LNG power recovery, Proceedings of the Institution of Mechanical Engineers Part A Journal of Power and Energy, June 1994;

[2] Estimation of the energy efficiency of cryogenic filled tank using in different system and devices, Applied Thermal Engineering, March 2016;

[3] On the recovery of LNG physical energy, Exergy, An International Journal, 2002;

[4] A combined power cycle utilizing low-temperature waste heat and LNG cold energy, Energy Conversion and Management, March 2009;

[5] Thermodynamic analysis and optimization of a two-stage organic Rankine cycle for liquefied natural gas cryogenic energy recovery, International Journal of Mechanical and Production Engineering, Aug. 2016;

[6] Thermodynamic analysis and optimization of an ammonia-water power system with LNG (liquefied natural gas) as its heat sink, Energy, February 2013;

[7] Advanced exergetic analysis of a novel system for generating electricity, Energy, February 2010;

[8] Thermodynamic analysis of a Brayton cycle and Rankine cycle arranged in series exploiting the cold exergy of LNG (liquefied natural gas), Energy, 1 March 2014;

[9] Thermodynamic analysis of a transcritical CO_2 power cycle driven by solar energy with liquified natural gas as its heat sink, Applied Energy, April 2012;

[10] Thermodynamic analysis an SOFC-GT-ORG integrated power system with liquefied natural gas as heat sink, Procedia Engineering, 2017;

[11] Thermodynamic analysis of direct expansion configurations for electricity production by LNG cold energy recovery, Frontiers in Energy, 2016;

[12] Energy analysis of a combined power cycle using low-grade heat source and LNG cold energy, International Journal of Exergy, January 2015;

[13] A transcritical CO_2 Rankine cycle with LNG Cold Energy Utilization and Liquefaction of CO_2 in gas Turbine exhaust, J. Energy Resour. Technol, Nov 12, 2009;

[14] Analysis of a combined cycle power plant integrated with a liquid natural gas gasification and power generation system, Proceedings of the Institution of Mechanical Engineers Part A Journal of Power and Energy, January 2010;

[15] Novel Gas Turbine Cycle with Integration of CO_2 Recovery and LNG Cryogenic Exergy Utilization, ASME International Mechanical Engineering Congress & Exposition, November 17-22, 2002, New Orleans, Louisiana.

IOP Conf. Series: Materials Science and Engineering **692** (2019) 012027 doi:10.1088/1757-899X/692/1/012027

Calculation of the Energy Complex Based on a Steam Gas Installation Assessment of the Parameters of the Contour of the Auxiliary Steam Power Plant Installation

E.S.Gaev, Panshin R.A., Shimanov A.A., Tereshchenko O.V. and Shatochin E.V. and D.A. Uglanov.*

Samara University, Department of Heat Engineering and Heat Engines, 443086, Moskovskoyeshosse, 34, Samara, Russia

*E-mail: dmitry.uglanov@mail.ru

Abstract. In this paper, we calculate the parameters of the energy complex on liquefied natural gas (LNG), which serves for the gasification of LNG with its further utilization in the circuit of a steam turbine plant, during operation of which additional electrical energy is generated. We evaluate the effectiveness of regasification of cryoproducts and evaluate the potential of using Cold Energy.

Various schemes of power plants using Cold Energy are considered. The parameters and characteristics of the combined-cycle plant consisting of the gas-turbine circuit and the steam-turbine circuit are calculated. The coefficient of cryoproduct energy return (liquefied natural gas) of the considered schemes is calculated and and based on the data obtained, the most optimal one is selected.

1.Introduction

In many systems of aerospace technology, transport, as well as in energy complexes for various purposes, cryogenic liquids are currently widely used.

When storing a cryoproduct, losses of the cryoproduct occur due to the lack of thermal insulation and heat leakage from the environment. During storage of the cryoproduct due to the ablation of the mass, its heat content changes. You can calculate the loss of low-temperature heat cryogenic product per day, lost during storage, according to the following formulas:

$$\Delta m = m_{\Sigma} \cdot \frac{\Theta}{100} \tag{1}$$

$$Q_{eva} = \Delta m \cdot r + c_p \cdot \Delta m \cdot (T_{\kappa} - T_{н}) \tag{2}$$

where Δm - mass of evaporated cryoproduct per day, m_{Σ} - mass of stored cryoproduct; Θ - loss of product from evaporation per day, % Q_{evo} - low potential heat of evaporated cryoproduct; r - cryoproduct evaporation heat; T_{κ} – cryoproduct temperature; T_o – environment temperature.

He evaporating cryoproduct has a rather high certain potential of low-temperature energy [1-2], which can be used as a “refrigerator” in a heat engine operating in the Carnot cycle (Figure 1). The loss of available energy L_{ava} per day, provided that the low-grade heat of the cryoproduct is used according to the Carnot cycle, is calculated as follows:

$$L_{ava} = \frac{Q_{evo} \cdot \eta}{1-\eta} \quad (3)$$

$$\eta = 1 - \frac{T_c}{T_e} \quad (4)$$

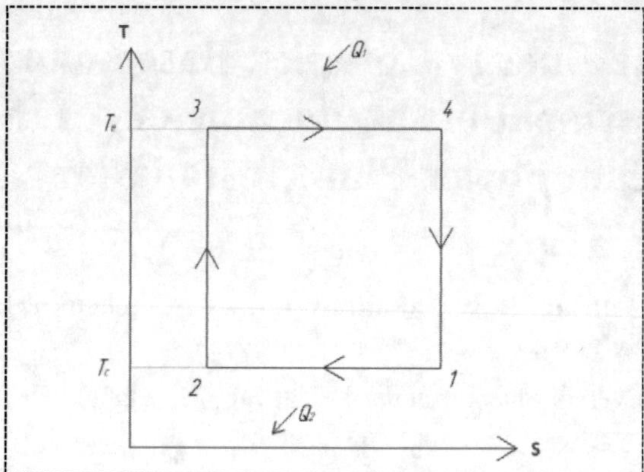

Figure 1. Thermodynamic cycle of heat engine.

The solution to the problem of the loss of a cryoproduct (for example, for LNG) during its storage and, accordingly, the loss of its disposable low-grade heat can be a low-temperature power plant based on liquefied natural gas, during operation of which additional electrical energy is produced [3-4].

The proposed power plant can provide consumers with electrical and thermal energy. The combined-cycle plant includes the contour of the gas turbine and steam turbine plants. The use of this power plant allows for higher energy efficiency during storage of LNG [5].

The gas turbine unit contour includes a compressor (C), a combustion chamber (CC), a gas turbine (GT), and an electric generator (EG). The steam-turbine plant includes a waste heat boiler (HB), a steam superheater, a steam turbine (ST), a heat exchanger-condenser (EC), a condensate pump (CP), and an electric generator (EG).

The power plant is shown in Figure 2.

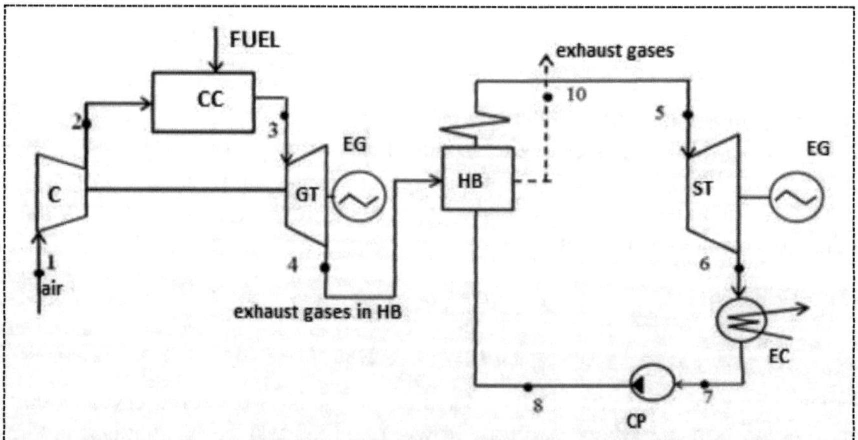

Figure 2. Scheme of power plant generating additional energy.

2. Calculation of parameters of the cycle circuit gas turbine installation

During evaporation, part of the LNG through the reducer enters the combustion chamber of a small-sized gas turbine unit. The compressor increases the air pressure and feeds it further into the combustion chamber. During the combustion of the air-fuel mixture in the combustion chamber, the gas temperature increases. Next, the gas enters the turbine, where its expansion occurs, with the result that we obtain effective (mechanical) work on the turbine shaft. One part of the efficient work is spent on driving the compressor. Another part of the effective work, with the help of the generator is converted into electrical energy.

The initial data for the calculation is as follows:

The flow rate of the working fluid G_{CH4}=0,06 kg/s. Point 1: air pressure at the inlet p_{in} = 0,1 MPa, temperature T_{in} = 288 K.

The cycle of the gas turbine is shown in Figure 3:

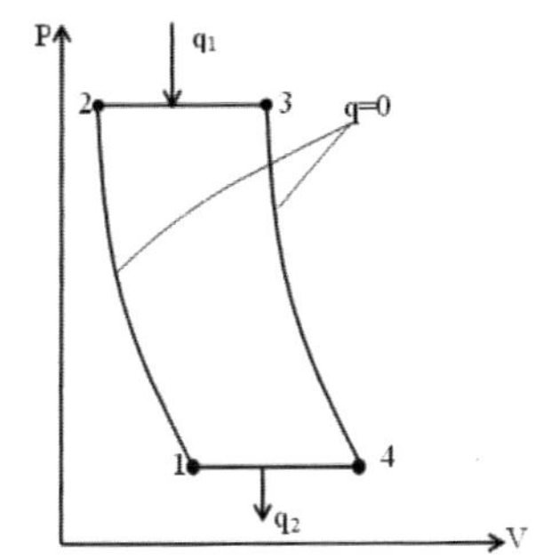

Figure 3. The gas turbine installation cycle in pv - coordinates.

The specific volume is found from the equation of state of an ideal gas:

$$p_1 \cdot v_1 = R \cdot T_1 \tag{5}$$

where it comes from: $v_1 = \frac{RT_1}{p_1}$,

where R – gas constant:

$$R = \frac{\mu \cdot R}{\mu}, \tag{6}$$

$$v_1 = \frac{RT_1}{p_1}. \tag{7}$$

Point 4: (4-1) - isobar process of heat removal

Specific volume will be found from the equation of state of an ideal gas at point 4:

$$v_4 = \frac{R \cdot T_4}{p_4}. \tag{8}$$

Point 3: (3-4) – adiabatic expansion of the working fluid in a gas turbine GTP;

$T_3 = 1500$ K;

The ratio of the parameters of the adiabatic process:

$$\frac{T_3}{T_4} = \left(\frac{p_3}{p_4}\right)^{\frac{k-1}{k}}, \tag{9}$$

where k - adiabatic index:

$$k = \frac{c_p}{c_v}, \tag{10}$$

c_p – is the mass heat capacity at constant pressure, c_v – is the mass heat capacity at constant volume.

According to the molecular kinetic theory:

$$c_p = \frac{i+2}{2} R, \tag{11}$$

$$c_v = \frac{i}{2} R, \tag{12}$$

where i – is the number of degrees of freedom of the molecule ($i = 5$ for a diatomic gas);

R – air gas constant.

Point 2: (2-3) – Isobaric process of heat supply in the combustion chamber of the GTP.
$p_3 = p_2 = 1{,}57MPa;$

(1-2) – adiabatic air compression in a gas turbine compressor. The ratio of parameters in the adiabatic process:

$$\frac{T_2}{T_1} = \left(\frac{p_2}{p_1}\right)^{\frac{k-1}{k}} \rightarrow T_2 = T_1\left(\frac{p_2}{p_1}\right)^{\frac{k-1}{k}}, \tag{13}$$

Specific volume will be found from the equation of state of gas at point 2:

$$v_2 = \frac{R \cdot T_2}{p_2} \tag{14}$$

The values of average isobaric heat capacities are selected from tabular data.
Specific amount of heat applied per 1 kg of working fluid in a gas turbine unit:

$$q_1 = c_p \cdot T_3 - c_p \cdot T_2, \tag{15}$$

Specific amount of heat abstracted from 1 kg of working fluid in a gas turbine unit:

$$q_2 = c_p \cdot T_4 - c_p \cdot T_1, \tag{16}$$

Theoretical work of 1 kg of working fluid in a gas turbine unit:

$$l_0^{GTP} = q_1 - q_2, \tag{17}$$

Fuel flow $G_{fuel} = 0{,}06\ kg/s$, air flow $G_{air} = 15 \cdot G_{\text{т}} = 0{,}9\ kg/s$.
Then the total flow rate $G_{\Sigma} = 0{,}06 + 0{,}9 = 0{,}96\ kg/s$.
GTU power:

$$W_{GTU} = l_0^{GTU} \cdot G_{\Sigma}. \tag{18}$$

3. Calculation of parameters of the cycle loop of a steam turbine installation

Exhaust gases leaving the gas turbine plant are fed to the waste-heat boiler, where part of the heat of the gases escapes with the flue gases and the other part enters the steam turbine. In the steam turbine, the expansion of gases occurs, resulting in effective mechanical work on the turbine shaft. This work with the help of an electric generator is transformed into electrical energy. The exhaust gases leaving the steam turbine enter the heat exchanger-condenser, where they are condensed — they change from a gaseous state to a liquid state. Next, from the heat exchanger, with the help of a pump-condenser, the liquid flows back into the waste heat boiler.

The cycle of the steam turbine installation is presented in Figure 4.

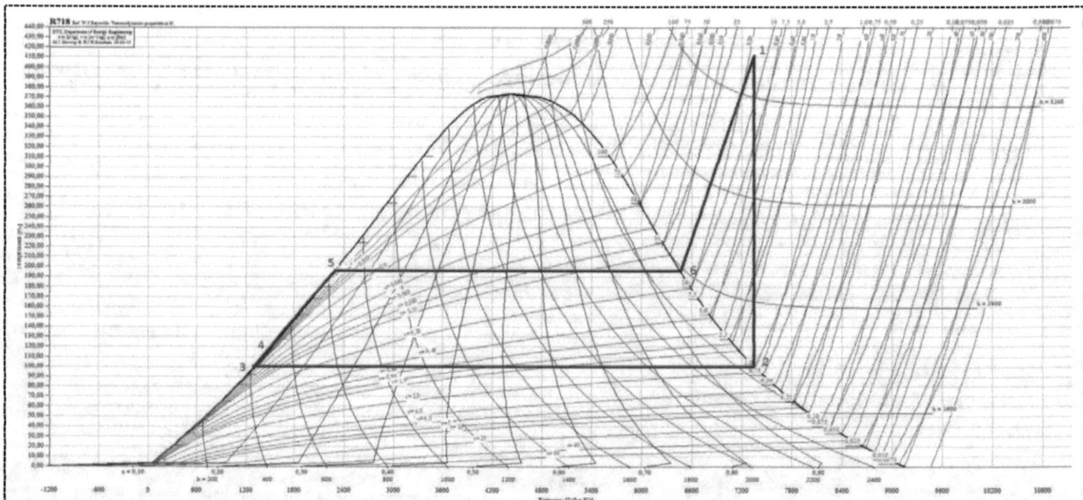

Figure 4. Steam turbine installation cycle.

IOP Conf. Series: Materials Science and Engineering **692** (2019) 012027 doi:10.1088/1757-899X/692/1/012027

Then the turbine is equal to:

$$L_T = i_1 - i_2 \tag{19}$$

Inlet and outlet heat:

$$q_1 = i_1 - i_4 \tag{20}$$

$$q_2 = i_2 - i_3 \tag{21}$$

Heat balance of the heat recovery boiler:

$$G_\Sigma \cdot (c_p \cdot T_4 - c_p \cdot T_{fg}) = G_п \cdot (c_{pв} \cdot (T_{bw} - T_{in}) + r), \tag{22}$$

where T_{fg} - is the flue gas temperature, T_{bw} – temperature of the boiling water, r - is the specific heat of vaporization, T_{in} - is the feed water temperature, c_{pw} - is the specific heat capacity of water,
Then steam flow:

$$G_{ST} = \frac{G_\Sigma \cdot (c_p \cdot T_4 - c_p \cdot T_{fg})}{c_{pw} \cdot (T_{bw} - T_{in}) + r}, \tag{23}$$

Steam turbine power:

$$W_{ST} = L_T \cdot G_{ST} \tag{24}$$

Combined-cycle power:

$$W_{CC} = W_{GTP} + W_{ST}, \tag{25}$$

Thermal efficiency steam cycle:

$$\eta_{tCC} = \frac{W_{GTP} + W_{ST}}{G \cdot q_{1GTP}} \tag{26}$$

The installation diagram is shown in Figure 5:

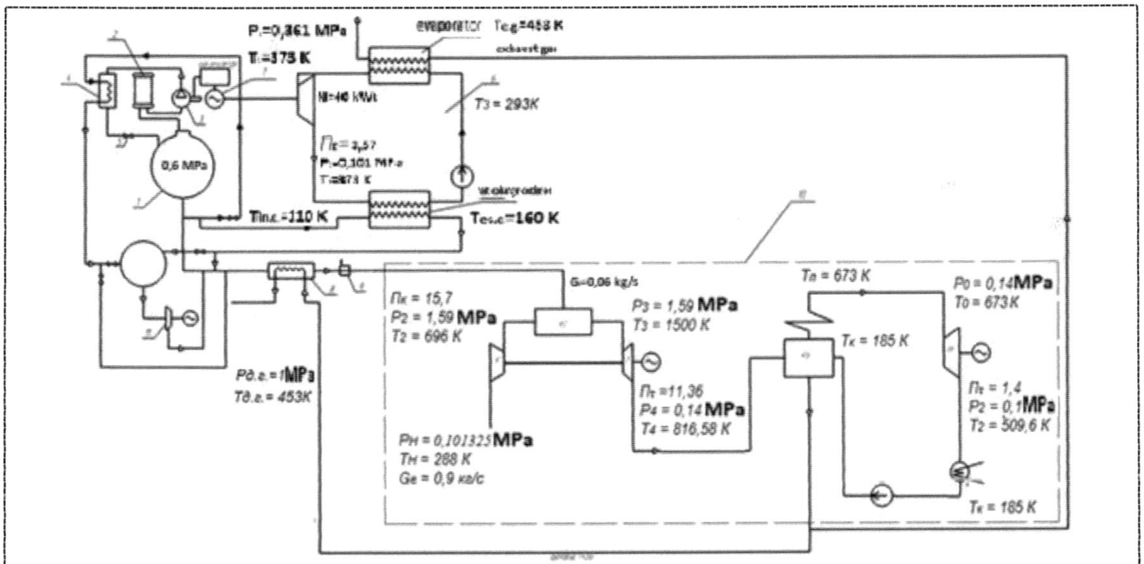

Figure 5. Schematic diagram of the installation based on thermal power plant and combined-cycle plant.

1 - Tank with liquefied natural gas (LNG); 2 - Gas receiver; 3 - Compressor; 4 - Heat exchanger-evaporator; 5 - Throttle valve; 6 - Auxiliary steam power plant; 7 - electric generator; 8 - Heat exchanger; 9 - Reducer; 10 - Steam-gas unit (SGU); 11 - Expansion turbine.

4. Alternative problem-solving schemes

Another scheme to reduce the loss of cryoproduct is the installation with a contour based on a Stirling engine (Figure 6).

IOP Conf. Series: Materials Science and Engineering **692** (2019) 012027 doi:10.1088/1757-899X/692/1/012027

The principle of operation of the circuit based on the Stirling engine is as follows. The compressor cavity of the Stirling engine is cooled with liquefied natural gas to a temperature of 158 K, and the expander cavity is heated from the exhaust gases of a gas piston installation to a temperature of 400 K. The actual engine power is 4.8 kW. Liquefied gas, passing through the heat exchanger of the compressor cavity of the Stirling engine, enters the heat exchanger-heat exchanger, where it is heated by exhaust gases from the combined-cycle plant. After complete gasification of the gas it is used to the combustion chamber of the combined-cycle plant.

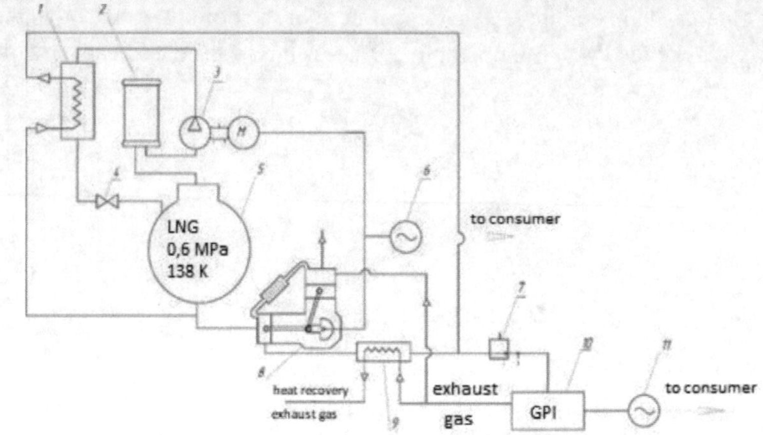

Figure 6. Schematic diagram of a low-temperature power plant based on a Stirling engine.
1 - Heat exchanger-evaporator; 2 - Gas receiver; 3 - Centrifugal compressor; 4 - Throttle valve; 5 - Tank with liquefied natural gas (LNG); 6 - electric generator; 7 - Reducer by 0.1 MPa; 8 - Stirling engine; 9 - Heat exchanger-utilizer; 10 - Gas piston installation (GPI); 11 - Electric generator.

It is possible to use a thermo-acoustic motor in the installation circuit as the main energy converter (Figure 7).

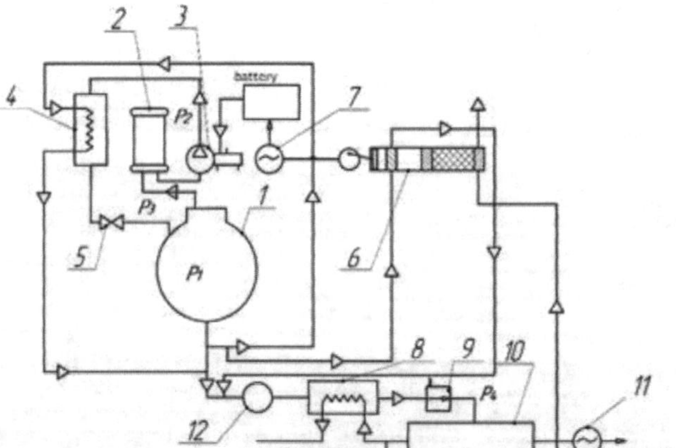

Figure 7. Schematic diagram of a low-temperature power plant based on a thermo-acoustic engine.
1 - Tank with liquefied natural gas (LNG); 2 - gas receiver; 3 - centrifugal compressor; 4 - evaporator heat exchanger ;; 5 - throttle valve; 6 - Thermo-acoustic engine based on traveling wave; 7 - electric generator; 8 - heat exchanger; 9 - gearbox; 10 - gas piston installation (GPI); 11 - electric generator; 12 - receiver.

This circuit works as follows: the cold heat exchanger of the thermo-acoustic engine 6 is cooled with liquefied natural gas from the tank 1, and the hot heat exchanger is heated by the exhaust gases of the hcp. Thus, the engine 6 generates energy, which is converted into electrical energy by an electric generator 7, which drives the compressor 3. LNG, passing through the cavity of the thermoacoustic engine 6, enters the heat exchanger 8, where it is heated by exhaust gases from the gas piston unit 10, then enters in the gearbox 9, which maintains the pressure at the inlet to the gas piston installation 10 constant and equal to p4. After complete gasification of the gas, it is fed to the gas piston unit 10, which produces electrical energy using an electric generator 11. The actual engine power is 1.2 kW.

The use of such an installation allows you to compensate for losses during the evaporation of LNG from the tank, as well as to obtain additional electrical energy that can be used for various purposes, including solving problems arising from the operation of a gas piston installation.

5. Calculation of energy recovery coefficient

For the energy assessment of the use of the low-temperature potential of the cryogenic product, an indicator is used that characterizes the fraction of the return energy from that which was previously spent on the liquefaction of the cryoproduct: (calculation results for the different schemes are shown in Table 1):

$$k = \frac{N_\Sigma}{W_{liq}} = \frac{N}{l \cdot G} \tag{27}$$

Table 1. The calculation results of the main parameters of the the power plants.

	Installation with a combined-cycle circuit	Installation with a Stirling engine	Installation with a thermo-acoustic engine
N, kWt	40	4,8	1,2
W_{liq}, kWt	129,6	129,6	129,6
k	0,308	0,037	0,009

6. Conclusion

As a result of this work, the amount of energy losses in storage systems of cryogenic products along cylindrical horizontal and vertical tanks was analyzed, energy potential losses were determined, as a result of which it was found that LNG has the greatest potential of cryogenic energy lost during storage. The proposed layouts for the organization of the LNG storage park, consisting of two, four and eight tanks, and determined that the scheme with two tanks provides the least mass loss of the cryoproduct. The weight loss of the cryoproduct for 28 days of work amounted to 1061.42 kg. The schemes of low-potential power plants that compensate for the loss of the cryoproduct during storage are considered. The calculation of the axial compressor of the gas turbine installation was carried out; as a result, the following were determined: parameters and geometrical dimensions of the compressor; the absolute speed at the exit from the impeller of the first stage is 185.73 m/s; the flow rate at the outlet of the first stage guide vane is 134.4 m/s; the absolute speed at the exit from the impeller of the last stage is 168.77 m/s; the flow rate at the exit from the guide vane of the last stage is 67.5 m/s; velocity triangles for the first and last stages are constructed; compressor power is 363.6 kW. The parameters of the gas turbine of the gas turbine unit were calculated, with the result that the specific operation of the turbine is 710.07 kJ/kg, the power of the turbine is 632.37 kW. The parameters of the heat exchanger – condenser was calculated and determined: the heat exchange surface area — 0.33 m^2, heat transfer coefficient — 95.13 $\frac{W}{(m^2 \cdot K)}$; the total pressure drop in the annular space is 13.366 kPa. As a result of the economic evaluation, it was determined that the payback period of a low-grade installation is 6 years.

References

[1] Filin N.V. Liquid cryogenic systems [Text]: reference book / N.V. Filin A.B. Bulanov - L .: Engineering, Leningrad Branch, 1985. - 247 p.

[2] Xiong, Y. A novel CO2-capturing natural gas combined cycle with LNG cold energy utilization [Text]: Xiong, Y., Luo, P., Hua, B. // Energy Procedia.Volume 61. - 2014, Pages 899-903

[3] E.V. Blagin. Estimation of the energy efficiency of cryogenic filled tank using in different system and devices [Text]: E.V. Blagin, A.I. Dovgyallo, S.O. Nekrasova, D.V. Sarmin, D.A. Uglanov//Applied Thermal Engineering. - 2016. Volume 101, Pages 537-544

[4] Novotelnov V. N. Cryogenic machines [Text]: textbook for high schools / Novotelnov V. N., Suslov A. D., Poltaraus V. B. - St. Petersburg: Polytechnic, 199.- 335 s

[5] J. Wang. Thermodynamic analysis and optimization of a transcritical CO2 geothermal power generation system based on the cold energy utilization of LNG [Text]: J. Wang, J. Wang, Y. Dai, P. Zhao.// Applied Thermal Engineering Volume 70, Issue 1, 5 September 2014, Pages 531-540

Calculation and Determination of Energy Parameters of the Mini-CHP on the Basis of the Heat Pump

Grinuk U.I., Panshin R.A., Shimanov A.A., Tereshchenko O.V. and D.A. Uglanov*

Samara University, Department of Heat Engineering and Heat Engines, 443086, Moskovskoyeshosse, 34, Samara, Russia

*E-mail: dmitry.uglanov@mail.ru

Abstract In this paper, the calculation and determination of the parameters of a mini-CHP on the basis of a heat pump is carried out, using four different plant diagrams as the power supply for the heat pump. Thermodynamic calculation of the heat pump. Calculation of the efficiency of the heat pump. Analysis of structural schemes for generating electricity using micro-thermal power plants and based on heat pumps A comparative analysis of the effectiveness of various schemes with mini-CHP. An assessment of the capital costs of installation, operation and determination of the payback period of the mini-CHP on the basis of a heat pump with various plant diagrams is carried out and the most economically advantageous scheme is selected. The practical application of the studied structural schemes is evaluated.

1. Introduction

The proposed power plant can provide thermal energy settlement Priozerny, located in the Stavropol Territory, Shpakovsky district. A feature of this mini-CHP is that its autonomy is ensured through the use of four different schemes for providing a heat pump with electricity [1-9]:

- Gas piston installation on diesel fuel;
- Gas piston installation on liquefied natural gas (LNG);
- Wind power installation (wind turbine);
- Photoelectric converter (FEP).

2.Thermodynamic calculation of the basic system parameters

The purpose of thermodynamic calculation is to determine the efficiency indicators of the heat pump cycle. The section proposes a method for calculating heat pump cycles.

Initial data for calculation:

– heat load Q_{hl}...1000 kWt;
– temperature of low potential coolant (brine) at the heat pump inlet t_{p1}...................................5 °C;
– temperature of low potential coolant (brine) after the heat pump t_{p2}......................................2 °C;
– temperature of low potential coolant (hot water) at the heat pump inlet t_{w1}...........................45 °C;
– temperature of hot water after heat pump t_{w2}...55 °C;
– temperature of environment t_0...–1 °C;
– temperature drops at the outlet of heat exchangers: evaporator Δt_e, condenser Δt_c, subcooler Δt_{sc}..5°C;

IOP Conf. Series: Materials Science and Engineering **692** (2019) 012028 doi:10.1088/1757-899X/692/1/012028

– temperature of steam superheat in the intermediate heat exchanger Δt_{he}.............................20 °C.

The refrigerant used is R152a freon, which is ozone-safe.

Temperature of evaporation of freon:

$$t_{и} = t_{н2} - \Delta t_{и} = 2 - 5 = -3^{0}C\ ; \tag{1}$$

By evaporation temperature t_e = –3°C from the tables of thermodynamic properties of R152a refrigerant in the saturated state or p, h-diagram (Figure 1, 2) defines the parameters at point 1 - enthalpy on the right of the boundary curve h'' and the pressure p:

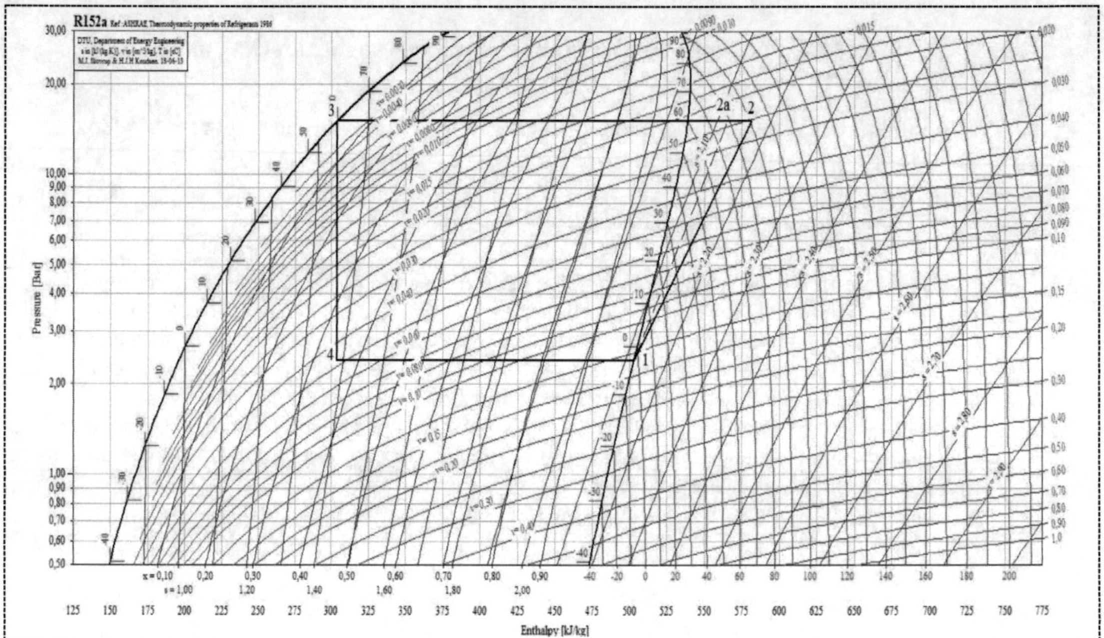

Figure 1. The cycle of vapor compression heat pump in the ph-diagram of the refrigerant R152a.

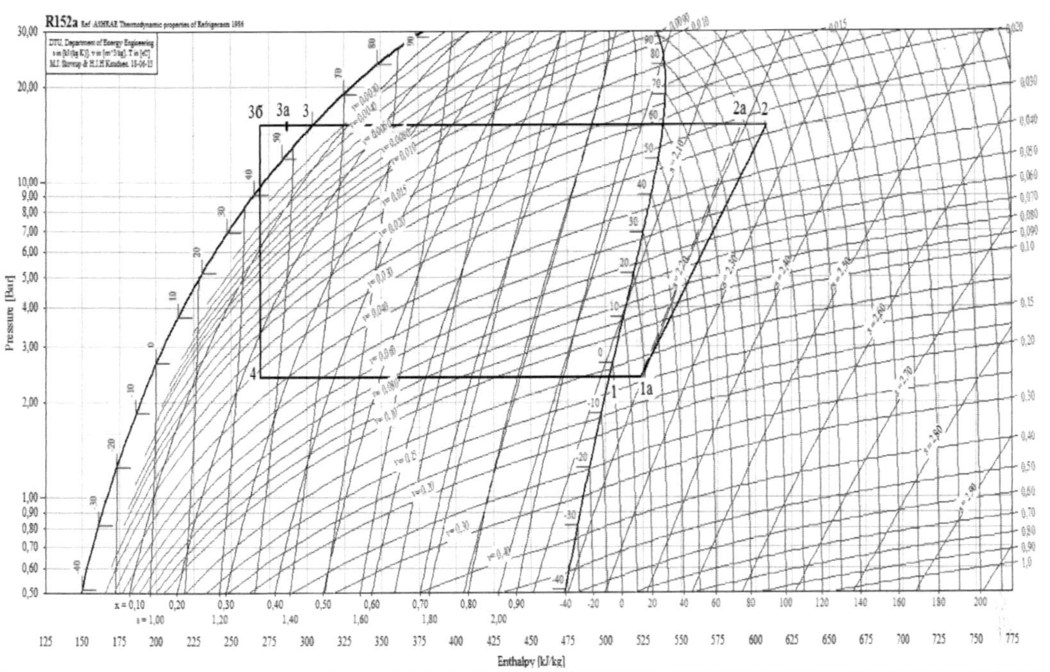

Figure 2. Heat pump cycle with intermediate heat exchanger and subcooler in ph-diagram of the refrigerant R152a.

$$h_1 = 502{,}73 \text{ KJ/kg}; \tag{2}$$

$$p_e = 0{,}24 \text{ MPa}; \tag{3}$$

point 1 is marked on the ph-diagram.

The condensation temperature of freon is determined as follows:

$$t_c = t_{w2} - \Delta t_c = 55 + 5 = 60^0 C; \tag{4}$$

1. By temperature of condensation t_c parameters at point 3 are determined from the tables of thermodynamic properties or from the ph-diagram — the enthalpy on the left boundary curve h' and the pressure p:

$$H_3 = 304{,}14 \text{ kJ/kg}; \tag{5}$$

$$p_c = 1{,}51 \text{ MPa}; \tag{6}$$

point 3 is marked on the ph-diagram.

2. On the ph-diagram, at the intersection of the constant entropy line S_1 passing through point 1 and the isobar line p_c, passing through point 3, point 2a is determined, then the enthalpy at this point is determined by the diagram:

$$H_{2a} = 565{,}56 \text{ kJ/kg}; \tag{7}$$

3. The adiabatic efficiency of the compressor η_a is determined as follows:

$$\eta_a = 0{,}98 \cdot \frac{273 + t_0}{273 + t_c} = 0{,}98 \cdot \frac{273 - 1}{273 + 60} = 0{,}8; \tag{8}$$

The enthalpy of freon after compression with allowance for losses is determined as follows:

$$h_2 = h_1 + \frac{h_{2a} - h_1}{\eta_a} = 502{,}73 + \frac{565{,}56 - 502{,}73}{0{,}8} = 581{,}3 \text{ kJ/kg}; \tag{9}$$

According to the value of enthalpy h_2=581,3 kJ/kg and pressure p_c = 1,51 MPa point 2 is marked on the diagram. Temperature at this point:

$$t_2 = 90 \text{ °C}; \tag{10}$$

IOP Conf. Series: Materials Science and Engineering **692** (2019) 012028 doi:10.1088/1757-899X/692/1/012028

4. According to the enthalpy $h_3 = h_4 = 304{,}14$ kJ/kg and pressure $p_e = 0{,}24$ MPa, point 4 is marked on the diagram.

5. Specific heat loads in the nodes of the heat pump:

$$q_e = h_1 - h_4 = 502{,}73 - 304{,}14 = 198{,}59 \text{ kJ/kg}; \quad (11)$$

$$q_c = h_2 - h_3 = 581{,}3 - 304{,}14 = 277{,}16 \text{ kJ/kg}; \quad (12)$$

$$l_{com} = h_2 - h_1 = 581{,}3 - 502{,}73 = 78{,}57 \text{ kJ/kg}. \quad (13)$$

The correctness of the calculation is determined by checking the heat balance:

$$198{,}59 + 78{,}57 = 277{,}16 \text{ kJ/kg}; \quad (14)$$

The heat load of the heat pump is determined as follows:

$$q_{hl} = q_c = 277{,}16 \text{ kJ/kg}; \quad (15)$$

The energy consumed by the electric motor W is determined by the formula:

$$W = \frac{l_{com}}{\eta_{el.m} \cdot \eta_{el}} = \frac{78{,}57}{0{,}95 \cdot 0{,}8} = 103{,}4 \text{ kJ/kg}; \quad (16)$$

6. Energy efficiency indicators of the heat pump:

– heat conversion factor:

$$\mu = \frac{q_c}{l_{com}} = \frac{277{,}16}{78{,}57} = 3{,}53; \quad (17)$$

– electric power conversion coefficient:

$$\mu_{el} = \eta_{el.m.} \cdot \eta_{el} \cdot \mu = 0{,}95 \cdot 0{,}8 \cdot 3{,}53 = 2{,}68; \quad (18)$$

– specific primary energy consumption:

$$EC = \frac{1}{\eta_{el.m.} \cdot \eta_{el} \cdot \eta_{el.s.} \cdot \eta_{prim}} \cdot \frac{1}{\mu} = \frac{1}{0{,}95 \cdot 0{,}8 \cdot 0{,}4 \cdot 0{,}95} \cdot \frac{1}{3{,}53} = 0{,}98; \quad (19)$$

Since EC <1, from the energy point of view, heating with the use of a heat pump is more profitable than when burning fossil fuel used for electricity generation.

7. The degree of pressure increase in the compressor is determined by the expression:

$$\pi_c^* = \frac{p_c}{p_e} = \frac{1{,}51}{0{,}24} = 6{,}29; \quad (20)$$

8. Produced exergy calculation circuit:

– logarithmic average temperature of the cold coolant

$$T_{mid.p} = \frac{t_{p1} - t_{p2}}{\ln \frac{t_{p1} + 273}{t_{p2} + 273}} = \frac{5 - 2}{\ln \frac{5 + 273}{2 + 273}} = 276{,}5\,K; \quad (21)$$

The results of the calculated options are shown in table 1.

Table 1. "Indicators of energy efficiency of the calculated options".

Scheme №	1	2
Specific heat load of the heat pump q_{hl}, kJ/kg	277,16	318,25
Specific energy consumed by an electric motor W, kJ/kg	103,4	111,84
The degree of pressure increase in the compressor $\pi_к^*$	6,29	
Heat conversion coefficient μ	3,53	3,74
electric power conversion coefficient μ_{el}	2,68	2,84
specific primary energy consumption EC	0,98	0,93
Exergetic efficiency η_{el}	0,411	0,444

For further calculations, choose the scheme № 2.

The mass flow rate of the refrigerant G_{ref} determined by the formula:

$$G_{ref} = \frac{Q_{hl}}{q_{hl}} = \frac{1000}{318,25} = 3,14 \, {}^{kg}\!/_{s} \, ; \tag{22}$$

Full load nodes heat pump is determined:

– in the evaporator:

$$Q_e = q_e \cdot G_{ref} = 233,22 \cdot 3,14 = 732,31 \, kWt \, ; \tag{23}$$

– in the condenser:

$$Q_{cond} = q_{cond} \cdot G_{ref} = 304,11 \cdot 3,14 = 954,91 \, kWt \, ; \tag{24}$$

– in the subcooler:

$$Q_{subc} = q_{subc} \cdot G_{ref} = 14,14 \cdot 3,14 = 44,4 \, kWt \, ; \tag{25}$$

– in the intermediate heat exchanger:

$$Q_{ihe} = q_{ihe} \cdot G_{ref} = 20,49 \cdot 3,14 = 64,34 \, kWt \, ; \tag{26}$$

9. Proposed installations for solving the problem of autonomy and their schemes[1-9].

3.GPI on diesel fuel

To ensure the autonomous operation of the mini-CHP one of the installations was chosen as a GPI on diesel fuel, this circuit arrangement is shown in Figure 3.

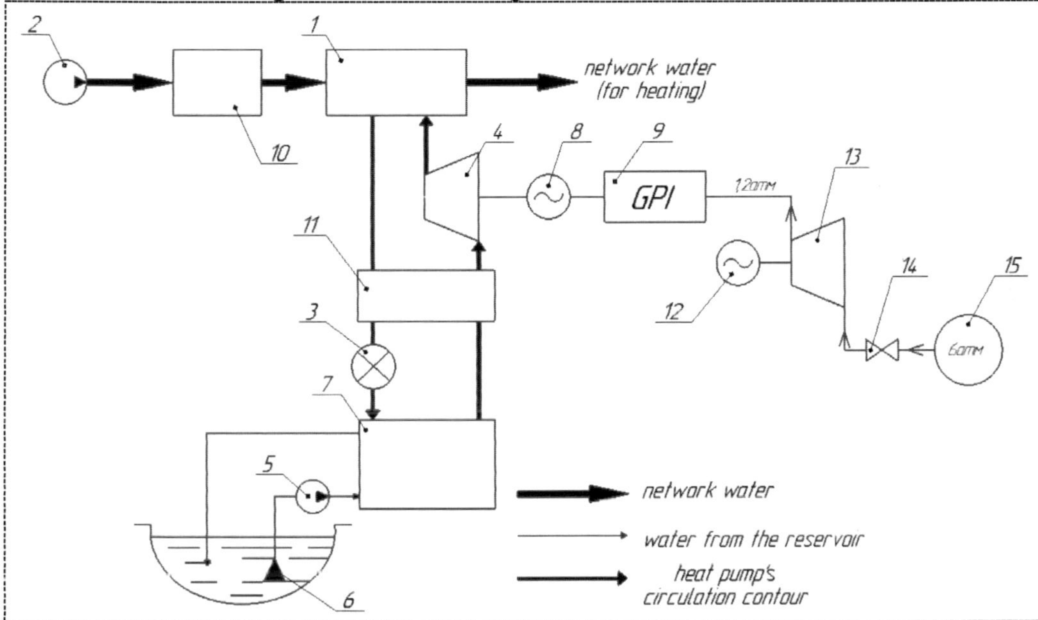

Figure 3. Scheme of mini CHP on the basis of a heat pump with the provision of compressor autonomy by a gas piston installation on diesel fuel.
1 - evaporator, 2 - circulation pump, 3 - throttle, 4 - compressor, 5 - circulation pump, 6 - intake device located in the non-freezing part of the reservoir, 7 - evaporator, 8 - electric generator, 9 - gas piston installation (GPI) operating on diesel fuel, 10 - subcooler, 11 - intermediate heat exchanger.

4.Gas piston installation on liquefied natural gas (LNG)

To ensure the autonomous operation of the mini-CHP one of the installations was chosen as a natural gas GPI, the scheme of this unit is shown in Figure 4.

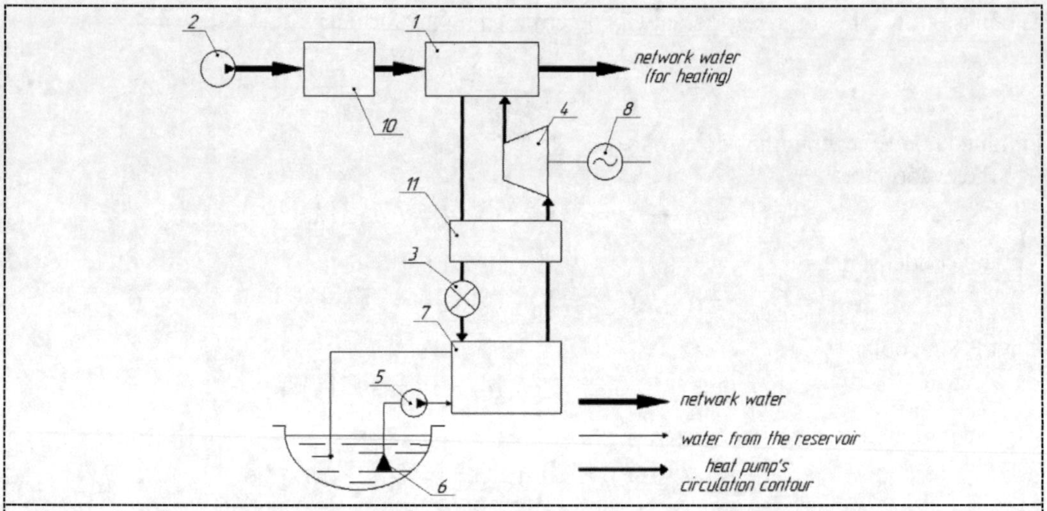

Figure 4. Scheme of mini CHP on the basis of a heat pump with the provision of compressor autonomy by a gas piston installation on liquefied natural gas (LNG).
1 - evaporator, 2 - circulation pump, 3 - throttle, 4 - compressor, 5 - circulation pump, 6 - intake device located in the non-freezing part of the reservoir, 7 - evaporator, 8 - electric generator, 9 - gas piston instalations (GPI) operating on liquefied natural gas (LNG), 10 - subcooler, 11 - intermediate heat exchanger, 12 - electric generator, 13 - expansion turbine, 14 - throttle, 15 - LNG cylinder.

5.Wind power installation (wind turbine)

To ensure the autonomous operation of the mini-CHP one of the installations was chosen wind turbine with a horizontal axis of rotation, the scheme of this installation is shown in Figure 5.

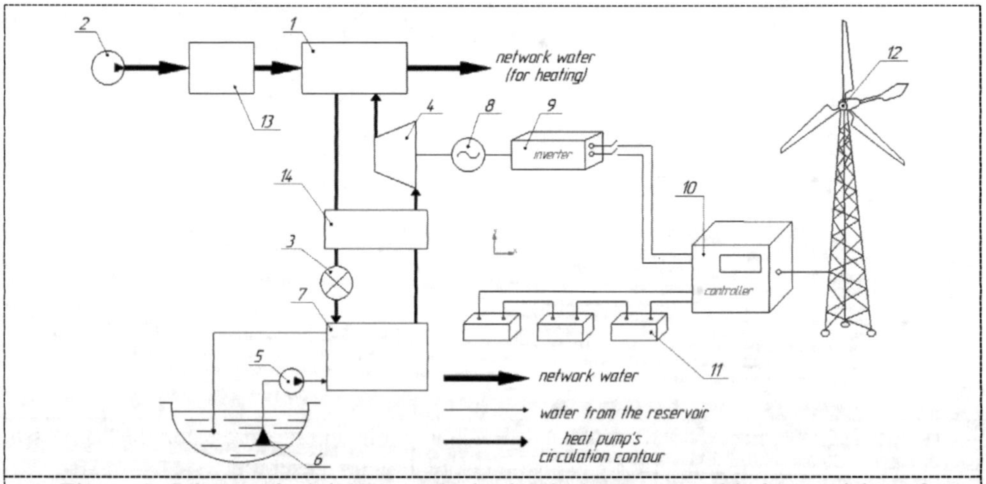

Figure 5. Scheme of mini CHP on the basis of a heat pump with the provision of autonomy of the com-pressor by a wind power installation.
1 - evaporator, 2 - circulation pump, 3 - choke, 4 - compressor, 5 - circulation pump, 6 - intake device located in the non-freezing part of the reservoir, 7 - evaporator, 8 - electric generator, 9 - inverter, 10 - controller, 11 - battery, 12 - wind power installation, 13 - subcooler, 14 - inter-mediate heat exchanger.

6. Photoelectric converter (FEP)

IOP Conf. Series: Materials Science and Engineering **692** (2019) 012028 doi:10.1088/1757-899X/692/1/012028

To ensure the autonomous operation of the mini-CHP one of the installations was chosen solar power plant based on a photoelectric converter, the scheme of this installation is shown in Figure 6.

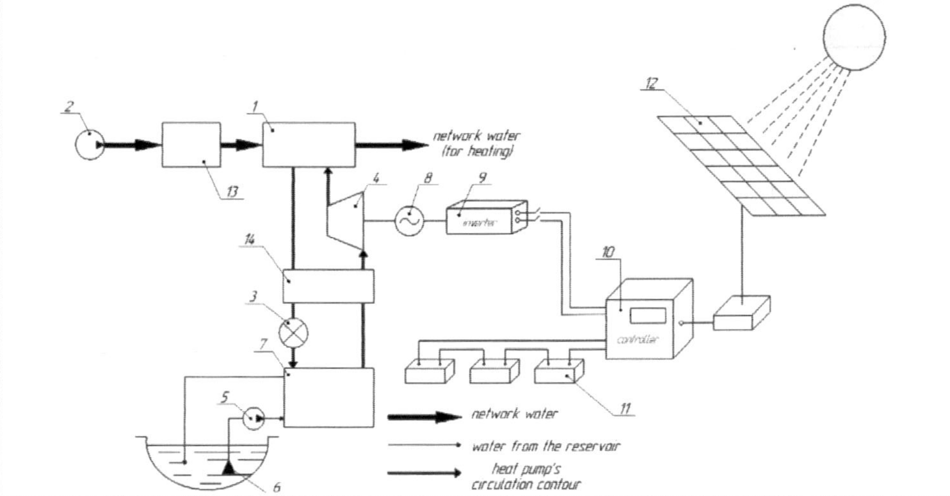

Figure 6. Scheme of mini CHP on the basis of a heat pump with the provision of compressor autonomy by a solar power plant based on photoelectric converters. 1 - evaporator, 2 - circulation pump, 3 - choke, 4 - compressor, 5 - circulation pump, 6 - water intake device located in the non-freezing part of the reservoir, 7 - evaporator, 8 - electric gen-erator, 9 - inverter, 10 - controller, 11 - battery, 12 - solar power station, 13 - subcooler, 14 - intermediate heat exchanger.

7. CONCLUSION

Thus, it is proposed to use mini-CHP on the basis of a heat pump, used as a heat supply system in the village of Priozerny, located in the Stavropol Territory, Shpakovsky district. Four systems are considered as a system for providing a heat pump with electric power: a liquefied natural gas combined-cycle plant, a diesel-fired combined-cycle plant, a wind power plant, a solar power station based on photoelectric converters.

For the heat pump in the Coolpack program, Carnot cycles were constructed, the cycle parameters were calculated for the vapor compression heat pump and for the heat pump with an intermediate heat exchanger and a subcooler. Refrigerant R152a was selected for this installation. Based on the obtained results, the optimal cycle was chosen - a heat pump with an intermediate heat exchanger and a subcooler with optimal values of the heat pump specific heat ($q_{TH} = 318,25 \, {}^{kJ}\!/_{kg}$) and heat conversion coefficient ($\mu = 3,74$).

The same was carried out calculation of the cost and payback period of these installations.

It is impossible to pay back the mini-CHP on the basis of the heat pump in combination with the GPI on diesel fuel, due to the high cost of diesel fuel and the further rise in prices for it.

The recoupment of the mini-CHP on the basis of the heat pump in combination with the GPI on liquefied natural gas will be 5 years.

The recoupment of the mini-CHP on the basis of a heat pump in combination with a wind power installation will be 27 years.

The recoupment of the mini-CHP on the basis of a heat pump in combination with a solar power station on the basis of solar cells will be 25 years.

REFERENCES

[1] Trubaev P.A. Heat pumps [Text]: Textbook. allowance / P.A. Trubaev, B.M. Grisho-ko. - Belgorod: Publishing House of BSTU. V.G. Shukhov, 2009. -- 142 p.
[2] Tsvetkov O.B., Laptev Yu.A., Barantsov A.A. Calculation of the properties of refrigerants [Text]: Textbook. Method. allowance. SPb .: ITMO University, 2016. 72 p.
[3] Alexandrov A.A., Grigoriev B.A. Tables of thermophysical properties of water and water vapor [Text]: Reference. Rec. Gos. Standard Data Service. GSSSD R-776-98 - M .: Publishing House MPEI. 1999. - 168 p.
[4] Savelyev N.I. Calculation and design of shell-and-tube heat exchangers [Text]: textbook. allowance / N.I. Saveliev, P.M. Lukin. - Cheboksary: Publishing house of Chu-vash. un-that. 2010. - 80 p.
[5] Yantovsky E.I., Levin L.A. Industrial heat pumps [Text]. - M.: Energoatomizdat, 1989.- - 128 p.
[6] Kharitonov V.P. Autonomous wind power installations [Text]. - M.: GNU VIESH, 2006. - 280 p.
[7] Zysin L.V., Sergeev V.V. Unconventional and renewable energy sources. Part 1. Renewable energy sources [Text]: Textbook. Allowance. - SPb .: Publishing house of the Polytechnic. University, 2008. - 192 p.
[8] Kirpichnikova, I.M. Wind power plants. Calculation of the parameters of the components [Text]: textbook / I.M. Kirpichnikova, E.V. Solomin. - Chelya-Binsk: Publishing Center of SUSU, 2013. - 71 p.
[9] Khakhaleva. L. V. Non-traditional and renewable energy sources [Text]: A manual for practical training. / Comp. Hakhaleva L.V. –Ulyanovsk, 2008. - 32 p.

Design and Analysis of Vegetable Transplanter Based on Five-bar Mechanism

Kai Sun[1], Rongyu Ge[1,*], Taoyuan Li[2] and Jiatai Wang[2]

[1]School of Mechanical Engineering, University of Jinan, Jinan, 250022, China
[2]Department of Mechanical Engineering, Zaozhuang Vocational College of Science and Technology, Jinan, 277599, China

*me_gery@ujn.edu.cn

Abstract. As one of the important components of automatic transplanter, duckbilled planting device, is a significant factor affecting the efficiency of the transplanter. Aiming at the problems of low transplanting rate, poor upright angle and poor stability of the transplanter, a five-pole transplanting mechanism is designed. Firstly, the parameters of the five-pole transplanting mechanism are analyzed and confirmed by MATALAB software. After virtual prototype model is established by SolidWorks software, simplified model is imported into ADAMS software for kinematics simulation, and then the trajectory, velocity and acceleration curve of the duckbilled tip are obtained. The simulation results verify the rationality of the structure and parameters and show that the trajectory of duckbilled planter meets the agronomic requirements of vegetable cultivation.

1. Introduction

Planting industry accounts for more than half of the total agricultural production value in China [1]. Transplanting is an extremely important production link in planting industry. At present, most of the transplanting operations of dryland crops in China adopt manual transplanting. Because of the increase in the proportion of rural employees going out to work and the adjustment of the overall industrial structure, the actual labor cost has been significantly higher than that of mechanized operations. Therefore, it is necessary to develop the automation and machinery and imperative to mechanize transplanting and improve its efficiency. Dryland transplanter is an agricultural equipment mainly for maize, cotton, cabbage and other dryland crops transplantation. It can replace manual operation to achieve automatic and semi-automatic transplanting work, and has been more and more widely used in different regions of China [2].

Since planting mechanism is an important component of dryland transplanter, a five-bar planting mechanism is designed to improve the overall stability of the planting mechanism and ensure the accuracy of the overall movement through the transmission of the five-bar mechanism. In order to overcome the difficulty of designing five-bar mechanism, the parameters are designed by MATLAB software, 3D (three-dimensional) model is built by SolidWorks software, and the planting mechanism is simulated and analyzed by ADAMS software. Finally, all the structural parameters of the five-bar planting mechanism can be obtained.

2. Structure and working principle of the planter

Five-bar mechanism is a multi-degree-of-freedom mechanism with the least number of links and rings. It is relatively stable and has more dimension and angle parameters. It can realize abundant curve trajectories. Therefore, five-bar mechanism is used as the executing mechanism of duckbilled tip trajectory [3].

The planting structure of transplanter is mainly composed of five parts: fixing plate, driving sprocket, upper cranks, lower cranks and duckbill, as shown in figure 1.

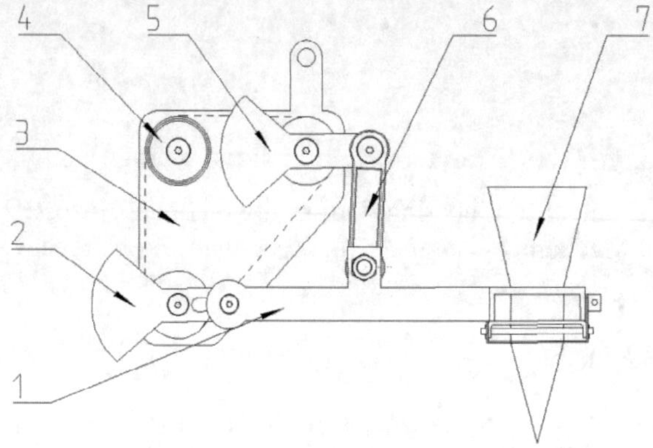

1-guide rod 2-lower crank 3-fixed plate 4-driving sprocket
5-upper crank 6-connecting rod 7-duckbill

Figure 1. Structural diagram of planting mechanism

In figure 1, one end of the upper and lower cranks are connected with the bearing seat, the other end of the upper crank is hinged with the guide rod and the other end of the lower crank is hinged with the slider embedded in the guide rod, which form a double crank guide rod mechanism. During operation, the upper and lower cranks of the planter rotate at equal speed under the action of driving sprockets. At the same time, the planting mechanism moves in a uniform straight line in the horizontal direction.

Under the compound movement, the duckbill can be planted vertically by adjusting the structural parameters of the mechanism, so as to ensure that the pot seedlings have a good upright angle after planting. The lower crank has a sliding groove, and the planting depth can be adjusted by changing the length, which can meet the requirements of different crops for planting depth and ensure the planting quality of seedlings.

3. Establishment of mathematical model

In order to study the motion law of the five-bar mechanism, the motion of the planter mechanism is analyzed by analytic method, and the closed vector position equation of the mechanism is established. The law of speed and acceleration could be obtained by further derivation. The schematic diagram of the transplanting device with five-bar mechanism is shown in figure 2.

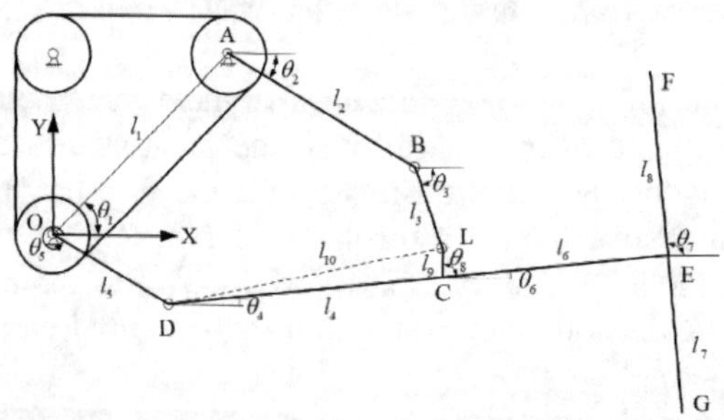

Figure 2. Mathematical model of planter mechanism

Cartesian coordinate system is established with O point as origin, horizontal direction as X axis and vertical direction as Y axis. The displacement equation of point D can be listed by analytic method as

$$\begin{cases} X_D = X_O + l_5 \cos\theta_5 \\ Y_D = Y_O + l_5 \sin\theta_5 \end{cases} \tag{1}$$

Similarly, the displacement equation of point B is defined as

$$\begin{cases} X_B = X_A + l_2 \cos\theta_2 \\ Y_B = Y_A + l_2 \sin\theta_2 \end{cases} \tag{2}$$

Since the displacement vector **OL** can be represented as

$$\boldsymbol{l}_{OA} + \boldsymbol{l}_{AB} + \boldsymbol{l}_{BL} = \boldsymbol{l}_{OD} + \boldsymbol{l}_{DL} \tag{3}$$

the coordinate of point L can be calculate by

$$\begin{cases} X_L = l_1\cos\theta_1 + l_2\cos\theta_2 + l_3\cos\theta_3 = l_5\cos\theta_5 + \cos\left(\theta_4 + \arctan\dfrac{l_9}{l_4}\right)\sqrt{l_4^2 + l_9^2 - 2l_4 l_9\cos\theta_8} \\ Y_L = l_1\sin\theta_1 + l_2\sin\theta_2 + l_3\sin\theta_3 = l_5\sin\theta_5 + \sin\left(\theta_4 + \arctan\dfrac{l_9}{l_4}\right)\sqrt{l_4^2 + l_9^2 - 2l_4 l_9\cos\theta_8} \end{cases} \tag{4}$$

To simplify the operation, a definition is written as

$$\begin{cases} a = l_1\cos\theta_1 + l_2\cos\theta_2 - l_5\cos\theta_5 \\ b = l_1\sin\theta_1 + l_2\sin\theta_2 - l_5\sin\theta_5 \\ c = \dfrac{l_4^2 + l_9^2 - 2l_4 l_9\cos\theta_8 - a_2 - b_2 - l_3^2}{2l_3} \end{cases} \tag{5}$$

The parameter θ_3 can be given as

$$\theta_3 = 2\arctan\frac{b \pm \sqrt{a^2 + b^2 - c^2}}{a - c} \tag{6}$$

Referring equation (4) and equation (6), parameter θ_4 can be obtained. So the displacement equation of point C is defined as

$$\begin{cases} X_C = X_L + l_9\cos(\pi - \theta_8 + \theta_4) \\ Y_C = Y_L + l_9\sin(\pi - \theta_8 + \theta_4) \end{cases} \tag{7}$$

The displacement equation of point E is expressed as

$$\begin{cases} X_E = X_D + (l_4 + l_6)\cos\theta_4 \\ Y_E = Y_D + (l_4 + l_6)\sin\theta_4 \end{cases} \tag{8}$$

The displacement equation of point F is described as

$$\begin{cases} X_F = X_E + l_8\cos(\theta_7 - \theta_4) \\ Y_F = Y_E + l_8\sin(\theta_7 - \theta_4) \end{cases} \tag{9}$$

The displacement equation of point G is defined as

$$\begin{cases} X_G = X_E + l_7\cos(\theta_7 - \theta_4) \\ Y_G = Y_E - l_7\sin(\theta_7 - \theta_4) \end{cases} \tag{10}$$

IOP Conf. Series: Materials Science and Engineering **692** (2019) 012029 doi:10.1088/1757-899X/692/1/012029

The displacement equation of the five-bar mechanism is solved. If the first derivative and the second derivative of the corresponding displacement equation is accomplished, the velocity and acceleration equations of the five-bar mechanism can be obtained respectively.

4. Parameter design of five-bar mechanism

According to the above formulas and the existing parameters, combined with the agronomic conditions of vegetable planting, the design objectives of the planting mechanism are set as follows:

(1) the efficiency of duckbill planting mechanism is 60-80 plants per minute.

(2) pot seedling spacing requirements is 200-350 mm;

(3) bowl seedling planting depth is 130-150 mm;

(4) the angle between bowl seedling and ridge is more than 80 degree.

Because it is difficult to complete the design of five-bar mechanism by traditional calculation method, the amount of calculation is too large and the efficiency is low[4]. If the parametric design of five-bar mechanism is completed by computer-aided method. Therefore, according to the established kinematics model, the parametric model of the five-bar planting mechanism is established using MATLAB software, so that the relevant parameters can be adjusted at any time to observe the movement law of the planting mechanism. Figures 3 and 4 show the relationship between the length of L2 and L5 and the end trajectory, respectively. Repeat this process to obtain a set of better parameters: l_1=300mm, l_2=100mm, l_3=175mm, l_4=215mm, l_5=50mm, l_6=265mm, l_7=280mm, l_8=170mm, l_9=45mm, θ_1=45°, θ_2=90°, θ_5=45°, θ_7=45°, θ_8=90°.

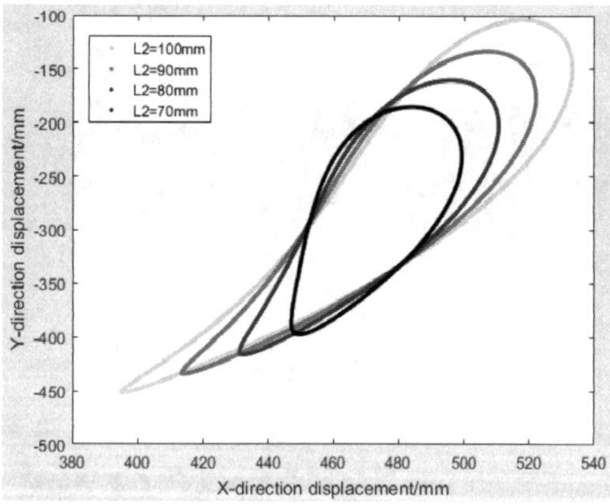

Figure 3. The relationship between L2 length and endpoint trajectory

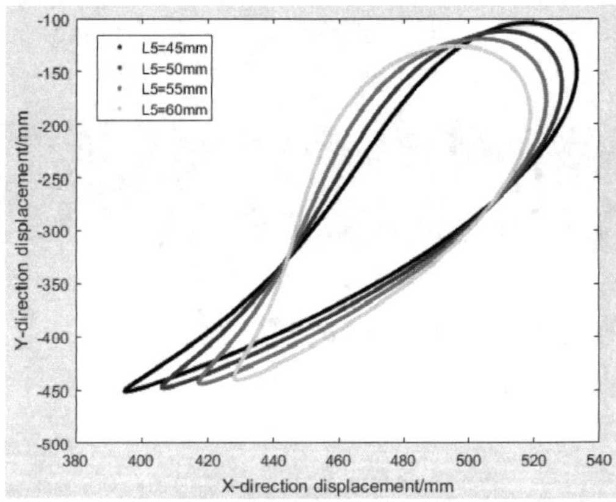

Figure 4. The relationship between L5 length and endpoint trajectory

5. Construction and simulation of virtual prototype for the transplanting mechanism

5.1 3D modeling

The parameters obtained by MATLAB software are modeled by SolidWorks software to form an assembly and check the interference, as shown in figure 5.

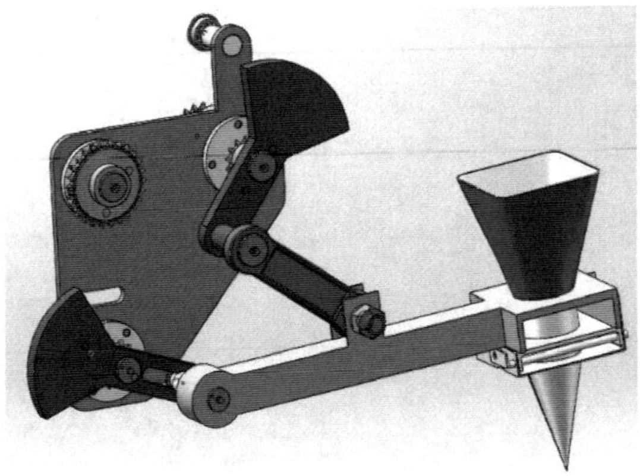

Figure 5 3D model of planter

5.2 Modeling of virtual prototype

ADAMS has powerful dynamic simulation function and can simulate the virtual mechanical system in real time [5], but its 3D modeling ability is relatively poor. With parameters designed using MATLAB Software, the virtual prototype model of planting mechanism is established by SolidWorks software, and the model is assembled and checked for interference. At the same time, simplification of components with less influence on mechanism motion is made and then it is imported into ADAMS software. Attention should be paid to the coordinate system of SolidWorks and ADAMS software when importing, so that the coordinate system is consistent and the trajectory coordinates of duckbill can be accurately collected.

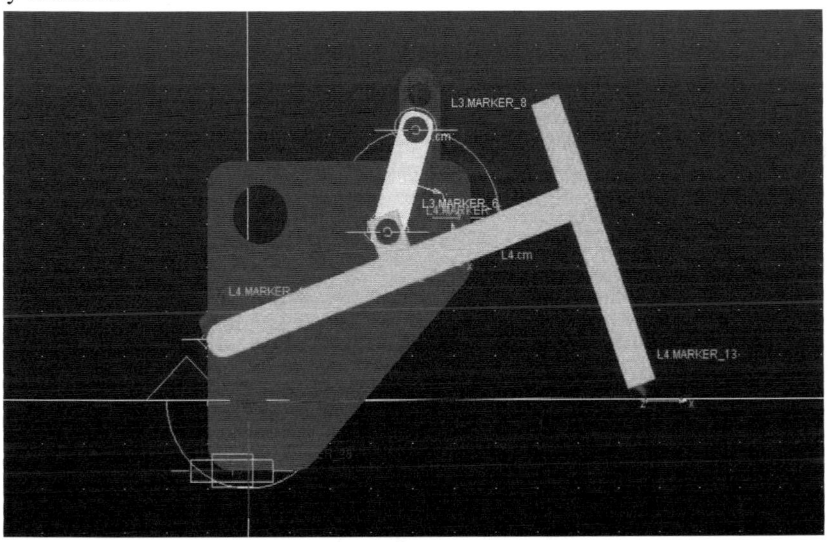

Figure 6 Planting mechanism model

A rotating pair is added to the hinge of the crank, a moving pair is added to the slider of the guide rod and a fixed pair is added to the fixed end of the connecting rod. As a result the whole virtual prototype model is completed as shown in figure 6.

5.3 ADAMS virtual simulation

According to the actual design requirements, adding driver to virtual prototype in ADAMS software, the end of the duckbill will be used as a measuring point for the trajectory of the duckbill planting mechanism. The simulation time is set to 10 seconds and the simulation steps are 500. As a result the trajectory curve of planting mechanism is shown in figure 7.

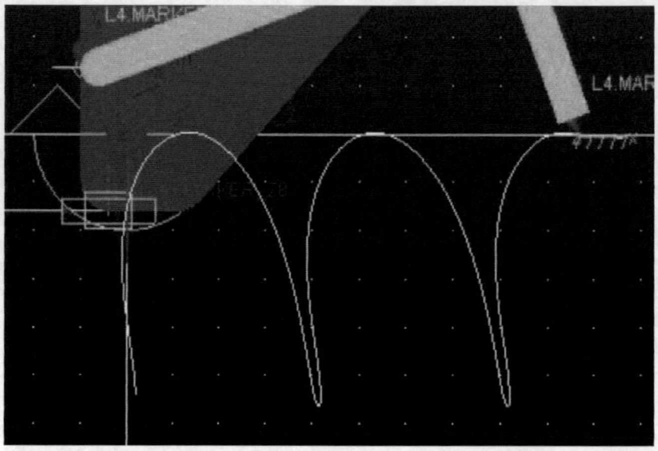

Figure 7 Trajectory of planting mechanism

Figure 7 shows the motion track of the duckbill tip. It can be seen from the graph that the trajectory of the tip of the duckbill is nearly perpendicular to the ground. The distance between the trajectory of entering soil and the trajectory of unearthing is small and the horizontal displacement is close to zero when planting, which improves the survival rate of potted seedlings and guarantees the quality of transplanting.

Based on the analysis of the simulation results above, the parameters can meet the requirements of the design objectives and the transplanting operation requirements.

6. Conclusion

A five-bar planting mechanism consisting of a fixed plate, a driving sprocket, a five-bar mechanism and a duckbill is described in this paper. The structure and working principle of the mechanism are introduced. The parameters of the mechanism are selected through mathematical modeling and MATLAB simulation. A virtual prototype of the five-bar planting mechanism is established by SolidWorks software. After simplification, the kinematics is simulated by ADAMS software. The results show that the parameters are reasonable.

Acknowledgement

The authors gratefully acknowledge that the present research is supported by the Key Research Project of Shandong Province in China (No. 2019GNC106123). In addition, valuable suggestions from the reviewers in improving the quality of the paper are highly appreciated.

Reference

[1] Han C J, Zhang X J, Yang W Z and Yin W Q 2011 Present situation and analysis of automatic transplanting technology of potted seedlings in Dryland *Agricultural mechanization research* **33** 238-240

[2] Wang G 2017 Research on the development process and current situation of vegetable transplanting machines *Agricultural science and technology and equipment* 35-36+38

[3] Xu G W, Liu H X, Jiang S C, Shi S and He T F 2018 Design and experiment of transplanting mechanism of Salvia miltiorrhiza on film based on five-bar mechanism *Journal of Agricultural Machinery* **49** 55-65

[4] Junhua Tong, Chuanyu Wu, Huanyu Jiang, Yaxin Yu and Xiuqin Rao 2017 Optimizing the path of seedling low-density transplanting by using greedy genetic algorithm *Computers and Electronics in Agriculture* **142** 356-368

[5] G.V. Prasanna Kumar and H. Raheman 2011 Development of a walk-behind type hand tractor powered vegetable transplanter for paper pot seedlings *Biosystems Engineering* **110** 189-197

IOP Conf. Series: Materials Science and Engineering **692** (2019) 012030 doi:10.1088/1757-899X/692/1/012030

Structural Optimization for Stiffness and Stochastic Fatigue Life Improvement of a Sport Utility Vehicle Chassis

Wenlin Wang[1,*], Bao Huang[1,2], Jianming Du[2] and Siyuan Cheng[3]

[1]School of Mechanical Engineering, Dongguan University of Technology, Dongguan 523808, China.
[2]College of Mechatronics and Control Engineering, Shenzhen University, Shenzhen 518060, China
[3]Jiangling Motors Co., Ltd, Nanchang 330001, China

[*]Corresponding Author: E-mail: pianowwl@163.com

Abstract. To further improve the structural dynamic performance of the Land-Wind X-6 chassis, structural optimization was performed: An X-shape cross member with high stiffness is added to the chassis, in addition, the cross member connecting the two torsion bar attachments is also strengthened. Extensive Finite Element Analysis (FEA) demonstrates that the inadequate torsion stiffness has been greatly increased by 102.6% after optimization, and the adequate bending stiffness has also been slightly improved by 3.8%, in addition, the level of local stresses on the chassis has been considerably declined; After optimization, the fatigue life coefficient when the vehicle is running on A-class concrete road at a speed of 110 km/h has been improved by about 40.36%, and that when the vehicle is running on D-class macadam road at a speed of 60 km/h has been improved by about 33.43%. Followed industrial implementation, field test result and product feedback information verify that the structural optimization performed in this study is effective and significant.

1. Introduction

Structural fatigue and fracture usually occur in the chassis system of a sport utility vehicle (SUV) although the fatigue strength allowance of the chassis system is well-designed in the product development process, and many trouble-shooting works indicate that the reason might due to the unpredictable actual load spectrum under severe in service conditions.

Finite Element Analysis (FEA) based approaches [1] are usually used in automotive structural development, fatigue life prediction and optimization. Vijayan et al. [2] used FEA modelling in the structural analysis of automotive chassis when considering the cross-section and material, Lee et al. [3] and Radzieński et al. [4] respectively used the stress relaxation approach and the experimental modal analysis approach in the automotive structural strength improvement, Thomas [5] characterized cut-edges in structural improvement and fatigue life prediction, Kang et al. [6] and Song et al. [7] both put efforts in structural durability improvement by optimizing the suspension elements.

Wang et al. [8] performed static and local stress analysis in the structural failure trouble-shooting of the Chinese Land-wind X6 chassis by establishing and validating the FEA model of the chassis, and based on the findings, Wang et al. [9] optimized the chassis using a novel dynamic force counteracting approach, and the preliminary industrial implementation proved the effectiveness of their research. However, the stiffness and the stochastic fatigue life of the chassis still needs to be improved.

IOP Conf. Series: Materials Science and Engineering **692** (2019) 012030 doi:10.1088/1757-899X/692/1/012030

This study continues the research of literatures 8 and 9 in furthering the structural optimization of Land-wind X6 chassis for its stiffness and stochastic fatigue life improvement. An X-shape cross member with high stiffness is added to the chassis and the cross member connecting the two torsion bar attachments is also strengthened. Extensive FEA analyses indicate that the inadequate torsion stiffness of the chassis has been greatly increased, the adequate bending stiffness has also been slightly improved, the level of local stresses on the chassis has been considerably declined, and the stochastic fatigue life has also obviously improved. Followed industrial implementation verifies the effectiveness of the structural optimization.

2. Structural optimization

Figure 1 shows the loading and calculation approach in stiffness FEA analysis of the chassis. So referring to figure 1(a), the bending stiffness $K_{bending}$ of the chassis can be formulated by

$$K_{bending} = \frac{Fa^3}{48f} \tag{1}$$

where F is the load, a is the automotive wheel-base and f is the vertical deflection of the chassis.

Similarly, referring to figure 1(b), the torsion stiffness $K_{torsion}$ of the chassis is formulated by

$$K_{torsion} = \frac{L^2\pi}{180} \cdot \frac{F}{h} \tag{2}$$

where L is the arm of load and h is the twist deflection of the chassis.

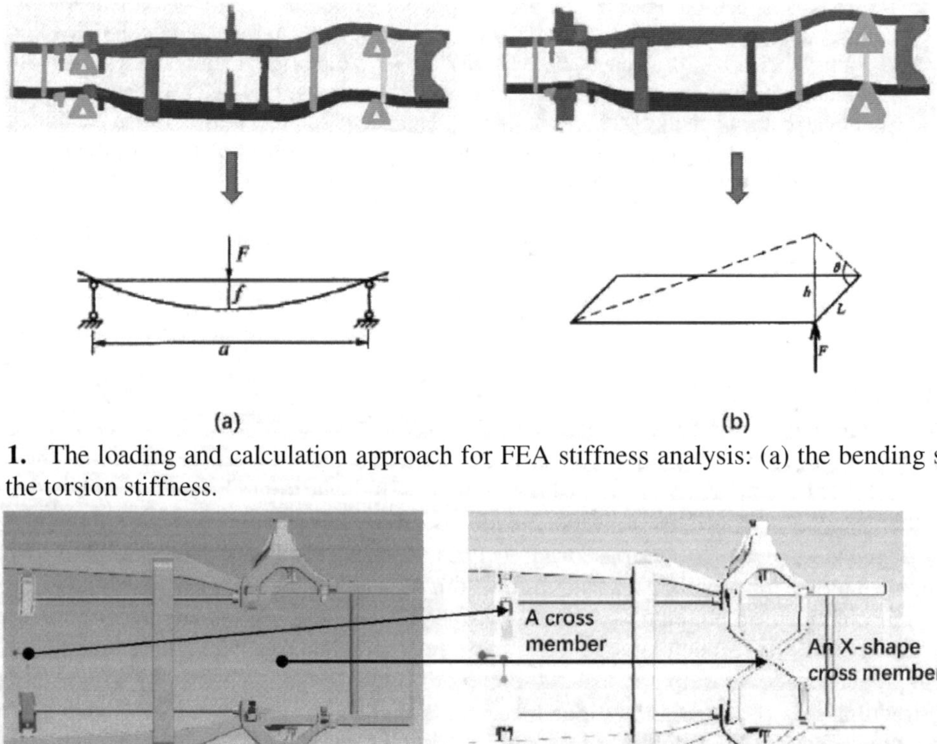

Figure 1. The loading and calculation approach for FEA stiffness analysis: (a) the bending stiffness and (b) the torsion stiffness.

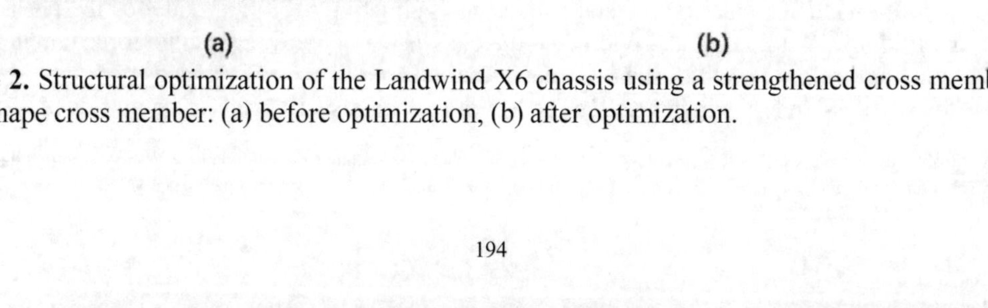

Figure 2. Structural optimization of the Landwind X6 chassis using a strengthened cross member and an X-shape cross member: (a) before optimization, (b) after optimization.

Assisted by FEA stiffness analysis, structural optimization of the Landwind X6 chassis using a strengthened cross member and an X-shape cross member is performed, as shown in figure 2. The current optimization is different from the initial structural refinement performed in [9]: An X-shape cross member with high stiffness is added to the chassis, in addition, the cross member connecting the two torsion bar attachments is also strengthened.

3. Structural dynamics improvement

3.1. The effectiveness of stiffness and local stress improvement

Figure 3 compares the FEA-based stiffness analysis results of the chassis before [9] and after the structural optimization. Figure 3(a) shows that under full load conditions the maximum vertical deflection drops from 0.561 mm to 0.542 mm after optimization, which leads to the bending stiffness of the chassis increases from 1.56×10^6 N m^2 to 1.62×10^6 N m^2, i.e., a 3.8% improvement after optimization.

Figure 3(b) illustrates that under full load conditions the maximum twist deflection drops from 49.36 mm to 24.36 mm after optimization, which leads to the torsion stiffness of the chassis greatly increases from 346.3 N m/° to 701.8 N m/°, i.e., a 102.6% improvement after optimization.

Thus, after adding an X-shape cross member and strengthening the cross member connecting the two torsion bar attachments, the inadequate torsion stiffness of the chassis has been greatly increased, and the adequate bending stiffness has also been slightly improved.

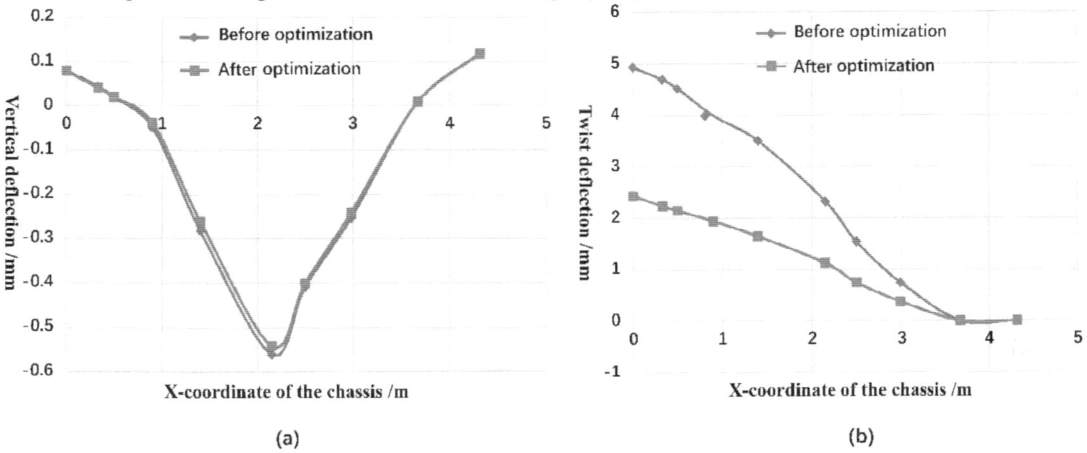

Figure 3. (a) A comparison of vertical deflections of the chassis before and after structural optimization under full load conditions; (b) A comparison of twist deflection of the chassis before and after structural optimization under full load conditions.

Figure 4 compares the local stress distribution of the chassis before and after structural optimization under full load conditions, and it demonstrates that the local stresses have been globally and greatly reduced. Particularly, the maximum local stress in the torsion bar attachment reduces from 446.34 MPa to 251.068 MPa, the maximum local stresses in the arm attachments respectively reduce from 404.41 MPa, 485.34 MPa and 549.88 MPa to 110.339 MPa, 190.766 MPa and 203.311 MPa. Thus, an obvious reduction of the local stresses on the key suspension components indicates that the original structural design imperfections of the chassis have been overcome.

3.2. The effectiveness of stochastic fatigue life improvement

Transient response FEA results show that after optimization a 1σ maximum stress of 120.652 MPa occurs at the attachment of the third cross member and the beam, when the vehicle is running on A-class concrete road at a speed of 110 km/h. Figure 5 shows the stochastic response spectrums at the point with 1σ maximum stress, it indicates that both of the displacement and the velocity responses at

17.5 Hz and between 35-40 Hz are larger and obvious for those frequencies are tightly related to the bending and twisting natural frequencies of the chassis.

Similar analysis is performed when the vehicle is running on D-class macadam road at a speed of 60 km/h, the 1σ maximum stress is 185.647 MPa.

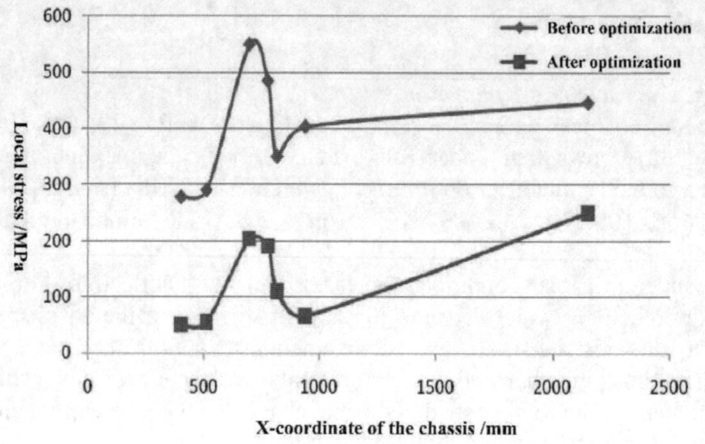

Figure 4. A comparison of local stress distribution of the chassis before and after structural optimization under full load conditions.

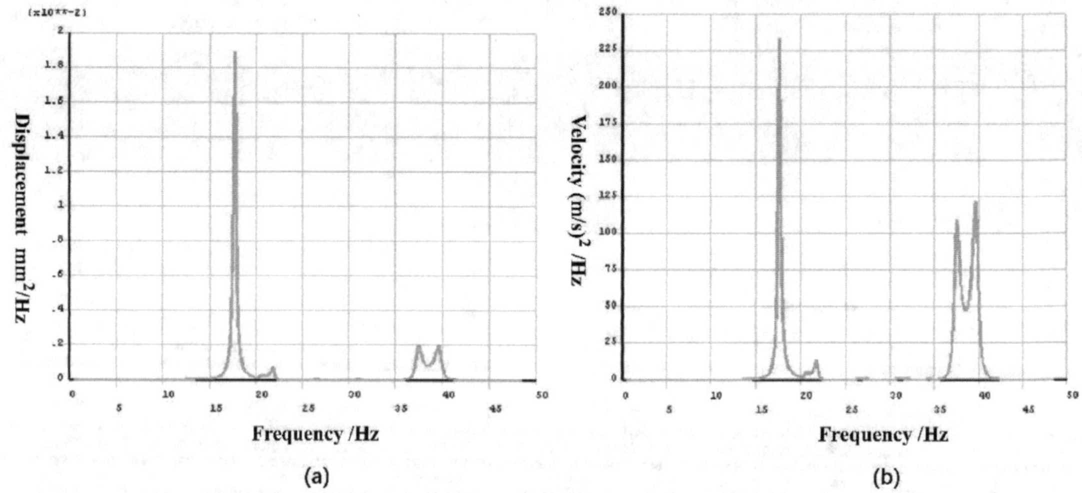

Figure 5. (a) Displacement response spectrum and (b) Velocity response spectrum at the point with 1σ maximum stress when the vehicle is running on A-class concrete road at a speed of 110 km/h.

Miner has proposed the linear fatigue cumulative damage law [10] as

$$\frac{n_1}{N_1} + \frac{n_2}{N_2} + \cdots + \frac{n_n}{N_n} = 1 \tag{3}$$

where n_1, N_1 are respectively the actual cycle number and the damage cycle number of a component when it undergoes the first level of stress, and so on.

Thus, the fatigue life coefficient of the chassis under stochastic vibrations can be calculated by using the three-region estimation approach [11], which is based on the Gaussian distribution and Miner fatigue cumulative damage law. The fatigue life coefficient is formulated by

$$D = \frac{n_{1\sigma}}{N_{1\sigma}} + \frac{n_{2\sigma}}{N_{2\sigma}} + \frac{n_{3\sigma}}{N_{3\sigma}} \tag{4}$$

where $n_{1\sigma}$, $n_{2\sigma}$ and $n_{3\sigma}$ are respectively the cycle numbers when the stresses fall within the scopes of 1σ, 2σ and 3σ; $N_{1\sigma}$, $N_{2\sigma}$ and $N_{3\sigma}$ are respectively the allowed cycle numbers corresponding to the stress scopes of 1σ, 2σ and 3σ, $N_{1\sigma}$, $N_{2\sigma}$ and $N_{3\sigma}$ can be obtained by looking up the *S-N* curve of a particular material in the materials handbook. For this research, the chassis is made of 16Mn.

Calculating by equation 4, the fatigue life coefficients of the chassis under different stochastic vibrations conditions are obtained. When the vehicle is running on A-class concrete road at a speed of 110 km/h, the fatigue life coefficient after optimization decreases from 0.4841 to 0.2887, which means that the fatigue life after structural optimization and under that vibration condition has been improved by about 40.36%.

Similar analysis shows that when the vehicle is running on D-class macadam road at a speed of 60 km/h, the fatigue life coefficient after optimization decreases from 0.8356 to 0.5562, which means that the fatigue life after structural optimization has got a 33.43% improvement.

3.3. A summary

After adding an X-shape cross member and strengthening the cross member connecting the two torsion bar attachments, the dynamic performance of the new chassis has been greatly improved, the concrete indices are summarized in Table 1.

Table 1. A comparison of performance indices of the chassis before and after structural optimization

Performance indices	Before optimization	After optimization	Change
Maximum stress under full bending load (MPa)	286.3	272.3	↓ 4.9%
Maximum stress under full torsional load (MPa)	238.9	199.9	↓ 16.3%
Bending stiffness of the chassis (N m^2)	1.56×10^6	1.62×10^6	↑ 3.8%
Torsion stiffness of the chassis (N m/°)	346.3	701.8	↑ 102.6%
Maximum local stress around the attachment of arms (MPa)	549.88	203.311	↓ 63%
Maximum local stress around the attachment of torsion bar (MPa)	446.34	251.068	↓ 43.75%
Fatigue life coefficient on A-class road (concrete)	0.4841	0.2887	↓ 40.36%
Fatigue life coefficient on D-class road (macadam)	0.8356	0.5562	↓ 33.43%

4. Conclusions

- Structural optimization was performed on the Land-Wind X-6 chassis: An X-shape cross member with high stiffness is added to the chassis, in addition, the cross member connecting the two torsion bar attachments is also strengthened.
- After optimization, the inadequate torsion stiffness has been greatly increased by 102.6%, and the adequate bending stiffness has also been slightly improved by 3.8%, in addition, the level of local stresses on the chassis has been considerably declined.
- After optimization, the fatigue life coefficient when the vehicle is running on A-class concrete road at a speed of 110 km/h has been improved by about 40.36%, and that when the vehicle is running on D-class macadam road at a speed of 60 km/h has been improved by about 33.43%.

- Followed industrial implementation, field test result and product feedback information verify that the structural optimization performed in this study is effective and significant.

Acknowledgements

The authors are grateful for financial support from the National Natural Science Foundation of China (NSFC) under Grant No. 11572123 and the Research Fund for High-level Talent of Dongguan University of Technology under Project No. GC200906-30.

References

[1] Ghazaly N M 2014 Applications of finite element stress analysis of heavy truck chassis: survey and recent development *J. Mech. Design Vib.* **2, 3** 69–73

[2] Vijayan S N and Sendhilkumar S 2015 Structural analysis of automotive chassis considering cross-section and material *Int. J. Mech. Eng. Auto.* **2** 370–6

[3] Lee H, Seo H and Park G J 2003 Design enhancements for stress relaxation in automotive multi-shell-structures *Int. J. Solids and Struct.* **40** 5319–34

[4] Radzieński M, Krawczuk M and Palacz M 2011 Improvement of damage detection methods based on experimental modal parameters *Mech. Syst. Signal Proc.* **25** 2169–90

[5] Thomas D J 2012 Characterization of cut-edges for improved automotive chassis and suspension fatigue CAE life predictions *J. Failure Analy. Preven.* **12** 408–18

[6] Kang B J, Kim N W, Kim J H, Lee Y H and Sin H C 2008 Improving the durability of automobile suspension systems by optimizing the elastomeric bushing compliance *Proc. IMechE, Part D: J. Auto. Eng.* **222** 469–84

[7] Song B C, Park Y C, Kang S W and Lee K H 2009 Structural optimization of an upper control arm, considering the strength *Proc. IMechE, Part D: J. Auto. Eng.* **223** 727–35

[8] Wang W L, Cao X C and Chen S Y 2018 Structural failure analysis of a sport utility vehicle chassis *Proc. Int. Conf. on Mechatronic Syst. and Robots* (*Singapore*), p 100

[9] Wang W L, Hou Z C, Zhou Z R and Chen S Y 2018 Stress relaxation of a sport utility vehicle chassis using a dynamic force counteracting approach *MATEC Web of Conferences* **249**, 03001

[10] Macdonald K A 2011 *Fracture and fatigue of welded joints and structures* (Cambridge: Woodhead Publishing Limited)

[11] Zhang Y H 2013 *Fatigue analysis of welded joints* (Beijing: Chemical Industry Press) (In Chinese)

Effect of Dynamic Balance Parameters on Vibration of Vertical and Horizontal Rotor System

Zhong Yan, Wu Ji and Ding Kejin

Shanghai Marine Equipment Research Institute, Shanghai, China

Corresponding author's e-mail address: zhongyan704@163.com

Abstract. In this paper, a rotor test device for vertical-horizontal conversion is designed. The effects of unbalanced mass, dynamic balance position and other dynamic balance parameters on the vibration response of the rotor system are experimentally analyzed, and the difference of the vibration response of the rotor system between vertical and horizontal States is compared. The test results show that the vibration of the bearing base foot will increase after the dynamic balancing rotor is installed in the vertical shafting in the horizontal state; in order to ensure the dynamic balancing accuracy, the dynamic balancing of the vertical rotor system should be carried out on the vertical system as far as possible, and the aggravation should be carried out far away from the driving end. The research results in this paper can provide necessary support for the determination of dynamic balancing process and the study of dynamic characteristics of vertical rotors.

1. Introduction

Vertical rotor system is widely used in marine auxiliary equipment, such as fuel pump, lube pump, fire pump, condensate booster pump and so on. However, due to the difference of layout and load-bearing conditions, the design calculation and test methods of horizontal rotor system can not be applied to vertical structure [1,2].

In engineering practice, the dynamic balancing of the vertical rotor is completed in the horizontal state, which results in a large deviation between the vibration response of the vertical rotor system and the theoretical value [3]. In this paper, a rotor test-bed for vertical-horizontal conversion is tested, the difference of vibration response between vertical and horizontal rotor systems is analyzed, and the relationship of vibration response between horizontal and vertical conversion process is established to provide necessary support for the study of dynamic characteristics, state assessment, fault diagnosis and life prediction of vertical rotor systems.

2. Difference of Vertical and Horizontal Rotors

For the vertical single-stage cantilever pump, the dynamic model of the rotor-bearing system can be established as follows (without considering the axial vibration of the rotor)[4,5].

$$[M]\{\ddot{U}\}+[G]\{\dot{U}\}+[C]\{\dot{U}\}+[K]\{U\}=[F] \tag{1}$$

Among them, [M] and [K] are mass matrix and stiffness matrix, [G] is rotation matrix, [C] is damping matrix, {F} is system force, for vertical rotor system, {F}={Fe}+{Ft}, for horizontal rotor system, {F}={Fe}+{Ft}+{W}, where {W} is rotor gravity, {Fe} is unbalanced force and {Ft} is fluid exciting force.

It can be seen from the above formula that the most obvious difference between vertical and horizontal rotors lies in the difference of system forces, that is, the gravity in horizontal rotors plays a dominant role in the rotor forces, while the unbalanced force and fluid-induced force have less influence; while for vertical rotors, the influence factors of gravity can be neglected, and the unbalanced force and fluid-induced force play a dominant role in the rotor system.

These differences make the vibration responses of vertical and horizontal rotors significantly different under the same unbalanced excitation, and the vibration response characteristics are the basis for calculating the unbalanced state in the dynamic balancing process. Therefore, the fundamental improvement of the dynamic balancing process of vertical rotors lies in the analysis of the dynamic characteristics of the rotor bearing system under the same unbalanced excitation.

3. Design of Vertical and Horizontal Conversion Test Device

In order to analyze the difference of vibration response between two rotor systems from the point of view of experiment, a kind of rotor test bench which can be used for vertical-horizontal conversion is designed. Vertical and horizontal conversion test-bed is mainly composed of driving motor, coupling, aligning roller bearing, simulation shaft, adjustable bearing seat, support, adjustable base, test system and hydraulic lifting system. The bearing-rotor system of the model vertical pump is taken as the model of the test-bed, and the geometric dimensions and relevant operating parameters of the test-bed are determined according to the similarity criterion. The structure sketch is shown in Figure 1.

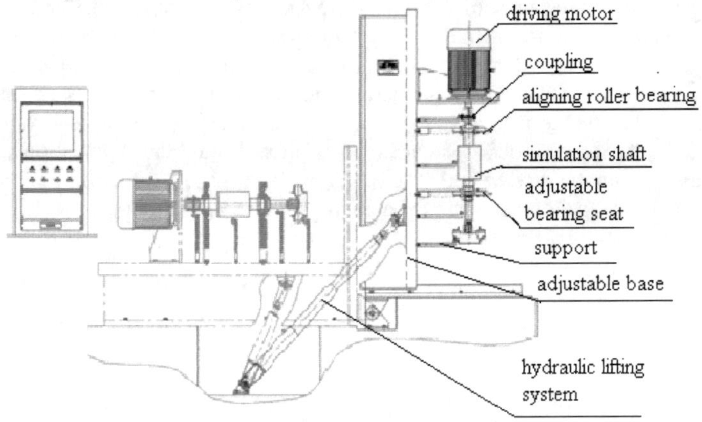

Figure 1. Schematic diagram of the test device

The test device has the following characteristics: (1) the parameters of the test-bed can be adjusted to meet the vibration response test and analysis of the rotor system under different working speeds, different balancing conditions, different support stiffness and non-span conditions; (2) the installation mode of the test-bed can be adjusted, and the vertical and horizontal conversion of the test-bed can be achieved through the hydraulic device at the base; (3) the installation and disassembly is convenient and the positioning is reliable.

The dynamic analysis of the rotor system shows that the first-order critical speed of the rotor is 3218 r/min and the second-order critical speed is 9468 r/min, which can meet the operation requirements of the test-bed. The calculation model and the Campbell diagram of the rotor system are shown in Figure 2 and Figure 3.

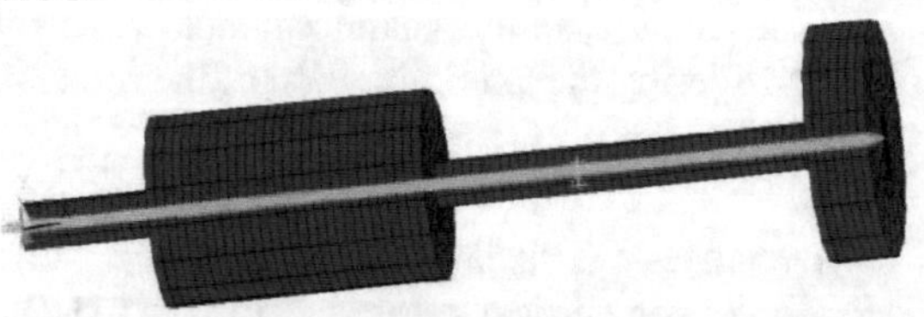

Figure 2. Computational finite element model

IOP Conf. Series: Materials Science and Engineering **692** (2019) 012031 doi:10.1088/1757-899X/692/1/012031

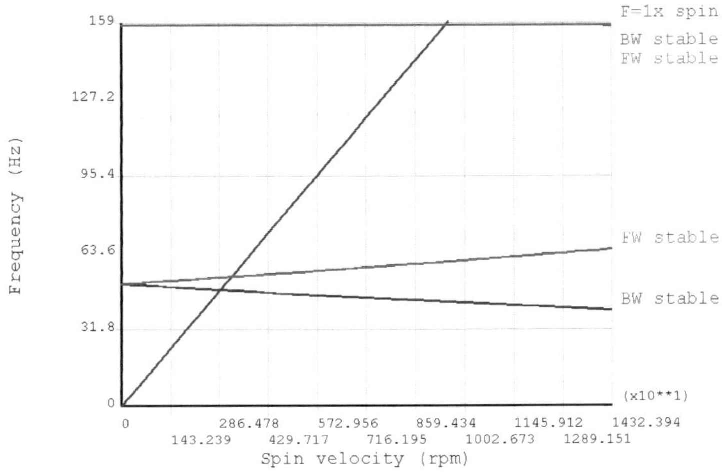

Figure 3. Campbell diagram of rotor system

4. Comparison of Vibration of Vertical and Horizontal Rotors under Unbalanced Excitation

The influence of unbalance on the vibration of the bench was tested on a typical test bench of vertical pump. By adding bolts of different mass on the balance plate, the influence of unbalance on the fundamental frequency vibration of the bearing base was analyzed, and the characteristics of the bench under different rotational speeds were compared and analyzed. According to the characteristics of the bench, four kinds of unbalanced mass are set up, which are 0.8g, 6.8g, 12.4g and 24.6g, respectively. The rotational speed of the bench is 1200 r/min. The test results are shown in Figure 4.

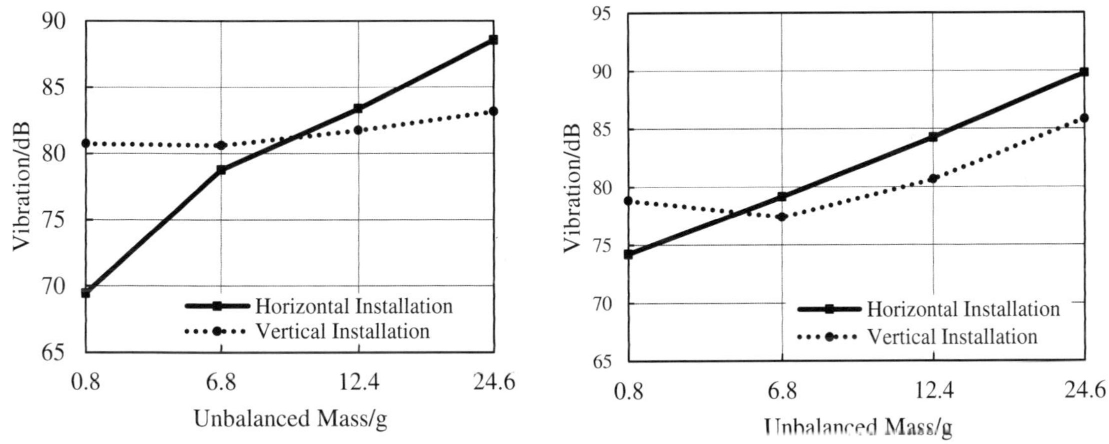

1) Vibration of front Baering 2) Vibration of After Baering

Figure 4. Comparison of Vibration of Vertical and Horizontal Rotors

It can be seen from the test results that: (1) in the ideal unbalanced state (unbalanced mass 0.8g), the vibration at the foot of the vertical bearing is higher than that at the horizontal state. With the deterioration of the balance state, when the unbalanced amount increases to 12.4g, the vibration at the foot of the vertical bearing is obviously lower than that at the horizontal state; (2) in the horizontal state, the vibration at the front and rear bearing seats is all along with the unbalanced amount. The vibration of bearing seat increases linearly with the increase of unbalance, but the gradient of variation decreases obviously when the bench changes from horizontal to vertical.

The test data show that the vibration of the bearing base foot will increase after the dynamic balancing rotor is installed in the vertical shafting under the influence of support stiffness and installation conditions. When the unbalance of the rotor increases after the operation of the vertical

IOP Conf. Series: Materials Science and Engineering **692** (2019) 012031 doi:10.1088/1757-899X/692/1/012031

shafting, the vibration of the bearing seat will increase, but the increase is less than that of the horizontal shafting.

5. Analysis of the Effect of Balanced Position on Vibration

The influence of uneven position on the vibration of the test bench is tested on a typical test bench of vertical pump. By applying the same unbalanced counterweight blocks at five different positions of the test rotor shafting, the influence of the unbalanced mass loading position of the vertical rotor on the vibration characteristics of the shafting is studied. The aggravating position is shown in Figure 5. Starting from the side of the cantilever rotor, the test is marked as position 1 to position 5, respectively. The weights are 12.4 g and the working speed is 1200 r/min. The test results are shown in Figure 6.

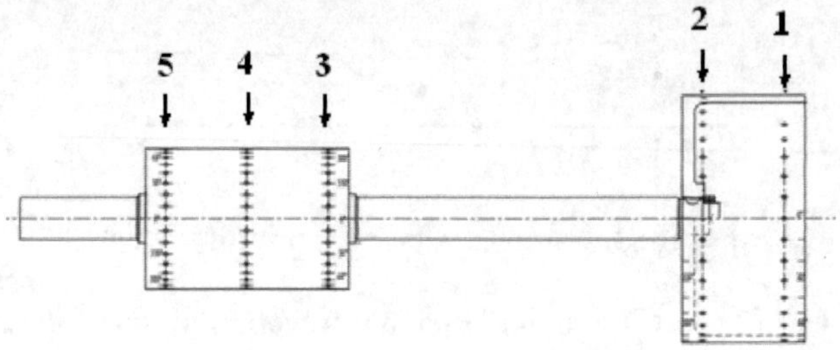

Figure 5. Schematic diagram of unbalanced mass applied position

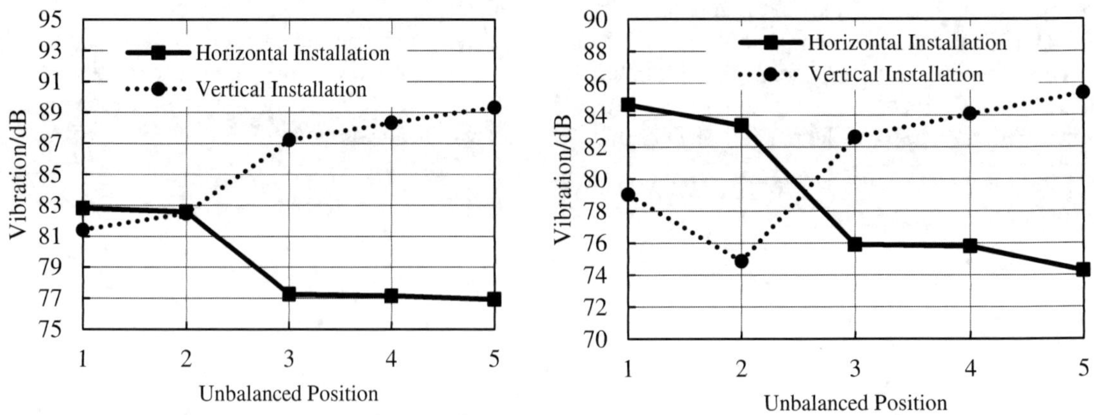

1) Vibration of front Baering 2) Vibration of After Baering

Figure 6. Influence of unbalanced position on vibration

From the test data, it can be seen that in the horizontal state, the aggravating position of dynamic balance is close to the whole drive, the smaller the vibration of the bearing seat, and in the vertical state, the closer the processing position is to the driving end, the greater the vibration of the bearing seat. The test data show that the dynamic balancing of the horizontal rotor system should be carried out near the driving end, and the dynamic balancing of the vertical rotor system should be carried out far away from the driving end.

6. Conclusion

In this paper, the effects of unbalanced mass, dynamic balance position and other dynamic balance parameters on the vibration response of the rotor system are experimentally analyzed on a rotor test device capable of vertical-horizontal conversion. The following conclusions are drawn:

(1) When the dynamic balancing rotor is installed in the vertical shafting, the vibration of the bearing base foot will increase.

(2) In order to ensure the accuracy of dynamic balancing, the dynamic balancing of the vertical rotor system should be carried out on the vertical system as far as possible.

IOP Conf. Series: Materials Science and Engineering **692** (2019) 012031 doi:10.1088/1757-899X/692/1/012031

(3) The aggravation of dynamic balance of vertical rotor system should be carried out far away from the driving end.

Reference

[1] Wang Zhengxian. 2009 *Research on Dynamic Characteristics of Vertical Rotor-Bearing System* (Harbin: Harbin Institute of Technology)

[2] Shi Xiangwei. 2016 *Design on Vibration Test Stand of Rotor System* (Xi`an: Xi`an University of Architecture and Technology)

[3] Sinou Jean-Jacques, Faverjon Béatrice. The vibration signature of chordal cracks in a rotor system including uncertainties [J]. *Journal of Sound and Vibration*, 2012,**(331)**:138-154.

[4] Yao-Yuhu, De Zhongwang, Yuan Fu. Numerical study on rotordynamic coefficients of the sealof molten salt pump[J]. *Nuclear Science and Techniques*, 2016,**(5)**:114-124.

[5] Sun Liquan, Liu Zhansheng, Qian Dashuai. Stability of Vertical rotor bearing system[J]. *Journal of Vibration and Shock*, 2010,**(29)**:41-43.

IOP Conf. Series: Materials Science and Engineering **692** (2019) 012032 doi:10.1088/1757-899X/692/1/012032

Research on Performance of Terrain Matching Algorithm for Underwater Autonomous Vehicle Based on Particle Filter

ShenJian[1], Xiong Lu[2] and Ba Ning[2]

[1] Post-doctoral program on control science and engineering, Naval University of Engineering, Wuhan 430033, China
[2]Department of Radar system, OrdnanceNCO Academy of Army engineering university, Wuhan 430075, China

Sterrain@126.com

Abstract. AUV is a powerful tool for exploration and development of marine resources. Underwater terrain matching navigation is an emerging passive autonomous navigation method, which improves the long-term and high-precision navigation capability of AUV. The underwater terrain matching algorithm based on particle filter (PF) use a large number of particle sampling to approximate the system state, which can avoid the Taylor expansion truncation error caused by the terrain linearization method, and has good matching precision and terrain adaptability. However, due to the strong nonlinearity and the uneven distribution of the underwater terrain features, it is difficult to achieve the full-range terrain matching optimize with a single configuration parameter of PF underwater terrain matching algorithm. For this reason, this paper has comprehensively studied the PF terrain matching algorithm (referred to as PF matching algorithm). The principle, accuracy and implementation steps of PF algorithm are analyzed. With the UKF algorithm as a reference, the advantages and disadvantages of PF-based algorithm are studied through computer simulation of terrain matching navigation. The matching precision and terrain adaptability of PF-based algorithm has been tested. From the simulation results, it is concluded that the particle diversity decline is one of the main factors affecting the matching performance, which provides a basis for algorithm performance improvement and engineering adaptability improvement.

1. Introduction

The navigation system is one of the most basic components of the underwater autonomous vehicle(AUV). It provides real-time position, velocity and attitude information for the AUV, which is the important Information input for AUV's autonomous control and navigation [1]. Underwater terrain matching navigation is a new type of passive, autonomous navigation method. The AUV's current position can be estimated by comparing the real-time measured terrain data with the underwater terrain database. From the perspective of estimation theory, the terrain matching algorithm essentially uses the observable measurement data to estimate the AUV's state variable by the estimation method [2]. Therefore, the recursive filtering is considered to be one of the effective ways to solve the underwater terrain matching problems.

In the 1970s, the terrain matching algorithm based on extended Kalmanfilter(EKF) has been successfully applied on the missiles. As the first-order linear truncation of Taylor expansion is used for extended Kalman filter's (EKF) terrain linearization[3], the state variables estimated precision of EKF algorithm is low. In addition, the Jacobian matrix solution of the system state equation in engineering application is usually complicated, which affects the system response speed to some extent[4].

With the continuous development of filtering technology, PF nonlinear filtering method is used the as the filtering kernel of the terrain matching algorithm, which improves the filtering precision of the matching algorithm. However, due to the strong nonlinearity of the underwater terrain and the randomness of the terrain feature distribution, it is difficult to achieve the ideal matching accuracy and robust effect by using the single PF based terrain matching algorithm in a region with varied terrain[5,6]. For this reason, it is necessary to optimize the PF parameters and algorithm structure in order to adapt the different terrain features. This paper fully analyze the performance of PF based terrain matching algorithm, firstly, the system state approximation principle of PF algorithm is introduced, the precision and characteristics of the PF algorithm are analyzed compared with the UKF algorithm. Secondly, the simulation of univariate nonstationary growth model (UNGM) is used to study the two algorithms' estimation performance. Thirdly, the AUV underwater terrain matching framework based on PF algorithm is given. The underwater terrain matching model is established and the underwater terrain matching simulation has carried out by using the two matching algorithms. Finally, the positioning effect of the terrain matching algorithms are studied to provide reference for the AUV navigation under different application conditions.

2. PF algorithm analysis

2.1. Principle of PF algorithm

Particle filtering is based on Bayesian estimation, Markov process and Monte Carlo theory. The core idea of PF is to simulate the probability distribution of system parameter by large random sampling from state sample set [7]:

$$p(x_{0:k} \mid z_{1:k}) \approx \sum_{i=1}^{N_s} \omega_k^i \delta(x_{0:k} - x_{0:k}^i) \tag{1}$$

The $\{x_{0:k}^i, \omega_k^i\}_{i=1}^{N_s}$ is defined as the particle set of the posteriori probability density function $p(x_{0:k} \mid z_{1:k})$, the $\{x_{0:k}^i, i=0,\cdots,N_s\}$ corresponding to the weight coefficient set $\{\omega_k^i, i=0,\cdots N_s\}$ is representing the particle set of system states, $x_{0:k}=\{x_j, j=0,\cdots k\}$, representing the system state set from the initial moment to the k-th moment, $z_{1:k}=\{z_j, j=0,\cdots k\}$ representing all observations from the 1-th moment to the k-th moment, $p(x_{0:k} \mid z_{1:k})$ representing the probability density under the observations sequence .

It can be seen from equation (1) that the more the number of particles, the better the probability approximating by the particle set $\{x_{0:k}^i, \omega_k^i\}_{i=1}^{N_s}$. The posterior probability density of k-th moment can be approximated by using the particle set $\{x_k^i, \omega_k^i\}_{i=1}^{N_s}$ at the corresponding moment, that is, by calculating the particle set at each moment, the recursive law of the corresponding system state can be obtained.

2.2. Comparison of PF algorithm and UKF algorithm

2.2.1. UKF algorithm. UKF is a nonlinear Kalman filtering framework method based on the Unscented transformation theory. The core idea is to achieve nonlinear approximation of the state distribution by deterministic sampling. Compared with the EKF, UKF's nonlinear estimation precision can reach above 2nd-order Taylor expansion truncation precision, and the approximate calculation is a Non-Jacobian matrix operation process, which is suitable for engineering realization[4,8].

The sigma point sampling of UKF is the basics for nonlinear state approximation, Commonly used sampling methods is:

- symmetric distributed sampling, the representative algorithm is the symmetric set sampling (Standard UKF)[8];

- hyper-spherical sampling, the representative algorithm is volume Kalman filter (Cubature Kalman Filter, CKF);
- Gaussian distributed volume sampling, the representative algorithm is Gauss-Hermite Kalman Filter (GHKF) [9,10].

If the system transfer equation is developed by Taylor expansion representation, the UT transformation can ensure the correct recursive of system state not higher than the 2nd order. When the distribution of Independent variables is symmetrical (symmetric distributed sampling), the UT transformation can achieve the precision of 3rd order. When the independent variable satisfies the Gauss distribution (Gaussian distributed volume sampling), for the one-dimensional case, the sampling method can achieve 4th-order precision, while for multi-dimensional variables, the UT transform can only guarantee the 4th -order axis directional precision.

By definition, the minimum mean square error can be classified as criterion of the norm minimum, which is not robust in statistics. Therefore, when the assumptions are deviated from the actual parameters, the estimation results will be greatly affected [4]. Based on the above reasons, the UKF-based underwater terrain matching algorithm will also face the lack of robustness.

2.2.2. PF algorithm. From the sampling principle of particle filtering, if the particles number is sufficient, the particle set can approximate the System state variables with arbitrary precision. Unfortunately, with the increasing of recursion time k, most particle weights are concentrating on a few particles, while the other particles' weights are too small to effective. This phenomenon is called the particle degradation which seriously affects the algorithm computational efficiency. Therefore Gordon.N.J proposed the resampling method, which removes the small weight particles and preserves the larger weight particles, This sampling method can avoid the repeated calculation of small weight particles in recursion, which ensures the computational efficiency[7].

Compared with the N=2n+1(n is the dimension of system state) sampling points of UKF algorithm, the PF system need all particles to participate the states /variance estimation in each recursive step, the calculation amount is much larger than the UKF algorithm. Therefore, reasonable optimization must be made between the precision of the algorithm and the calculation speed according to the actual application. In addition, as a fixed component of the particle filter framework, resampling can effectively solve the particle degradation problem, however, since high-weight particles are copied many times in the iterative process, a large number of particles are clustered the high-weight state region. It reduces the diversity of particle types, affects the state approximation ability of particle set , reduces the accuracy and robustness of the algorithm.

3. Univariate nonstationary growth model simulation

In this section, the univariate nonstationary growth model (UNGM) [4,5,7] is used to test the performance of the UKF and PF. The UNGM has the characteristics of high nonlinearity and bimodal state distribution, so it is suitable for verifying the nonlinear estimation ability of filter algorithm. The UNGM discrete dynamic space model can be expressed as:

$$x_k = 0.5x_{k-1} + 25\frac{x_{k-1}}{1+x_{k-1}^2} + 8cos\left(1.2(k-1)\right) + u_{k-1} \tag{2}$$

$$y_k = \frac{x_k^2}{20} + v_k \quad k = 1,2,\cdots,N \tag{3}$$

Where, $u_n \sim N\left(0,\sigma_u^2\right)$ and $v_n \sim N\left(0,\sigma_v^2\right)$ respectively represent the system noise and observed noise which satisfy the Gaussian distribution. Taking $\sigma_u^2 = \sigma_v^2 = 1$. In order to increase the contrast, the UKF algorithm is used as the comparison algorithm in the simulation. The sampling scale factor of the

UKF is $\alpha = 0.5$, $\beta = 2$, $\kappa = 1$, and the PF particle number N_s=500, the simulation time is t=100, the Monte Carlo simulation time is N=500 . The Mean Square Error (MSE) is define as:

$$\mathrm{MSE} = \frac{1}{N}\sum_{i=1}^{N}\left(x_i - \bar{x}_i\right)^2, N = 500 \tag{4}$$

The simulation results are shown in Figure 1.

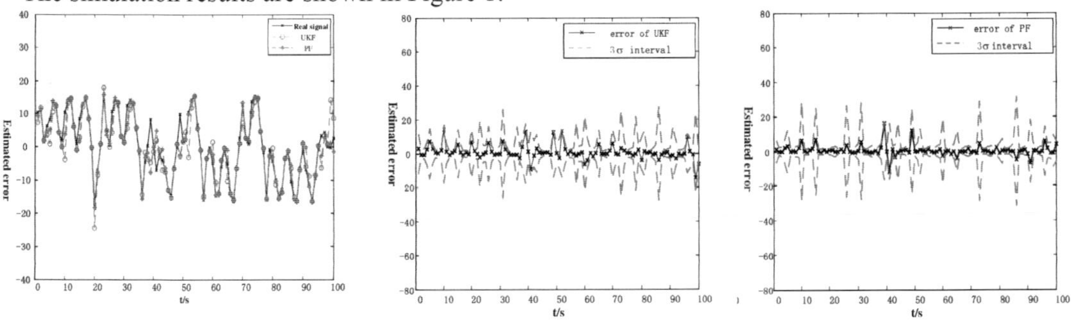

(a) Estimated error of UKF2 and PF

(b) Estimated MSE and 3σ confidence interval of UKF

(c) Estimated MSE and 3σ confidence interval of PF

Figure 1. Estimated error and Estimated MSE of UKF and PF

In the simulation result, the PF shows a better estimate ability than that of the UKF. This because the state distribution approximates of UKF is just first/second order, while PF achieves an overall approximation of the conditional probability distribution through a large number of particles sampling, and thus has higher precision than that of UKF. Comparing Figure 1(a) and Figure1(b), the estimation error of PF algorithm almost falls within the 3σ confidence interval, which reflects a high estimation result confidence of algorithm. In addition, the $MSE^{UKF2} = 19.1991$, $MSE^{PF} = 9.3434$ are also prove a better estimation accuracy of the PF algorithm. However, the average Monte Carlo simulation time of UKF is t^{UKF}=10.32ms, while the average simulation time of PF is t^{PF}=48.16ms (QudaCore Inter Core i7 7700HQ,3800MHz, None GPU computing mode). It indicates that better estimation accuracy of particle filter is at the expense of calculation amount increasing.

4. PF-based underwater terrain matching algorithm

4.1. PF-based underwater terrain matching algorithm

The AUV underwater navigation mode includes fixed depth navigation and terrain following navigation. Since the pitching motion of the AUV easily causes the depth measurement error, the AUV usually uses fixed-depth navigation in the matching process. To this end, different terrain matching algorithms can be studied by the two-dimensional underwater terrain matching model [6]:

$$\begin{cases} X_{t+1} = X_t + \Delta s + \omega_t \\ z_t = h\left(X_t\right) + \zeta_t \end{cases} \tag{5}$$

Where, $X_t = \left(x_t, y_t\right)$ is the current position of the AUV, Δs indicating the distance between the two matching positions, the system noise ω and the measured noise ζ are zero mean white Gaussian noise, Q and R are the variances are respectively.

4.2. Basic framework of PF underwater terrain matching algorithm

The PF underwater terrain matching algorithm includes four parts: Initialization, time update and measurement update, and resampling[6,7].

- Initialization: Passing the navigation information at the AUV starting matching position to the particle set, So that the AUV navigation information can be continuously corrected in the subsequent recursion;
- Time update: PF predicts the position and its variance of the next matching time through the navigation equation and the state particle set;
- Measurement update: The predicted position and its variance are adjusted according to the difference between the measured depth and the estimated depth.
- Resampling: Eliminate small weighted particles i and copy large weighted particles n the particle set to obtain a new posterior particle set

4.3. Algorithm implementation

- Particles Initialization : Particles initialization is performed using the AUV navigation initial position $\overline{X}_0 = (\overline{x}_0, \overline{y}_0)$ and its variance P_0. The $\{\chi_0^i, \omega_0^i\}_{i=1}^{N_s}$ is the particle set of the prior probability P_0, where, $\chi_0^i = (x_0^i, y_0^i)$ is the AUV positional information ofthei-th particle, The $\omega_0^i = \frac{1}{N_s}$ is Weight of the i-th particle.
- Time update: The particles $\chi_{k|k}^i$ $(i = 1, 2 \cdots N_s)$ is updated using equation (5) to obtain the predicted particles set $\{\chi_{k+1|k}^i, \omega_{k+1|k}^i\}_{i=1}^{N_s}$ in time k+1.

$$w_k^i = w_{k-1}^i p\left(z_k \mid x_k^i\right) = w_{k-1}^i p_{e_k}\left(z_k - h\left(x_k^i\right)\right) \; i = 1, 2, \cdots N_s \tag{6}$$

The predicted position of AUV at time k+1 can be calculated using equation(7):

$$\hat{x}_k \approx \sum_{i=1}^{N_s} w_k^i x_k^i \tag{7}$$

- Measurement update: Let $\chi_{k+1|k+1}^i = \chi_{k+1|k}^i$, The predicted positions of the particles $\chi_{k+1|k+1}^i$ $(i = 1, 2 \cdots N_s)$ can be find in the underwater digital map. The depth information h_{k+1}^i $(i = 1, 2 \cdots N_s)$ of these positions are obtained by terrain interpolation algorithm. Compare these predicted depths with the measured depth z_k to complete the update of particle set weights:

$$\omega_{k+1|k+1}^i = \omega_{k+1|k}^i \bullet p_{\zeta_k}\left(z_k - h_{k+1}^i\right) \; i = 1, 2, \cdots N_s \tag{8}$$

The AUV Posterior position and its estimate variance at time k+1 can be obtained:

$$\hat{X}_{k+1|k+1} = \sum_{i=1}^{N_s} \omega_{k+1|k+1}^i \bullet \chi_{k+1|k+1}^i \tag{9}$$

$$P_{k+1|k+1} = \sum_{i=1}^{N_s} \omega_{k+1|k+1}^i \left(\chi_{k+1|k+1}^i - \hat{X}_{k+1|k+1}\right)\left(\chi_{k+1|k+1}^i - \hat{X}_{k+1|k+1}\right)^T \tag{10}$$

- Resampling: The particle set resampling threshold is:

$$N_{eff} = \frac{N_s}{1 + Var(\omega_{k+1}^i)} \approx \frac{1}{\sum_{i=1}^{N_s} (\omega_{k+1}^i)^2} \tag{11}$$

When $N_{eff} \le N_{off}$,it indicates that the particle degradation is severe and needs to take the resampling measures. Otherwise go to time update step. By repeating the above recursive process, the AUV position can be estimated in real time, and the matching track of the AUV is obtained.

5. Underwater terrain matching simulation research

In this section, the computer simulation of underwater terrain matching navigation based on PF will be carried out in order to test the matching effect of algorithm.

The terrain matching digital map is constructed by using the depth data of a certain sea area (2500m×2500m). The map resolution is 6m, the symbols ★ and ● are the starting position and the ending position of the terrain matching routes, and the matching route of two simulations are shown in Figure 2. The terrain matching route shows that the terrain features at 0-1500m is obviously undulating, while the terrain at the 1500m-2500m gradually becomes flat.

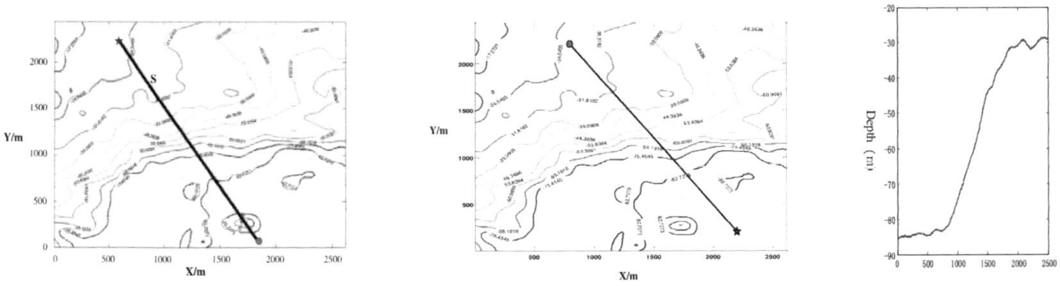

(a) Terrain matching route I (b)Terrain matching route II (c) Depth (route II)

Figure 2. The two simulation routes and the depth along the route II .

For comparison, the terrain matching algorithm based on UKF is simulated and analyzed under the same navigation conditions. The PF underwater terrain matching algorithm model is shown in (5)-(11) above, and the other simulation condition is show in Table 1.

Table 1. Terrain matching simulation parameters setting

Navigation parameter	System parameters
Initial deviation of position: (55m,55m)	Measurement error variance: $0.2m^2$
Position deviation (east): 0.1m/s	Measurement interval :5s
Position deviation (north) : 0.1m/s	Particle number: n=1000
Initial velocity error : 0.2m/s	UKF factors: $\alpha = 0.5$
Initial heading error : 0.3°	UKF factors: $\beta = 2$
AUV velocity: 2m/s	UKF factors: $\kappa = 1$
Initial variance of position: $\text{diag}\left[(50\text{m})^2, (50m)^2\right]$	

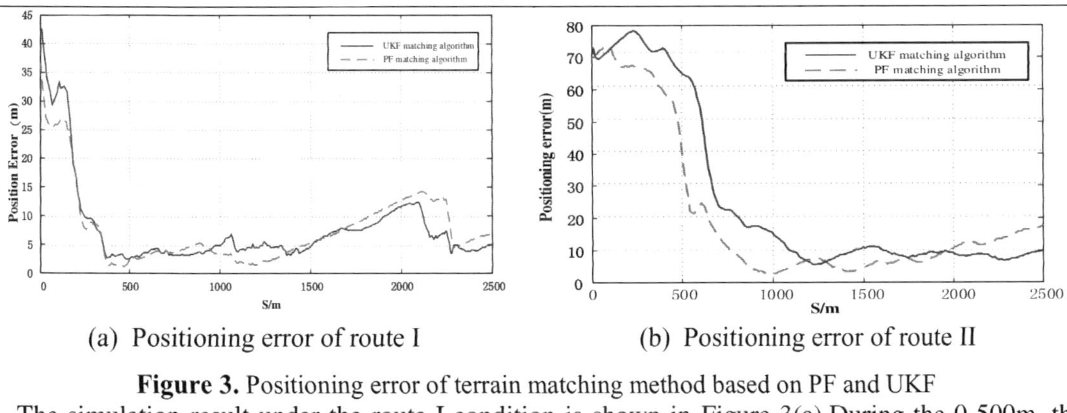

(a) Positioning error of route I (b) Positioning error of route II

Figure 3. Positioning error of terrain matching method based on PF and UKF

The simulation result under the route I condition is shown in Figure 3(a).During the 0-500m, the positioning error reducing speed of the PF-based terrain matching algorithm is faster than the UKF ones. And during the route of 0-1500m, the terrain feature along the route is sufficient, the PF-based algorithm shows a better terrain matching accuracy. The above simulation results illustrates that the state distribution approximation by a large number of particles is more accurate than the approximation of only 2-nd order Taylor expansion.

However, in the 1500m-2500m, as the terrain gradually becomes flat, the positioning error of the PF-based algorithm increases faster than the UKF-based algorithm. In order to prove that the above phenomenon is not caused by the terrain distribution, the starting position and the ending position of the terrain matching routes is exchanged, the AUV is navigate from the terrain flat area into the Terrain-rich area, as shown in Figure 2 (b).

It can be seen from the figure that under the condition of matching route II, the terrain near the AUV starting point is relatively flat, and the terrain features gradually become obvious with the extension of the route. Keep the simulation parameters unchanged and simulate again. The result is shown in Figure 3(b).

Comparing the Figure 3(a) with Figure 3(b), it can be found that the matching performance of the PF-based algorithm is better than the UKF algorithm regardless of the terrain distribution, and the positioning error can be kept within 10m. However, as the matching process progresses, the positioning error of the PF algorithm gradually deteriorates. At 2000m, the positioning error is greater than the UKF matching algorithm.

The simulation results above show that the PF terrain matching algorithm has better matching accuracy and terrain adaptability than the UKF-based ones, but its good matching performance is weakened with the passage of time because of the particle diversity degradation.

6. Conclusion

This paper starts with the principle of PF algorithm, the principle, matching precision and implementation steps have been studied. the advantages and disadvantages of PF matching algorithm are deeply studied through terrain matching simulation. The simulation results show that the PF-based terrain matching algorithm has the better matching accuracy and terrain adaptability, but the particles diversity degradation is one of the main factors affecting the performance of the algorithm. In the follow-up, the research team will study the degradation of particles diversity, find an effective method to reduce the degradation rate, and improve the engineering practicability of the algorithm.

Acknowledgments

ShenJian and Xiong Lu contributed equally to the writing of this article.

References

[1] Shen Gongxun,Sun Jianfeng. Application of information fusion theory in inertial/astronomical/GPS integrated navigation sSystem [M]. Beijing: national defense industry press, 1998: 56-81.

[2] Tian Fengmin. Research on underwater terrain navigation model solving and navigation cell selecting[D]. Harbin: Harbin engineering university.

[3] Kalman R,Bucy R.New Results in Linear Filtering and prediction Theory [J].Journal of Basic Engineering, 1961, 82(1):35-46.

[4] Chang Guobin. Nonlinear Kalman filtering algorithm and its application in attitude estimation problems [D]. Wuhan: Naval University of Engineering, 2011.

[5] MerweR.Sigma-point Kalman filters for probabilistic inference indynamic state-space models[D]. Portland: Oregon Health & Sci.Univ, 2004.

[6] Andrea L.Estimation Methods for Terrain Navigation[D].Stockholm:School of Electrical Engineering, 2006.

[7] Gordon N,Salmond D.Novel approach to non-linearand non-Gaussian Bayesian state estimation[J]. Proceedings of Institute Electric Engineering, 1993, 140(2):107-111.

[8] Karlgaard C.Huber-based unscented filtering and its application to vision-based relative navigation[J]. Radar, Sonar, Navigation, 2010, 4(5):744-745.

[9] Wu Y,Hu D.A Numerical-Integration Perspective on Gaussian Filters [J].Transactions on Signal Processing, 2006, 54(8): 2910-2921.

[10] IenkaranA, SimonH. CubatureKalman Filters[J]. IEEE Transactions on automatic control, 2009, 54(6):1254-1269.

Simulation Study on Gap Flow Field in EDM Micro-hole Machining Based on Gas-liquid-solid Three-phase Flow

Yu Liu, Chuangye Liu, Wenchao Zhang, Shengfang Zhang*, Fujian Ma, Dapeng Yang and Zhihua Sha

School of Mechanical Engineering, Dalian Jiaotong University, Dalian, China 116028

*E-mail: zsf@djtu.edu.cn

Abstract: In the process of micro-hole machining in EDM, if the debris and bubbles are not removed from the gap flow field, the concentration of debris and bubbles will increase, which will cause arcing discharge and seriously affect the processing efficiency and surface quality. In this paper, the geometric model of three-dimensional cylindrical gap flow field was established. The influence of bubble generation on the gap flow field and the debris under single pulse discharge condition was analyzed by Fluent. Then using the secondary development function of Fluent, the process of the debris and bubble on the bottom surface is continuously generated and the influence of bubble on the movement of debris under continuous discharge conditions were simulated and analyzed. Finally, by considering side flushing, the motion process of bubbles and debris under the conditions of flushing was analyzed.

Keywords: EDM; micro-hole; debris; bubble; gap flow field

1. Introduction

EDM is a kind of non-contact machining. Most of the machining process occurs in the dielectric working fluid [1]. The local high temperature generated by the instantaneous pulse discharge is used to remove the workpiece. When machining micro holes, compared with traditional processing methods, it has strong applicability, no cutting force, and low processing cost, etc.[2]. However, the micro-holes in EDM also have certain limitations. During the processing, if the debris generated by the high-temperature etching materials and the bubbles generated by the cracking working fluid cannot be removed in time, the accumulation of debris and bubbles in the bottom gap will cause arcing discharge, thereby reducing processing efficiency and processing quality. For the movement of the micro-hole gap flow field in EDM. Zhang Jie et al. established a three-dimensional simulation model of the gap flow field, simulating the movement of the working fluid and the debris in the gap flow field during the lifting motion of the electrode. It is believed that the reciprocating tool movement can promote the movement of working fluid and debris in the gap flow field [3]. Koening W et al. simulated the flushing process in the EDM process, established a mathematical model in the flow field of the bottom machining gap during the flushing process, and calculated the pressure field and velocity field in the machining gap by mathematical methods [4]. Masanori Kunieda et al. simulated the flow field in the EDM process. In the course of the research, the effects of bubbles and working fluids in the flow field were considered, and the motion of the debris and the velocity field of the debris was simulated [5]. At present, scholars of the world have only considered the solid-liquid two-phase flow for the simulation analysis of the gap flow field of the EDM, the effect of bubble generation on the movement of the debris in the gap flow field is not considered. Some studies have considered bubbles, but the bubble model does not reflect well the effects on debris and flow fields. In this paper, the gas-liquid-solid three-phase flow simulation model in the process of EDM micro-hole machining is established by analysing the characteristics of the gap flow field. The influence of bubble generation on the distribution of debris in the gap flow field is analysed, and the side flushing was considered to analyse the movement process of the debris and bubbles.

2. Gap flow field simulation mode

In the process of machining micro-holes in EDM, when electrode moves down to a position close to workpiece, the pulse voltage between the two electrodes breaks down the working medium to form an electrical discharge. As shown in Fig 1, The generation of the discharge channel converts most of the energy of the pulsed power source into thermal energy. Under the action of high temperature, two electrodes were melted and thrown into the working fluid medium to form debris, at the same time, the working fluid generates high-pressure gas under the high-temperature cracking.

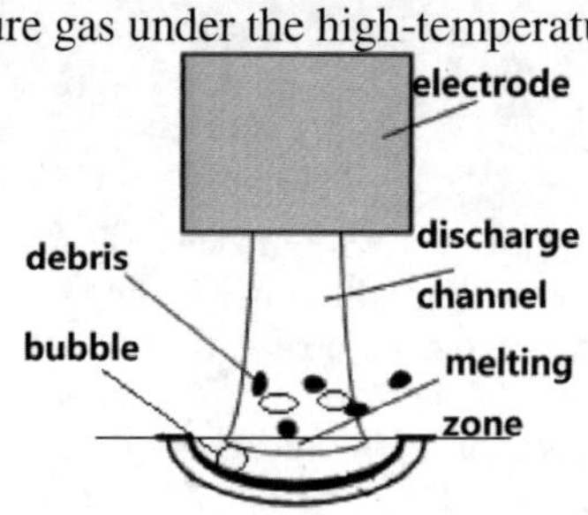

Figure 1. Physical model of EDM

2.1. Simulation parameter

When machining micro holes in EDM, different electrical parameters impact on the machining gap and the size of the bubbles. This paper does not consider the influence of electrical parameters. According to the existing experimental equipment, the simulation parameters are shown in table 1 .

Table 1. Acquisition simulation parameters in the experiment

Discharge voltage	Peak current	Discharge cycle	Working fluid	Flushing speed	Electrode	Workpiece	Electrode diameter
70V	2A	40μs	Deionized water	4m/s	Copper	TC4	1mm

2.2 Gap flow field geometry modeling

When machining micro holes in EDM, the shape of the electrode is cylindrical, which can simplify the gap flow field into a three-dimensional cylindrical assembly. As shown in Fig 2, based on previous research results, this paper estimates the bottom machining gap under the simulated discharge parameters is 50μm[6], through the experimental hole measurement, the single side machining gap is 100μm.

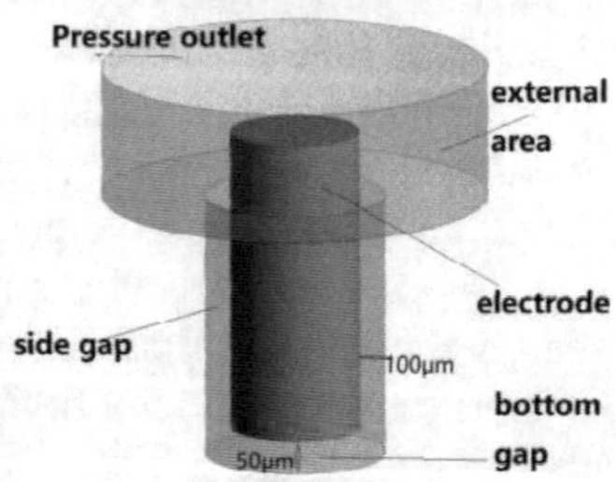

Figure 2. Geometric simulation model of gap flow field

IOP Conf. Series: Materials Science and Engineering **692** (2019) 012033 doi:10.1088/1757-899X/692/1/012033

2.3 Bubble and debris generation model

The high temperature generated by the pulse discharge can crack the working fluid to generate bubbles. Since the pulse discharge process is short, the generated gas has a small density and a large volume, so in the process of machining micro-hole in EDM, the pressure of the bubble in the bottom surface gap is much larger than the surrounding fluid medium, so under the action of the pressure difference, the bubble will keep the surrounding working fluid away from the bubble, and the flow of the working fluid can drive the movement of the debris, realizing the coupling of the three phases. Since the movement of the bubble is formed under the action of the pressure difference, and the discharge time is short, considering the discharge efficiency, it can be considered that a small volume of high-pressure bubbles is formed in the bottom processing gap in every two discharge cycles, bubbles move freely under the influence of pressure difference. The VOF model is mainly used to describe this process. For the initial size and pressure of the bubble, it can be estimated by the observing the bubble. The diameter of the bubble simulated in this paper is 0.04mm and the pressure is 1.5×108pa.

In the process of EDM, each pulse discharge will produce debris. The following assumptions are made on the gap flow field: (1) The working fluid in the gap flow field is a constant temperature and incompressible medium. (2) During the processing, after the flow field is stable, it is in steady state. (3) The debris are spherical. It is estimated by experiments that about 38 debris are produced every pulse discharge. Through the DPM model in Fluent, the secondary development function can be used to periodically generate particles randomly in the bottom machining gap, so that the movement state of the debris in the gap flow field can be visually seen.

3. Simulation result analysis

The distribution of bubbles and debris under single-pulse discharge was simulated. Based on the single-pulse model, the distribution of bubbles and debris under continuous discharge with or without side flushing was simulated.

3.1 Bubble and debris generation model

In order to better observe the motion of bubbles generated by pulse discharge, the first bubble was set at the bottom center, and then simulated and observed. As shown in Fig 3, the blue phase is the working fluid and the red phase is the gas. It can be clearly seen that the bubbles will expand and accumulate in the bottom gap.

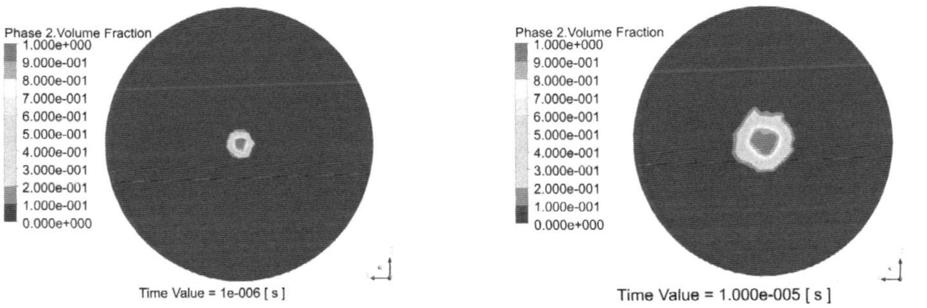

(a) Phase contour of bubble at 1μs (b) Phase contour of bubble at 10μs

Figure 3. Bubble phase at bottom gap under single pulse discharge

The liquid velocity contour and velocity vector contour of the bottom gap are shown in Fig. 4.

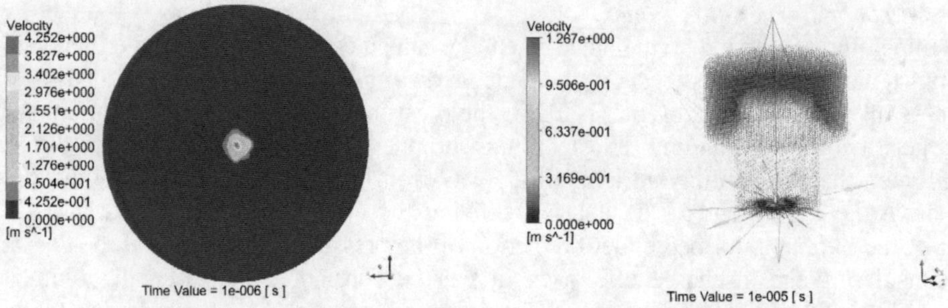

(a) Bubble velocity contour (b) Bubble velocity vector

Figure 4. Velocity contour and velocity vector contour at bottom gap

It can be seen that when the bubble is generated, a velocity vector will be generated from the inside of the bubble to the outside of the bubble, and the velocity value will decrease with the increase of the volume of the bubble.

In order to study the effect of single pulse bubbles on the distribution of debris, a layer of debris is simulated in the bottom machining gap. As shown in Fig. 5 (a), the distribution of debris is shown in Fig. 5 (b) after a period of single pulse.

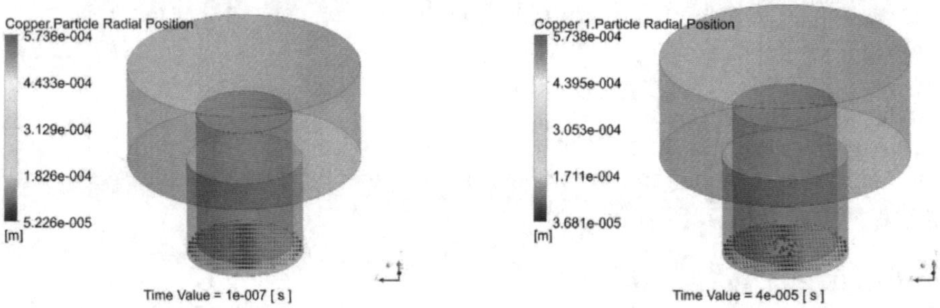

(a) Initial distribution of debris (b) Distribution after bubble formation

Figure 5. Distribution of debris under the action of bubbles generated by single pulse

3.2 Effect of bubbles on debris under continuous discharge

Through UDF function in Fluent, a bubble is randomly generated in bottom gap every two discharge cycles. On the base of the initial distribution of a layer of debris, the randomly generated debris are added. The bubble distribution in bottom gap with processing time is shown in Fig. 6.

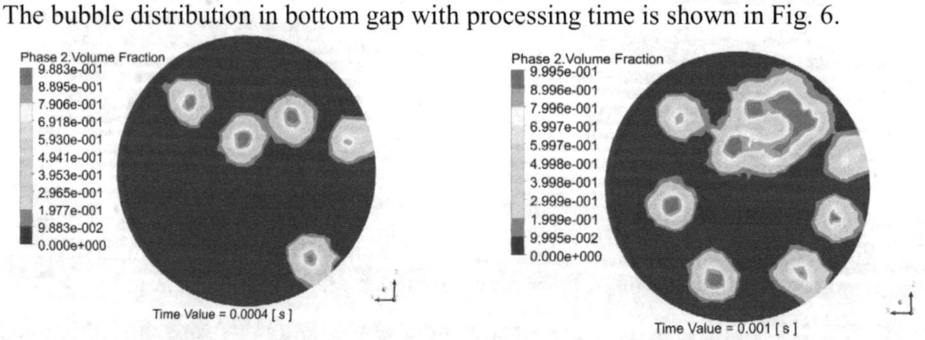

(a)Phase distribution of bubbles at 0.4ms (b) Phase distribution of bubbles at 1ms

Figure 6. Bubble phase contour at bottom gap under continuous discharge

As can be seen from Fig. 6, single bubble was generated randomly at the bottom. As the processing time goes on, the number of bubbles increases, bubbles merge and accumulate in the processing gap at the bottom. The effect of bubbles generated on the distribution of debris under continuous discharge was shown in Fig. 7.

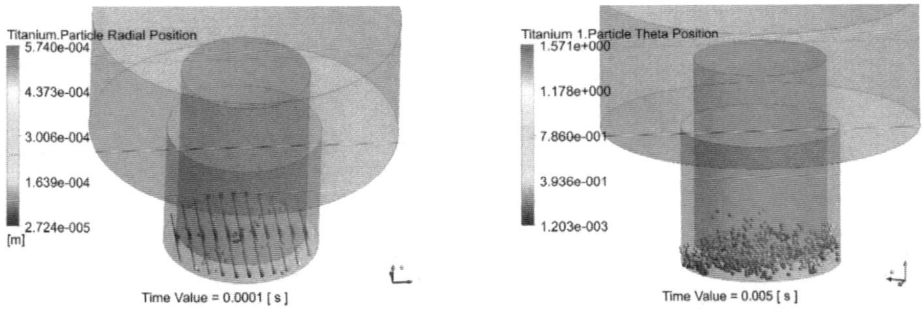

(a) Debris distribution at 0.1ms (b) Debris distribution at 5ms

Figure 7. Effect of bubble generation on the distribution of debris under continuous discharge

From Fig. 7 (a), it can be seen that when the processing time is 0.1ms, the pulse discharge occurs once. It can be seen that the randomly generated particles of the debris are added to the bottom gap, but bubbles are fewer at this time. It is not obvious that bubbles impact on the distribution of the debris. With the processing time going on, at 5ms, it can be seen that the distribution of the debris is affected by bubbles. Part of the debris are removed from bottom gap to side gap under the action of bubbles. By counting the number of particles, 342 particles can be obtained in the side gap. At this time, 3235 particles are generated in the bottom gap, the removal rate is only 10.5%. Therefore, it can be concluded that bubbles can make part of the debris remove from the bottom gap to the side gap, but it is difficult to ensure the continuous processing.

3.3 Distribution of bubbles and debris under unilateral flushing

Flushing is a common debris removal method when machining micro holes in EDM, the flow state of gap flow field under side flushing was studied in this paper. Fig 8 shows the velocity contour of gap flow field under flushing and the distribution of debris.

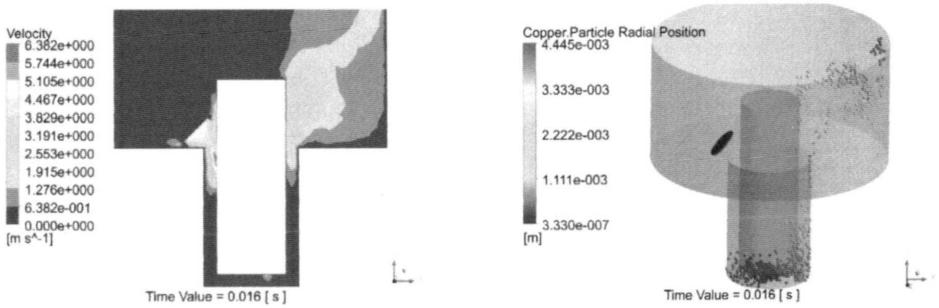

(a) Velocity contour of gap flow field (b) Distribution of debris in gap flow field

Figure 8. Velocity contour and distribution of debris in gap flow field

Fig. 8 shows that when flushing fluid is added, a flow field will be formed in the gap flow field from the left side to the bottom to the right side. The debris will be removed from the gap flow field quickly under the action of flushing fluid. The removal effect of flushing fluid is much greater than that of bubbles, so that it can avoid the accumulation of debris in the bottom gap and ensure continuous processing. From the side of the electrode, the bubble phase distribution in the gap flow field can be seen as shown in Fig 9.

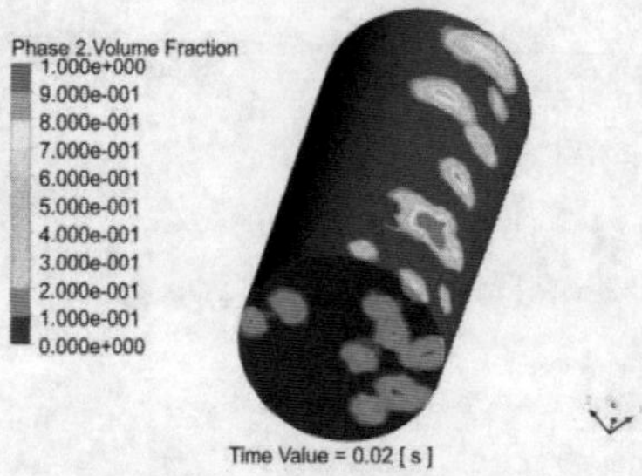

Figure 9. Bubble phase distribution in gap flow field

From Fig. 9, it can be seen that there are still some bubbles remaining in the bottom gap, but under the action of flushing, bubbles are continuously removed from the bottom gap, the concentration of bubbles in the bottom gap are reduced and ensuring the stability of processing.

4. Conclusion

In this paper, the characteristics of gap flow field in EDM were analyzed, the process of machining micro-holes in EDM was simulated. The gas-liquid-solid three-phase coupling model was established by Fluent, which showed the movement of debris and bubbles in gap flow field. Through the analysis of gap flow field, the following conclusions are obtained.

(1) The bubbles generated by high temperature cracking working fluid produce a decreasing velocity from inside to outside, and is reduced in a very short time. The bubbles will drive the debris to move away from the bubbles.

(2) During continuous machining, some of the debris will be removed from the bottom gap under the action of bubbles, but the effect of bubbles on the removal of debris is limited. If debris removal method is not considered, the debris will still accumulate in the bottom gap, which will affect the processing stability.

(3) Side flushing can form a flow field from the left to the bottom to the right in the gap flow field. The debris and bubbles can be removed from the processing gap under the action of flushing fluid.

Acknowledgments

The financial support from the National Natural Science Foundation of China under grant no. 51875074, Scientific Research Platform Foundation of Liaoning Province under grant no. JDL2016006, and General Program of Natural Science Foundation of Liaoning Province under grant no.20180550425 is acknowledged.

References

[1] Gaofei, Gu Lin. Research status and prospect of micro-hole special processing technology *J. Aviation manufacturing technology*, 2016, 518(23)108-111.

[2] Wang Feng, Zhao Changhui, *et al.* High-speed EDM micro hole processing technology and its application *J. New technologies and products in China*, 2016(20)68-68.

[3] Zhang jie, Isago Soichiro, *et al.* Three-dimensional simulation of gap flow field in EDM *J. Electric machining and mold*. 2008(2)12-15.

[4] Koenig W, Weill R, *et al.* The flow fields in the working gap with electro-discharge-machining *J. Annals of the CIRP,* 1977, 25(1): 71-76.

[5] Takeuchi Haruka, Masanori Kunieda. Relation Between Debris Concentration and Discharge Gap Length in EDM Process *J. International Journal of Electrical Machining*. 2007 (12): 17-22.

[6] Kiyoshi Inoue. Principle of electrical discharge machining *M.* Shuai Yuanlun, Yu Xuewen, *Translation. Beijing: National Defense Industry Press*, 1983.

IOP Conf. Series: Materials Science and Engineering **692** (2019) 012034 doi:10.1088/1757-899X/692/1/012034

Rolling Bearings Fault Diagnosis Using VMD and Multi-tree Mahalanobis Taguchi System

Jun Zhan, Longsheng Cheng* and Zhaiming Peng

School of Economics and Management, Nanjing University of Science and Technology, 210094, Nanjing, China

*Corresponding author's email: cheng_longsheng@163.com

Abstract. In order to effectively identify the various faults of rolling bearings and the severity of the faults, an intelligent fault diagnosis system based on variational mode decomposition and multi-tree Mahalanobis Taguchi system with multiple mahalanobis distance was proposed. Vibration signals were decomposed into multiple BLIMFs by variational mode decomposition, and time domain and frequency domain features were extracted. The multiple mahalanobis distance method was used to solve the problem of many features in the diagnosis system. By using the advantages of Mahalanobis Taguchi system in features optimization, sensitive modal components for diagnosis and recognition were selected. Multi-tree Mahalanobis Taguchi system was constructed for intelligent identification of multiple fault states. Finally, rolling bearings fault data was tested to verify the accuracy of the algorithm and compared with other algorithms. The results show that the algorithm can simplify the diagnosis system, reduce training time and improve the recognition accuracy.

1. Introduction

Rolling bearings are key components of rotating machinery. Their running state directly affect performance, running efficiency and useful life of equipment. The failure of rolling bearing usually occurs in the components of the bearing, including inner race, outer race and rolling elements. Mechanical equipment vibration signals are mostly non-stationary and nonlinear signals. Early fault signals are very weak, which may encounter attenuation and be easily interfered by background noise in the transmission path. Therefore, extracting fault information from such signals is the key to accurate diagnosis of mechanical equipment [1].

Empirical mode decomposition (EMD) proposed by Huang et al. [2] can adaptively decompose a nonstationary signal into multiple intrinsic mode functions (IMFs). However, EMD has some disadvantages like mode mixing problem and sensitivity to noise. Ensemble Empirical Mode Decomposition (EEMD) eliminates the noise of the original signal by adding Gaussian random white noise, and restrain mode mixing problem in EMD [3]. However, they still have defects in anti-noise, end effect, and incorrect separation of signals with similar frequencies.

Variational modal decomposition (VMD) has been proposed by Dragomiretskiy and Zosso [4]. It can adaptively decompose vibration signals into a number of band-limited intrinsic mode functions (BLIMFs) and get rid of a series of problems caused by recursive decomposition methods such as EMD. Many scholars have successfully applied VMD to signal decomposition of rolling bearing fault diagnosis [5-6]. VMD needs to set the number K of BLIMFs, and many scholars have discussed the setting of K value [7-8]. If we setup initial K values firstly, then sensitive modal components in every fault condition can be intelligently selected based on data drive. It not only won't cause excessive decomposition, but also be favourable for removing redundant modal components and accelerating the identification efficiency. It is the point of this study.

On the other hand, selecting an appropriate classifier to intelligently identify the bearing fault is important. Many scholars have tried various intelligent algorithms to identify rolling bearing fault, including support vector machine (SVM), artificial neural network (ANN) and other intelligent algorithms. However, all kinds of algorithms also have their own defects. For example, SVM has low classification efficiency when solving multi-classification problems. In ANN, network structures are difficult to determine and easy to fall into local minima when training data.

Mahalanobis Taguchi system (MTS) has been proposed by quality engineering expert Dr.Genichi Taguchi in 1990s, which is a multi-system pattern recognition method [9]. MTS is successful because it has many advantages. MTS is based on data driven without relying on the data distribution assumption. It can realize dimension reduction based on OAs and SN ratios. Diagnosis and prediction are simple and fast. In recent years, many researchers have used MTS into rolling bearing fault diagnosis.

The paper decomposed the vibration signal into multiple BLIMFs through VMD, and extracted multiple fault features in each component. At this time, for each vibration signal, the number of fault features is huge. Due to the limitation of OAs in designing orthogonal experiments, MTS is difficult to optimize the classification system containing many variables. Therefore, firstly, this paper adopted Mahalanobis Taguchi system based on multiple mahalanobis distance. Feature subsets replaced individual features in participating in the construction of the classifier to solve the problem of large number of features. At the same time, OAs and SN ratios were used to select important feature subsets in each signal. Thus, VMD was combined with MMD-MTS classifier. Secondly, due to MTS is designed to solve binary classification problem, this paper built Multi-tree MTS (MT-MTS) based on training data to solve the fault diagnosis of rolling bearings. And the diagnosis results using other algorithms were compared with improved MT-MTS based on VMD and MMD to verify this algorithm's rationality and effectiveness.

2. Variational mode decomposition and features extraction

2.1. Variational mode decomposition

VMD obtains adaptive signal decomposition thought searching optimal solution of constrained variational model. Accordingly, the input signal is decomposed into a series of modal component with sparse properties.

In VMD, the square norm L^2 of the gradient of the demodulated signal is calculated, and the bandwidth of each mode component is estimated. The expression for the corresponding constrained variation model is as follows:

$$\min_{\{u_k\},\{\omega_k\}}\left\{\sum_k\left\|\partial_t[(\delta(t)+\frac{j}{\pi t})u_k(t)]\mathrm{e}^{-j\omega_k t}\right\|^2\right\} \quad \text{s.t} \quad \sum_k u_k=f \tag{1}$$

Where, $\{u_k\}=\{u_1,\cdots,u_k\}$ represent the K BLIMF after decomposition, and $\{\omega_k\}=\{\omega_1,\cdots,\omega_k\}$ represent central frequency of each component.

The second-order penalty factor ∂ and Lagrange multiplying operator $\lambda(t)$ are introduced to solve the optimal solution of the constrained variational problem. The extended Lagrange expression is as follows:

$$L(\{u_k\},\{\omega_k\},\lambda) = \alpha\sum_k\left\|\partial_t[(\delta(t)+\frac{j}{\pi t})u_k(t)]\mathrm{e}^{-j\omega_k t}\right\|_2^2+\left\|f(t)-\sum_k u_k(t)\right\|_2^2+<\lambda(t),f(t)-\sum_k u_k(t)> \tag{2}$$

The "saddle point" of extended Lagrange expression is calculated via the alternating direction method of multiplier algorithm. The detail algorithm can be found in [4].

2.2. Choosing sensitive BLIMFs in the diagnostic process

When signals are decomposed using VMD, K value need to be determined. Many scholars have studied on determining K value based on the principle of similar center frequency, kurtosis selection criteria, particle swarm optimization, etc. However, in order to unify measurement in all states, the number K of modal components in each state are required to be equal to participate in the diagnosis and identification of the classifier.

Firstly, the initial K value is given to decompose the vibration signal by VMD base on the principle of similar center frequency. Then, because MTS can optimize features, invalid modal components are eliminated in MTS' diagnosis process. It can reduce computational complexity and improve the diagnosis accuracy. See the following section for specific elimination methods. This method can not only obtain modal components that affect the diagnosis result and achieve the best diagnosis effect, but also eliminate the invalid modal components based on the diagnosis process, making the signal decomposition method closely combined with the intelligent classifier.

2.3. BLIMFs' features extraction

Vibration signals contain a large amount of information during mechanical operation. The success of fault diagnosis depends on the selection of BLIMFs' features. Time domain features are intuitive and easy to understand, which the original basis for signal fault diagnosis is. Extracting time domain features can judge the running state. Therefore, this paper extracts 10 time domain features of each modal component to reflect the weak change of vibration signal in time domain.

However, time domain features are difficult to determine the location, type and severity of the faults. Therefore, this paper continues to extract two frequency domain features from each modal component through fast Fourier transform. Frequency domain features can distinguish different faults. To sum up, a total of 12 features in time domain and frequency domain are extracted for each modal component. Then the total number of features in the system should be $12 \times K$. K is the number of components in modal decomposition. All selected features are presented in table 1.

Table.1 Selected features.

Time domain features		Frequency domain features
Mean	Wave index	Center frequency
Root mean square	Peak index	Center amplitude
Standard deviations	Impulsion index	
Skewness	Tolerance index	
Kurtosis	Kurtosis index	

3. The improved multi-tree Mahalanobis Taguchi system

3.1. Mahalanobis Taguchi system

MTS includes Mahalanobis space (MS) and Taguchi method. MS is constructed as a reference space by using normal samples of standardized data. Its core is Mahalanobis distance (MD), which is used to measure the similarity between unknown samples and known samples. In Taguchi method, orthogonal arrays (OAs) and signal-to-noise (SN) ratios are used to measure the contribution of each variable in system. Selecting important variables from the original variables is essential to MTS. Finally, unknown samples are identified by threshold. More details of the MTS algorithm can be found in [9]. In this paper, use the threshold T [10] to differentiate between normal and abnormal samples:

$$T = \mu + 2.66\overline{R} \tag{3}$$

Where μ is mean of the normal samples MDs, $\overline{R}$ is mean of normal samples MD moving ranges.

3.2. Multiple Mahalanobis distance

In the process of multiple pattern recognition, the selection of features not only determines the recognition ability of the system, but also determines the recognition speed of the system. However, as the number of features increases, classifiers become very difficult to process large data sets. Also, MTS is more difficult to choose the important features through OAs and SN ratio. Therefore Dr.Genichi Taguchi designed MMD method.

In MMD method, a large number of features are divided into feature subsets. MD of each feature subset is calculated to obtain a new measurement for measuring the huge features, which is recorded as MMD. MMD is introduced to compute SN ratios and thereby select suitable feature subsets. MMD method is suitable for processing a large number of features, and it can effectively reduce the complexity of the problem. The specific steps of MMD-MTS method are as follows:

Step 1: Defines feature subsets of the original features. These feature subsets may be extracted in different modes. Meanwhile, the number of variables in these feature subsets can be inconsistent.

Step 2: MD corresponding to each feature subset of normal samples and abnormal samples is calculated and recorded as MD_{nor} and MD_{abn}. Then the square root of the MD_{nor} and MD_{abn} are calculated and recorded as MMD_{nor} and MMD_{abn}.

Step 3: These feature subsets are regarded as feature variables (control factors). The MD_{nor} and MD_{abn} would provide required data for these subsets. If there are k feature subsets, the problem will be transformed into a new multi-system pattern recognition problem with k variables. At this time, MMD_{nor} and MMD_{abn} calculated according to MD_{nor} or MD_{abn} will be used as the new measurement to participate in the construction and validity analysis of MTS.

Step 4: For each experiment of OA, the SN ratio is calculated by MMD_{abn} and the important feature subsets are identified.

Step 5: According to the optimized feature subsets of the system, the unknown samples are identified and predicted thought the new measurement and threshold.

3.3. MMD-MT-MTS

MTS is used to solve the problem of binary classification in multivariate pattern recognition. However, there are a lot of multi-classification problems in practice. Therefore, constructing reasonable multi-classification rules are an important means to improve the classification accuracy of classifiers.

In computer science, a multi-tree is an ordered tree in which each node contains multiple subtrees. Different hierarchical structures of multi-tree have great influence on classification accuracy. If a classification error occurs at a certain node, the error will be extended to the next node, and the error rate of subsequent classification will be higher and higher. Therefore, it is necessary to select the appropriate multi-tree structure. Firstly, we divide the classes that are easy to be divided (which are not easy to generate misclassification), and then divide the classes that are not easy to be divided. Thus the possible errors can be kept away from the roots as far as possible. MMD-MT-MTS is divided into the following four steps:

Step 1: Each class executes MMD-MTS. Calculate recognition accuracy of each class as reference space for the same testing set. The class with the highest recognition accuracy is selected as the reference space of root node. The root node samples data are removed in training set and testing set. The MMD-MTS is executed for every remaining category again and the recognition accuracy is calculated. The class with the highest recognition accuracy is taken as the reference space of the node in the second level. And so on until every level node is found.

Step 2: If a class contains many small classes, it is first regarded as a large class to execute step 1. When the class is divided as "leaf" node, the small categories begin to be divided. Calculate recognition accuracy of each small categories as reference space for the same testing set. The class with the highest recognition accuracy is selected as the reference space of the node. Chen [11] introduces MTS to identify Fisher's irises data based on MD range and demonstrates the effectiveness of MTS in multi-classification. Therefore, choose a special category as the reference space and the MMDs range are used as the selection path. Then the nodes are divided into endpoints of small categories. Finally, the multi-tree structure is constructed.

Step 3: Once the multi-tree is established, the important feature sets are selected using OAs and SN ratios for each node. At the same time, threshold and MMD range is calculated as branch path.

Step 4: The testing samples are identified thought the MMD-MT-MTS model.

4. Fault diagnosis of rolling bearings based on MMD-MT-MTS

The vibration signal of rolling bearing is decomposed into K modal components by VMD, and features are extracted for each component. The features are extracted from this K modal components are regarded as K feature subsets and used to construct the initial feature subsets of MMD-MT-MTS. Then a multi-tree structure is constructed and MMD-MT-MTS is executed to realize intelligent diagnosis of multiple types of faults. The rolling bearing fault diagnosis method based on VMD and MMD-MT-MTS includes three stages. The specific steps and flow chart are shown as follows:

Step 1: For rolling bearing vibration signals with multiple states, VMD is used to decompose the original vibration signals, and the time domain and frequency domain features of each component are extracted to construct the initial feature subsets of MMD-MT-MTS.

Step 2: The collected features data are divided into training sets and testing sets according to cross-validation rules. The training samples are used to construct a proper multi-tree, and the reference space of each node and the branch paths based on threshold or MMD ranges are determined to construct an MMD-MT-MTS classifier.

Step 3: Through MMD-MT-MTS classifier, fault diagnosis is carried out on testing samples to identify fault types.

The algorithm flow chart is shown in figure 1.

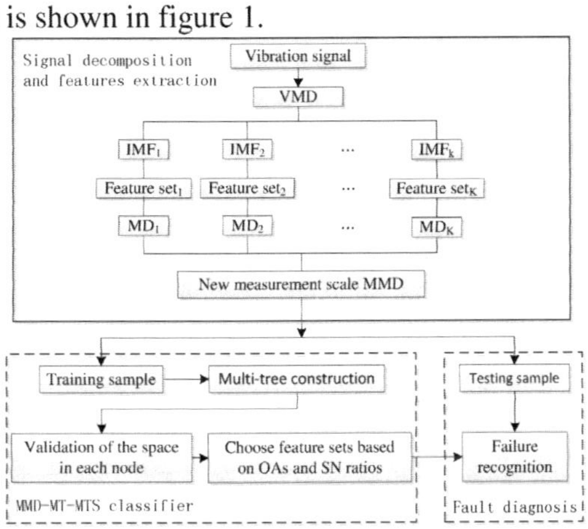

Figure 1. The flow chart of rolling bearing fault diagnosis.

5. Results and discussion

In order to verify the validity of VMD and MMD-MT-MTS in mechanical fault diagnosis, bearing data were collected from the experimental rig of rolling bearing failure simulation in the Electrical Engineering Laboratory of Case Western Reserve University. The experimental equipment is shown in Figure 2.

Figure 2. Experimental apparatus

Motor bearings were seeded with faults using electro-discharge machining (EDM). Faults ranging from 0.007 inches in diameter to 0.021 inches in diameter were introduced separately at the inner raceway (IR), rolling element (i.e. ball) (B) and outer raceway (OR). In this paper, the diameter of damage is 0.007 inches for early failures, 0.014 inches for moderate failures and 0.021 inches for severe failures. According to this classification, bearing failure types are determined to be 10 categories, and the specific category labels are shown in Table 2.

Table 2. The labels of bearing fault categories.

Bearing state	Label	Fault diameter (inches)
N	1	0
IR	2/3/4	0.007/0.014/0.021
B	5/6/7	0.007/0.014/0.021
OR	8/9/10	0.007/0.014/0.021

The original vibration signal is divided into a plurality of sample signals by a window of the same length. The vibration signal images in the normal state and the other three early fault states are plotted, as shown in Figure 3. The 100 vibration signals data in every state were selected, then the sample size is 1000. The validity of VMD and MMD-MT-MTS in rolling bearing fault diagnosis was verified by 5-fold cross-validation, that is, 80 groups of samples in every state were taken as training set and the remaining 20 groups were taken as testing set in each experiment.

5.1. Signal decomposition and feature subsets construction

1000 signals in the sample were decompose by VMD. Based on the principle of centre frequency approximation [7], the number $K=6$ of BLIMFs in VMD was preset. Figure 4 shows the BLIMFs of a signal decomposed by VMD in the normal state. At the same time, the time domain and frequency domain characteristics of each modal component in the signal are extracted based on Matlab software, and the Mahalanobis distance of each modal component is calculated as characteristic parameter of this components.

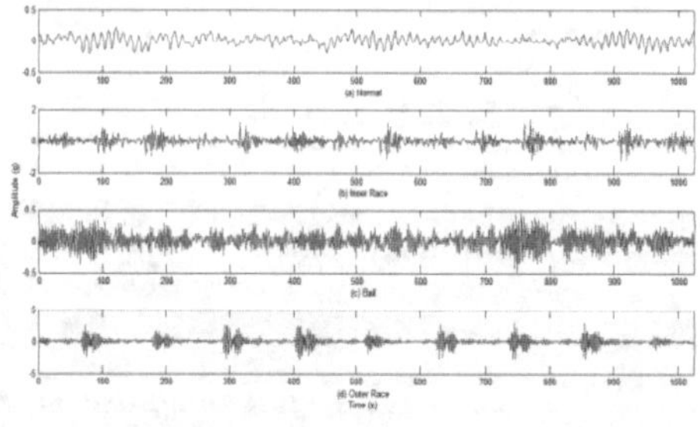

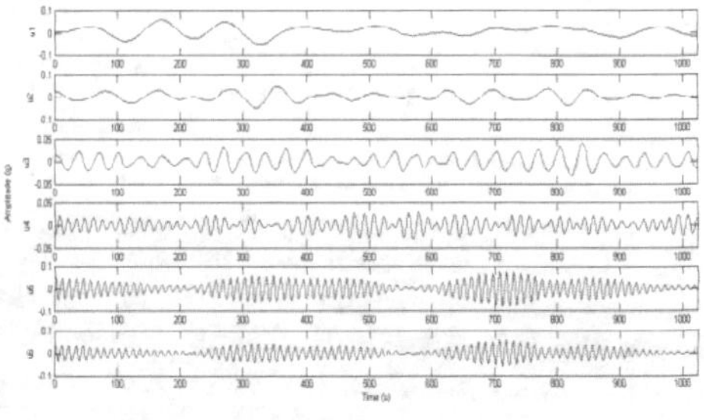

Figure 3. Rolling bearing vibration signals. **Figure 4.** Component signals decomposed by VMD.

5.2. Construction and fault identification of MMD-MT-MTS classifier

This section takes a cross-validation as an example to illustrate the experiment.

Step 1: Based on the principle of 5-fold cross-validation, the training samples were selected, totalling 800 samples. The features of each modal component are extracted as a feature subset of each sample. The MDs of these feature subsets are calculated and the root mean square of these MDs is calculated as a new measurement scale to participate in the construction of the classifier.

Step 2: Build a multi-tree structure. Firstly, the different damage degree of each state as a class participates in the selection of root nodes. Calculate the accuracy of the testing set with the four states as the reference space, as shown in Table 3. As can be seen from Table 3, in the first level, the classifier has the highest accuracy when the normal state as the normal sample and the other three fault states as the abnormal sample. Therefore, the root node selects the MMD-MTS classifier with the normal state as the reference space.

After the root node is determined, the normal state samples are removed. The training set only has samples in IR, B and OR fault state. Continue with the MMD-MTS for these three states and calculate the accuracy of the testing set, as shown in Table 3. The level 2 node selects the MMD-MTS classifier with IR fault sample as the reference space. By analogy, the level 3 node selects the MMD-MTS classifier with B failure sample as the reference space.

After each fault state is divided into node, the different damage degree of each state began to be divided. It is interesting to find that the recognition accuracy is highest when the early failure as the reference space. Thus, the reference space is constructed based on the samples of the early failure, and MMD-MTS is executed to classify the damage degree and form leaf nodes, so as to construct a complete multi-tree structure, as shown in Figure 6.

Table 3. Executed accuracy of MMD-MTS algorithm.

Multi-tree levels	N	IR	B	OR
The level 1	100%	97.5%	97.5%	97%
The level 2	removed	97.7%	96.7%	96.1%
The level 3	removed	removed	99.2%	97.5%

After confirming the structure of multi-tree, each node executes MMD-MTS. Calculate the MMDs of normal and abnormal samples in training samples, and verifies the validity of the reference space of each node. The calculated MMDs are shown in Figure 5.

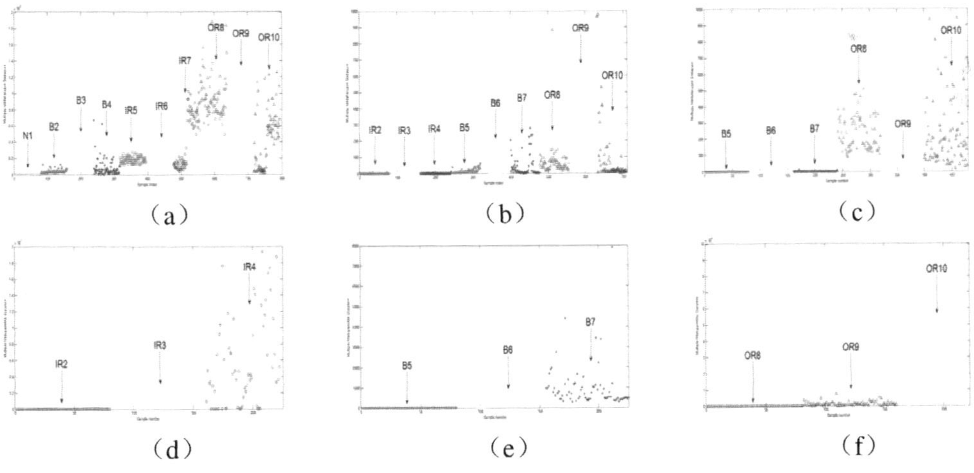

Figure 5. MMD calculated with N, IR, B, IR2, B5 and OR8 samples as reference space respectively.

According to the constructed multi-tree, MMD-MT-MTS is performed to select the feature subsets of the four states. The OAs and SN ratios response values of each component in the four states are calculated by MMD, and the sensitive modal components are selected as shown in Table 4. At the same time, the path conditions of each branch of the multi-tree are determined according to the thresholds or MMD ranges calculated by the MMD-MT-MTS. The results and multi-tree construction are shown in Figure 6.

IOP Conf. Series: Materials Science and Engineering **692** (2019) 012034 doi:10.1088/1757-899X/692/1/012034

Table 4. The selected BLIMFs of fault signals.

State	The selected BLIMFs
N	IMF1,IMF2,IMF4,IMF6
IR	IMF1,IMF2,IMF4,IMF5,IMF6
B	IMF2,IMF3,IMF5
OR	IMF4,IMF5,IMF6

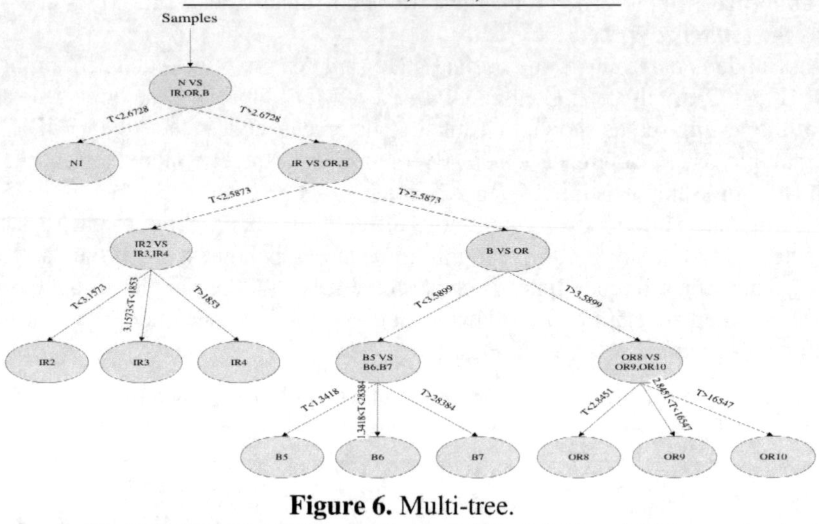

Figure 6. Multi-tree.

Step 3: The optimized MMD-MT-MTS is performed. The MMDs of 200 samples in the testing set are calculated with a different optimized reference space in each level node. Then these MMDs compared with the branch path. Diagnostic results and accuracy of the testing samples are shown in Table 5.

Table 5. The result of MMD-MMTS fault diagnosis.

		Testing results										
		N1	IR2	IR3	IR4	B5	B6	B7	OR8	OR9	OR10	Accuracy
Actually class	N1	20	0	0	0	0	0	0	0	0	0	100%
	IR2	0	20	0	0	0	0	0	0	0	0	100%
	IR3	0	0	20	0	0	0	0	0	0	0	100%
	IR4	0	0	1	19	0	0	0	0	0	0	95%
	B5	0	0	0	0	20	0	0	0	0	0	100%
	B6	0	0	0	2	0	18	0	0	0	0	90%
	B7	0	0	0	0	0	1	18	1	0	0	90%
	OR8	0	0	0	0	0	0	0	20	0	0	100%
	OR9	0	0	2	0	0	0	0	0	18	0	90%
	OR10	0	0	0	0	0	0	0	0	0	20	100%
Accuracy		96.5%										

5.3. Discussion

After 5-fold cross-validation, the identification results of various states of rolling bearings are shown in Table 4. The accuracy of the 200 test data predictions was 96.5%. The experimental results show that the rolling bearing fault diagnosis algorithm based on VMD and MMD-MMTS is effective and accurate.

In the MMD-MMTS algorithm, the effective feature subsets of each state can be effectively identified by OAs and SN ratios. The system selected four sensitive BLIMFs in the normal state and 3-5 sensitive BLIMFs in other states. It realizes the components reduction based on the diagnosis process. The effective modal components obtained through the screening can be used to assist the subsequent fault diagnosis work.

In addition, Table 6 lists the accuracy of the diagnostic recognition based on the same database with other algorithms. As can be seen from Table 6, the VMD and MMD-MMTS algorithms have higher recognition accuracy than other algorithms. This also shows that the improvement of the method in this paper has practical significance.

Table 6. The accuracy of algorithms.

Signal decomposition and classification algorithm	Accuracy(%)	The number of features
VMD+MMD+MT+MTS	96.5	3-5
EEMD+AMTS	94	9
VMD+SVM	95.5	3-5

6. Conclusion

Based on VMD and MMD-MT-MTS method, an intelligent fault diagnosis method is presented in this paper. The vibration signal is decomposed into several modal components by VMD, and the extracted features of each component construct the feature subsets. The MDs of the subsets are regarded as the characteristic value of the signal. Taking the root mean square of the MDs as a new measurement scale, we participate in the construction of the MMD-MTS. At the same time, a multi-tree is constructed for multi-class fault identification, so as to realize the intelligent fault diagnosis of rolling bearings. Compared with other research methods, the advantages of this paper are shown as follows:

(1) Based on the diagnosis process, the sensitive modal components in each state are selected reasonably through OAs and SN ratios of the MMD-MTS. The redundant modal components are eliminated, and the diagnosis efficiency is improved. The VMD signal decomposition method is closely combined with the MMD-MT-MTS classifier based on finding the key modal components in the diagnosis process.

(2) In order to solve the problem of the large number of features, MMD method is used to construct MMD-MT-MTS classifier in the form of feature subsets and new measurement scale. It not only fully reflects the fault characteristics, but also optimizes the subsets of features based on the advantages of the classifier itself. It realizes the real dimension reduction of the system, and the recognition effects are very good.

(3) A multi-classification MTS based on multi-tree is proposed. By constructing MMD-MT-MTS for multi-classification fault diagnosis of rolling bearings, the severity of rolling bearings under different fault states can be identified effectively. This method not only solves the limitation that MTS is binary classifier, but also makes the multi-classification recognition based on multi-tree, which is easy to operate and has high classification accuracy.

In the future, MMD-MT-MTS will be applied to more equipment for mechanical fault diagnosis. At that same time, the method can be extended to more computer technology field, just like speech recognition and face recognition.

References

[1] Gu X and Chen C 2017 *International Journal of Rotating Machinery* pp 1–12.
[2] Huang N E, Shen Z, Long S R, Wu M C, Shih H H, Zheng Q, Yen N C, Tung C C and Liu H H 1998 *Proceedings of the Royal Society* 454 903-93.
[3] Wu Z H and Huang N E 2009 *Advances in Adaptive Data Analysis* 1 1-41.
[4] Dragomiretskiy K and Zosso D 2014 *IEEE Transactions on Signal Processing* 62 531-44.
[5] Yi C, Lv Y and Dang Z 2016 *Shock and Vibration* 1–10.
[6] Li K, Su L, Wu J J Wang H and Chen P 2017 *Applied Sciences* 7 1004.
[7] MA Z Q, LI Y C, LIU Z and Gu C J 2016 *Journal of Vibration and Shock* 35 134-9. (in Chinese)
[8] Tang G J and Wang X L 2015 *Journal of Xi'an Jiaotong University* 49 73-81. (in Chinese)

[9] Taguchi G and Jugulum R 2000 *Sankhyā: The Indian Journal of Statistics* 62 233-248.
[10] Sheng Z R, Cheng L S and Gu Y P 2017 *Journal of Applied Statistics and Management* 36 1059-68. (in Chinese)
[11] Chen X L, Gu Y P and Han Z J 2010 *Mathematics in practice and theory* 40 74-9. (in Chinese)

IOP Conf. Series: Materials Science and Engineering **692** (2019) 012035 doi:10.1088/1757-899X/692/1/012035

A Multi-Objective Model with Probability Constraints of Trans-Provincial Hydropower Consumption Considering Dynamic Stability Limit

Ke Yang[1], Rui Mao[1], Xu Wen[1,2,3], Chuncheng Gao[4,5], Miao Huang[3] and Wei Yan[2]

[1]Southwest Subsection of State Grid, Chengdu, China
[2]State Key Laboratory of Power Transmission Equipment & System Security and New Technology(Chongqing University), Chongqing ,China
[3]College of Automation, Chongqing University of Posts and Telecommunications, Chongqing, China
[4]Beijing Ke Dong Electric Power Control System Co., Ltd.Beijing, China
[5]Nari Group Corporation/State Grid Electric Power Research Institute, Nanjing, China

E-mail: wenxu@cqu.edu.cn

Abstract. With the rapid development of clean energy and serious delay of grid construction, water abandonment is becoming more and more severe. Under this condition, this paper proposed a multi-objective model with probability constraints of trans-provincial hydropower consumption considering dynamic stability limit. Firstly, a mathematical expression about the dynamic stability limit of trans-provincial channel is constructed. Secondly, a risk assessment index of abandoned water is established on the basis of semi-absolute deviation. Then, based on the framework of stochastic programming, a multi-objective model with probability constraints is proposed for hydropower transmission in trans-provincial trade. By means of genetic algorithm with embedded Monte-Carlo simulation technology, the model can be solved efficiently. Finally, data of Tibet power grid is calculated to verify the validity and universality of the research. The results show that the established model can achieve optimal economic benefit between grid companies and generation enterprises, and also effective management of water abandonment risk.

1. Introduction

With the gradual shortage of primary energy, clean energy such as hydropower has been rapidly developed[1]. However, the reverse distribution of electricity supply and demand in China determines that the large-scale hydropower resources in southwestern China must be optimized in the whole country. However, due to many factors, such as the incoordination between network and source planning, large-scale water abandonment in southwest region is becoming more and more serious[2]. Therefore, it is of great practical significance to fully explore the capacity of trans-provincial channels, maximize economic benefits and reduce abandoned water risks by means of market-oriented methods. Foreign researches emphasize more on the coordination of the income and economic risk from the perspective of purchasing or selling power[3,4,5]. However, resource endowment, power system, network structure and market reform process in foreign countries are different from ours. Moreover, their market subjects don't take the responsibility of electric quantity balance and safe grid operation, which make it is generally not necessary to consider the influence of network constraints on the decision-making of purchasing and selling. Therefore, foreign research results cannot be applicable to our "province as entity" transaction decision-making under China's electric power system.

Considering the contradiction between high hydropower supply and low demand, the requirement of abandoned water management, existing literatures mainly adopt definite indexes, such as installed capacity, peak shaving, to evaluate and abandoned water [2]. However, most hydropower outputs are stochastic, the above indexes are only suitable for "ex-post" deterministic assessment. They can' t meet the requirements of "ex-ante" risk assessment and management of abandoned waters in random environment.

In view of this, from the perspective of promoting the optimal allocation of clean energy across the province, under the unified national electricity market, this paper carries out pioneering research in the following three aspects:1) establishes the mathematical expression of dynamic stability limit of trans-provincial transmission channel, which realizes the transformation from "static" to "dynamic" management in trans-provincial transaction; 2) constructs a risk assessment index based on semi-absolute deviation for abandoned water, which extends the assessment from certainty to uncertainty and puts forward the concept of risk in advance.3) proposes a risk of management and control model of transferring hydropower considering dynamic stability limit, which realizes the effective management of abandoned water risk. Based on the dynamic stability limits processing technique of trans-provincial transmission channel, the multi-objective genetic algorithm of embedded monte-carlo simulation technique is used to solve the model. Take Tibet power grid as an example, this paper verifies the effectiveness of the above work. Finally, universality of the study is discussed.

2. Establishment of "dynamic" stability limit expression of trans-provincial channel and risk assessment index of abandoned water

2.1. Equation of "dynamic" stability limit of trans-provincial channel

In the existing trans-provincial trading model, it is assumed that the trans-provincial transmission channel is static[6,7]. In fact, the influencing factors of the stability limit of the trans-provincial transmission channel are extremely complex, including the number of starting units (or) boot capacity, the cutting capacity of key units, the output of key units, and the transmission power of associated coupling section, etc. At the same time, the influence factors of the same trans-provincial transmission channel are different in opposite power directions. In order to realize the "dynamic" management of trans-provincial transmission channel limit, the dynamic equation is constructed.

$$P_{\mathrm{ex},k,\max}=\begin{cases}P_{\mathrm{ex},k,1,\max} & \begin{array}{l}\text{if } N_G=1;\\ P_{G,i,1,\max}>P_G>P_{G,i,1,\min};\\ P_{G,\mathrm{Q},\max}>P_{G,\mathrm{Q}}>P_{G,\mathrm{Q},\min};\\ P_{gl,\max}>P_{gl}>P_{gl,\min};\\ \cdots\cdots\end{array}\\ \cdots\cdots & \\ P_{\mathrm{ex},k,n,\max} & \begin{array}{l}\text{if } N_G=n_G;\\ P_{G,i,n,\max}>P_G>P_{G,i,n,\min};\\ P_{G,\mathrm{Q},\max}>P_{G,\mathrm{Q}}>P_{G,\mathrm{Q},\min};\\ P_{gl,n,\max}>P_{gl}>P_{gl,n,\min};\\ \cdots\cdots\end{array}\end{cases} \tag{1}$$

In the formula, $P_{ex,k,n,\max}$ represents the dynamic quota value of the k-th inter-provincial channel under the influence of the nth factor; N_G represents the number of key units that affect the quota of inter-provincial channel n_G ; P_G , $P_{G,i,\max}$ and $P_{G,i,\min}$ respectively represent the starting output and corresponding upper and lower limits of key units that affect the inter-provincial channel limit; P_{gl} , $P_{gl,\max}$ and $P_{gl,\min}$ respectively represent the transmission power and corresponding upper and lower limits of the associated channels that affect the inter-provincial channel quota; $P_{G,Q}$, $P_{G,Q,\max}$ and $P_{G,Q,\min}$ respectively represent the cutting capacity and corresponding upper and lower limits that affect the trans-provincial channel limit.

2.2. Construction of a risk assessment index based on semi-absolute deviation for abandoned water

The factors affecting abandoned water are extremely complex.From the consumption of electricity market perspective, the existing assessment indicators are deterministic[2]. However, most hydropower output is characterized by randomness. Those indicators are only suitable for post-assessment, which can't meet the requirements of prior risk assessment and management in the random environment. Therefore, a risk assessment index considering the randomness of hydropower output and referring to the basic concept of semi-absolute deviation risk[6] is constructed as follows:

$$\mathrm{E}\left[\left|\sum_{k=1}^{N_{\mathrm{w}}}(W_{\mathrm{w},k}-W_{\mathrm{w},k,m})\right|_{-}\right] \quad (2)$$

In the formula, $E[\bullet]$ represents the expectation operator; N_W represents the number of hydropower units; $W_{w,k,m}$ represents the monthly random electric quantity of the k-th hydropower unit; $W_{w,k}$ represents the expected monthly electric quantity of the k-th hydropower unit.

It can be known from equation (2), when the actual electric quantity of the hydropower unit is lower than the expected value, that is, there is the risk in abandoned water. It can be seen that the constructed risk assessment index inherits the definition method of risk severity degree of semi-absolute deviation and at the same time, this index extends the existing assessment of abandoned water from certainty to uncertainty. Based on the above, this paper puts forward the concept of risk management of abandoned water, which can realize the pre-assessment and pre-control.

3. Establishment of a risk management and control model of transferring hydropower in electricity market

3.1. Modeling ideas and assumptions

Considering more random factors and also fully exploring the market optimization space and trans-provincial transmission channel capacity, the model is built within the framework of stochastic programming.

Referring to the literature[7], the influence of monthly electric power, electric quantity and unit output on power network security and market trading decisions are described in typical periods, which is divided into monthly peak, flat and valley, of monthly market. Meanwhile, the following assumptions are also made: 1) Random factors, such as load and electricity price, are all subjected to follow normal distribution; 2) Network losses are ignored in the trade; 3) The annual priority of power generation plan has been deducted from the total load of the monthly plan and is no longer shown separately; 4) Each hydropower station and thermal power plant is equivalent to one unit.

3.2. Model construction

3.2.1 Objective function. Aiming at minimizing the expectation of the total purchase cost and the abandoned water risk in monthly trans-provincial and inter-provincial market, the risk management and control model of abandoned water is established as follows:

$$\min \sum_{k=1}^{N_{\mathrm{w}}} u_{\mathrm{w},k} W_{\mathrm{w},k} + \sum_{k=1}^{N_{\mathrm{t}}} u_{\mathrm{t},k} W_{\mathrm{t},k} - \sum_{k=1}^{N_{\mathrm{ex}}} u_{\mathrm{ex},k} W_{\mathrm{ex},k} \quad (3)$$

$$\min \ \mathrm{E}\left[\left|\sum_{k=1}^{N_{\mathrm{w}}}(W_{\mathrm{w},k}-W_{\mathrm{w},k,m})\right|_{-}\right] \quad (4)$$

In the formula: $u_{w,k}$, $u_{t,k}$ and $u_{ex,k}$ respectively represent the unit price expectations of the k-th hydropower, thermal power and transferring power; N_t , N_{ex} respectively represent the number of thermal power units and transferring power units.

3.2.2 Constraints

1) Opportunity constraint of "dynamic" stable limit in trans-provincial channel

$$\alpha_{\mathrm{s,ex}} = \Pr\left\{P_{\mathrm{s,ex},k} \geq P_{s,\mathrm{ex},k,\max}\right\} \leq \alpha_{\mathrm{s,ex,max}} \quad s = 1,3 \tag{5}$$

In the formula: $\Pr\{\}$ represents the probability operator; When s is set as 1, 2 and 3 respectively, it corresponds to peak, plat and valley load; $a_{s,ex}$, $a_{s,ex,\max}$ represent the stability limit probability and corresponding limit value under the typical load s; $P_{s,ex,k}$ and $P_{s,ex,k,\max}$ respectively represent the k-th unit power of transmission under s load and the corresponding stability limit of transmission channel, as shown in equation (1).

2) Opportunity constraints of monthly electricity balance considering hydropower transmission

$$\gamma_1=\Pr\left\{W_{\mathrm{d,max}} \leq \sum_{k=1}^{N_{\mathrm{w}}} W_{\mathrm{w},k,m} + \sum_{k=1}^{N_{\mathrm{t}}} W_{\mathrm{t},k} - \sum_{k=1}^{N_{\mathrm{ex}}} W_{\mathrm{ex},1,k}\right\} < \gamma_{1,\max} \tag{6}$$

$$\gamma_3=\Pr\left\{W_{\mathrm{d,min}} \geq \sum_{k=1}^{N_{\mathrm{w}}} W_{\mathrm{w},k,m} + \sum_{k=1}^{N_{\mathrm{t}}} W_{\mathrm{t},k} - \sum_{k=1}^{N_{\mathrm{ex}}} W_{\mathrm{ex},3,k}\right\} < \gamma_{3,\max} \tag{7}$$

In the formula: γ_s is the risk probability value that beyond the limit in the monthly electricity balance (s=1 means it beyond the upper limit, s=3 means it beyond the lower limit); $W_{d,\max}$ and $W_{d,\min}$ represent the maximum and minimum monthly electric quantity respectively; $\gamma_{1,\max}$ and $\gamma_{3,\max}$ respectively represent the limit value of the probability of monthly electricity balance which beyond the limit.

In addition, the remaining constraints of the model include:peak shaving in the state of monthly peak and valley load, monthly typical branch power flow,power balance equation of dc power flow under monthly peak and valley load, power equation of branch power flow under peak and valley load, upper and lower limits of power of thermal power and hydropower units, upper and lower limits of electric quantity of thermal power and hydropower units, etc.

4. Algorithm for solving the model

In essence, the established model is a multi-objective model with probability constraints[8,9]. By referring to the literature[10], the model is solved within the framework of genetic algorithm, and the fitness of multi-objective function is calculated by the method of relative superiority of targets. The steps of the algorithm include input of basic data, generation of initial population, calculation of stability limit of trans-provincial channel, opportunity constraints test, calculation of fitness function, operation of selection, cross and mutation, acquisition of hydropower transmission scheme and water abandonment risk, etc. Due to the limitation of space, this paper focuses on the algorithm processing skills, such as calculating dynamic stability limit of trans-provincial channel, which cannot be processed in literature[6,7].

(1)Generation of each iteration population. On the basis of generating population with upper and lower constraints of hydropower and thermal power units, each external power transmitted can be obtained by the power balance equation of typical load in each month. This method guarantees non-random generation to avoid seeds that do not meet the equation constraints, and also improves the solving efficiency indirectly.

(2) Calculation of dynamic stability limit of trans-provincial channel.In the process of each iteration of the population, dc power flow equation is used to calculate the branch power flow. According to the output state of each unit, the key information, such as the number of key units starting up, the output of key units, and the transmission power of associated channels that affect the trans-provincial channel limit, and formula (1), the dynamic stability limit of trans-provincial channel corresponding to each iteration of the population is obtained.

(3)Inspection of opportunity constraints. In order to improve the inspection efficiency, it is suggested to use Latin hypercube sampling technology [11] to generate random samples, such as hydropower output in typical load, and to check whether the opportunity constraints meet the conditions.

5. Verification of calculation examples

5.1. Basic date

The basic data of July 2017 of Tibet power grid was used to verify the validity of the research. The basic data are set as follows: Monthly electric quantity of load is 400 million kilowatt-hours; Monthly maximum and minimum load is 1 million, 0.5millon kilowatts respectively; Standard deviation of electric quantity of each month is determined as 20% of expected value; The probability of each opportunity constraint is 10.0%. The maximum stability limit of Chaila DC is 430,000 kilowatts, and the maximum stability limit of Chuanzang AC 130,000 kilowatts. Latin hypercube sampling scale is 500 times; The population size of genetic algorithm is 80, the crossover probability is 0.60, the mutation probability is 0.10, and the maximum iteration number is 300. The criterion of iteration termination is that the optimal individual remains unchanged for 30 consecutive generations or reaches the maximum iteration number.

5.2. Analysis of the impact of dynamic stability limit of trans-provincial channel on clean energy consumption

Based on the simulation conditions in section 4.1, the proposed model is compared with the treatment method in stability limit of trans-province channel in literature [6, 7], and the results are shown in Table 1.

Table 1. Comparison of hydropower transmission considering trans-provincial stability limit.

Solution	Power purchase cost /One hundred million RMB	Electricity of abandoned water/one million kilowatt-hours	Consumption of transmission/one million kilowatt-hours
Literature[6]	1.40	0.53	1.81
Literature[7]	1.46	0.67	1.89
In this paper	1.58	0.83	1.61

As can be seen from table 1, because the model established in this paper takes the dynamic stability limit of the trans-provincial channel into account, the electricity purchase cost and electricity of abandoned water increase greatly, while the electricity consumption of transmission decreases. The essential reason is that this paper strictly considers the influence of the dynamic stability limit of trans-provincial channel on clean energy external consumption. The stability limit of the trans-provincial channel is no longer set as the fixed number, but is changed dynamically with the boot combination, cutting capacity and other factors. Traditional literatures do not consider the dynamic change of trans-provincial channel limit, which is not consistent with the actual operation of power grid.

5.3. Validation of the effectiveness of risk management and control of water abandonment and analysis of its impact on the external consumption of clean energy

Keep the simulation conditions unchanged, make comparisons with the two schemes: scheme 1 considers the index of water abandonment risk, while scheme 2 doesn't consider that index. The results are shown in Tab.2 which validates the management of model on abandoned water risk and its influence on the external consumption of clean energy.

Table 2. Comparison of consumption of hydropower transmission considering water abandonment risk

Scheme	Power purchase cost /One hundred million yuan	Electricity of abandoned water/one million kilowatt-hours	Risk of abandoned water/one million kilowatt-hours	Consumption of transmission/one million kilowatt-hours
1	1.58	0.83	0.32	1.61
2	1.46	0.77	0.18	1.82

As can be seen from table 2, after considering the risk assessment indexes of abandoned water, both the electricity purchase cost and the abandoned water risk are reduced, but the electricity consumption is increased. This result reflects that the introduction of abandoned water risk management promotes wider consumption of clean energy. The essential reason can be explained as follows: When the abandoned water risk assessment is introduced and the load demand is unchanged, the trading decision optimization space of the provincial grid will extend to external transmission in order to reduce the risk of abandoned water. These not only guarantees more benefits of power selling for the grid company, but also increases the total economic benefit and reduces the abandoned water risk. Therefore, both the grid companies and the generation enterprises can make profits.

5.4. Validation of the effectiveness of risk management and control of water abandonment and analysis of its impact on the external consumption of clean energy

Keeping the simulation conditions unchanged, three schemes of 5.0%, 10.0% and 15.0% of the probability of opportunity constraint were adopted to verify their influences on the risk of abandoned water and the consumption of clean energy. The results are shown in Tab.3.

Table 3. Comparison of hydropower transmission considering different probability of opportunity constraint

The probability of opportunity constraint	Power purchase cost /One hundred million RMB	Electricity of abandoned water/one million kilowatt-hours	Risk of abandoned water/one million kilowatt-hours	Consumption of transmission/one million kilowatt-hours
5.0%	1.65	0.89	0.42	1.53
10.0%	1.58	0.83	0.32	1.61
15.0%	1.49	0.77	0.28	1.72

As can be seen from Tab.3, with the increasing probability value of opportunity constraint, both electricity purchase cost and abandoned water risk are reduced, but the external electricity consumption is increased. This result reflects that different probability of opportunity constraint has an impact on promoting large-scale consumption of clean energy. The essential reason can be explained as follows: In the mathematical way, the increment of probability of opportunity constraint means the expansion of space for optimization model. In the physical way, it means that the trading decision optimization space of the provincial grid will extend to external transmission in order to get more profits when the load demand is unchanged. This extension leads to increment in the total economic benefit and reduction in the abandoned water risk.

6. Conclusion

1) Since the stability limit of trans-provincial channel involves many complicated factors, such as the boot capacity of the key unit, boot combination, generation output, associated channel limit etc., it means that the stability limit of trans-provincial channel will change dynamically with different grid operation. Therefore, it is more reasonable to extend stability limit of trans-provincial channel from static to dynamic.

2) Considering the uncertainty of hydropower output, the built risk assessment index of abandoned water extends the existing assessment of water abandonment from certainty to uncertainty, and effectively describes the severity of water abandonment risk. The proposed risk management concept of abandoned water is conducive to wider consumption of clean energy, and is more suitable for the prior control of water abandonment risk than the existing indexes.

3) The established risk control model of water abandonment consumption for hydropower transmission, which considers the dynamic stability limit of trans-provincial channel, achieves the optimal economic benefit and effective management of water abandonment risk. In addition, it makes profits for both grid companies and generation enterprises. Therefore, it is of practicability and promotion value.

Acknowledgments
The Project supported by National Natural Science Foundation of China（51677012）and scientific and technological research project of Chongqing municipal education commission (Grant No. KJ1704101).

References
[1] KAHRL F and WILLIAMS J H 2013 The political economy of electricity dispatch reform in China. *Energy Policy*,**9(8)** 361-369.
[2] CHEN Yuehui,XIANG Meng and ZHANG Bin 2018 Optimal dispatch modes of cross-provincial tie-line and early-warning strategy.*Power System Protection and Control,***46(18)** 100-107.
[3] Liuya,an and Guan X. H 2003 Purchase allocation and demand bidding in electric power marks.*IEEE Transactions on Power systems,***18(1)** 106-112.
[4] Dehua Zheng, Abinet Tesfaye Eseye, Jianhua Zhang , et al 2017 Short-term wind power forecasting using a double-stage hierarchical ANFIS approach for energy management in microgrids. *Protection and Control of Modern Power Systems,***2(2)** 33-39.
[5] Vilim M 2014 Wnd power biding in electricity markets with high wind Penetration.*Appllied energy,***118** 141-155.
[6] YAN Wei, WEN Xu, WANG Junmei, et al 2012 An optimal monthly power purchasing model considering differences between Internal and external grid.*Automation of Electric Power Systems,*36(17)：56-60.
[7] WEN Xu, WANG Junmei, GUO Lin 2013 A monthly Power Purchasing Risk Management Model for Provincial Power Grid Considering Wind Power Uncertainty.*Electric Power Automation Equipment,* **33(7)** 27-33
[8] LIU Gushuai, XIAO Yiyao, HE Yuqiang, et al 2017 Multi-objective optimal method considering types of grid connected new energy of electric power system.*Power System Protection and Control,***45(10)** 31-37.
[9] YAN Jun, YAN Feng 2018 Location method of charging and swapping service facilities based on a two-step planning.*Power System Protection and Control,***46(14)** 48-56.
[10] LIN Ye 2011 Studies on optimization of global generation and transmission maintenance scheduling and its security correction.Chongqing: Chongqing university.
[11] YU H, CHUNG C Y, WONG K P, et al 2009 Probabilistic load flow evaluation with hybrid Latin hypercube sampling and Cholesky decomposition.*IEEE Transactions on Power Systems,***24(2)** 661-667.

IOP Conf. Series: Materials Science and Engineering **692** (2019) 012036 doi:10.1088/1757-899X/692/1/012036

Coupling Analysis of Multi-Physical Field of High-Speed Motorized Spindle Based on Fractal Contact Theory

Shufei Xue[1, 2], Jianhua Wang[1, 2], Congbin Yang[1, 2,*] and Zhifeng Liu[1, 2]

[1]College of Mechanical Engineering and Applied Electronics Technology, Beijing University of Technology, Beijing 100124, China
[2]Institute of Advanced Manufacturing and Intelligent Technology, Beijing University of Technology, Beijing 100124, China

[*]Corresponding author: yangcongbin@bjut.edu.cn

Abstract. In order to improve the modeling accuracy of the motorized spindle, a comprehensive thermal contact resistance (TCR) modeling method was proposed. The fractal theory was introduced to calculate the joint thermal contact resistance. The W-M function and the M-B fractal model were established to characterize the microscopic contour of the rough surface. The bulk thermal resistance, constrained thermal resistance, micro-gap thermal resistance and overall thermal contact resistance of the rough surface of different materials were discussed. The finite element method(FEM)was used to establish the multi-physics coupling analysis model of the motorized spindle, adding the theoretical calculation of thermal contact resistance in the joint surface of the motorized spindle. The temperature rise and deformation variation of the motorized spindle were simulated under the condition of thermal contact resistance. The simulation results demonstrate the influence of thermal contact resistance on the temperature and thermal deformation of the motorized spindle. The results show that when the thermal resistance of the high-speed motorized spindle is considered, the temperature rises and the deformation increases, and the accuracy of the model is more accurate, which is closer to the actual processing.

1. Introduction

As a key functional part of the machine tool, the high-speed motorized spindle plays an important role in machining accuracy. The thermal error caused by the time variation of the spindle temperature is mainly affected by the machining accuracy of the machine tool. A lot of research shows that the thermal error is the main error source inside the machine tool. In the existing thermal error theory research, most of the effects of thermal contact resistance between rough solid surfaces are neglected, and the results cannot reflect the actual engineering application. Ma [1] analyzed the transfer relationship between multi-variables in the bearing balance equation system, and proposed a Newton-Raphson algorithm with contact angle as iteration variable. The Weierstrass-Mandelbrot (W-M) function in fractal geometry was used to characterize the rough surface morphology of the bearing, and the asperities contact deformation model was established to calculate the contact parameters in the thermal contact resistance model. Zhang [2] used a three-dimensional W-M function to simulate a rough surface, and considered three deformation states, namely complete plastic deformation, elastoplastic deformation and elastic deformation, and established a thermal contact conductance considering the bulk thermal resistance and constrained thermal resistance model. Liu [3] proposed a

IOP Conf. Series: Materials Science and Engineering **692** (2019) 012036 doi:10.1088/1757-899X/692/1/012036

closed-loop iterative modeling method for thermal characteristics, considering the influence of thermal contact resistance on the spindle system, analyzed the main heat sources of the spindle, simulated and calculated the boundary conditions of the spindle, and proposed a new geometric mechanical model of thermal contact resistance. Cui [4] proposed the temperature distribution and thermal deformation modeling of the motorized spindle system and the thermal-structural coupling simulation analysis method. The relevant thermal contact resistance of the machine tool was obtained by experiment and theory, and the validity of the method was verified. Zhao [5] proposed a three-dimensional fractal theory based on thermal contact conductance model. The effects of different fractal dimensions and roughness parameters on thermal contact conductance were compared. The results show that the deformation of micro-convex body and gap thermal conductance should not be neglected. In order to get a more accurate model, considering the influence of the thermal contact resistance on the heat transfer of the spindle joint surface, a comprehensive thermal contact resistance prediction model is established in this paper. In general, the thermal contact resistance is a very necessary and important factor in the thermal analysis of spindle.

2. Fractal contact theory

2.1 Microscopic morphology of joint surface

The Weierstrass-Mandelbrot function [6] (W-M function) is often used to characterize the micro-contour of rough surface topography. On this basis, a fractal model of elastic-plastic contact of a single asperities body on rough surface, namely M-B model [7] is established. Its two-dimensional expression is as follows:

$$z(x)=G^{D-1}\sum_{n=n_1}^{\infty}\gamma^{(D-2)}n\cos(2\pi\gamma^{n}x) \tag{1}$$

$$1<D<2;\gamma>1$$

Where, $z(x)$ is a two-dimensional rough surface contour height function; x is the displacement coordinate of the rough surface contour; D is the contour fractal dimension; G is the contour fractal roughness parameter; f_1 is the ordinal corresponding to the lowest cut-off frequency; n is the spatial frequency ordinal; γ is greater than 1 Constant, usually taken $\gamma=1.5$.

As shown in Figure 1. For the two-dimensional microscopic topography of the rough surface, it can be clearly seen that the complexity of the curve increases with the increase of the fractal dimension D. The larger the D, the more complex the contour structure and the richer contour details.

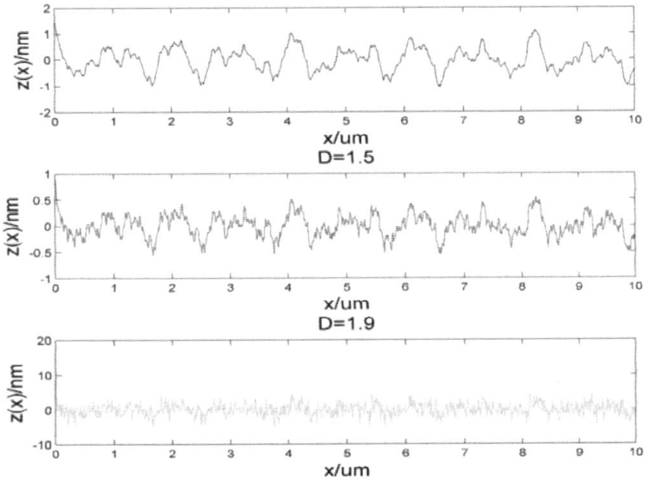

Figure 1. Two-dimensional topography of random rough surface.

Considering the elastic deformation, plastic deformation and elastoplastic deformation of asperities, an M-B modified model based on three-dimensional W-M function is proposed [8]:

$$z(x,y)=L\left(\frac{G}{L}\right)^{(D-2)}\left(\frac{\ln\gamma}{M}\right)^{1/2}\sum_{m=1}^{M}\sum_{n=0}^{n_{\max}}\gamma^{(D-3)n}\times\left\{\cos\phi_{m,n}-\cos\left[\frac{2\pi\gamma^{n}\sqrt{x^{2}+y^{2}}}{L}\cos\left(\arctan\left(\frac{y}{x}\right)-\frac{\pi m}{M}\right)+\phi_{m,n}\right]\right\} \quad (2)$$

$$2<D<3,\gamma>1$$

Where, $z(x,y)$ is a three-dimensional rough surface contour height function; M is the overlap number of the surface; $\phi_{m,n}$ is a random phase, and $\phi_{m,n}\in[0,2\pi]$; $n_{\max}$ is a frequency ordinal, and $n_{\max}=\text{int}[\log(L/L_s)/\log\gamma]$; L is the sampling interval.

The MATLAB software is used to simulate the generation of three-dimensional fractal rough surface and contour map. As shown in Figure 2.the larger the G value of the contour fractal roughness parameter, the larger the surface contour amplitude and the rougher the contact surface.

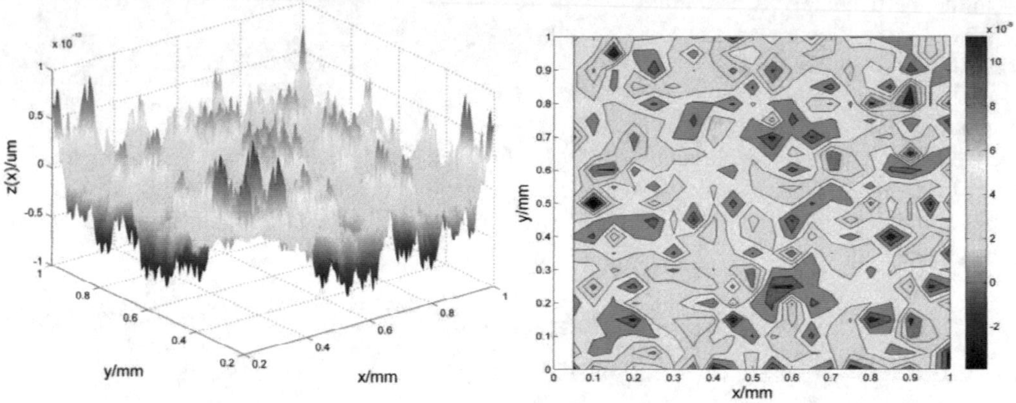

Figure 2. Three-dimensional shape and contour map of random rough surface.

2.2 Jointing surface thermal contact resistance mechanism

Thermal contact resistance plays an important role in heat transfer and has a large impact on the spindle system. Ignoring the thermal contact resistance results in inaccurate temperature distribution or thermal deformation calculations. As shown in Figure 3.when the two solid rough surfaces are in contact with each other, asperities of different sizes are formed. When heat flows through the solid surface, the heat flux density shrinks at these discrete points. In addition, the rougher the contact surface, the smaller the contact area, and the more severe the heat flux density shrinks. Therefore, the heat flux is limited by the thermal contact resistance of the solid, resulting in a sharp drop in temperature at the solid contact surface [9].

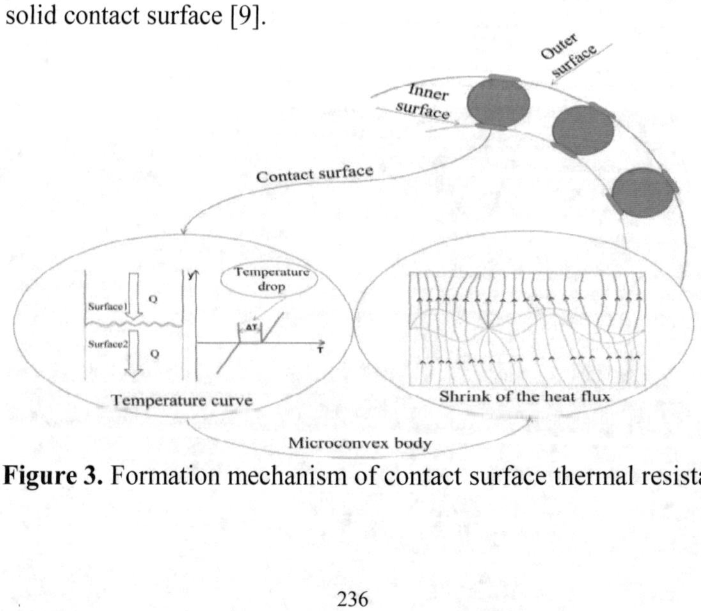

Figure 3. Formation mechanism of contact surface thermal resistance.

The rough surface contact between the main shaft and the bearing is calculated as follows according to the thermal contact resistance [10]:

$$R = \frac{L_g(\lambda_1+\lambda_2)}{A\left[2A_r^*\lambda_1\lambda_2+\lambda_f\left(1-A_r^*\right)(\lambda_1+\lambda_2)\right]} \tag{3}$$

Where, L_g is the thickness of the joint surface void; λ_1, λ_2, λ_f respectively, the joint surface material and the intermediate medium conduction coefficient; A is the nominal contact area; A_r^* is the actual contact area.

When the fractal dimension of the rough surface $1 < D < 1.5$,

$$A_r^* = \left(\frac{3\sqrt{2\pi}}{4}\right)^{2/(3-D)}\left(\frac{L_u}{G}\right)^{(2D-2)/(3-D)} g_4(D)\left(\frac{F}{E}\right)^{2/(3-D)} \tag{4}$$

When the fractal dimension of the rough surface $1.5 < D < 2$,

$$A_r^* = \left\{\frac{4}{3\sqrt{2\pi}}\left(\frac{G}{L_u}\right)^{D-1}[-g_1(\mathrm{D})]\left(\frac{2a_c}{L_u^2}\right)^{(3-2D)/2} + \frac{H}{E}g_2(D)\left(\frac{2a_c}{L_u^2}\right)^{(2-D)/2}\right\}^{-2/D} g_3(D)^{-2/D}\left(\frac{F}{E}\right)^{2/D} \tag{5}$$

Where

$$\begin{cases} g_1(D) = \left[\dfrac{2-D}{D}\right]^{D/2}\dfrac{D}{3-2D} \\ g_2(D) = \left[\dfrac{D}{2-D}\right]^{(2-D)/D} \\ g_3(D) = 2^{D/2}\psi^{\left(4-4D+D^2\right)/4} \\ g_4(D) = \left[g_1(D)g_3(D)\psi^{(D-2)(3-2D)/4}\right]^{2/(D-3)}\left[\dfrac{D}{4-2D}\right]^{(3-2D)/(3-D)} \end{cases} \tag{6}$$

Where , a_c is the critical micro contact area; H is the hardness of the material; L_u is the upper limit of the length of the fractal domain; F is the pressure; E is the equivalent elastic modulus, $E = \left(\frac{1-v_1^2}{E_1}+\frac{1-v_2^2}{E_2}\right)^{-1}$, E_1, E_2, v_1, v_2 are the Elastic modulus and Poisson's ratio of the two materials, respectively.

3. Thermal contact resistance model discussion

In this paper, four materials of Stainless steel, HT300, Copper and Aluminium alloy are taken as examples. The physical properties of the materials are shown in Table 1. Under normal temperature and pressure.

Table 1. Physical properties of four different materials.

Material	Elastic modulus $E \backslash Mpa$	Thermal conductivity $k\backslash W/(m \cdot K)$	Poisson's ratio v	Hardness coefficient $H\backslash$ Mpa
Stainless steel	1.9×10^5	16.2	0.29	3800
HT300	1.2×10^5	45	0.25	2300
Copper	1.2×10^5	398	0.35	1380
Aluminium alloy	1.9×10^5	154	0.33	1400

In this paper, the bulk thermal resistance, constrained thermal resistance and micro-gap thermal resistance of the rough surface of different materials are discussed. The thermal resistance of the four materials under certain fractal dimension D and roughness parameter G is shown in Figure 4.which

IOP Conf. Series: Materials Science and Engineering **692** (2019) 012036 doi:10.1088/1757-899X/692/1/012036

shows that under the same load. The materials have different thermal contact resistance. With the increase of load, the bulk thermal resistance, constrained thermal resistance and micro-gap thermal resistance and thermal contact resistance of the four materials are linearly decreasing. The thermal contact resistance of stainless steel material is greater than that of the other three materials, and that of copper material is the smallest. Due to the influence of physical properties of the material on the thermal resistance of the contact, the material with higher hardness has a larger contact gap as the load increases, so that the thermal contact resistance is large. On the contrary, the material with smaller hardness has smaller contact gap and smaller change of thermal contact resistance.

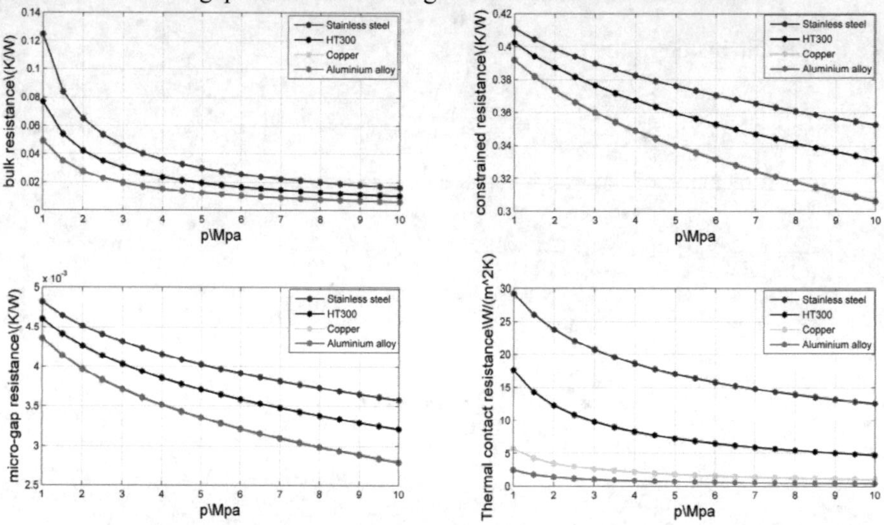

Figure 4. Relationship between thermal resistance and load of four materials.

4. Simulation analysis of motorized spindle

The finite element method is used to simulate and analyze the motorized spindle system. Adding the material properties of the motorized spindle system is the key to analyze the thermal characteristics. After importing the model, the material properties of the motorized spindle are set as shown in Table 2. The boundary conditions are set in the finite element modeling, and when the thermal contact resistance is considered, the calculated value is manually added to the contact surface of the spindle unit. Modeling and simulation are carried out without considering contact thermal resistance and contact thermal resistance respectively.

Table 2. Material properties of motorized spindle components.

Parts	Density Kg/m^3	Thermal conductivity W/m·k	Specific heat J/Kg·k	Modulus of elasticity GPa	Poisson's ratio	Linear expansion coefficient
Spindle	7800	45	400	210	0.23	1.20×10^{-5}
Bearing	7850	40	400	206	0.30	1.25×10^{-6}
Bearing block	7800	45	480	120	0.21	1.11×10^{-5}
Rolling element	3200	35	850	314	0.26	3.20×10^{-6}
Box body	7300	45	480	120	0.25	1.20×10^{-5}

The thermal-structural coupling analysis of high-speed motorized spindle system is carried out. The model is imported into the software to divide the CFD hexahedral mesh and add boundary conditions to solve the temperature field and thermal deformation changes when the thermal contact resistance is taken into account and the thermal contact resistance is not taken into account. As can be seen from Figure 5-6. The maximum temperature of the motorized spindle is 34.212°C without considering the thermal contact resistance of the joint surface. The temperature of the stator is lower than that of the rotor, which is due to the circulating cooling sleeve outside the stator, which can reduce the temperature in real time. Considering the thermal contact resistance of the interface, the maximum temperatures are 37.64 °C , respectively. The average temperature increased by 3.428 °C .

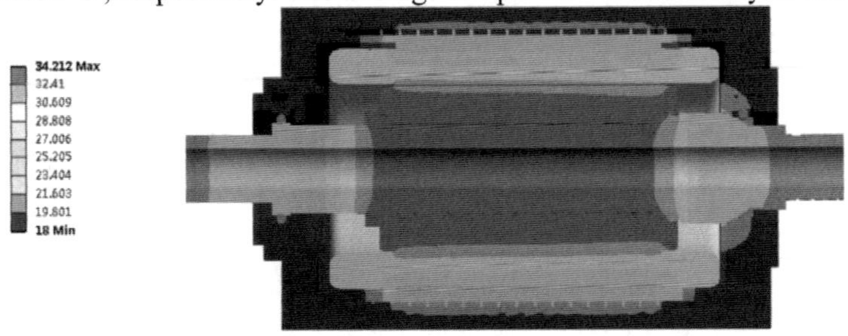

Figure 5. Without considering the joint surface thermal contact resistance temperature field cloud map.

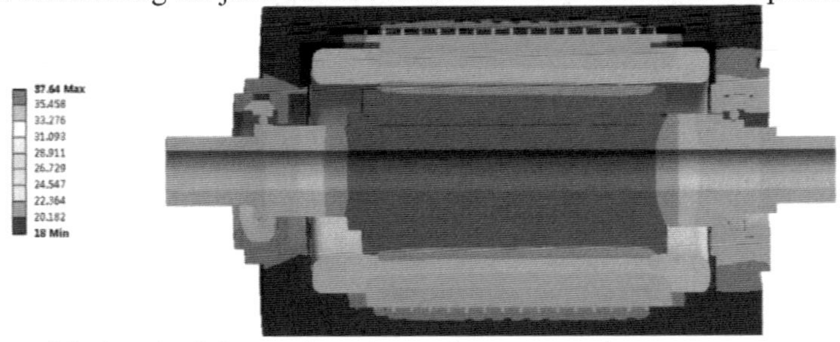

Figure 6. Considering the joint surface thermal contact resistance temperature field cloud map.

As can be seen from Figure 7-8. The maximum axial thermal deformation of the spindle is 0.01554 mm without considering the thermal contact resistance of the joint surface, and 0.020364 mm with considering the thermal contact resistance of the joint surface. The maximum axial thermal deformation of the spindle with considering the thermal contact resistance of the joint surface is 31% higher than that without considering the thermal contact resistance of the joint surface. Therefore, considering the thermal contact resistance, the spindle temperature rises rapidly and the thermal deformation increases, which is closer to the actual temperature and deformation in the process of processing.

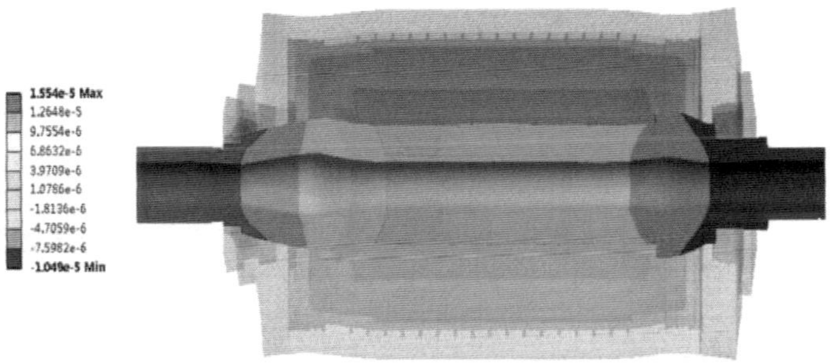

Figure 7. Without considering the joint surface thermal contact resistance heat deformation cloud map.

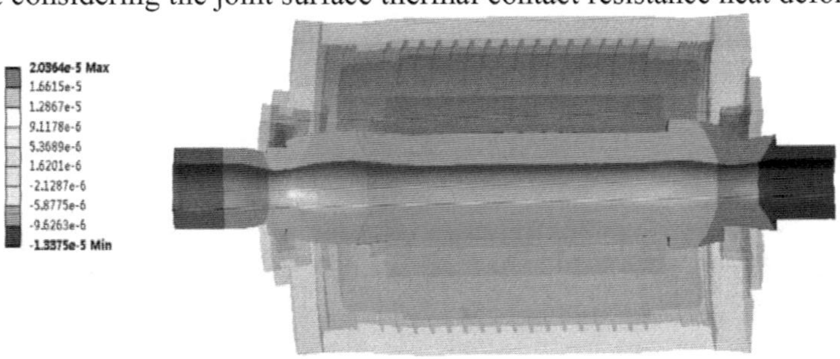

Figure 8. Considering the joint surface thermal contact resistance heat deformation cloud map.

5. Conclusion

Traditionally, the influence of thermal contact resistance on temperature and thermal deformation is often neglected in the thermal analysis and modeling of spindle, which leads to inaccurate modeling. In practice, considering the thermal contact resistance caused by asperities of different sizes between rough contact surfaces of high-speed motorized spindle, the simulation results of motorized spindle are closer to the actual processing situation. The specific conclusions are as follows:

A comprehensive thermal contact resistance analysis model was established, and the influence of thermal contact resistance on spindle temperature field and thermal deformation was analyzed. The simulation results of multi-physical field coupling of the motorized spindle system show that the temperature value of the thermal contact resistance of the spindle considering the joint surface is 3.428 °C higher than that without considering the joint surface, and the thermal deformation value is increased by 31%. Therefore, the thermal contact resistance must be fully considered in the thermal analysis of the spindle.

Acknowledgments

The authors gratefully acknowledge the National Natural Science Foundation of China (No.51805012), National Science and Technology Major Project (No.2018ZX04032002) and Beijing Science and Technology Plan Project (No.K2001011201801).

References

[1] Ma C, Yang J, Zhao L, Mei X M and Shi H 2015 Simulation and experimental study on the thermally induced deformations of high-speed spindle system *Appl. Therm. Eng.* **86** pp 251–268

[2] Zhang J H, Liu Y L and Yan K 2019 A fractal model for predicting thermal contact conductance considering elasto-plastic deformation and base thermal resistances *J. Mech. Sci. Technol.* **33** pp 475–484

[3] Liu J L, Ma C , Wang S L, Wang S B, Yang B and Shi H 2018 Thermal-structure interaction characteristics of a high-speed spindle-bearing system *Int. J. Mach. Tools Manuf.* **137** pp 42–57

[4] Cui Y, Li H L, Li T J and Chen L 2018 An accurate thermal performance modeling and simulation method for motorized spindle of machine tool based on thermal contact resistance analysis Int. J. Adv. Manuf. Technol. **96** pp 2525–2537

[5] Zhao Y S, Fang C, Cai L G and Liu Z F 2018 A three-dimensional fractal theory based on thermal contact conductance model of rough surfaces *Proc. Inst. Mech. Eng. Part E-J. Process Mech. Eng.* **232** pp 528–539

[6] Mandelbrot B B and Aizenman M 1979 Fractals: form, chance, and dimension *Phys. Today.* **32** pp 65–66

[7] BHUSHAN M B 1991 Fractal model of elastic-plastic contact between rough surfaces *J. Tribol.-Trans. ASME.* **113** pp 1–11

[8] Yan W and Komvopoulos K 1998 Contact analysis of elastic-plastic fractal surfaces *J. Appl. Phys.* **84** pp 3617–3624

[9] Ji J D, Hong R J, Sun F Z and Huang X D 2018 Thermal characteristic analysis of Z-axis guideway based on thermal contact resistance *Adv. Mech. Eng.* **10** pp 1–10

[10] Liu Z F, Pan M H, Zhang A P, Zhao Y S and Ma C Y 2015 Thermal characteristic analysis of high-speed motorized spindle system based on thermal contact resistance and thermal-conduction resistance *Int. J. Adv. Manuf. Technol.* **76** pp 1913–1926

IOP Conf. Series: Materials Science and Engineering **692** (2019) 012037 doi:10.1088/1757-899X/692/1/012037

Distributed Architecture Optimal Design for Aircraft Thrust Management System with Safety Method

Wei Zhang[1] **and Yakui Gao**[2]

[1] Department of flight Control system, AVIC the First Aircraft Institute, Xi'an 710089, China
[2] Department of Chief Engineer's office, AVIC the First Aircraft Institute, Xi'an 710089, China

gilles923@163.com

Abstract. In order to ensure the safety and reliability of aircraft thrust management system design and reduce the risk of engineering development, the distributed optimization architecture design based on the aircraft safety method for thrust management system has been put forward in this paper. First, the thrust management function requirement is analysed; second, the distributed architecture design method for thrust management is presented by following the SAE ARP4754 and ARP4761; then the automatic take-off thrust control system function which is not available in Airbus and Boeing civil aircraft thrust management architecture is designed by using this method. The results show that the method proposed in this paper is feasible and effective. The usage efficiency of airborne computer is improved and the aircraft weight is decreased, which greatly enhanced the safety and reliability of aircraft thrust management system.

1. Introduction

Thrust management technology is generated based on aircraft performance optimization. In different flight phases such as take-off, climb, cruise, approach and landing etc, Thrust management system controls aircraft fly along an optimized flight profile by calculating right engine thrust target in order to reduce operating costs.

Thrust management was first adopted by Boeing Aircraft Corporation in the 1970s and successfully applied to its commercial airplanes such as the B737 and B747. In the late 1970s, thrust management technology was applied by US Department of Defense (DOD) to its large aircraft such as bombers, tankers, and transport planes. Later, DOD also tried to use this technology to the fighters [1-2]. The Airbus aircraft company in Europe has also adopted thrust management technology for the A300, A310, A320 and other series of aircraft. In recent years, many domestic scholars have also analyzed and studied the thrust management functions and architecture [3-8], but they did not discuss how to design the thrust management system.

Being one of the aircraft systems, thrust management system architecture should be built at the beginning of the design phase. Modern civil aircraft systems designs are generally more complicated. If the mature safety method can be applied, the thrust management system architecture design will be more reasonable and safe, and the risk of aircraft development will be reduced effectively. In order to decrease the system design weight on the condition of safety guaranteed, we design the thrust management system based on requirement by using the safety method and the distributed optimization architecture.

2. Function design based on requirement

2.1. Requirement Analysis

The function design of the thrust management system is mainly to meet three aspects of aircraft requirements: (1) Reduce the operation and maintenance cost; (2) Automatic control engine thrust during the flight phase (including the automatic take-off thrust control system, ATTCS); (3) Fan speed(N1) synchronous for noise reduction.

The main factors considered in the requirement analysis are:

(1) Flight phase

The function requirements for thrust management are vary from one flight phase to another flight phase. In the take-off phase, it is mainly considered that the take-off thrust cannot exceed the engine thrust limit. If there is one engine inoperative(OEI), it is necessary to automatically increase the remaining engine thrust to the emergency take-off state (ATTCS function); in the cruise phase, thrust management mainly synchronize fan speed for noise reduction, and control speed or thrust automatically by moving throttle for auto-navigation.

(2) Cost-effectiveness

The derated thrust technology in thrust management can extend engine life and reduce maintenance costs, but it needs to be considered comprehensively for different sizes of aircraft design. The large aircraft has more high efficiency in reducing thrust, and would generally design multiple take-off thrust levels, but the corresponding technical cost is also high. However, the small aircraft has only one take-off thrust level for less benefit.

2.2. Function design

The thrust management system function is designed according to the requirements of the aircraft. It is closely related to the flight management system, automatic flight control system, engine control system as well as the airborne computer importance level and safety index. Different aircraft have different thrust management system architectures, but functions are basically the same, which mainly include the thrust level / thrust limit / target thrust calculation, automatic throttle control, and engine N1 synchronization, and automatic take-off thrust control.

2.2.1. Thrust level / thrust limit / target thrust calculation. According to the different thrust requirements on different flight phase, the thrust level in an aero-engine such as take-off, maximum continuous, climb and cruise usually are designed. In order to fly along the optimal flight profile every time, the thrust management system needs to calculate the required thrust level and target thrust according to the flight phase and cost index, while also giving real-time thrust limits to prevent undesired engine thrust response.

2.2.2. Automatic throttle control. The automatic throttle control includes the thrust control mode and speed control mode, different flight phase has different mode. In the climb phase, thrust control mode is selected to keep the thrust target; When in the cruise phase, the speed control mode is selected to hold the airspeed or Mach number which chosen by the pilot using the automatic flight control unit or given by the flight management computer; During the landing phase, the automatic throttle holds a certain safe speed or enters the idle state.

2.2.3. Engine N1 synchronization. The radiated noise for high bypass engine mainly comes from the discrete single-tone noise of the fan components. When the multiple working engines N1 are close, it is easy to generate acoustic wave interference called beating noise in cabin. In order to eliminate the beating noise effect, the engine N1 will be adjusted automatically to keep consistent. Engine N1 synchronization is only valid during the cruise phase and invalid during take-off, climb or idle engine state. Because this function is mainly to eliminate the beating noise, the engine N1 synchronization can not significantly affect the aircraft three-axis motion, so the N1 speed adjustment values need to be limited in a small range.

2.2.4. Automatic take-off thrust control system (ATTCS). The take-off thrust level plays an important role during aircraft safety take-off. Most of civil commercial aircraft have the OEI take-off capability (after the decision velocity). Some aircrafts may not have sufficient OEI take-off thrust under full load, the ATTCS function in the thrust management system is trigged when an OEI has happened. Aircraft may have two or four engines, each engine thrust control is independent and cannot communicate each other, when the OEI has happened, the ATTCS function cannot be triggered by themselves. So this function requires the aircraft system to transfer the OEI signal from one engine control systems to the other in order to trigger the ATTCS.

3. Safety design method

In the modern aircraft design process, the safety design for complex systems need to follow the SAE ARP4754A and ARP4761 standards. Otherwise the aircraft can not obtain the airworthiness certification issued by the FAA and EASA [9-10].

The ARP4754A establishes a compliance demonstration method for airworthiness regulation to ensure the development process. This SAE standard distinguishes the aircraft system function development assurance level (FDAL) and item development assurance level (IDAL) in detail, and stipulates the system FDAL needs to be implemented by the corresponding IDAL, or formed with the non-similar redundancy architecture by the degraded IDAL. The ARP4761 describes the safety analysis process including Function Hazard Analysis (FHA), Preliminary System Safety Analysis (PSSA), System Safety Analysis (SSA), and Common Cause Analysis (CCA). The main safety analysis methods include Fault Tree Analysis (FTA), Failure Mode and effect Analysis (FMEA) and etc (Table 1). This paper mainly uses the FTA method when optimizing the thrust management architecture.

The top system function in Figure 1 can be composed of multiple sub-functions. If any one sub-function fails, the top system function is considered invalid. For function 1, since the non-similar devices 1A and 1B form redundancy architecture, so the item assurance level of devices 1A and 1B can be lower than that of function 1; for function 2, there is no other device to form redundancy architecture, and then device 2 item development assurance level must be at least the same as the function 2.

Table 1. System architecture development assurance level allocation requirements.

Top-level failure	Failure probability	Corresponding FDAL	Function failure sets with members	
			option 1	option 2
Catastrophic (Ⅰ)	Less than 10E-9	FDAL A	At least one member should reach IDAL A	At least two members should reach IDAL B
Hazardous (Ⅱ)	Less than 10E-7	FDAL B	At least one member should reach IDAL B	At least two members should reach IDAL C
Major(Ⅲ)	Less than 10E-5	FDAL C	At least one member should reach IDAL C	At least two members should reach IDAL D
Minor(Ⅳ)	Less than 10E-3	FDAL D	At least one member should reach IDAL D	
No safety effect(Ⅴ)	More than 10E-3	FDAL E	IDAL E	

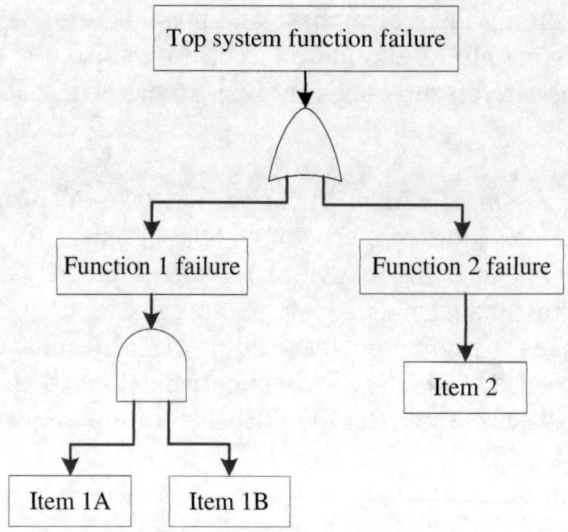

Figure 1. Fault tree analysis for FDAL and IDAL.

4. Application

Different types of aircraft have different thrust management system architectures because of their differences in functions and airborne computers. However, the safety and reliability should be considered in a better thrust management system architecture design.

4.1. Current mainstream thrust management architecture

Although the thrust management functions of Boeing and Airbus aircraft are basically same (except ATTCS), their architectures are quite different. Boeing aircraft has an independent thrust management system; Airbus aircraft's thrust management function is embedded in flight management system.

4.1.1. Boeing series aircraft thrust management architecture. The Boeing series aircrafts have the thrust management system (TMCS) concept (figure 2), which includes automatic throttle (AT) control law, N1 trimming, thrust limit calculation and so on. When the high integrated system design is popular, TMCS' hardware is often integrated with the flight management system (FMS) in order to conveniently visit the FMS database information. The FMS function includes navigation, performance, and guidance modules, where the performance module refers to calculating the aircraft's most economical speed, optimal flight altitude and top decent point for the whole flight profile, and the FMS needs to calculate the thrust limits Value (N1 limits) and thrust target value (N1 targets) for each flight phase. On Boeing's early airplanes, the thrust limit calculation was done by the FMS; while on later airplanes, the thrust limit calculation was handled by the TMCS.

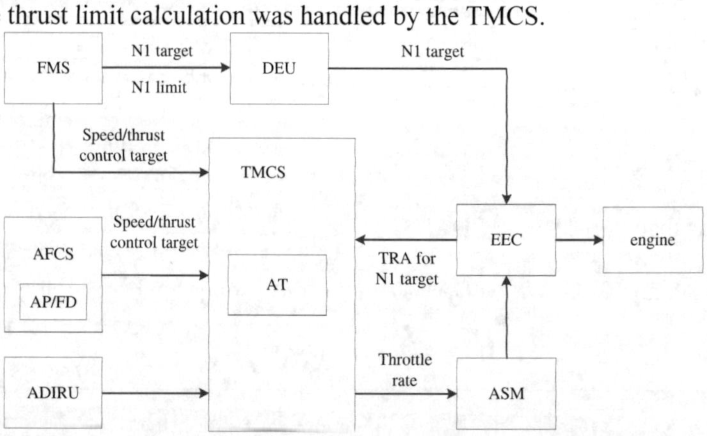

Figure 2. Boeing typical airplanes thrust management architecture diagram.

The Boeing 787 integrates all thrust management functions into an integrated modular avionics (IMA), which also integrates the flight management system and the automated flight system, and will help to interact deeply and reduce aircraft system weight. The Boeing 787 thrust management with more perfect function includes automatic throttle control laws as well as thrust levels, thrust electronic trim and thrust monitoring. The thrust level calculates the data such as the reduced take-off and the reduced climb. The electronical trim thrust completes fan speed trimming and small thrust adjustment without moving the throttle.

4.1.2. Airbus series aircraft thrust management architecture. Airbus series aircraft does not have an independent thrust management system, which is included in the flight management guidance envelope computer (FMGEC). The engine electronic controller (EEC) has also played an important role in the thrust management function realization (figure 3). FMGEC includes flight management, flight guidance and flight envelope protection modules. The automatic flight control and automatic throttle are integrated in the flight guidance module. Therefore, FMGEC only provides automatic control thrust target, thrust level and thrust limit value to EEC. The automatic throttle commands connect or disconnect is placed in the EEC. The FMGEC provides data to EEC needs to pass through the flight control computer (FCU) and the engine Interface vibration monitoring unit (EIVMU).

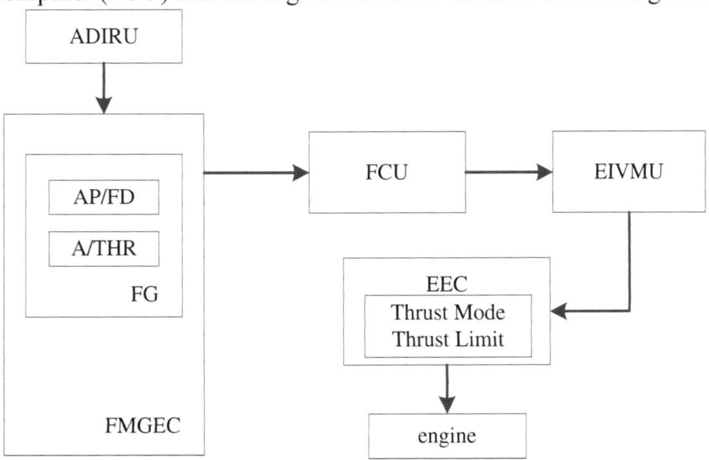

Figure 3. Airbus typical airplane thrust management architecture diagram.

4.2. Thrust management architecture design process

4.2.1. Function hazard analysis. The FHA of the thrust management function is analyzed by referring Table 1, and the results are shown in table 2.

Table 2. Function hazard analysis.

No.	Function hazard	Level
1	Loss of thrust level / thrust limit / thrust target calculation capacity	III
2	Automatic throttle control failure	III
3	Engine N1 synchronization failure	IV
4	ATTCS failure	I

4.2.2. Safety assessment analysis. Each function hazard item in Table 2 is analysed by using FTA method, and take No.4 ATTCS failure as example. Generally, the flight management system (FMS) can reach FDAL III, and the automatic flight control system (AFCS) can reach FDAL II, but the No.4 function FHA is the level I, single aircraft system cannot meet the safety requirement. According to

IOP Conf. Series: Materials Science and Engineering **692** (2019) 012037 doi:10.1088/1757-899X/692/1/012037

SAE ARP4754, the question can be solved by adding a new IDAL I item, or by optimizing the design based on the existing item. The later is selected in this paper.

4.2.3. Distributed architecture design. ATTCS function failures include engine control system failure and OEI signal transferring failure, both of them can lead to ATTCS function failure. The engine control system can reach FDAL I, and the OEI signal transferring failure also needs to reach FDAL I. According to the ARP4754, the FDAL I can be formed by two different systems or devices with FDAL II, so the architecture can be designed by using dual engine interface unit (EIU) and AFCS for transferring the OEI signal, especially there is no common mode fault between them. The fault tree analysis is shown in Figure 4.

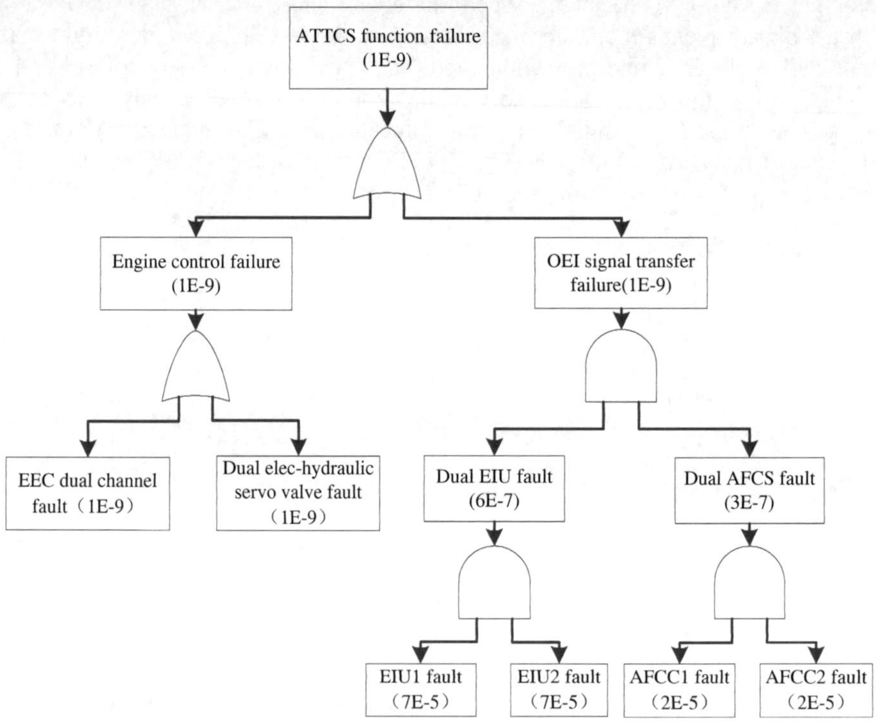

Figure 4. ATTCS fault tree analysis.

4.2.4. Optimized design. Weight is an important index in the aircraft design. The numbers of airborne computer decrease not only reduce the equipment's weight, but also reduce aircraft cable's weight. Although the Boeing aircraft's independent thrust management system can enhance system functions and the Airbus aircraft's thrust management system embedded in FMGEC can easily visit the flight guidance data, the distributed thrust management architecture has the advantage of saving resources and decreasing aircraft weight. In this paper, aircraft design weight can be decreased by 18.7kg by using distributed design method for the ATTCS function (Table 3).

The distributed thrust management architecture is optimized without increasing the new development equipment, which will significantly improve the efficiency of airborne computer.

Table 3. Aircraft weight optimization analysis for ATTCS function.

Solution	Equipment Weight	Cable Weight	Total Weight
Increase new development equipment	8.7kg	15kg	23.7kg
Optimize system architecture	0kg	5kg	5kg

IOP Conf. Series: Materials Science and Engineering **692** (2019) 012037 doi:10.1088/1757-899X/692/1/012037

5. Conclusion

In this paper, a new distributed thrust management system architecture design method is presented during the aircraft design. This method includes function requirement analysis and architecture safety design, it can realize complex thrust management system function design. The ATTCS function which is not available in Airbus and Boeing civil aircraft thrust management architecture is simulated by using this method, and the results show that the aircraft design weight can be decreased by 18.7kg. Meanwhile, the safety and reliability of the thrust management system can be guaranteed in the design stage. Therefore, the decreased weights will have the benefits for aircraft loading and extended-range operations.

References

[1] Charlers A 1982 *Aircraft thrust/power management can save defense fuel reduce engine maintenance costs and improve readiness* (Washington D.C.: United States General Accounting Office)

[2] Huynh and Kiet T 2003 *Engine thrust management-new design architecture* US patent 6886786

[3] Payne C and Lewis A 1970 *Concorde thrust management* S.A.E. Tech. Paper 700817

[4] Zong Junyao 2011 Research on the large civil aircraft flight management system function *Sci. Tech. Inno. Herald* **32** 210

[5] Zong Junyao and Zheng Zhiming 2014 Analysis for the architecture of civil aircraft thrust management system *Sci. Tech* **10** 155

[6] Yao Zhichao 2012 Research on civil aircraft for a new type thrust management architecture *The 9th Yangtze River Delta Sci. Tech. Conf. Aerospace Science Technology Innovation Sub-Conf.* (Nanjing) 269-71

[7] Ke Jie 2012 Design and analysis for advanced civil aircraft integrated flight and thrust management function *Soft. G* **11** 59-61

[8] Sun Peng Liu and Chao Ke Jie 2012 Research on thrust management function for large civil aircraft *Civ. AC. Desn & Res* **106** 18

[9] SAE ARP4754A 2010 Guidelines for Development of Civil Aircraft and Systems *S. A. E. International* 44

[10] SAE ARP4761 1996 Guidelines and Methods for Conducting the Safety Assessment Process on Civil Airborne Systems and Equipment *S. A. E. International* 12

IOP Conf. Series: Materials Science and Engineering **692** (2019) 012038 doi:10.1088/1757-899X/692/1/012038

Research on Characteristics of Spacecraft Transport Dynamics Ambient Excitation Based on Conditions of Highway and Railway Transportation

Liu Guangtong, Xukai, Guotao, Zhang Qiang

104#You Yi Road Haidian District Beijing China, Beijing Institute Of Spacecraft Environment Engineering, China Academy Of Space Technology

liugt2007@qq.com

Abstract. Based on the highway and railway transportation conditions of several spacecraft from Beijing to launch site, the dynamics environment data acquisition scheme is designed. The response values of sensors before vibration reduction of transport packaging box are collected throughout the whole process. The power spectral density is obtained by filtering and Fourier transform, and the peak response frequency of the excitation source of spacecraft transportation environment is analyzed. It can be used as the basis for the design of vibration reduction system for spacecraft transportation, and some suggestions for the optimization design of vibration reduction system are put forward.

1. Introduction

Spacecraft transport is a significant link during the manufacturing process of spacecraft, which mainly contains 3 kinds of conditions, namely deck transport, cabin transport and whole-spacecraft transport, especially under the whole-spacecraft transport condition, which is a quasi-launching condition, vibration environment evoked by transport will have potential influences on the satellite, especially large-scale structural components, so it is a complex condition with high requirements on mechanical environment[1]. After all works of the development stage at the developing area, spacecrafts need to be sent to the launching site for final works before launching, different and complex operating conditions like automobiles and trains, etc. will be experienced during the transportation processes, so it is significant to guarantee a safe mechanical environment during the transportation processes of spacecrafts, as it is directly related to the development of works in the satellite launch site.

With the development of spacecraft development technologies, the transportation states become increasingly complex, and requirements on transportation mechanical environment keep being higher. Generally, vibration transmissibility is reduced by installing an absorber at the pedestal of packaging boxes, which can realize a damping effect. At present, during the designing process of spacecraft transportation packaging boxes, due to insufficiencies in characteristic analysis on external expiation sources, amplification phenomenon may happen to some points of spacecrafts after vibration attenuation through shock mitigation system during the transportation process, which may have certain influences on transportation safety of the spacecrafts, so it is in urgent demands to research the ambient excitation characteristics during spacecraft transport based on operating conditions of highway and railway transportation[2,3].

2. Research on Necessity of Excittion Characteristic Analysis

Damping effect of spacecraft transportation system is directly reflected by the vibration transmissibility, which is defined as the specific value of response amplitude X_0 and input excitation amplitude X_i, and the standard model of single steel wirerope absorber is composed of a linear spring with rigidity of K and a damper system with damping of C, as shown in Figure 1.

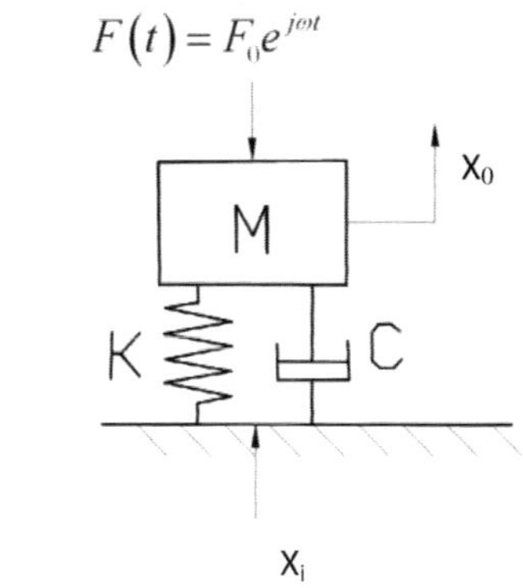

Figure 1. Unidirection model.

Motion equation of vibration isolation system is:

$$M \cdot \ddot{X} + C \cdot \dot{X} + K \cdot X = F_0 \cdot e^{j\omega t} \tag{1}$$

The most common physical quantity to describe and evaluate vibration isolation effect is the vibration transfer coefficient T, and under steady-state sinusoidal excitation, transmissibility T is:

$$T = \frac{X_0}{X_i} = \sqrt{\frac{1+(2\lambda\zeta)^2}{(1-\lambda^2)^2+(2\lambda\zeta)^2}} \tag{2}$$

$\lambda=\omega/\omega_0$, λ is the frequency ratio, ω is the forced vibrational frequency, ω/ω_0 is the inherent frequency of natural vibration, $\zeta=C/C_c$, ζ is the damping ratio of vibration system, and C_c is the critical damping coefficient.

It can be known that vibration transmissibility is influenced by frequency ratio λ, when frequency ratio $\lambda<\sqrt{2}$, T>1, and the vibration is amplified; when frequency ratio $\lambda=1$, T is at its maximum value, and there is resonance; frequency ratio λ continues amplifying, then frequency ratio T will gradually decrease; when frequency ratio $\lambda>\sqrt{2}$, then T<1, and the vibration damper will come into play[4].

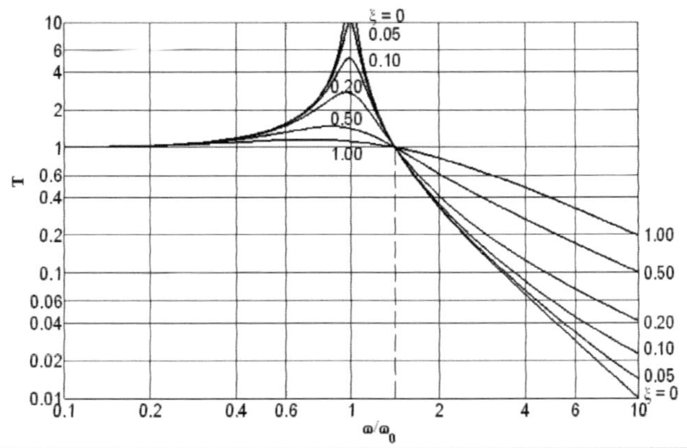

Figure 2. Reelationship between transmissibility, frequency and damping.

Based on Figure 2, when $\lambda=\omega/\omega_0>\sqrt{2}$, the shock mitigation system can be confirmed to keep away from the resonance region and meanwhile has a good vibration damping performance, if the interfering frequency ω of the system is low, the system can hardly fulfill the requirement of $\omega/\omega_0>\sqrt{2}$ when designing, so vibration response of the system must be restrained through increasing the damping of vibration isolation system. It is in urgent demand to know about the ambient excitation frequency during the transportation process of spacecrafts with complex operating conditions, while as spacecrafts are installed in the spacecraft transportation packaging boxes, which are fixed to the automobiles or trains during transportation process, existing road spectrums of highways and railways with different classifications have no referential significance on the shock mitigation system of spacecraft packaging boxes, excitation sources of shock mitigation system should be the response of upper surfaced of automobiles and train carriages, thus it will be of significant meanings on optimization of vibration damping performance of spacecraft packages and improvement of reliability of spacecraft transport to have deep analysis on excitation sources of spacecraft transportation packaging boxes under different road conditions and modes of transportation.

3. Brief Introduction of Spacecraft Transportation System

Spacecraft transportation system mainly contains 3 parts: spacecraft, packaging box and packaging box shock mitigation system, transport vehicle and vehicle shock mitigation system, as shown in Figure 3. The spacecraft is installed in the packaging box through an adapter bracket, a shock mitigation system is arranged between the adapter bracket and baseplate of the packaging box, and the whole packaging box is fixed to the carrier vehicle by a fixing device, during the transportation process, vibrations and impacts caused by road surface on the spacecraft during transportation process will be reduced by the double-layer effect of the shock mitigation system of the vehicle and packaging box.

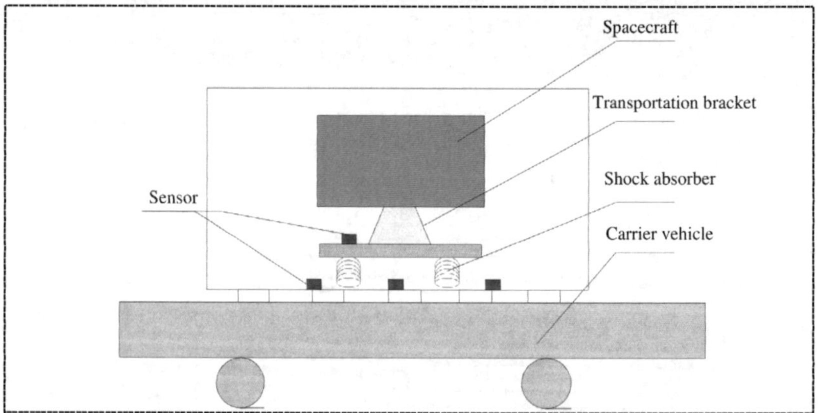

Figure 3. Sketch map of spacecrraft transportation system.

3.1. Road condition

Road conditions during spacecraft transportation process mainly contain those of highways and railways. Highways contain expressways, first-class highways and second-class highways, on which speed of transport vehicles is generally 30~60Km/h[5]; railway classifications contain Class-I or Class-II passenger-freight co-Line railways, on which speed of transport vehicles is generally no higher than 90Km/h[6], and traveling distance of the whole course is 800km.

3.2. Transport vehicles

Transport vehicles contain highway and railway transport vehicles. Highway transport vehicles are large-scale carrier vehicles purposely made by some brand, which come with an air-fuel mixture shock mitigation system that can effectively isolate the vibrations and impacts from road surface, as shown in Figure 4. Railway transport vehicles are railway flatcars, whose main vehicle model is NX17 series, it is with a full-wood floor structure, and there is no special vibration damper inside it[7] , as shown in Figure 5.

Figure 4. Motor transport of packaging box.

Figure 5. Railway transport flatcar of packaging box.

3.3. Spacecraft transportation packaging box

Spacecraft packaging box is the carrier of spacecraft transport, integrated design is conducted on packaging boxes integrated with demands of satellite development and transport based on universalization and integration theories, which can not only take whole-course highway and railway transportation of spacecraft into consideration, meanwhile it also realizes functions like controllable inside-box temperature and whole-course real-time monitoring of environmental data, etc. for the spacecraft under extreme environmental conditions, thus guaranteeing the environmental safety of spacecraft transport.

Built-in vibration attenuation mode is adopted by the packaging box shock mitigation system, the shock absorber is arranged between the adapter bracket and packaging box of the spacecraft, whose significant characteristics lie in its small volume, simple arrangement and good universality. The spring most widely used by built-in vibration attenuation mode is the steel wirerope spring, which is basically the same as the vibration attenuation mode of spacecraft transportation packaging boxes home and abroad. Figure 6 shows the sketch map of transportation packaging box and shock mitigation system of a foreign spacecraft with a certain model.

Figure 6. Transportation mode of JWST and GOCE satelite of ESA.

4. Excitation Analytic Data Acquisition

Excitation source of spacecraft transport is the response of upper surface of the vehicle, so the shock mitigation system of transport vehicle is not needed, only shock mitigation system of packaging boxes needs to be considered. In the packaging box shock mitigation system during spacecraft transport, data measured through the vibrating sensor at the bottom of packaging boxes is used as the input source of the shocking mitigation system, computation study is made on the shock excitation spectral characteristics under different operating conditions with data of highway and railway transportation with some model as analytic basis, so as to analyze the characteristics of ambient excitation of spacecraft transport.

4.1. Distribution of measure points of transportation packaging boxes

Totally 3 mechanical sensors are set before vibration attenuation during the transportation process, and responses of the 3 points before vibration attenuation (piont 1, piont 2 and piont 3) are mainly analyzed. Figure 7 shows the measuring points of sensors.

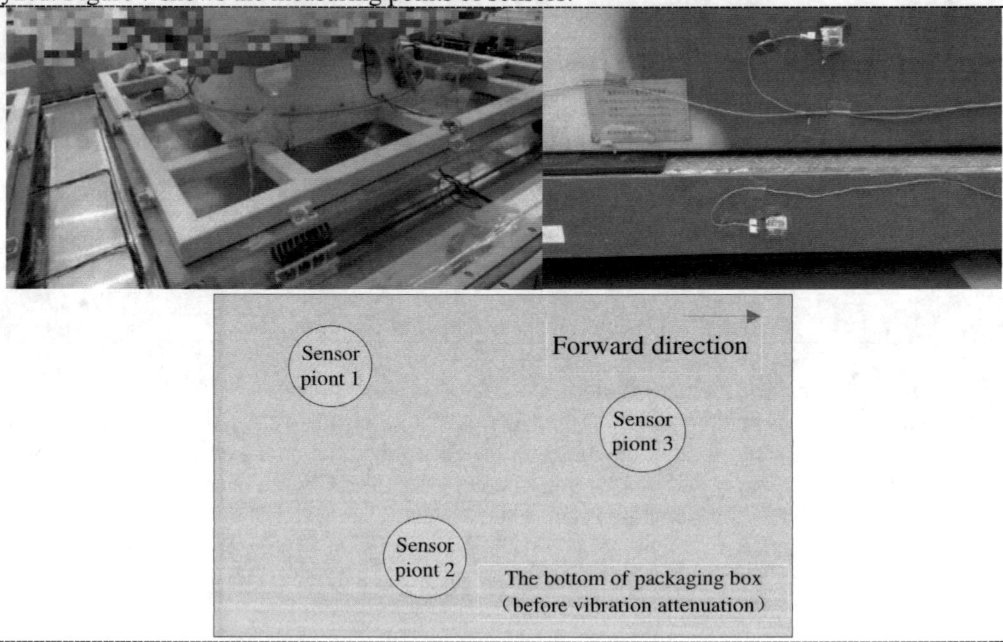

Figure 7. The measuring points of sensors on packaging box.

4.2. Test system conditions

The test system can conduct real-time monitoring on information like vibration, running speed and position, etc. inside the packaging boxes, the piezoelectricity acceleration sensor and signal conditioner of the 3 three-axial direction built-in ICs are connected by low-noise coaxial cables, data of all sensors collected by the main engine is stored in the memory spaces of each channel, meanwhile it is connected with the laptop through the form of Ethernet TCP/IP, which can display the measured real-time data on the computer of the carriage. Monitors conduct curvilinear playback, spectral analysis and power spectral density analysis, etc. on the collected data through data acquisition and analysis software. Figure 8 shows the sketch map of monitoring system composition.

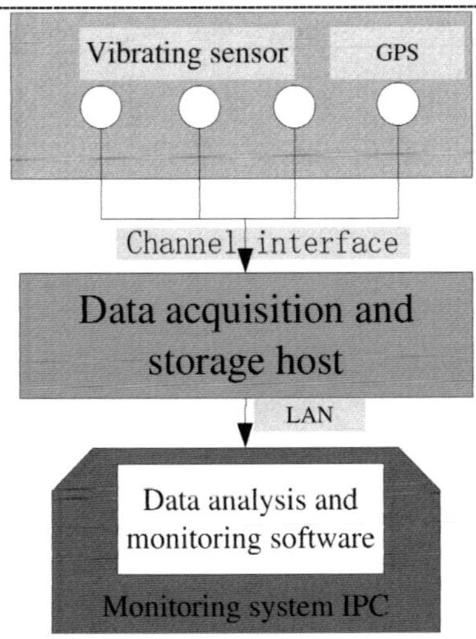

Figure 8. Sketch map of monitoring system composition.

4.3. Data processing

During the test process, packaging box data acquisition system is used to collect the signals during the transportation process, a high-pass filter is adopted to filter the mingled low-frequency noise ingredients in the signals, and FFT is adopted to acquire the PSD curve during transportation process, then analyze the PSD curve and acquire the peak value response frequency, which is the peak value response frequency of ambient excitation source during spacecraft transport. Detailed treatment scheme is shown in Figure 9.

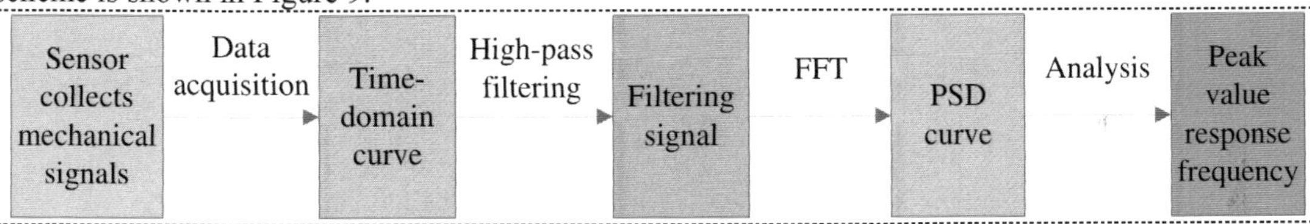

Figure 9. Data processing flow.

4.4. Analysis on applicability of excitation source data towards different spacecraft transportations

Excitation source of packaging box shock mitigation system is the response of upper surface of transport vehicles, when transport vehicles convey spacecrafts with different weights, excitation sources of packaging box shock mitigation system will also change relevantly, to guarantee the universality of the excitation source on different spacecraft transportations, influences of changes in spacecraft weight on changes of excitation need to be researched.

Generally the total weight of a middle-large packaging box of spacecraft is over 10t, and the weight of a spacecraft (remote sensing satellite) is generally 2t~3t, difference among weights of different spacecrafts is about 1t, when transport vehicles are conveying spacecrafts with different models, variation in the weight of carrier is about 7% of the total weight, thus the weight of spacecraft takes a small proportion in the whole transportation system, and it is thought that change in weight of spacecrafts will not affect the ambient excitation of spacecraft transportation.

5. Excitation Analysis of Highway Transportation

Conduct spectral analysis on the time-domain curves of the 3 measure points collected through operating conditions of highway transportation, vibration acceleration spectrum density curves of measure point 1-3 can be seen in Figure 10-12, where vertical acceleration is far higher than that in the forward and lateral direction, and frequency components are concentrated on 2.2Hz, 4.2Hz and 11Hz; difference value of three-way acceleration PSD at measure point 1 and 2, namely the power density spectrum of angular acceleration on baseplate of packaging box is shown in Figure 13, and angular acceleration generated through difference in Z-direction acceleration is far larger than the angular velocity generated through difference in Y-direction angular acceleration.

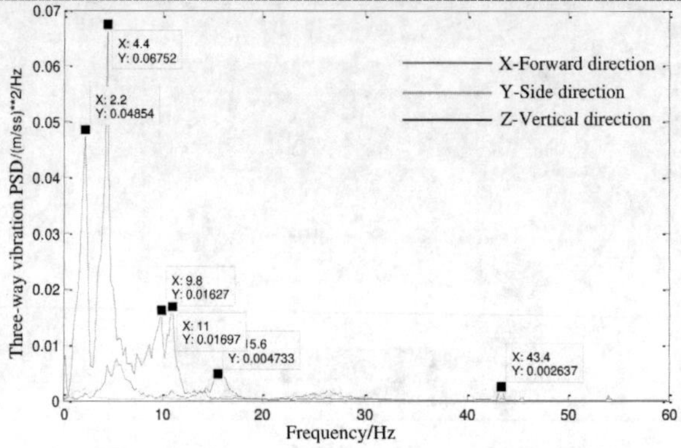

Figure 10. Vibration acceleration PSD spectral line of measure point 1 during highway transportation.

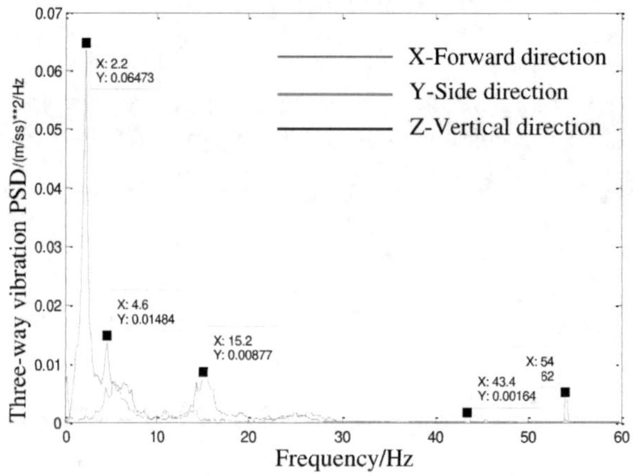

Figure 11. Vibration acceleration PSD spectral line of measure point 2 during highway transportation.

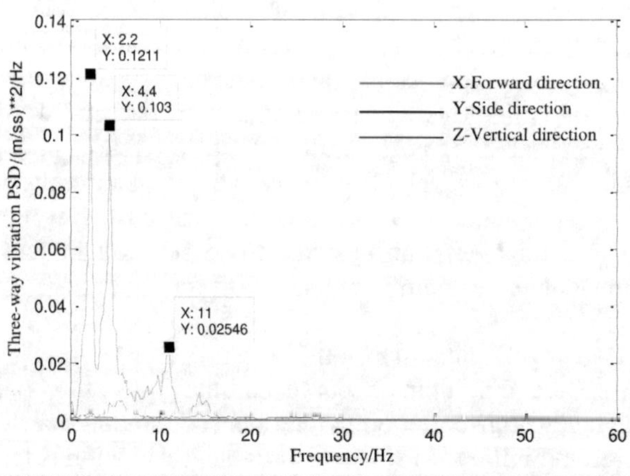

Figure 12. Vibration acceleration PSD spectral line of measure point 3 during highway transportation.

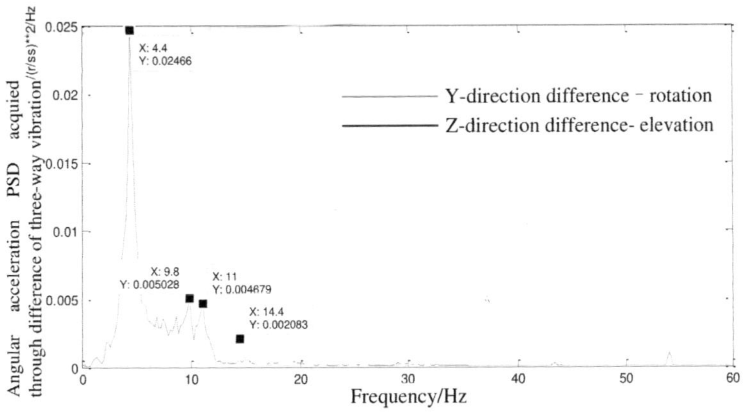

Figure 13. Angular velocity PSD spectral line of packaging box baseplate.

6. Railway transport excitation analysis

The vibration acceleration spectrum density curves at the 3 measure points of operating conditions during railway transport are shown in Figure 14-16, the vertical acceleration is far larger than that of forward and side direction, and the frequency components are concentrated on 1.8Hz and 6.2Hz. Difference value of three-way acceleration PSD of measure point 1 and 2, namely the angular acceleration power density spectrum of packaging box baseplate is shown in Figure 17. Angular acceleration generated through difference in Z-direction acceleration is far larger than that generated through difference in Y-direction angular acceleration, and frequency components are concentrated on 2.2Hz, 6.6Hz, 8.6Hz, 9.8Hz and 11.4Hz.

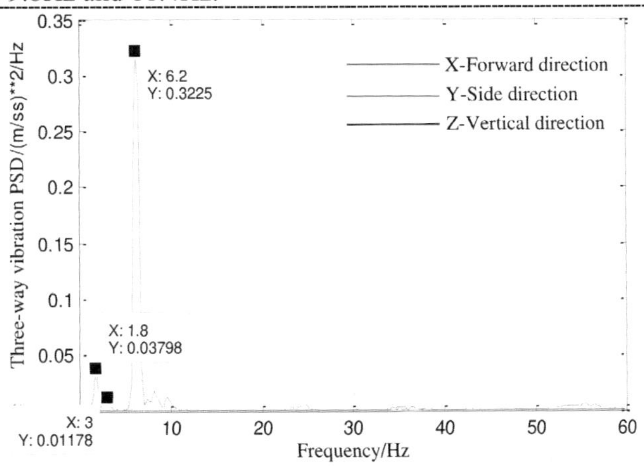

Figure 14. Vibration acceleration PSD spectral line of measure point 1 during railway transportation.

IOP Conf. Series: Materials Science and Engineering **692** (2019) 012038 doi:10.1088/1757-899X/692/1/012038

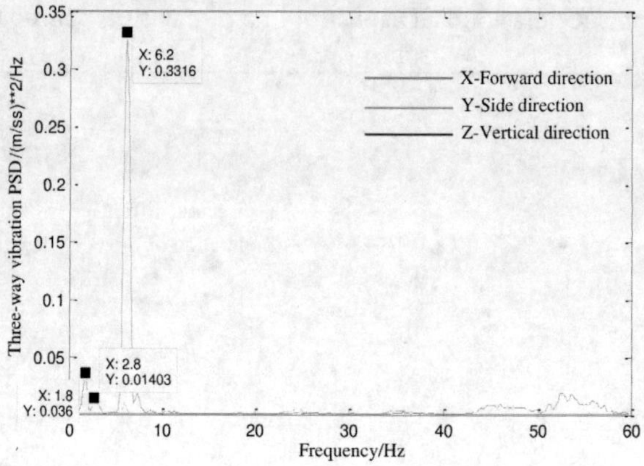

Figure 15. Vibration acceleration PSD spectral line of measure point 1 during railway transportation.

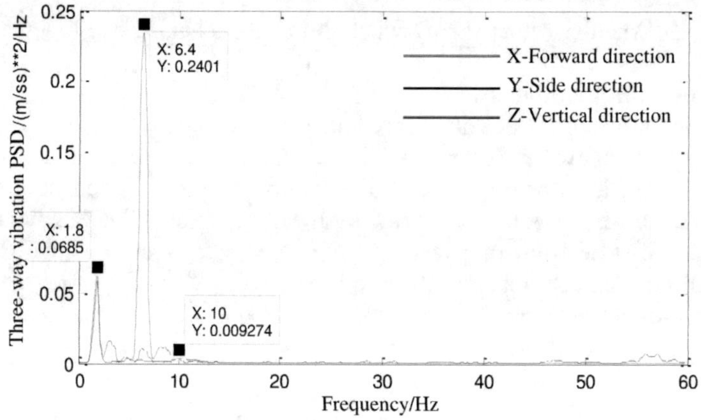

Figure 16. Vibration acceleration PSD spectral line of measure point 1 during railway transportation.

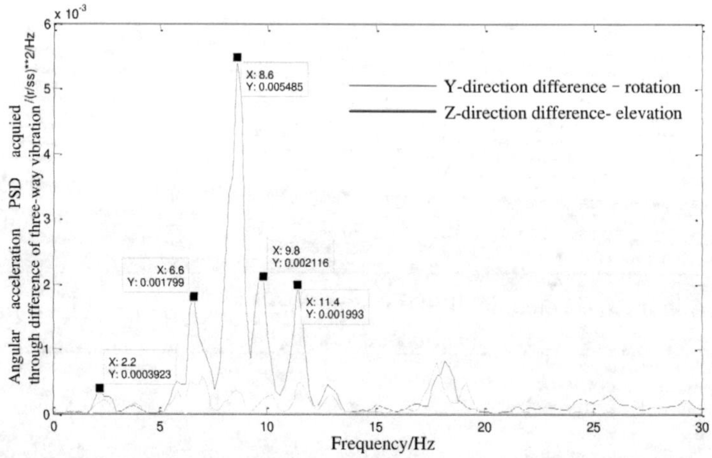

Figure 17. Angular velocity PSD spectral line of packaging box baseplate.

7. Suggestions on Shock Mitigation System for Spacecraft Transport

- Shock excitation of automobile transport is concentrated on: 2.2 Hz, 4.4 Hz and 10- 11Hz, shock excitation of railway transport is concentrated on: around 1.8 and 6.2Hz, and design of inherent frequency of packaging box shock mitigation system should keep away from areas concentrated with energies;
- Maximum energy contribution frequency of excitation is between 2.2Hz and 4.4Hz, to effectively reduce the excitation of this frequency band, the inherent frequency of the system can be reduced to below the maximum contribution frequency, while as the transformation of damper is large, attention should be paid to guaranteeing the stability and system shock resistance capacity, and enough capacitance should be guaranteed for the system under the most rigorous conditions required for the system input environment.
- If the system interfering frequency ω is low, it is difficult to reach the requirement $\omega/\omega_0>\sqrt{2}$,when designing the system, so vibration response of the system must be restrained through increasing the resonance of vibration isolation system, do not simply pursue a best vibration isolation effect without consideration on the influence that large amplitudes will have on spacecraft safety;
- In view of the stiffness characteristics of wire rope dampers and rubber dampers commonly used in spacecraft transportation, the optimum operating conditions of the damper system are fixed when the quality of the transporting products is the same, and the natural frequencies of the damper system can not be adjusted flexibly in the transportation process, which has limitations. The stiffness auto-adjusting vibration absorber system based on air spring has certain advantages, and it is one of the directions for the optimization of the vibration absorber system for spacecraft transportation.

8. Conclusion

Through collecting mechanical data of highway and railway transportation with a certain model, computation and research were made on shock excitation PSD under different transportation conditions to analyze the ambient excitation characteristics during spacecraft transport and to provide basis for the confirmation of inherent frequency of shock mitigation system for packaging box shock mitigation system during spacecraft transportation, and suggestions were put forward for the design of shock mitigation system.

Reference

[1] Zhang Hua, Zong Yiyan. Vibration environment and reliability analysis of solar array in transportation [J]. Spacecraft Environmental Engineering, 2014, 31(5): 536-542

[2] Zhu Hongya, Liu Guangtong. The acquisition, storage and analysis of mechanical data in spacecraft transportation [J]. Spacecraft Environmental Engineering, 2014, 31(5): 564-567

[3] Zhang Li, Qi Xiaojun, Fu Guoqing, Response Analysis on Vibration during Satellite Transportation [J], Spacecraft Environmental Engineering, 2009, 26(z1): 55-61

[4] Zhao Jigang, Li Yufa, Han Quanyou. The Principle and Application of Engine Mount for Vibration Isolation [J]. Agricultural Equipment and Vehicle Engineering, 2012, 31(5): 6-9

[5] Shao Chhunfu, Wang Ying, Zhou Zhixiang. Comparative Study of Highway Engineering Technique Standards [J]. Academic Journal of Traffic Engineering and Information, 2005, 03: 16-28

[6] Feng Dequan. Comparative Analysis on Railway Design Standards between China and Russia [J]. Railway Standard Design, 2012, 6: 28-31

[7] Dai Anguo, Yu Shiming, Zhang Zhong. Exploration of Loading and Consolidating Techniques of Flatcars on Chinese Railways [J]. Railway Freight Transport, 2013, 03: 53-57

IOP Conf. Series: Materials Science and Engineering **692** (2019) 012039 doi:10.1088/1757-899X/692/1/012039

Reconstruction on Rotor Profile of Twin Screw Compressor Based on Cubic B-spline Curve

Yinglei Chen[1], Rongyu Ge[1,*], Beibei Kong[2] and Xiaomei Ni[2]

[1]School of Mechanical Engineering, University of Jinan, Jinan, 250022, China
[2]School of Mechatronics Engineering, Qilu Institute of Technology, Jinan, 250200, China

Corresponding author: me_gery@ujn.edu.cn

Abstract. In order to overcome the inconvenience of adjusting the rotor profile of twin screw compressor, the method of using cubic B-spline curve to express the rotor profile is put forward. CFD simulation analysis of the rotor compressor formed by the modified cubic B-spline curve shows that the performance of the rotor profile described by the cubic B-spline curve has little difference with that of the original profile line, and the rotor profile described by cubic B-spline curve is more convenient to modify.

1. Introduction

Twin screw compressor plays an important role in many fields such as machinery, electric power, chemical industry, pharmaceutical industry, energy and military industry due to its simple structure, good power balance and high reliability [1]. Through the research on the rotor profile of twin screw compressor, the curve types of the profile mainly include arc, involute, cycloid and straight line [2]. The composition curve of rotor tooth profile is expressed by cubic B-spline curve, and the design results show that this method is feasible, which provides an idea for the design of rotor profile on twin screw compressor.

2. Rotor profile design of twin screw compressor

As the core component of twin screw compressor, the rotor is particularly important for the performance of the compressor. The design method of rotor profile adopted in this paper is to define the tooth profile of a female rotor in advance, use cubic B-spline curve to describe the female rotor profile, and then model it in 3d software, and then solve the conjugate tooth profile of the corresponding male rotor by means of three-dimensional solid modeling [3].

2.1. Design of female rotor profile

A unilateral asymmetric female rotor profile formed by a combination of straight line, arc and cycloid is shown in figure 1. This profile is modified by radial line and chamfering, and the cycloid forms a point moving inward. In addition, the arc tooth curve is extended to a certain angle to form a protection angle. Its rotor engagement line seal and continuous contact line can be better applied in production [4]. Where, AB and DE are straight lines, BC and EF are arcs, CD is cycloid and D is cycloid forming point, which constitute a complete and continuous female profile line

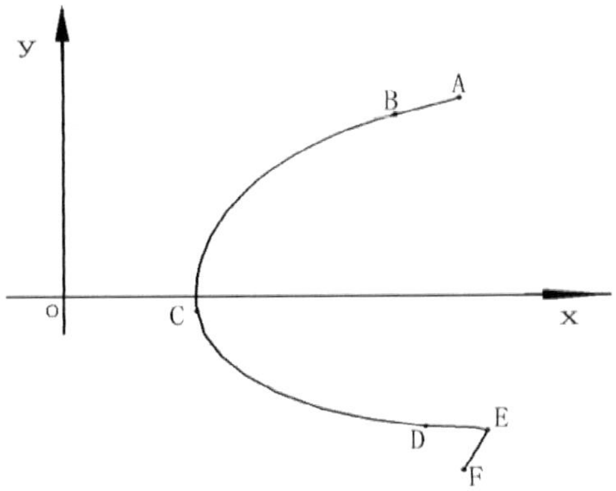

Figure 1. Composition of female rotor profile line.

2.2. Cubic B-spline curve expression of rotor profile

Cubic B-spline curve has a series of excellent properties, such as local properties, variation reduction and convexity, etc., which are conducive to realizing the control of spline curve shape, especially the local shape control of spline curve. By adjusting the control points rather than modifying data points, spline curves can be smoothed and convex preserving requirements can be achieved [5]. When using cubic B-spline to describe the tooth curve of female rotor, the basic curve segments AB, BC, CD, DE and EF are defined first. The five-segment curve is described by a complete cubic B-spline curve, and the parametric equation of cubic B-spline curve $p(u)$ is shown as follows:

$$p(u)=\sum_{j=0}^{n} d_j N_{j,3}(u) \quad u\in[u_i,u_{i+1}]\subset[u_3,u_{n+1}] \quad i=3,4\cdots,n+1 \tag{1}$$

Where d_j is the control vertex and $U=\{u_0,u_1, \cdot\cdot\cdot,u_n,u_{n+4}\}$ is the node vector.

Since the rotor profile line is an open curve, the fixed support condition of the endpoints of quadruple nodes is taken, and the standard definition domain is taken, so $u_0=u_1=u_2=u_3=0$, and $u_{n+1}=u_{n+2}=u_{n+3}=u_{n+4}=1$. A number of coordinate points of rotor profile line were selected, and the parametric method of normalizing accumulated chord length was adopted. As shown in equation (2), the data points were parameterized.

$$\begin{cases} u_3=0 \\ u_{3+i}=u_{i+2}+|\Delta q_{i-1}|, i=1,2,\cdots n \end{cases} \tag{2}$$

Where u is node vector, $\Delta q_{i-1}=q_i-q_{i-1}$ is string vector, and q is coordinate point of rotor profile line.

In combination with the conditions of the initial and terminal points, the additional equation of the head and end points is given by using the tangent vector conditions. The arc length parameter is usually the tangent vector equivalent to the unit tangent vector under the parametric method of cumulative chord length, that is, the module length is set as $p_0=|\Delta p_0|(\Delta_0)^{-1}$, $p_{n-1}=|\Delta p_{n-1}|(\Delta_{n-1})^{-1}$, so that the additional equation of the initial and terminal points is shown in equation (3) :

$$\begin{cases} \boldsymbol{p}_0=d_0 \\ \boldsymbol{p}_n=d_n \end{cases} \tag{3}$$

Where d_0 and d_n are given initial and terminal tangent vectors.

By successively substituting the node values in the defined domain $u\in[u_i,u_{i+1}]\subset[u_3,u_{n+1}]$ into cubic B-spline equation (1), the linear equations can be obtained as shown in equation (4):

IOP Conf. Series: Materials Science and Engineering **692** (2019) 012039 doi:10.1088/1757-899X/692/1/012039

$$\begin{bmatrix} N_{0,3}(u_3)N_{1,3}(u_3) & & \\ N_{0,3}(u_4)N_{1,3}(u_3)N_{2,3}(u_3) & & \\ \bullet \quad \bullet \quad \bullet & & \\ & \bullet \quad \bullet \quad \bullet & \\ & & \bullet \quad \bullet \quad \bullet \\ & & N_{n-3,3}(u_3)N_{n-2,3}(u_3)N_{n-1,3}(u_3) \\ & & N_{n-2,3}(u_3)N_{n-1,3}(u_3) \end{bmatrix} \begin{bmatrix} d_0 \\ d_1 \\ \bullet \\ \bullet \\ \bullet \\ d_{n-2} \\ d_{n-1} \end{bmatrix} = \begin{bmatrix} P_{n-1} \\ P_0 \\ \bullet \\ \bullet \\ \bullet \\ P_{n-3} \\ P_{n-2} \end{bmatrix} \tag{4}$$

The elements in the coefficient matrix are the values of the B-spline basis function, which are only related to the node values. It is more convenient to use the matrix equation as follows.

$$\begin{bmatrix} b0 & c0 & & & \\ a1 & b1 & c1 & & \\ & \bullet & \bullet & \bullet & \\ & & \bullet & \bullet & \bullet \\ & & & \bullet \quad \bullet & \bullet \\ & & a_{n-2} & b_{n-2} & c_{n-2} \\ & & & c_{n-1} & b_{n-1} \end{bmatrix} \begin{bmatrix} d_0 \\ d_1 \\ \bullet \\ \bullet \\ \bullet \\ d_{n-2} \\ d_{n-1} \end{bmatrix} = \begin{bmatrix} e_0 \\ e_1 \\ \bullet \\ \bullet \\ \bullet \\ e_{n-2} \\ e_{n-1} \end{bmatrix} \tag{5}$$

Where

$$\begin{cases} a_i = \dfrac{(\Delta_{i+2})^2}{\Delta_i + \Delta_{i+1} + \Delta_{i+2}} \\ b_i = \dfrac{\Delta_{i+2}(\Delta_i + \Delta_{i+1})}{\Delta_i + \Delta_{i+1} + \Delta_{i+2}} + \dfrac{\Delta_{i+1}(\Delta_{i+2} + \Delta_{i+3})}{\Delta_{i+1} + \Delta_{i+2} + \Delta_{i+3}} \qquad i = 0,1,\cdots,n-1 \\ c_i = \dfrac{(\Delta_{i+1})^2}{\Delta_i + \Delta_{i+2} + \Delta_{i+3}} \\ e_i = (\Delta_{i+1} + \Delta_{i+2})q_{i-1} \end{cases} \tag{6}$$

$$\Delta_i = u_{i+1} - u_i$$

Tangent vector conditions for the first and end.

$$b_1 = 1, c_1 = a_1 = 0,\ e_1 = q_0 + \frac{\Delta_3}{3}\boldsymbol{q}_0 \tag{7}$$

$$c_{n-1} = a_{n-1} = 0, b_{n-1} = 1, e_{n-1} = q_{n-2} + \frac{\Delta_n}{3}\boldsymbol{q}_{n-2} \tag{8}$$

According to equations (5), (6), (7) and (8), the control point di can be obtained by solving the equation with the chase method. Then, the equation and shape of cubic B-spline curve can be determined by substituting the obtained parameters into equation (4). In this paper, Matlab is used to inversely solve cubic B-spline curves, as shown in figure 2, to obtain the control points and curve equations of cubic B-spline curves of the female rotor. In the three-dimensional software SolidWorks, by using the spatial solid envelope, the profile line of the corresponding conjugate curve segment of the female rotor is obtained, that is, the profile line of the male rotor.

IOP Conf. Series: Materials Science and Engineering **692** (2019) 012039 doi:10.1088/1757-899X/692/1/012039

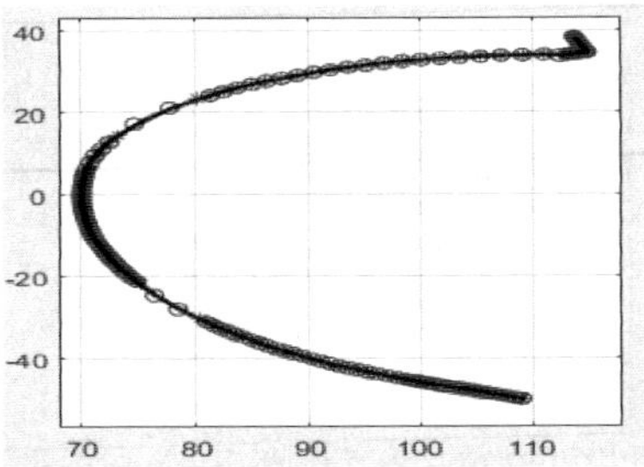

Figure 2. Cubic B-spline curve calculated by Matlab.

3. Performance analysis of rotor profile

In order to judge the performance of the rotor profile described by cubic B-spline curves, the reconstructed cubic B-spline female rotor profile data points and the calculated male rotor profile data points are imported into SolidWorks to generate the three-dimensional solid model of the rotor, as shown in figure 3. Then, the performance of the new rotor profile line is judged by CFD simulation analysis. Since the volume of the rotor teeth and the volume of the working cavity of the compressor are continuously and periodically changing, the gas in the compressor cavity is compressed and flowing, and the process is unsteady. In order to study the internal pressure distribution and fluid velocity of the compressor, Fluent software was used to establish the fluid models of the unilateral asymmetric linear rotor compressor and the new linear rotor compressor, and the dynamic numerical simulation of the internal flow field of the compressor was carried out by using the dynamic grid technology. Finally, the characteristics of flow field inside compressor are compared and analyzed.

Figure 3. Three-dimensional solid model of the rotor.

Figure 4 and figure 5 respectively show the pressure cloud diagram of the new-type profile line compressor and the unilateral asymmetric profile line compressor. By comparing the two figures, it can be found that the axial upward pressure of the female and male rotors of the new-type profile line rotor compressor and the unilateral asymmetric profile line rotor compressor has increased significantly. The air pressure at the exhaust end of the new-type profile line rotor compressor is up to 0.7989MPa, while that at the exhaust end of the prototype profile line rotor compressor is 0.809MPa. There is no significant difference between the two, indicating that the compression performance of the new profile line is little different from that of the prototype profile line in the design of the rotor profile using the cubic B-spline curve as the rotor tooth curve.

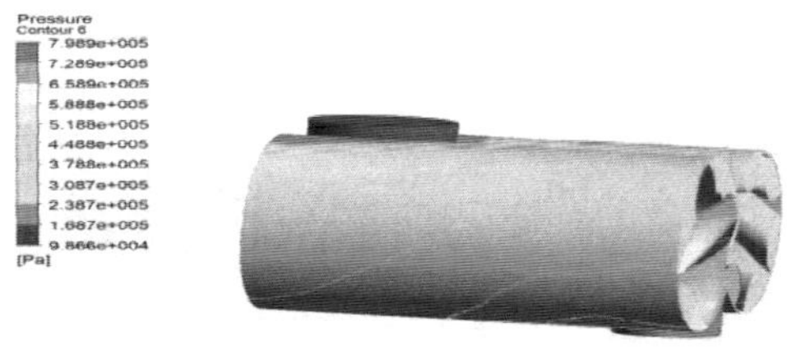

Figure 4. Pressure cloud diagram of the reconstructed profile line compressor.

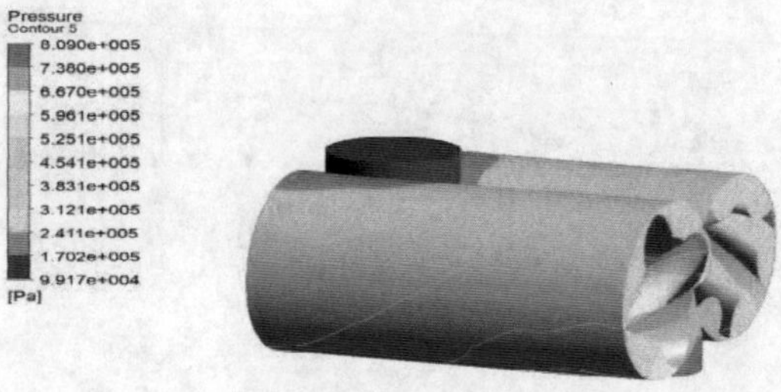

Figure 5. Pressure cloud diagram of unilateral asymmetric line compressor.

Figure 6 and figure 7 are respectively the fluid pressure cloud images of the middle section of the new-type profile line rotor compressor and unilateral asymmetric profile line rotor compressor. By comparing the two figures, it can be found that the pressure has an obvious change trend near the contact line, indicating that the sealing effect of the engagement line is good, which is consistent with the actual situation. At the same time, the gas pressure changes on both sides of the two contact lines are obvious, and the gas pressure on the low pressure side is relatively low, which indicates that when the cubic B-spline curve is used as the rotor tooth curve to design the rotor profile line, the sealing performance of the engagement line obtained is not much different from that of the prototype profile line.

Figure 6. The section pressure diagram of profile after reconstruction.

Figure 7. The section pressure diagram of unilateral asymmetric profile.

Figure 8 and figure 9 show the fluid velocity of the new-type profile line rotor and unilateral asymmetric profile line rotor compressor respectively. As there is a certain gap between the female and male rotors, the gas on the high-pressure side will leak to the low-pressure side through the gap. The maximum fluid velocity of the new linear rotor compressor is 274.3m/s, and that of the unilateral asymmetric rotor compressor is 294.3m/s, which indicates that the new rotor line contact line can effectively prevent fluid leakage from the high-pressure side to the low-pressure side, which is not much different from the sealing performance of the original contact line.

Figure 8. Internal velocity diagram of rotor compressor after reconstruction.

Figure 9. Internal velocity diagram of unilateral asymmetric rotor compressor.

4. Conclusions

Using cubic B-spline curve, the control point of tooth curve of female rotor is obtained, and the reverse process of cubic b-spline curve section is introduced in detail. The simulation results show that the performance of the newly designed rotor profile is little different from that of the original profile under the same parameters, and the rotor profile composed of cubic b-spline curve is more convenient to modify.

Acknowledgement

The authors gratefully acknowledge that the present research is supported by the Key Research Project of Shandong Province in China (No. 2019GNC106123). In addition, valuable suggestions from the reviewers in improving the quality of the paper are highly appreciated.

References

[1] Liu H 2014 Current situation and prospect of the market of screw compressor for aerodynamics. *G Mach* **10** pp31-33
[2] He X, Shi G, Wu M and Zhang R 2018 Application of B-spline in the reverse design method of screw rotor Profile. *Mech. Sci. Tech. A. E.* **37** pp43-54
[3] Xu L 2014 Mutual design and simulation of female and male rotors *J. Mech. Trans.* **38** pp24-26
[4] Xing Z 2000 *Screw compressor–theory design and application* (Beijing: Mechanical Industry Press) p30
[5] Shi F 2001 *Computer-aided geometric design and non-uniform rational B-spline* (Beijing: Higher Education Press) pp304-325

Investigation of Blocking Effect of Aviation High-Speed Towing Tank Based on CFD and EFD

Zhijian Xiao[1,2], Rong Jiang[1,2], Bin Wu[1,2], Mingzhen Wang[1,2] and Wenjing Yang[3*]

[1] Key Aviation Scientific and Technological Laboratory of High-Speed Hydrodynamic, Jingmen 448035, China
[2] China Special Vehicle Research Institute, Jingmen 448035, China
[3] Huazhong University of Science and Techonolgy, Wuhan 430074, China

[*]E-mail address: yangwenjing10@hust.edu.cn

Abstract. Taking the unpowered model of a seaplane as the object, the blocking effect of aviation high-speed tank was studied by experimental and numerical methods in this paper. Firstly, the experiment of air field and drag were carried out in the bottom and front of the trailer. The experimental results of air field showed that the blocking effect of the air field was obvious in the bottom of trailer, while ignored in the front of trailer. The results of drag experiment under two conditions showed significant difference at mid-high speed, which proved the influence of blocking effect was obviouse. RANS method and overset grid were adopted in numerical simulation, and the accuracy of numerical method adopted in this paper was verified by comparing computed results of the model in the bottom of trailer with the experimental results. The computed results of hydrodynamic performance of the seaplane in finite and infinite domain showed that the influnence of the blocking effect of water could be nearly ignored, but the influnence of the blocking effect of air field was dominated. Finally, the hydrodynamic performance of the seaplane under the trailers with different sizes of chassis were carried out. The computed results showed that within the certain range, the degree of blocking effect of air field increased with the increase of geometrical variables of trailer chassis, further, which reached the maximum and kept constant when the variables exceeded a certain value.
Keywords: Seaplane; Hydrodynamic performance; RANS; Overset mesh; Blocking effect

1. Introduction

The boundary effect is inevitable in any towing test due to the limited boundary. There are three kinds of boundary effect, containing blocking effect, shallow water effect and wall effect. Generally speaking, for the ship test in towing tank, blocking effect is the largest one while the shallow water effect and wall effect can be ignored[1-3]. The wind speed has a small effect on the test of low-speed surface ship, but the change of wind speed has a greater impact on hydrodynamic results of high-speed vehicles such as seaplane, ground-effect aircraft and planing boat. Ideally, the wind speed at the bottom of the high-speed trailer was equal to the speed of the trailer for the aviation high-speed towing tank. However, when the high-speed trailer travels fast in the tank, the wind speed in the bottom of the high-speed trailer will increase because of the blocking effect caused by the limited air flow along the trailer chassis, tank wall and water surface. To reduce the impact of increment of wind speed in the bottom of high-speed trailers, some high-speed tanks, such as the one at Russia's central institute of air and fluid dynamics, placed the models in the front of trailer. Therefore, it is necessary to consider the blocking effect of both water and air fields when study the blocking effect in aviation high-speed towing tanker. At present, there were many researches on the blocking effect of tank test focusing on the medium and low speed ship instead of high-speed vehicle. A modified formula of blocking effect was given in the 13th international towing tank conference (ITTC)[4]. Sheng Zhenbang pointed out that the shallow water effect and wall effect of ship model test in towing tank could be generally ignored while blocking effect was the main part of boundary effect[5]. Comparing the modified formulas of blocking effects in the drag test of ten ship models in three tanks with differente sections, Xie Kezhen pointed out that the influence of the tank width on the drag was greater than that the tank depth and the influence of the width-depth ratio should be considered when selecting the modified formula[6]. Studying the influence of different widths and depths of tank on the attitude of KVLCC2 ship model, Evert Lataire proposed the equivalent width mathematical model for solving the attitude change of ship model[7].

The unpowered model of seaplane was adopted in this paper to study the blocking effect of aviation high-speed towing tank by experimental and numerical methods. The air field and drag test in the bottom and front areas of the trailer were carried out. The numerical simulation was mainly including two parts. On the one hand, the hydrodynamic performance of the seaplane in the finite and infinite domain was computed to study the blocking effect of water and air field. On the other hand, the hydrodynamic performance of the seaplane in the bottom of the trailers with different sizes of chassis to study the influence of trailer chassis geometry parameters on blocking effect.

2. Geometry and Gird

2.1 Geometry

The unpowered model of seaplane was taken as the object of the experiment and numerical simulation. The angles of flaps and elevators in this model was 0 degree, and the buoy was above the water surface during the whole test. The geometrical model of seaplane was shown in figure 1. All experiments were carried on aviation high-speed towing tank with the size of 510m×6.5m×6.8m as figure 2 shown. The size of high-speed trailer is 9.6m×6.5m×2.5m, which is 2.1m above the water surface. The accuracy of speed is between 0.1% and 0.3% in aviation high-speed towing tank.

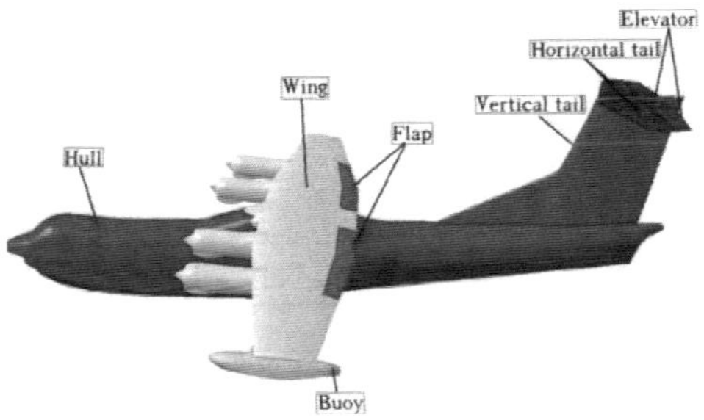

Figure 1. Geometrical model of seaplane.

Figure 2. Aviation high-speed towing tank.

2.2. gird generation and boundary condition

The trimmed mesh adopted in this paper was generated the high quality grid for complex surface of seaplane, and the refined mesh density was achieved by using volumetric controls applied to the area with serious flow separation and near-wall surface. The overset mesh was used to solve the problem of model motion. The grid distributions of finite domain and infinite domain were shown in figure 3 and figure 4. The mesh distributions of the seaplane in the front and bottom of the trailer were same except the difference of the model position. Since the model and trailer were geometrically symmetric, the half model was adopted in all simulation. The boundary condition of model and trailer surface was set as no-slip wall. As shown in figure 5 and 6, velocity-inlet boundary condition was imposed on the inlet, top, side and bottom, and the boundary conditions of outlet was set as pressure-outlet in the infinite domain. While the condition of side and bottom were set as wall in the infinite domain where the influence of side wall and bottom should be considered. The size of finite domain was set according to the actual tank, and the size of infinite domain were set as follow: $-5.0L \leqq x \leqq 1.5L$, $0L \leqq y \leqq 2.5L$, $-2.0L \leqq z \leqq 2.0L$.

Referring to previous work and literature [8], RANS method and SST K-ω turbulence model were adopted for simulation in this study. The governing equation was discretized by the second order upwind finite volume method (FVM). The convection term was discretized by the second order upwind scheme, and the diffusion term was discretized by the second order scheme. Free surface was captured by two-phase fluid volume (VOF) technology.

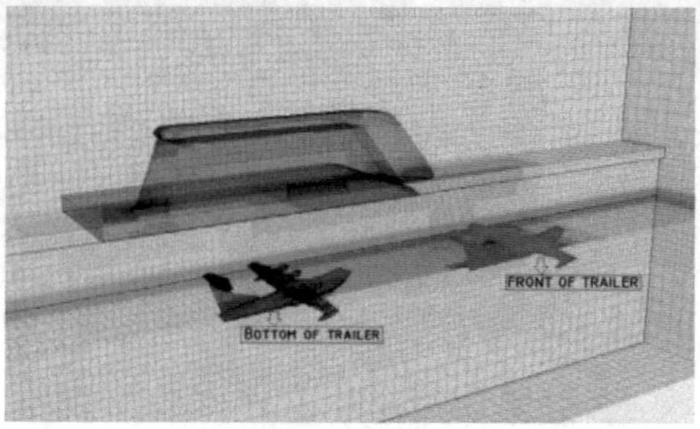

Figure 3. Mesh distributions of seaplane in the finite domain.

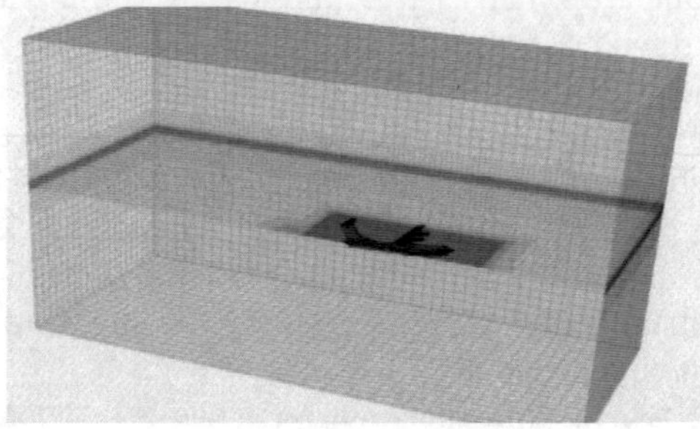

Figure 4. Mesh distributions of seaplane in the infinite domain.

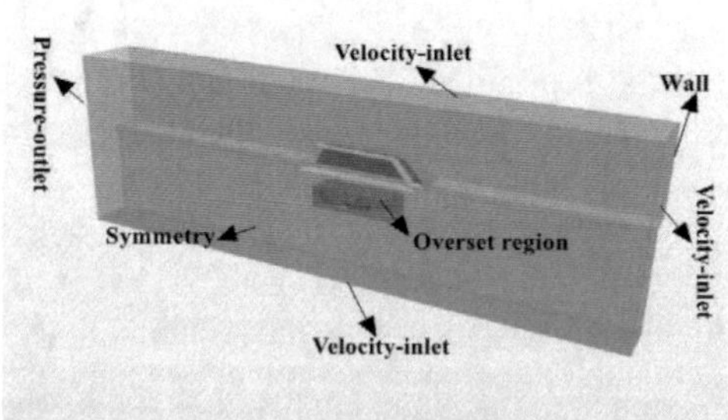

Figure 5. The applied boundary conditions in the finite domain.

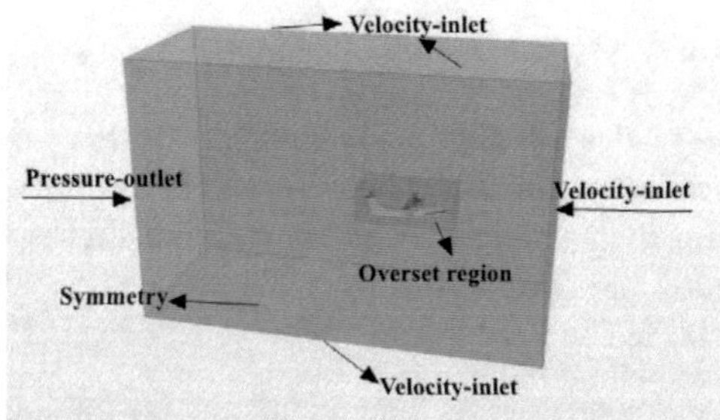

Figure 6. The applied boundary conditions in the infinite domain.

3. Experimental methods

3.1. the design of air field experiment

To research the variation of air field around the trailer, wind speed measurements were installed in the front and bottom of the trailer separately. The five-hole pitot tube, five micro-pressure sensors with amplification and temperature compensation and a data acquisition system were installed in the bottom of trailer. The five-hole point tube had been calibrated in the wind tunnel of the Fluid Mechanics Laboratory, and the computer data acquisition software can display and record the wind speed in real time. Due to the installation problem of the five-hole pitot tube, the wind field of the front of trailer were tested by the ultrasonic anemometer. The measuring points were arranged around the seaplane, and the positions of the test points were as shown in figure 7, and the layout of ultrasonic anemometer was shown in figure 8.

3.2. the design of resistant experiment

The experiment in the bottom of the trailer was conventional, while it was the first time to carry out the model test in the front of the trailer for the hydrodynamic research center, so only the experimental device of the latter was described here. The experiment device mainly included trailer, forward extension device, movement device and restrainer. The forward extension device was consolidated with the trailer for extending the position of the test model to the front of the trailer. The movement device guaranteed the degree of freedom of the model, including a small pulley, a lifting rod and Barycentric connecting rod. During the test, the lifting rod was vertically translated with the model. To ensure that the model can rise and fall freely, the lower end of the Barycentric connecting rod was articulated with the model at the center position. The restrainer mainly limited the yaw motion of the model. The front device of trailer was shown in figure 9.

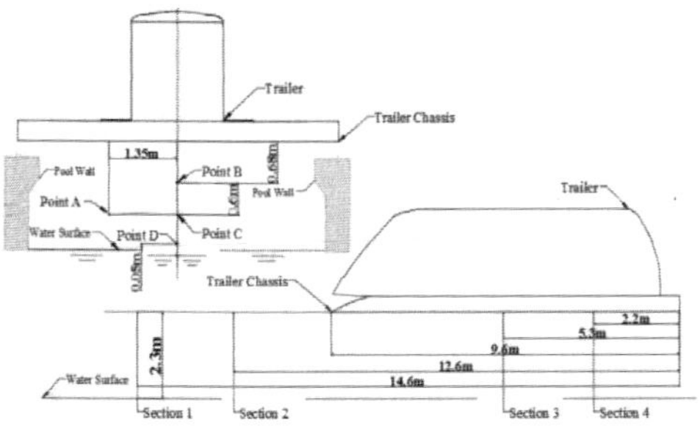

Figure 7. Location of measuring points.

Figure 8. The layout of ultrasonic anemometer.

Figure 9. The typical layout of device in the front of trailer.

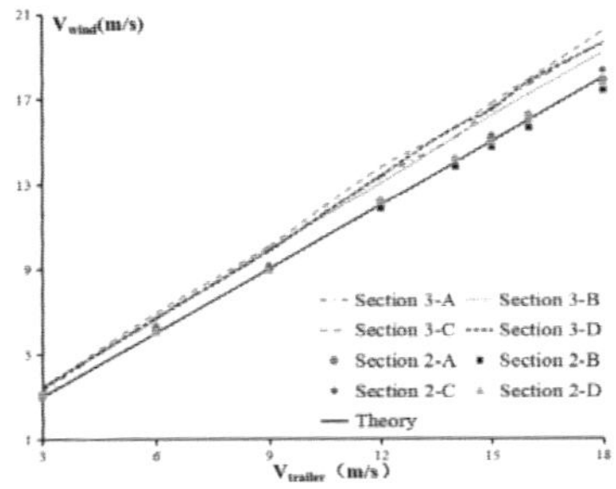

Figure 10. The experimental results of measuring points.

4. Results and Discussion

4.1. wind speed measuring experimental results analysis

Since the experimental results of section 2 and section 1, section 3 and section 4 were consistent respectively, the experimental results of measuring points in the section 2 and section 3 which shown in figure 10 were compared to analyze the wind speed increment problem. The angles of wind on the horizontal and vertical direction of the measuring points varied within ±2 degrees, which was within the errors generated by wind pulsation and installation, so it was considered that the wind was not deflected for all experimental condition. The experimental results of measuring points in section 2 differ less than 2% from the speed of trailer, indicating that the blocking effect of the area in the front of trailer could be ignored. The experimental results of measuring points in the section 3 were 9.5% larger than the speed of trail in average, and the wind speed of all measuring points were close to each other, forming a stable wind speed increment interval, indicating that the wind field in the bottom of the trailer had obvious blocking effect. Therefore, in order to evaluate the influence of blocking effect on the experiment, resistant experiment should be carried out in section 1 and section 2 (front of the trailer), section 3 and section 4 (bottom of the trailer) respectively.

4.2. Research on the experimental results of the forward and trailer bottom

Figures 11-13 showed the experimental results in the front and bottom of the trailer, and figure 14 showed the relative errors of experimental results under two conditions. Error (%) = (forward-bottom)

/ bottom * 100%. It should be noted that Δ(displacement) and *L*(length) adopted in this paper were constant values for the dimensionless treatment of experimental and numerical results. The results of the low speed experiment ($V \leqq 6.5$ m/s) showed that the error of drag, the heave and the trim angle of model in the front and bottom of the trailer were within 2.1%. At mid-high speed ($V \geq 8$ m/s), comparing the results of model in the bottom of trailer, the model in the front of the trailer showed greater drag, bigger trim angle and smaller heave amplitude. In the figure 11, the trim angle of model in the front of trailer was maintained at about 8.5°, while the trim angle obtained from the model in the bottom of trailer decreased with the increase of the speed. The relative error of trim angle under the two conditions increased from 5.1% gradually with the increasing speed. As the speed was greater than 12m / s, the tail of the model in the bottom of the trailer began to leave the water, while the tail of the model in the front of the trailer was still under the water, so the errors of trim angle and drag (shown in figure 13) under the two conditions were significant when the speed was above 12m/s. In the figure 12, the heave amplitude obtained from model in the front of the trailer was 5.5% smaller than the bottom of the trailer in average. In the figure 13, the drag obtained from model in the front of the trailer was7.2% bigger than the bottom of the trailer in average.

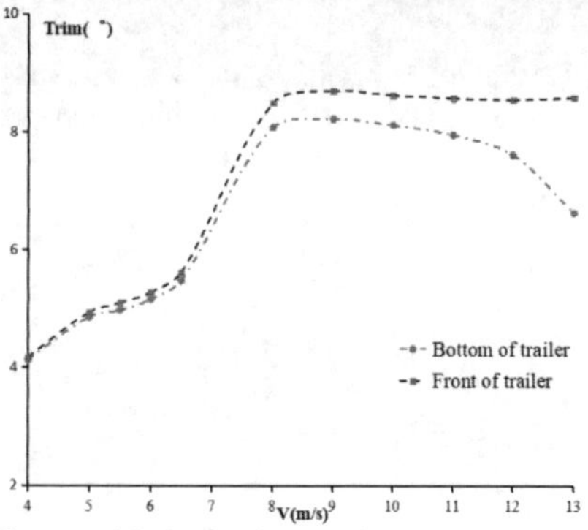

Figure 11. Experimental results of trim angle under the two conditions.

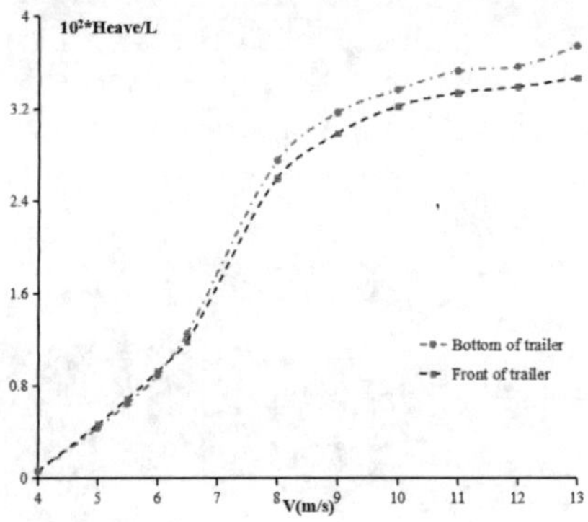

Figure 12. Experimental results of heave amplitude under the two conditions.

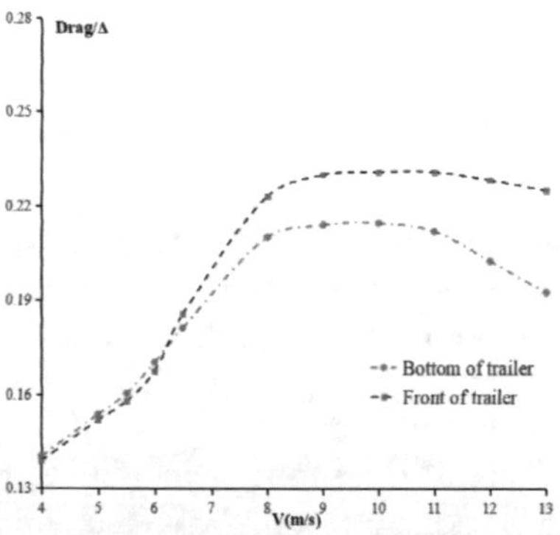

Figure 13. Experimental results of drag under the two conditions.

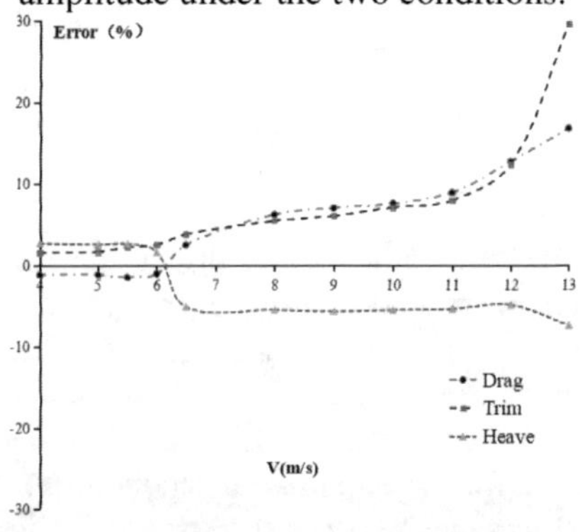

Figure 14. Relative errors of experimental results under the two conditions.

The experimental results under the two conditions were in good agreement at low speed, while the error of the experimental values under the two conditions was more than 5% at mid-high speed. It could be seen that the influence of blocking effect was obvious at the mid-high speed.

4.3. Blocking effect of aviation high-speed towing tank

Unlike the general ship test, the effect of air field on the seaplane couldn't be ignored at the mid-high speed. Therefore, the degree of blocking effect of water and air on the model test during the mid-high speed was necessary to be further studied by CFD method. The experimental results of air field showed that the speed of wind in the front area of the trailer was consistent with the theoretical value. By comparing the computed results of the model in the front of the trailer and infinite domain, the blocking effect of water in the model towing was studied. According to figure 15 (a) and (b), the difference of drag, trim angle and heave amplitude of the model in the front of the trailer and infinite domain was nearly ignored, the average of which was less than 2%. figure 16 showed that the distribution of wave system under the two conditions were basically the same at the speed of 11 m/s, indicating that the hydrodynamic of model weren't affected by the side wall and shallow water effects. Analyzing as the energy transfer, the larger the wave fluctuation of the model, the more energy it needed to consume, then the energy was ultimately reflected as the form of drag. In this respect, the similarity of drag results computed from the two conditions were rationality. Li Guangnian pointed out that when the blocking ratio was less than 0.2%, the blocking effect could be ignored[9]. The blocking ratio between the model and tank adopted in this paper was 0.098%. The computed results of the model in the infinite domain and the front of the trailer were very similar, which was consistent with Li Guangnian's theory, accordingly, the blocking effect of water on the seaplane in aviation high-speed towing tank can be ignored.

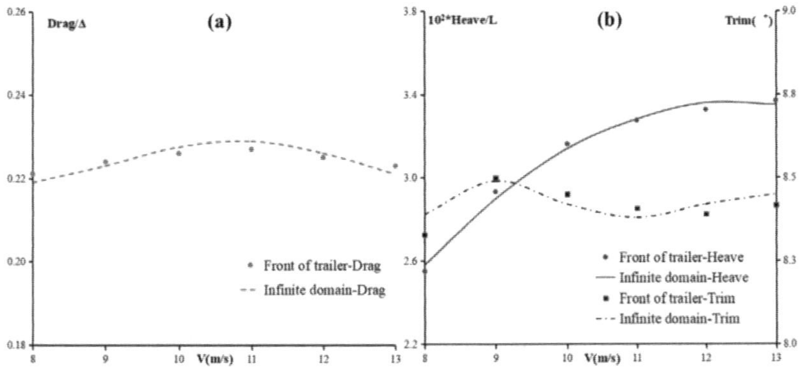

Figure 15. Computed drag (a) and attitude (b) under the different conditions

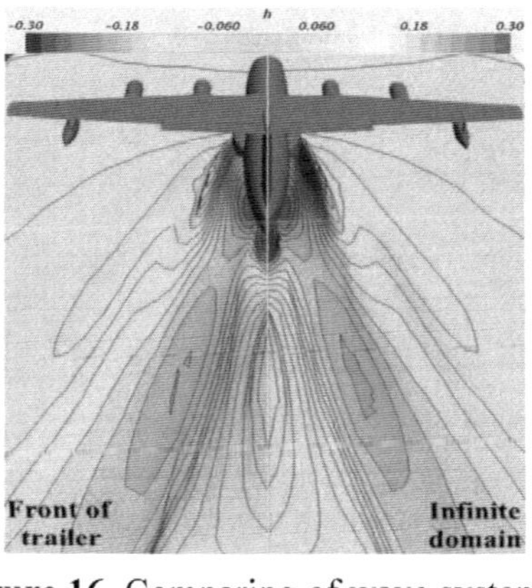

Figure 16. Comparing of wave system at different conditions

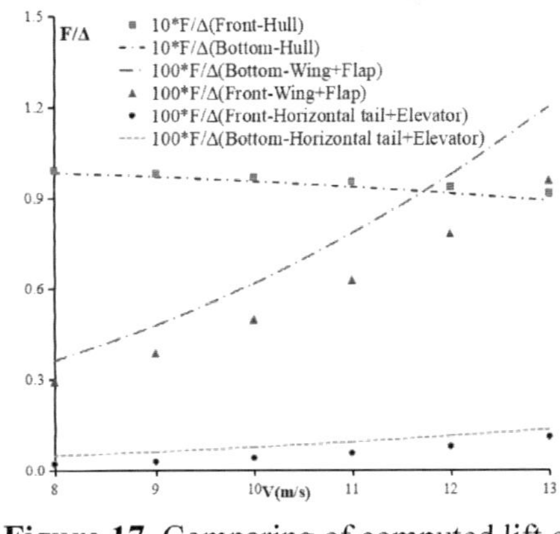

Figure 17. Comparing of computed lift of different parts.

The drag and attitudes were given by the towing test of seaplane in the front and bottom of the trailer, and numerical simulations was conducted under the same conditions. By studying the computed results of each part of the model under the two conditions, the influence of the blocking effect of the air field on the hydrodynamic performance of the model had been further studied. In figure 17, considering that the speed of wind in the bottom of the trailer was 9.5% higher than that in the front of the trailer in average, the aerodynamic lift of the wing+flaps in the bottom of the trailer was obviously greater than that in the front of the trailer. In figure 18, it can be seen that the hydrodynamic moments of the hull were opposite to the aerodynamic moments generated in horizontal tail+elevator and wing+flaps, and the aerodynamic moment was significantly bigger when the model was located at the bottom of the trailer than in the front of the trailer, and the difference increased with the increase of the speed. In summary, due to the blocking effect of the air field, compared with the model in the front of the trailer, the model in the bottom of the trailer suffered more aerodynamic lift and bigger bow moment, resulting in larger heave amplitude, smaller pitch angle, in further smaller wet surface of model and drag.

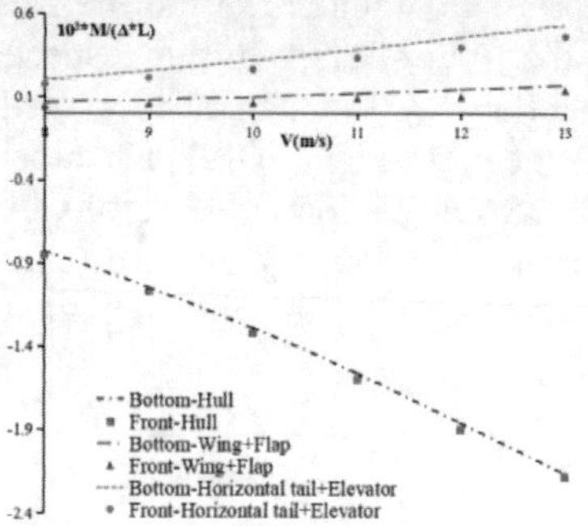

Figure 18. Comparing of computed moment of different parts.

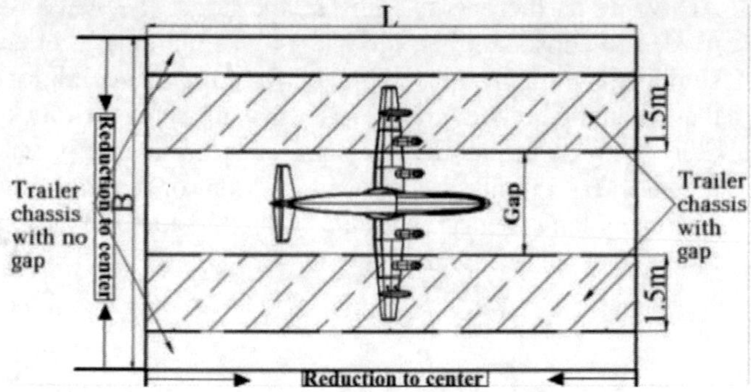

Figure 19. Geometry parameter of trailer chassis.

4.4. Analysis of influence of geometric parameters of trailer chassis on blocking effect

The hydrodynamic performance of seaplane under different shape of trailer chassis were computed at the speed of 11m/s by changing the length, breadth and gap of the bottom of trailer, studying the influence of geometric size of trailer chassis on blocking effect. The sketch of the geometric parameters of the trailer chassis was shown in the figure 19. It should be noted that the geometry of the trailer chassis was symmetrically changed by taking the geometric center of chassis as the origin, and the geometric center of the model was always directly below the geometric center of the trailer chassis.

Figure 20(a) showed the relationship between the hydrodynamic performance of the model and the width of the trailer chassis when $L = 9.6$m and B_{gap}=0m. Figure 20(b) showed the relationship between the hydrodynamic performance of the model and the breadth of the trailer chassis when $B = 6.5$m and B_{gap}=0m. Figure 20(a) showed the relationship between the hydrodynamic performance of the model and the breadth of the gap when $L = 9.6$m and $B = 6.5$m.

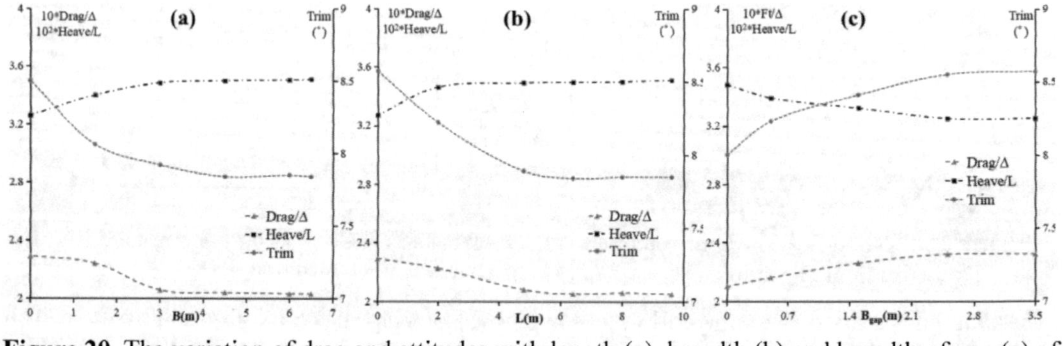

Figure 20. The variation of drag and attitudes with length (a), breadth (b) and breadth of gap (c) of trailer chassis.

We can observed from these figures that when the breadth was larger than 3 m and the length was longer than 4.5m, the computed results of the model remained basically unchanged, which were basically consistent with the computed results obtained from the actual size of the trailer chassis. When the breadth was less than 3m and the length was less than 4.5m, the drag and trim angle decreased obviously with the increase of geometric parameters, but the heave amplitude increases accordingly. With the increase of the breadth of gap, the drag and trim angle increased, and the heave amplitude decreased. When the breadth of gap exceeded 2.5m, the computed results basically remained unchanged.

In summary, when the breadth of trailer chassis was less than 3m, the length of trailer chassis was less than 4.5m or the breadth of gap was less than 2.5m, the degree of blocking effect of air field increased with the increase of geometric variables, but when the size of chassis exceeded these specific values, the degree of blocking effect remained its maximum, and kept basically unchanged.

5. Conclusion

Experiment and CFD simulation were conducted in this paper to study the blocking effect of aviation high-speed towing tank. From the results of these works, the following conclusions could be derived:

1. The experiment of air field showed that the blocking effect of area in the front of the trailer could be ignored, while the blocking effect of area in the bottom of the trailer was obvious.
2. The experiment of the model in the front and bottom of the trailer showed that the experimental results obtained from two conditions were in good agreement at low speed, but the error of the results of model at two conditions obtained at mid-high speed was more than 5%. The blocking effect was ignored at low speed but obvious at mid-high speed.
3. Comparing the computed results of the model in the infinite domain and front of the trailer, it could be seen that the blocking effect of water on the model can be ignored. By analyzing the computed results of each part of the model in the front and bottom of the trailer, the blocking effect of air field had obviously influence on the hydrodynamic performance of the model in the bottom of trailer.
4. The computed results of hydrodynamic performance of seaplane under different trailer chassis shapes showed that when the breadth of trailer chassis was less than 3m, the length of trailer chassis was less than 4.5m or the breadth of gap was less than 2.5m, the degree of blocking effect of air field increased with the increase of geometric variables, but when the size of chassis exceeded these specific values, the degree of blocking effect remained its maximum, and kept basically unchanged.

Reference

[1] Hao Y P 1989 *Expeimental Technique of Ship Performance*. Beijing: National Defense Industry Press.
[2] Yu X S 1991 *Experimental Technique of Ship Performance*. Shanghai: Shanghai Jiao Tong University Press.
[3] Li Y B 2006 *Ship Resistance*. Harbin: Haibin Engineering University Pres.
[4] GROSS A 1972 On blockage correction. *Proc. of the 13th ITTC*. Hamburg&Berlin, Germany, 209-240.
[5] Sheng Z B 2003*Principle of Ship*. Shanghai: Shanghai Jiao Tong University Press.
[6] Xie K Z 1978 The experiment study of blockage effect of the tank. *Journal of Shanghai ship and shipping research institute*, 1 (2),1-27.
[7] Lataire E, Vantorre M & Delefortrie G 2012 A prediction method for squat in restricted and unrestricted rectangular fairways. *Ocean Engineering, 55*, 71-80.
[8] Lang Z, Yong-Sheng C & Fu-Xin W 2018 Coupled hydrodynamic and aerodynamic performance analysis of seaplane take-off process in calm water. *Science Technology and Engineering*.
[9] Guangnian L I, Xie Y & Guo X 2011 Design of towing tank. *Shipbuilding of China*.

IOP Conf. Series: Materials Science and Engineering **692** (2019) 012041 doi:10.1088/1757-899X/692/1/012041

Design of Adaptive Camouflage System Based on SOPC

LI Ce[1], ZHU Yimeng[2] and LI Ran[3]

[1] College of Mechanical Engineering, Donghua University, Shanghai 201620, China
[2] Office of Academic Affairs, Donghua University, Shanghai 201620, China
[3] College of Material Science and Engineering, Donghua University, Shanghai 201620, China

Email:ce_li@dhu.edu

Abstract. With the development of electrochromic devices, the requirements for the accuracy and real-time performance of control devices are constantly increasing. For the reliability problems of single-chip microcomputer control technology, through the SOPC development platform, based on the Cydone IV core chip, the hardware design is completed by itself. The Quartus II integrated development environment is used to tailor the NIOSII real-time operating system, to achieve data collection and control. The signal is filtered using an FIR filter. The designed system has the characteristics of real-time, stability, scalability, low power consumption, etc., and can adaptively control the MOFs-based flexible electrochromic device.

1. Introduction

Materials and devices that can change color according to their environment have broad application prospects in the field of visible stealth and artificial intelligence. However, the current color-changing response devices are usually based on stimuli or control means such as gas, light, heat, etc., the color change effect is not obvious, the coordination is poor, and it is difficult to adapt to the changing environment[1][2].The flexible electrochromic device based on Metal organic Frameworks (MOFs) has the dual function of photoelectric response[3]. By connecting external logic control, the adaptive change of photoelectric characteristics can be realized, so that the optical characteristic parameters of the camouflage body and the environmental optical characteristics can be highly matched and integrated. Commonly used control devices have non-linearity and parameter inconsistency. When replacing the device, the amplifier needs to be re-commissioned due to zero drift of the amplifier. In the signal acquisition and output control technology, the control system is mainly based on CPU or single chip microcomputer. The system speed of controlling operation and operation by software is much lower than that of pure hardware system, and the reliability is not high. System-on-a-Programmable-Chip (SOPC) is a special embedded system: first, it is a system-on-a-chip, which is the main logic function of the entire system by a single chip; second, it is a programmable system with a flexible design that can be reduced, scalable, upgradeable, and has hardware and software programmable functions in the system[4].This paper aims to develop the system hardware design based on Cydone IV E as the core chip SOPC system development platform. On the hardware platform, the data acquisition and control system is realized through the transplantation and tailoring of the Quartus II integrated development environment and the NIOSII real-time operating system, and the MOFs-based flexible electrochromic device is adaptively controlled.

2. System design

The system adopts MOFs electrochromic materials. The excellent electrochemical stability and porous structure of MOFs meet the requirements of electrochromic materials. After solvothermal reaction, a layer of MOFs film is attached to the conductive glass substrate. These films have good electrochemical stability and excellent electrochromic properties. When different colors of light are irradiated on the material, the material will produce corresponding small current signals. The signal is collected, amplified and filtered by the front-end adjustment circuit, and A/D analog-to-digital conversion is performed on it. The system performs real-time processing on the collected data, and completes the output control of the voltage according to the collected data. By controlling data acquisition, processing and feedback output, high quality acquisition, coordination and feedback of MOFs device signal are realized.

The overall structure design of the system is shown in Figure 1. It consists of front-end adjustment circuit, data acquisition module, data processing module, data transmission control module and embedded software system based on SOPC.

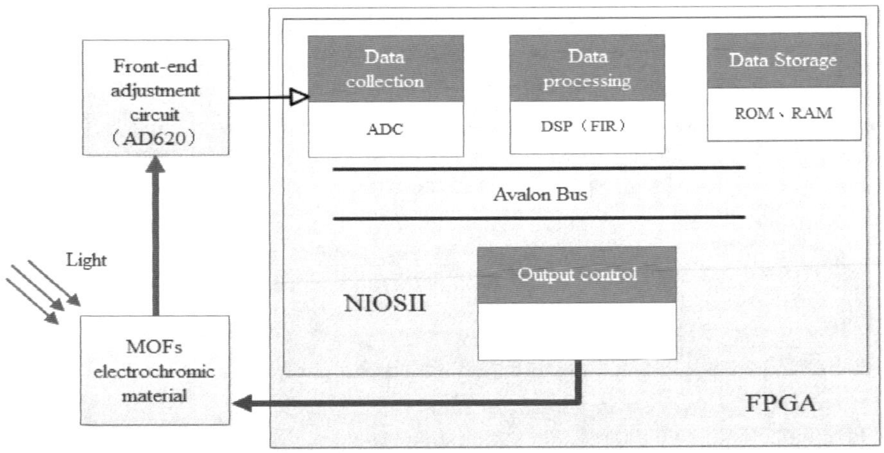

Figure 1. SOPC-based adaptive camouflage system

3. Function Module Design

A complete adaptive camouflage system needs to include sensing, analysis, control and discoloration mechanism. The key technology lies in the performance of sensing and discoloration materials and the adaptability of logical control system.

3.1 Data Acquisition Module Design

Since the current range of the input signal is between 0 and 0.5 μA, the measured signal cannot be directly collected and processed. First, the circuit amplifies the signal by AD620 amplifier. The AD620 features high accuracy, low offset voltage and low offset drift[5].And meets the system's requirements for small size, low voltage, and low power consumption.

The system designs a signal conditioning circuit to amplify and filter the analog signal of the input system. The circuit realizes low-pass filtering by programmable amplifier AD620, and filters the analog current signal input from the front end. The analog current signal is converted into analog voltage signal by using MAX4081 chip.

Then low power ADC128S022 chip is used to realize analog-to-digital conversion. ADC128S022 chip communicates with the controller through SPI interface. The communication sequence diagram is shown in the Figure 2.

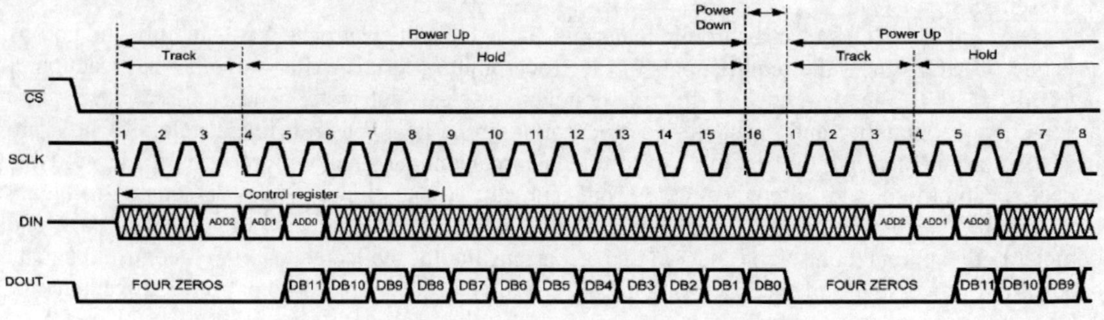

Figure 2. ADC128S022 operational timing diagram

3.2 Data Processing Module Design

The electrical signals generated by MOFs electrochromic materials under illumination have low noise, low drift and low common mode rejection ratio after passing through preamplifier circuit[6][7]. At this time, the electrical signal is mainly disturbed by power frequency, EMG and other signals. The system customizes the convolution operation unit with the DSP (Digital Signal Processor) multiplier module integrated by FGPA to generate FIR filter with pulsating array structure[8]. The filter adopts full pipeline structure, which can realize real-time signal processing at high speed and without delay.

The transition process of FIR filter has a finite interval, and its system function is equation (1):

$$\mathrm{H(Z)} = \sum_{n=0}^{N-1} h(n)z^{-n} \tag{1}$$

N-1 is the delay period of FIR filter, it can also be expressed by equation (2):

$$\mathrm{y(n)} = \sum_{m=0}^{N-1} h(m)x(n-m) \tag{2}$$

x(n) is the input sampling sequence, *h* (m) is the filter coefficient, N-1 is the delay period of FIR filter, and y (n) represents the output sequence of the filter. Direct FIR filtering is used as shown in Figure 3.

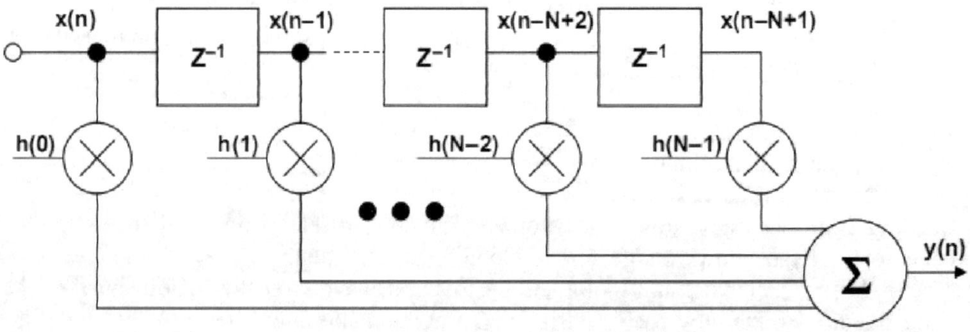

Figure 3. Direct form FIR filter

The FIR filter designed in this system is a low-pass filter, and the difference equation is equation (3):

$$y\ (n)= h(0)x(n) + h(1)x(n-1) + \cdots + h(13)x(n-13) \tag{3}$$

The sampling frequency is Fs, the CUT-OFF frequency of the filter is Fco, and the length of the filter is Nfir, as shown in the following Figure 4:

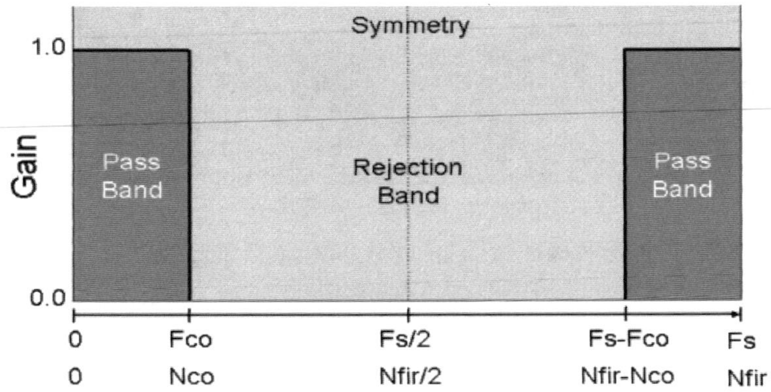

Figure 4. Filter length and frequency relationship

That Nfir = 128, Nco = 13 notes: Fco = Fs* (Nco / Nfir), h(t) is calculated as Figure 5:

```
CSp[0].real()=1.0:
CSp[0].imag()=0.0:
for (i==1;i<==Nco;i++){    //pass-band
      CSp[i].real()=1.0;    //Left page(Gain=1.0)
      CSp[i].imag()==0.0;
      CSp[Nfir-i].real()=1.0;    //Right page(Gain=1.0)
      CSp[Nfir-i].imag()==0.0;
}
for(i=Nco+1;i<==(Nfir-Nco-1);i++) { //Rejection  band
      CSp[i].real()=0.0;    //Gain=0.0
      CSp[i].imag()=0.0;
}
```

Figure 5. Program of h(t) calculation

In this FIR filter, 13 delay units and 14 multiplication units are needed. If this system uses ordinary digital signal processors, delay, multiplication and addition operations can be only performed sequentially in a serial manner. It needs a few instruction cycles to complete this process. By using the parallel structure of FPGA, the output of a FIR filter can be obtained in one clock cycle.

4. System acquisition Test

The system collects and processes the current signals produced by the material irradiated by red and blue light respectively. In this paper, Quartus II and ModelSim software are used to compile and simulate, and the simulation waveform is shown in the Figure 6.The working frequency of Clk is 50MHz, Div_PARM is the clock frequency division setting, SCLK is the clock signal of ADC serial data interface, Channel is the channel selection, Data is the result of ADC. The processed signal meets the design expectation.

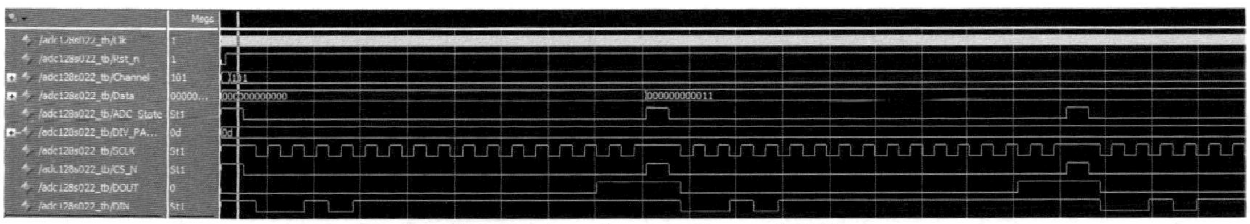

(a)Simulation data under red light

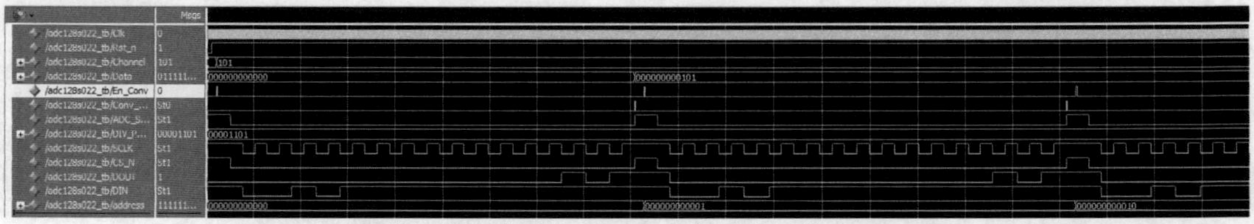

(b) Simulation data under blue light

Figure 6. Data acquisition simulation diagram

As the Figure 7 shows, the controller outputs the corresponding voltage. When the chromotropic device loads 2.5V voltage, the device turns red and the current changes from 2.5mA to 0.1mA. It indicates that the resistance of the device changes from the 1000Ω to 25000Ω. When 3 V voltage is applied, the device becomes blue and the current changes from 8 mA to 0.2 mA, which indicates that the resistance changes from 400Ωto 15000Ω.

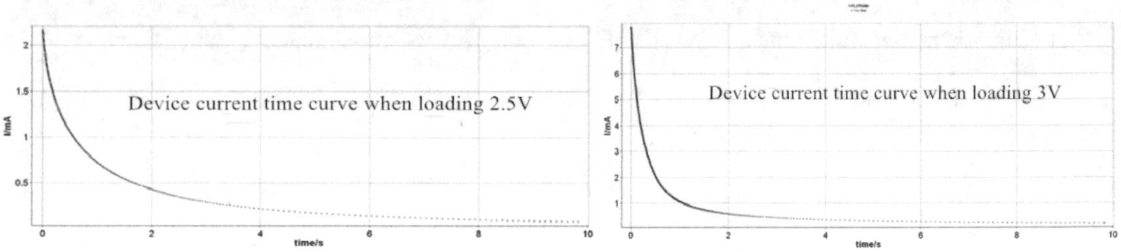

Figure 7. Device current varies with applied voltage

5. Conclusion

According to the application requirement of wearable material control system, FPGA is used as hardware platform and SOPC technology is used to carry NIOS II soft core in the design. The hardware and soft core are flexibly tailored according to the actual control requirements. Materials will produce different current signals under different colors of light. The system collects and analyses them, controls the output of the corresponding rated voltage and feeds back to the material, so that the material can be converted from transparent to red, green and blue. SOPC technology can be used to improve the real-time and accuracy of system data acquisition and control. At the same time, the system has the characteristics of flexible design space, simple debugging and small hardware volume. Reserve more space for subsequent system upgrades.

Acknowledgments

The paper supported by "the Fundamental Research Funds for the Central Universities", thank you very much.

References

[1] C. R. Wade, M. Li, M. Dincă, Facile deposition of multicolored electrochromic metal–organic framework thin films. Angew. Chem. Int. Edit., 2013, 52: 13377-13381.
[2] S. I. Park, S. Kim, J. O. Choi. Low-cost fabrication of WO_3 films using a room temperature and low-vacuum air-spray based deposition system for inorganic electrochromic device applications. Thin Solid Films, 2017, 589: 412-418.
[3] K. R. Li, H. Z. Wang, Q. H. Zhang, Y. G. Li. Lightweight, highly bendable and foldable electrochromic films based on all-solution-processed bilayer nanowire networks. J. Mater. Chem. C, 2017, 4: 5849-5857.
[4] Swarnalatha A, Shanthi A P. Complete hardware evolution based SoPC for evolvable hardware [J]. Applied Soft Computing, 2014, 18:314-322.
[5] Ain K, Soelistiono S, Wibowo R A, et al. Design and Development of Device to measure Body

Fat using Multi-frequency Bio-impedance Method[J]. Journal of Physics Conference Series, 2018, 1120:012043.

[6] Meher P K, Chandrasekaran S, Amira A. FPGA Realization of FIR Filters by Efficient and Flexible Systolization Using Distributed Arithmetic [J]. IEEE Transactions on Signal Processing, 2008, 56(7):3009-3017.

[7] Zhang Shulin. Research on Implementation Technology of FIR Filter Based on FPGA [D]. Harbin Engineering University, 2013.

[8] Liu Yanping. DSP Technology Principles and Application Tutorial [M]. Beijing Aerospace University Press, 2012.

IOP Conf. Series: Materials Science and Engineering **692** (2019) 012042 doi:10.1088/1757-899X/692/1/012042

Sliding Bearing Material Testing Machine under High Temperature Corrosion Conditions

Wenxun Jiang[1] **and Wen Wang**[1, *]

[1] College of Mechanical and Electrical Engineering and Automation, No. 99 Shangda Road, Shanghai University

[*] Correspondence author's E-mail: 770897570@qq.com

Abstract. Sliding bearings are widely used in submerged rolls of strip galvanizing line immersed in zinc bath. Submerged roller bearing pair materials include sleeves and bushings. These two parts are in high temperature, load and corrosive working environment. They are prone to wear and failure. The service life of bearing pair materials determines the operation time of the whole galvanizing line. In order to ensure long-term operation of galvanizing line, the most ideal method at present is to select suitable sleeve and bushing material matching pairs. To screen sleeve and bushing materials with good matching performance, a friction and wear tester simulating sleeve and bushing materials in high temperature molten zinc environment was designed and developed. The test machine is divided into three parts: driving system, loading system and heating system. The driving system drives the sample rotation of the sleeve; the loading system drives the bushing pattern to load the sample of the sleeve; the well heating furnace heats the zinc ingot to 480 °C to simulate the zinc pool environment; the vertical loading of bearing pattern and sleeve pattern, the alignment and insulation of the driving system and the loading system are comprehensively considered and solved. The stability of the test machine under test conditions is analyzed to ensure that the signal collected by the sensor has a high reference value.

1. Introduction

To solve the problem of surface corrosion of steel plate and to protect the surface of steel plate better, hot-dip galvanizing process of steel plate emerged as the times require. It has been more than 80 years since the appearance of hot-dip galvanizing production line in the 1930s. Generally speaking, galvanizing line can be divided into hot-dip pure zinc (GI) and its alloyed products (GA), which are widely used in automobile steel plate production in Japan, Europe and the United States [1-2]. The principle of hot-dip galvanizing anticorrosion relies on a dense layer of zinc liquid plated on the surface of the steel plate, so that the steel plate is completely separated from the external electrolytes and other substances, thus effectively avoiding corrosion of the steel plate. When the surface of the coating is not very dense, the external electrolyte, the zinc plated on the surface of the steel plate and the iron plated on the surface of the steel plate naturally form the primary battery. In this primary cell, zinc is the anode and iron is the cathode, so zinc is gradually dissolved, but iron is protected. Generally, the corrosion life of steel sheet after hot-dip galvanizing will be increased to about 10-25 times.

Figure 1.1 is a schematic diagram of the galvanized sheet production line. The sinking roll is located in the Zinc Bath. Figure 1.2 is a schematic diagram of the sinking roll working in the zinc pool.

The temperature of the zinc pool is 480 degrees. Sleeves and bushings are common consumables in the process of galvanizing strip steel. On the one hand, due to the load received by the sinking roll passing through; the axle sleeve acting on the bushing at both ends causes higher contact stress, and because the actual speed of the sinking roll of galvanizing line is not high, the liquid zinc is difficult to form hydrodynamic lubrication, and the contact fatigue between the bearing and the axle sleeve is very easy to occur; on the other hand, the bushing and the axle sleeve are all positioned. In high temperature molten zinc solution, it receives strong corrosion, which inevitably leads to chemical reactions and accelerates the failure of sleeves and bushings.

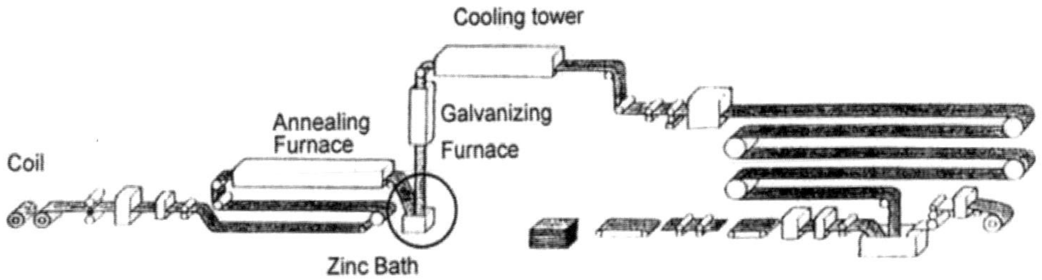

Figure 1.1 Galvanized sheet production line

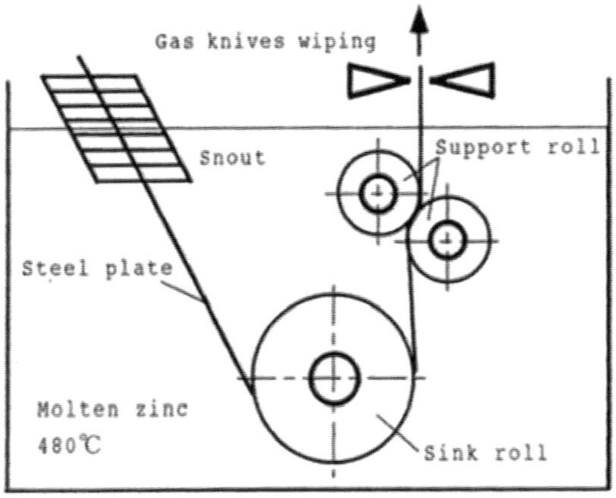

Figure 1.2 Working condition of sink roll

At present, the service life of bearing pair material in industrial galvanized line is about two weeks. Once bearing pair material fails, it is necessary to stop production line immediately and replace spare sleeve or bushing, which will bring huge economic losses. To prolong the service life of bearing pair materials in galvanizing line, the most effective method is to achieve satisfactory service life of bearing by matching contact pair materials and verifying and screening through simulation experiments. For this reason, a test machine was developed to simulate the friction and wear of bearing pairs of sink rolls in high temperature molten zinc solution.

2. Design of testing machine

Considering the actual working conditions of galvanizing line, the test machine is divided into driving part, loading part and heating part [3]. The driving part uses the frequency conversion motor to drive the sample rotation of the sleeve to simulate the rotation of the sunk roll sleeve. The loading part uses the hydraulic cylinder to simulate the bearing pressure caused by the tension of the strip, the buoyancy of zinc liquid and the gravity of the sunk roll itself in the actual working condition. In the heating part, a well-type heating crucible furnace is used to heat the zinc ingot in a graphite crucible to 480 °C, which simulates the high temperature molten zinc liquid in the galvanizing pool of the actual galvanizing line. The structure of the testing machine is shown in Figure 2.1.

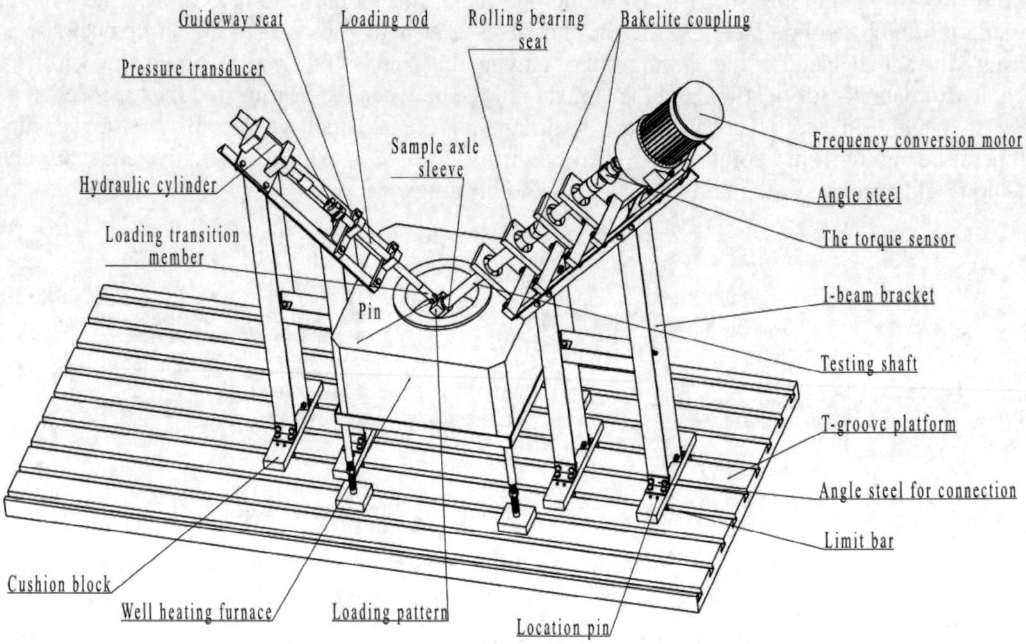

Figure 2.1 Integral structure of testing machine

2.1. Driving system

The driving system is powered by an integrated motor with frequency conversion and speed regulation. The motor shaft is connected with the torque sensor, the intermediate shaft and the testing shaft in turn through the coupling. The sample sleeve is fixed with the test shaft through bolts and rotates with the test shaft. Two copper cage deep groove ball bearings are assembled on the intermediate axle and the testing axle respectively (copper cage is adopted considering the requirement of heat insulation) and fixed in the bearing seat. All couplings use phenolic resin coupling, as shown in Figure 2.2. Thermal insulation is also considered to prevent the signal of the torque sensor from being disturbed by high temperature and the motor from burning out.

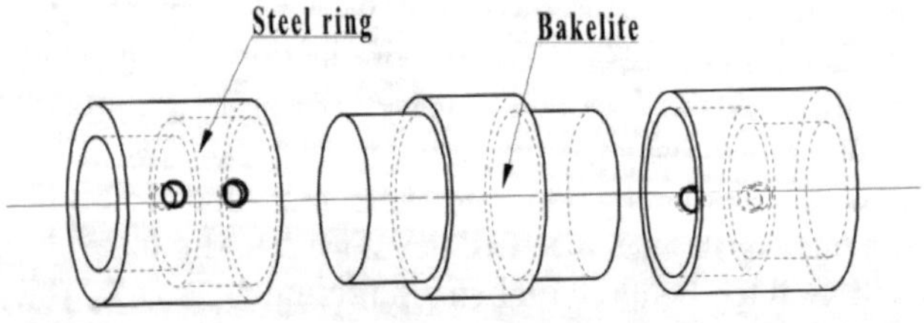

Figure 2.2 phenolic resin coupling

2.2. Loading system

The loading part of the loading system is a hydraulic cylinder. The outer thread of the piston rod end of the hydraulic cylinder is connected with the inner thread of one end of the pressure sensor through a connecting rod, and the other end of the pressure sensor is connected with the loading rod. In order to prevent the inclination of the loading rod in the process of loading, two guideway pedestals are designed to cooperate with the loading rod. During the loading process of the hydraulic cylinder, the test shaft of the driving system will be slightly deformed, leading to the impossibility of vertical loading. The loading structure shown in Figure 2.3 is designed. A loading transfer part is added between the loading rod and the bushing pattern. One end of the loading transfer member is a cylinder, which can be directly inserted into the loading rod and the cylindrical end face of the loading transfer part does not contact the end face of the loading rod hole; the other end of the loading transfer part is provided with a rectangular groove. The bushing pattern can be embedded therein, and the load transfer member can be connected with the bushing pattern through the pin. The rectangular end surface of bushing pattern and the end surface of loading transfer parts with rectangular grooves leave gaps. This is due to the deformation of the testing shaft section during actual loading. The loading rod can rotate through the pin on the loading transfer member leaving room for slight rotation, ensuring vertical loading and making the test results more convincing.

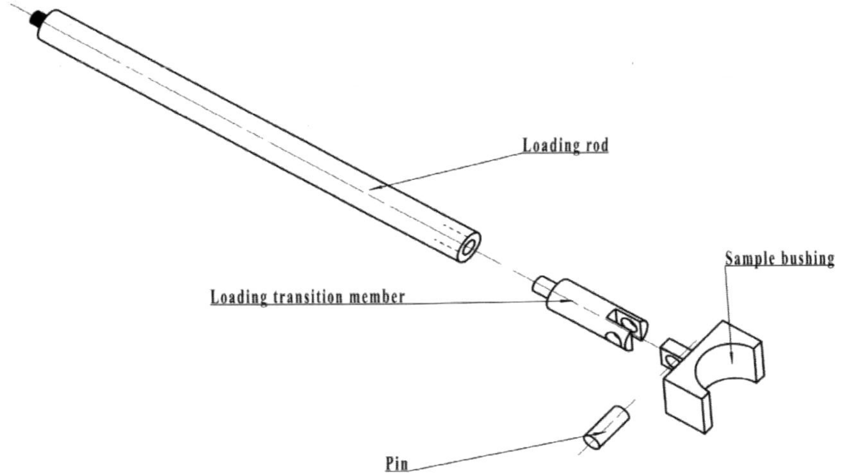

Figure 2.3 (a), Loading rod components

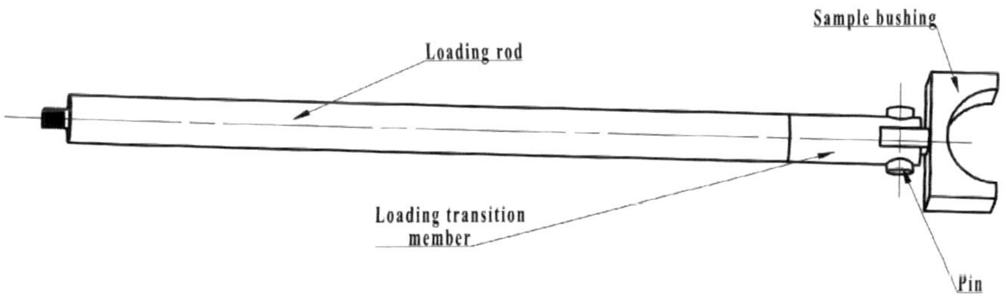

Figure 2.3. (b) Loading rod assembly structure

2.3. Heating system

The heating system is mainly composed of well crucible heating furnace and lifting platform. During the test, zinc ingot will be put into isostatic graphite crucible, graphite crucible will be put into stainless steel inner liner, together with stainless steel inner liner and graphite crucible, into well-type high temperature heating furnace. Finally, the heating curve and holding time can be set.

After the experiment, it is necessary to separate the pattern from the zinc liquid immediately, take off the pattern, and design the lifting platform as shown in Figure 2.4. The upper end of the bench is a supporting heating furnace bench welded by angle steel. Four corners of the bench are welded with four hollow steel tubes respectively. The screw rod is penetrated into the hollow steel pipe, the furnace is lowered to the appropriate position by the hydraulic jack at the bottom of the heating furnace, and then the screw rod and the hollow steel pipe are locked and fixed by the nut, so is the operation of the rise.

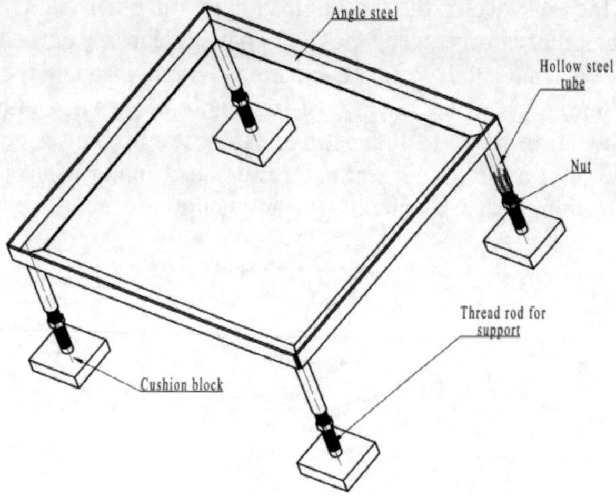

Figure 2.4. Heating furnace lifting platform

2.4. Assembly alignment of testing machine

As shown in Figure 2.1, the bottom of the test machine is a T-groove platform, and the T-groove has a high straightness. The brackets used in the test machine are all I-beams of the same size, and the size and processing accuracy of the bottom pads are exactly the same. Therefore, the better straightness of T-groove is used to locate the bracket of driving system and loading system, so that the bracket of the two parts is completely aligned. The following steps can be taken:

- Firstly, the limit bar is threaded with the cushion block, and the limit bar is pressed into the T-groove. The matching size and tolerance are shown in Figure 2.5.
- The positioning pin is used to limit the limit bar and cushion block to ensure that the bottom of the whole testing machine is fixed.
- Taking the same side of the four pads with the same size and processing precision as the datum level, the positioning grooves of I-beam steel are set up on the four pads to ensure that the distance between the I-beam support and the datum level is the same.
- Assemble the supporting angle steel and I-beam steel, take one side of the angle steel as the datum plane, and process symmetrical positioning pin holes on the large-scale machining center to ensure that the drive system and the shafting parts of the loading system are concentric.
- Insert the I-beam bracket into the positioning groove of the cushion block, and fix it.

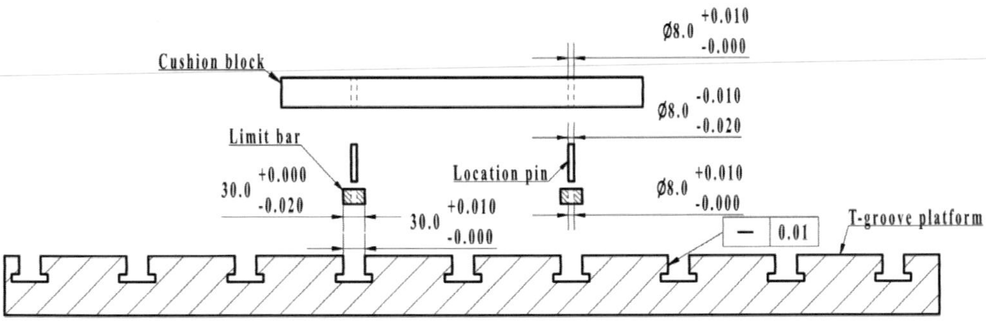

Figure 2.5 Bottom positioning assembly drawing

Through the above steps, the whole testing machine ensures higher requirements for alignment and concentricity, and the test results are more accurate. The overall assembly of the test machine is shown in Figure 2.6. The testeing machine can adjust the speed, load, temperature and change the sample material. Sleeves and bushings may be made of ceramic [2-3], cobalt-based [4] and iron-based alloys [5] or coated materials [6-8], detailed material appendix is shown in Table 4.1.

Figure 2.6 Physical object of testing machine

3. Reliability analysis of testing machine

3.1. Heat transfer analysis

To prevent heat transfer from burning out sensors and other components, it is necessary to simulate and analyze the heat transfer in the front part of the sample test shaft, the sleeve and the bushing immersed in molten zinc at 500°C. The driving system is powered by an integrated motor with frequency conversion and speed regulation. The motor shaft is connected with the torque sensor, the intermediate shaft and the testing shaft in turn through the coupling. The sample sleeve is fixed with the test shaft through bolts and rotates with the test shaft. Two copper cage deep groove ball bearings are assembled on the intermediate axle and the testing axle respectively (copper cage is adopted considering the requirement of heat insulation) and fixed in the bearing seat. All couplings use phenolic resin coupling, as shown in Figure 2.2. Thermal insulation is also considered to prevent the signal of the torque sensor from being disturbed by high temperature and the motor from burning out.
The heat transfer analysis of the drive system and loading system is carried out, and the results are shown in Fig. 3.1. The bearing temperature of the torque sensor is 80. The simulation results show that the temperature of the position of the torque sensor is less than 35.6°C, which shows that the torque sensor can work properly. The actual air-fired heating furnace uses infrared thermometer to measure the position and display the temperature value of 35°C (room temperature), this temperature is very close to the simulation results. Similarly, the temperature of the pressure sensor and the hydraulic cylinder in the loading system is close to the actual room temperature and can work normally.

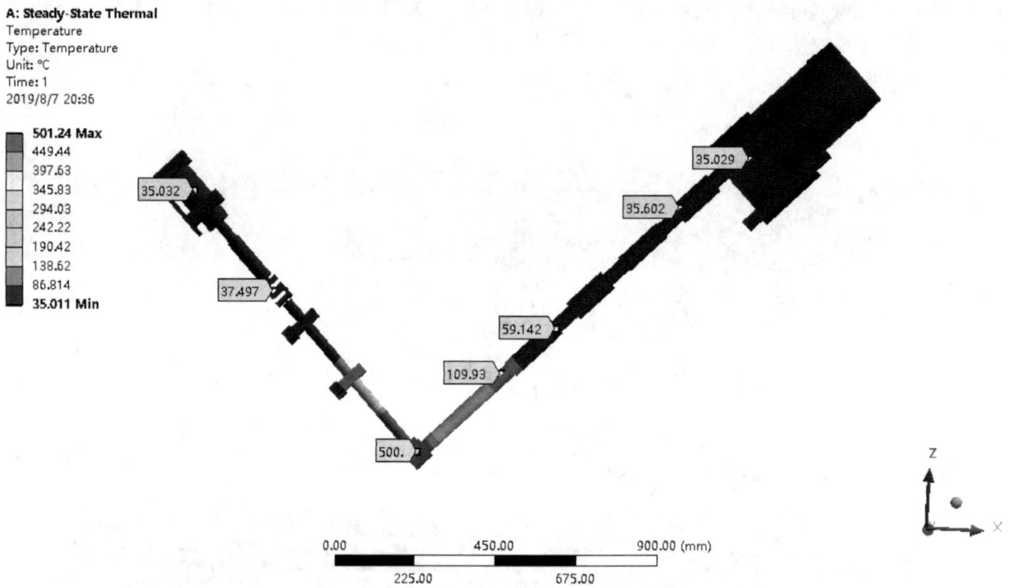

Figure 3.1. Heat transfer analysis of driving and loading system

3.2. Coupling Analysis of Thermodynamic Field

Under the dual action of high temperature and load, whether the strength of parts of loading system and driving system meets the need of test is an urgent problem to be solved. It is necessary to make corresponding analysis before experiment. All shafting components are steel by default. The results of deformation and stress distribution in the thermal field can provide reference for practical test. Firstly, load, speed and other boundary conditions are added. The overall boundary conditions are shown in Figure 3.2, in which ball bearings are replaced by equivalent constraints [9-10].

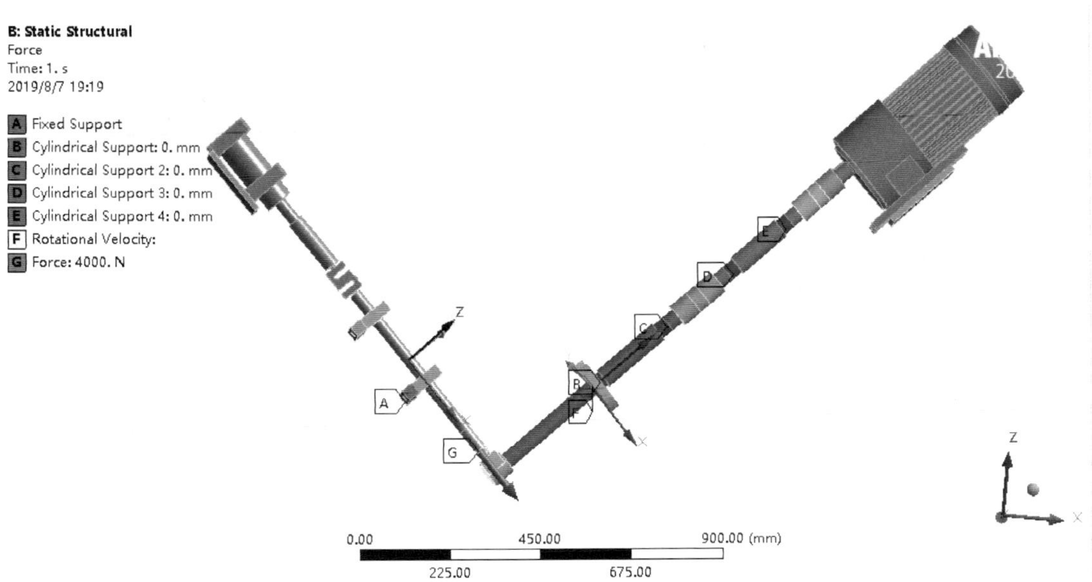

Figure 3.2. Boundary conditions of thermal stress field

After the corresponding parameters are set, the distribution of deformation and thermal stress is solved, as shown in Fig. 3.3 (a) and (b), respectively. It can be seen that in the thermal field, the maximum deformation of the shaft reaches 2 mm. In the shoulder position of the shaft near the rotating sleeve, there is a greater stress concentration, which is close to the yield limit of steel. In addition, the molten zinc at high temperature is corrosive to some extent. Considering comprehensively, the material of shaft can be replaced by Hot-Strength steel in practical test.

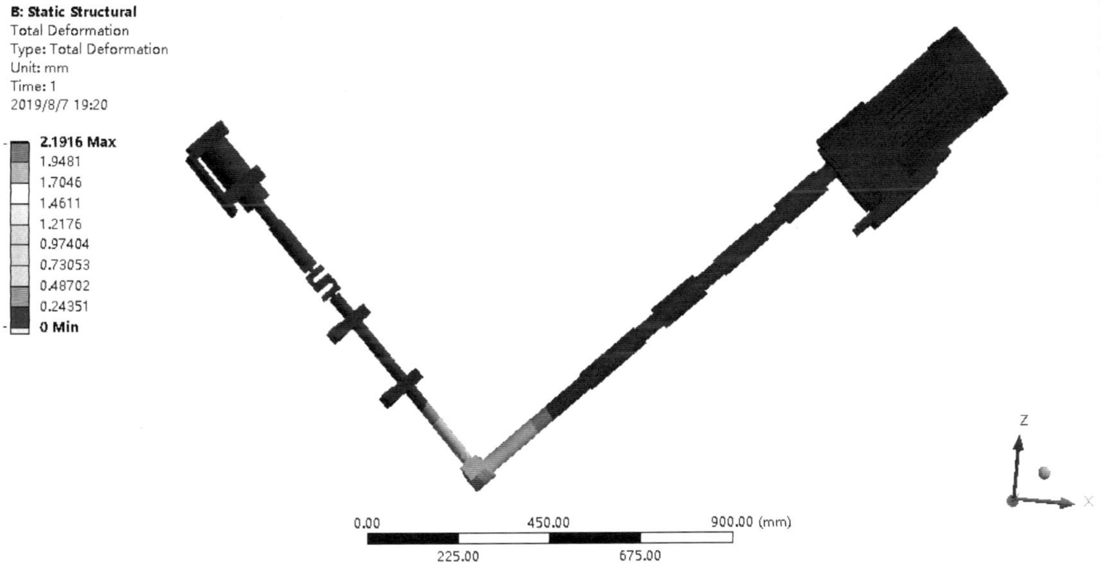

Figure 3.3. (a) Deformation under thermal stress field

IOP Conf. Series: Materials Science and Engineering **692** (2019) 012042 doi:10.1088/1757-899X/692/1/012042

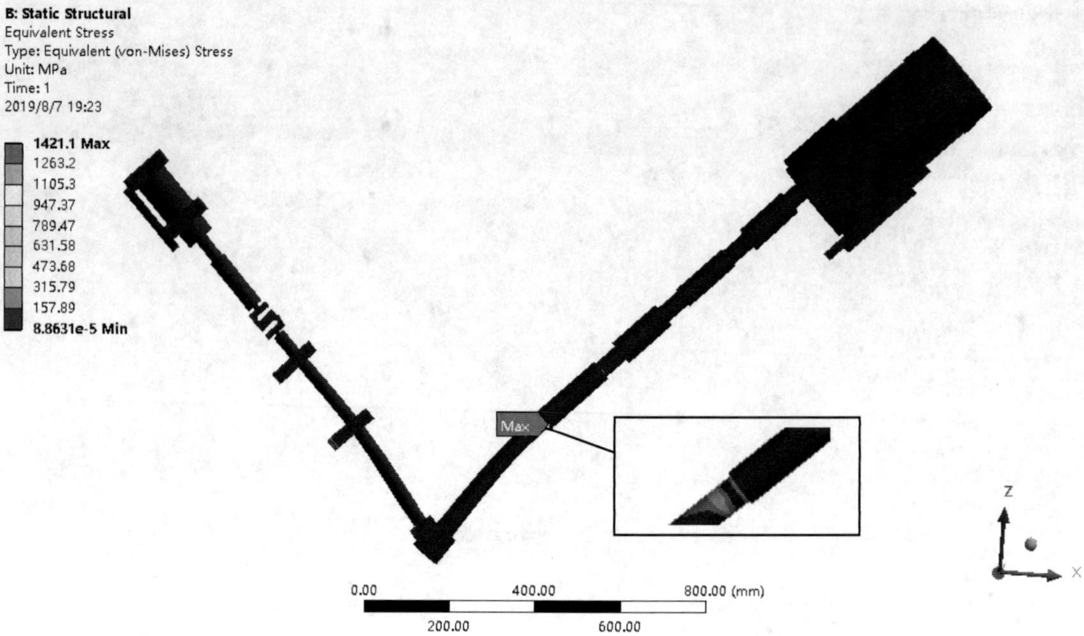

Figure 3.3. (b) Stress distribution under thermal stress field

4. Conclusion

The sliding bearing material testing machine can effectively evaluate the sleeves and bushings material under high temperature, corrosion and loading conditions. The maximum temperature can be adjusted to 1200 °C, the maximum loading force can reach 1 ton, the maximum speed of the pattern can reach 600 RPM, and it is easy to replace test materials, various material combinations are shown in Table 4.1. The matching materials with superior performance can be screened out through a large number of tests.

Table 4.1. Material combinations tested in molten zinc alloy bath.

Bushing	Sleeve
Stellite #6	WC-Co coating on 316L stainless steel
SiAlON	Laser clad WC coating on 316L stainless steel
SiC/graphite	Stellite #4
Stellite #6	Stellite #6

References

[1] Somjai Janudom,Vishnu Rachpech, Narisara Mahathaninwonga and Somkid Thiwong, Corrosion Behavior of Galvanized Steel for Porcelain Insulator's Pin in HVAC Transmission Line, *Key Engineering Materials* vol.803 pp 45-49

[2] K. Zhang and L. Battiston, Sliding wear of various materials in molten zinc, 2002 *J. Materials Science and Technology*. 18:12, 1551 1560

[3] K. Zhang and L. Battiston, Friction and wear characterization of some cobalt- and iron-based superalloys in zinc alloy baths, 2002 *J. Wear*. 252 (2002) 332–344

IOP Conf. Series: Materials Science and Engineering **692** (2019) 012042 doi:10.1088/1757-899X/692/1/012042

[4] K. Zhang, Wear of cobalt-based alloys sliding in molten zinc, 2003 *J. Wear.* 255 (2003) 545–555

[5] N.Y. Tang, Y.H. Liu, and K. Zhang: Proc. Int. Symp. on Zinc-CoatedSteel, Iron & Steel Society 44th Mechanical Working and Steel Pro-cessing, and 8th International Steel Rolling Conf., Orlando, FL, Sept.8–11, 2002, *The Iron and Steel Society*, Warrendale, PA, vol. XL, pp. 1285-97.

[6] K. Zhang, Effects of test conditions on the tribological behaviour of a journal bearing in molten zinc, 2005 *J. Wear.* 259 (2005) 1248–1253

[7] F.E. Goodwin, E.A. Silva, Zinc-based coating production capabilities for steel sheet: current technical topic and scientific needs, in: 44th MWSP Conference Proceedings, vol. XL, Orlando, FL, September, 2002, pp. 365–379.

[8] Florian Summer,Florian Grün, Martin Offenbecher and Stuart Taylor,Challenges of friction reduction of engine plain bearings-Tackling the problem with novel bearing materials, 2019 *J. Tribology International.* 131 (2019) 238-250

[9] Wei Kou, Li Yang and Yixun Hong, Dynamic simulation of infrared signature of deep groove ball bearing based on ANSYS/LS-DYNA, 2015 *J. Applied Optics and Photonics China.*

[10] K.R. Kadam, S.S. Banwait. Performance characteristics of two-axial groove journal bearing for different groove angles,2018 J. *Industrial Lubrication and Tribology*, 70(2).

IOP Conf. Series: Materials Science and Engineering **692** (2019) 012043 doi:10.1088/1757-899X/692/1/012043

Design and Evaluation of Scene-Based Product Device

Yanhua Wang* **and Dong Yan**

School of Construction Machinery, Chang'an University, Xi'an City, Shaanxi Province

*1303844110@qq.com

Abstract. The production of persimmon has a long history, however, the traditional production of persimmon depends mainly on manpower and is greatly affected by the natural weather. The yield and quality cannot be guaranteed which affects the taste, sales and word of mouth. According to the scenario design program, four kinds of user requirement models are constructed. After systematic analysis, four functional principle schemes are obtained, and five evaluation indexes are summarized. Finally, scheme 4 is selected as the most ideal scheme. Through the research and application of scenario design method in the process of product design, this paper provides an evaluation method for the evaluation of the product design, which helps designers to determine the direction of product development effectively.

1. Introduction

1.1. For Scenario design

In the product design, the user's experience is difficult to quantify by the statistical method, therefore, it is usually done by building a user scenario. Scenario construction can help the designer to take into account the different state of the product in different scenarios, and understand the way of life and behaviour of the user deeply [1]. In this paper, Scenario story method is used to analyze the story board and the user demand model. Through the combination of analytic hierarchy process and triangular fuzzy number method, optimum scheme can be selected, meanwhile, the innovative design method of the user's personalized needs is explored.

A scenario is a description of the actual situation of the past or the future. It is a collection of expected scenarios [2]. The paper uses the first three steps of raising questions, scenario segmentation, and scenario construction in scenario design program. Because of the research of persimmon suspension drying device, so "putting forward questions" is to solve the related problems of persimmon hanging drying. The factors of the scene model include the interaction among the environment, the user, the product and the three.

According to the design procedure of scenario story and the results of market investigation, the production of persimmon can be divided into four first-class scenarios, namely, persimmon drying is shown in figure1, persimmon hanging is shown in figure 2, persimmon suspension is shown in figure 3, manual drying is shown in figure 4. The four first-class scenarios are divided into a number of secondary scenarios such as cleaning, staging, placing, and so on. Through on-site investigation, analysis, data inspection and so on, scenario of persimmon production have been constructed. namely, persimmon drying, persimmon hanging, persimmon suspension, manual drying. The traditional production of the artificial persimmon depends on the non-stable nature. The quality of the high-quality products is low, however, the artificial drying method can overcome the drawbacks of the natural drying system. After analysis, the present situation of persimmon cake can be obtained as follows: (1) environmental factors: drying and hanging the natural environment, weather and air cannot be effectively controlled, however ,the drying environment is opposite, but the cost is higher; (2) user factors: consuming a lot of manpower, repetitive work, high physical requirements, etc, but less drying workload; (3) Product factors: drying only the baking yield and quality are guaranteed; the cycle is short; the equipment covers a small area; and the degree of mechanization is high.

1.2. User requirement model extraction

There are many decisive factors in product design, such as product function, style and so on. The user's demand for the product usually includes material, psychological and social needs. According to the story board analysis, we can get the needs of agricultural users, detailed user requirements as shown in Table 1.

Figure 1. Persimmon drying story board

Figure 2. Persimmon hanging story board.

Figure 3. Persimmon hanging story board

Figure 4. Manual drying story board

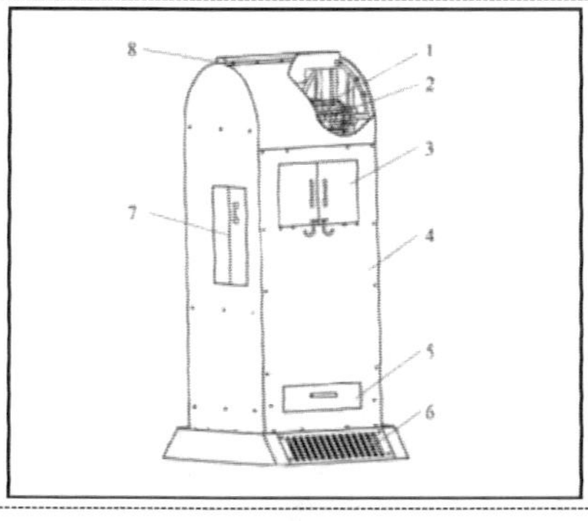

Figure 5. The whole structure of persimmon baking

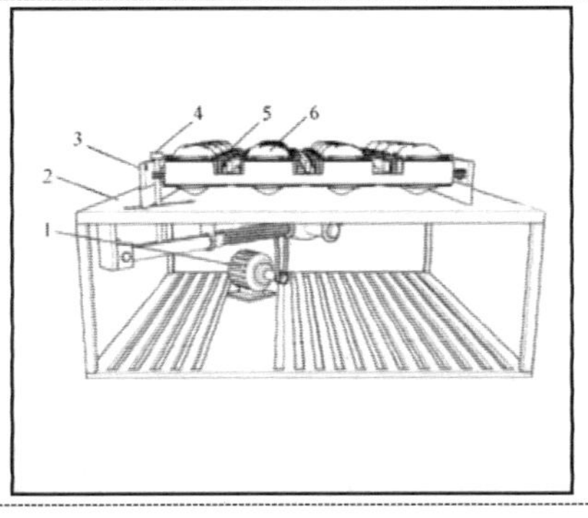

Figure 6. Automatic kneading machine for persimmon

IOP Conf. Series: Materials Science and Engineering **692** (2019) 012043 doi:10.1088/1757-899X/692/1/012043

2. Systematic Analysis of existing products

2.1. Total function analysis

Functional Analysis is a commonly used method in Product systematic Analysis when we adopt the function analysis method to carry on the scheme design, the step is to first determine the total function, abstract its design problem. Secondly, the total function of the system was decomposed into Sub-function. Finally, the combination of principle solution of general function was form a variety of principle design scheme.

The main structure of the connecting rod-conveying chain combined persimmon cake baking device [3] comprises a shell, a rotating device, a persimmon rack, a hot air device and an air guide device. As shown in figure 5, the front door 3 and the side door 7 is arranged on the shell 4, the rotation device 1 and the hanging frame 2 is arranged in the shell 4, the air outlet 8 is arranged at the top of the shell 4, the heating device 6 is arranged at the bottom of the shell 4, and the receiving box 5 is connected with the shell 4, A number of hangers 2 are connected to the rotation device 1. The function and layout of this device provide an effective reference value for the device designed in this paper.

The persimmon cake automatic kneading machine[4] is composed of a power device 1, a bracket 2, a grid device 3, a driving rod 4 and a rolling wheel device 5, the support 2 is used for supporting the rest of the structure and the grid device 3 is used for containing the persimmon, the driving rod is used for driving the kneading wheel device 5 to knead the persimmon 6. The device integrates the principle of shape change in the process of transformation between quadrilateral and rhomboid into the kneading design of persimmon cake so that the kneading device is combined with the oven tray, and the whole structure of the persimmon cake kneading device is shown in figure 6.

Table 1. User demand.

Demand factor categoy	Demand points	Demand factor categoy	Demand points
Social Demand	Solve the demanding conditions of weather to avoid fly and fly sand, etc.	Physiological Demand	Reduce labor costs Safe and reliable Unconstrained posture No high security risks
Material demand	Stable production and high quality Easy to recycle . . Batch, mechanized	Psychological demand	Reduce fatigue Self-realization

2.2. Morphological matrix calculation

The morphological synthesis method is based on the morphological matrix and finds out all kinds of answers through the decomposition and combination of the system [5]. According to the functional decomposition diagram and scenario map of persimmon making device, the functional elements of persimmon baking device are solved by the method of design catalogue. The morphological matrix of functional elements of persimmon baking device is constructed by means of morphological analysis[6]. The morphological matrix of the main functional elements is shown in figure 7.

According to the above principles and the user's demand model, the following several suitable schemes can be obtained. Scheme 1: semi-automatic persimmon hanging drying device (A1+B2+C2+D2+E2+F2+G1).Scheme 2: self-rotating kneading and kneading persimmon baking device (A1+B4+C4+D1+E2+G1) Scheme 3: drive type persimmon cake baking device (A2+B3+C4+D1+E1+G2) Scheme.4 (A2+B1+C3+D2+E2+F3+G1).

Table 2. Evaluation Index of persimmon hanging drying device

Classification	Primary indicator	Secondary indicator
Technical index of persimmon suspension device	Technical index A1	Temperature rise rate B1 Floor area B2 Persimmon space area B3 Observable area B4
	Human-machine relationship A2	Color B5 Modelling B6 Operation panel B7
	Practicability index A3	Security B8 Operational ease B9 Degree of automation B10 Maintainability B11
Economic index of persimmon suspension device	Economic index A4	Device cost B12 Work cost B13 Work period B14 Product period B15
Society Index of persimmon suspension device	Green index A5	Pollution level B16 Fuel B17 Security B18 Retrievability B19

Table 3. The first level index matrix and weight of the device.

A	A1	*A2*	A3	*A4*	A5	*W*
A1	1	5	2	1/7	1/2	0.151
A2	1/5	1	1/5	1/6	1/6	0.037
A3	1/2	5	1	1/2	1	0.157
A4	7	6	2	1	1	0.409
A5	2	6	1	1	1	0.243

Table 4. The assembly matrix of the evaluator's opinions.

	U1	*U2*	U3	U4	U5
X1	(0.21,0.27,0.32)	(0.16,0.20,0.28)	(0.20,0.22,0.25)	(0.25,0.29,0.33)	(0.32,0.37,0.42)
X2	(0.23,0.26,0.31)	(0.08,0.14,0.22)	(0.26,0.29,0.33)	(0.22,0.25,0.29)	(0.08,0.11,0.14)
X3	(0.16,0.20,0.25)	(0.14,0.18,0.24)	(0.19,0.22,0.26)	(0.18,0.21,0.24)	(0.22,0.24,0.27)
X4	(0.22,0.25,0.30)	(0.37,0.46,0.55)	(0.19,0.23,0.26)	(0.21,0.23,0.26)	(0.24,0.25,0.29)

3. Scheme Optimization based on the functional principle of user satisfaction

3.1. Determination of evaluation index system of product function principle scheme

The evaluation index system refers to the framework and system for the evaluation of industrial products. Because of its wide range and complexity, we should first establish an evaluation system to measure the comprehensive performance of products [6]. So we can get the evaluation index of persimmon hanging drying device. Which is shown in Table 2.

The analytic hierarchy process (AHP) is mainly to compare the elements in pairs, and then to sort and judge the overall weights of these elements, and finally to establish the weights of each element [7]. Through the expert consultation method, we compare the importance of the first-level index of the 1-9 scale method, and calculate the evaluation index of the judgment matrix. In order to achieve satisfactory consistency [8], we get the weight of the influencing factors, and the random consistency of the hierarchy total order order ($CR=0 < 0.1, P < 0.01$). Therefore, the matrix is effective, and the weights of the first-order indexes and the synthetic evaluation indexes are obtained, as shown in Table 3.

According to the final comprehensive evaluation of weight, five evaluation indexes of persimmon hanging drying device can be selected, including the work cycle, safety, device cost, persimmon space area and automation degree.

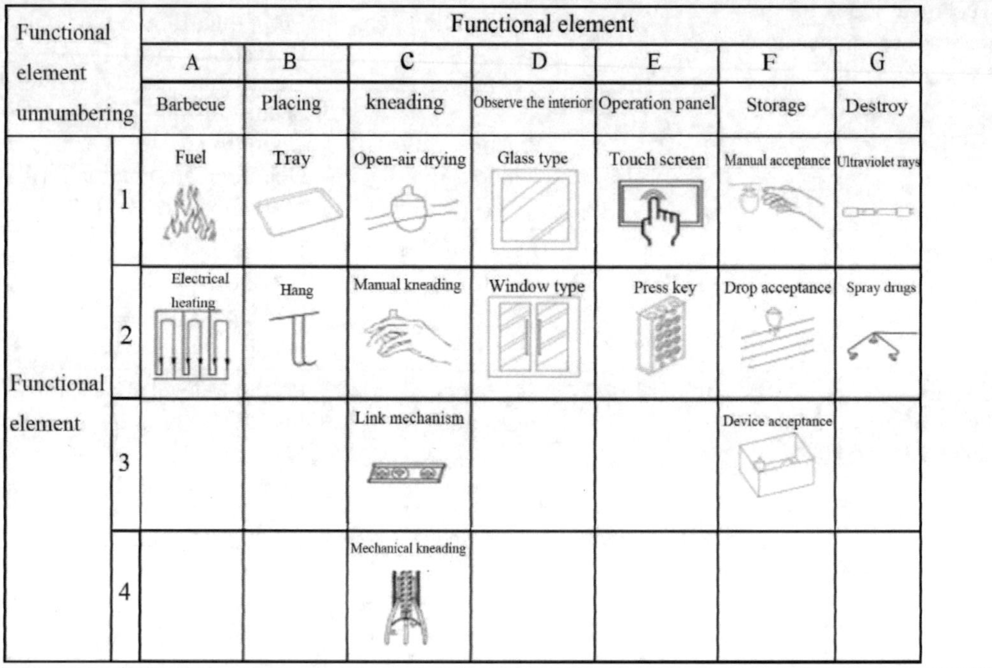

Figure 7. Morphology matrix of persimmon baking device

3.2. Determined by the group multi-evaluation index decision-making method of the triangular fuzzy number

Firstly, the fuzzy decision matrix of the Kth expert is constructed $\tilde{A}=[\tilde{a}_{kj}^{k}]_{m*n}$, $\tilde{a}_{ij}^{k}=\left(a_{ij}^{ku},a_{ij}^{kl},a_{ij}^{km}\right)$ Represents the value of the K evaluator's evaluation of the scheme X_i under the target U_j, which In order to eliminate the influence of different physical dimensions on decision-making results. The normalized formula is as follows[8].

$$b_{ij}^{kL}=\frac{a_{ij}^{kL}}{\sum_{i=1}^{m}a_{ij}^{kL}},b_{ij}^{kM}=\frac{a_{ij}^{kM}}{\sum_{i=1}^{m}a_{ij}^{kM}},b_{ij}^{kU}=\frac{a_{ij}^{kU}}{\sum_{i=1}^{m}a_{ij}^{kL}} \quad (1)$$

The aggregation of the evaluator's opinions is to aggregate the individual evaluation value of each evaluator under the subjective evaluation index. The evaluation value of the evaluator e^k under the evaluation index u_j is set as the evaluation value of the evaluation index uj $\tilde{b}_{ij}^{k}=\left(b_{ij}^{kL},b_{ij}^{kM},b_{ij}^{kU}\right)$.

This paper holds that the weights of the three experts are the same, that is, $\omega_k=(1/3,1/3,1/3)$. Therefore, the assembly matrix of evaluators' opinions can be obtained, as shown in Table 4.

The $U^* = (\tilde{u}_1^+ \tilde{u}_2^+ ... \tilde{u}_n^+)$ is ideal Schema. We are based on the distance between the two triangular fuzzy numbers, that is, the average value of the alternative scheme evaluation to the evaluation value of the positive ideal scheme and the negative. The European distance of the evaluation value of ideal scheme and the European distance are shown in the formulas.

$$d_i^+ = \sqrt{(d_{i1}^+)^2 + (d_{i2}^+)^2 + ... + (d_{in}^+)^2} \tag{2}$$

$$d_i^- = \sqrt{(d_{i1}^-)^2 + (d_{i2}^-)^2 + ... + (d_{in}^-)^2} \tag{3}$$

$$L(X_i) = \frac{d_i^-}{d_i^- + d_i^+} \tag{4}$$

$$L(X_4) = 0.781 \tag{5}$$

4. Conclusion

In this paper, he first three steps of scenario design program can be used to construct the feature model of product user group effectively. The scheme of constructing innovative product can be determined by the method of systematic product analysis. Comprehensive analytic hierarchy process (AHP) and trigonometric fuzzy number (TFA) re used to evaluate the product scheme the innovative design of persimmon hanging drying device, it shows that the innovative design of mechanical and electrical products based on theory and technology is effective, which provides an innovative approach.

References

[1] Liu Lanlan, "Application of situational storytelling in Product Design and Development." Journal of Packaging Engineering, 2007(9).

[2] Hu Kang, "Design of Fire helmet based on situational Construction." Journal of Packaging Engineering, 2018(6).

[3] Liu Jianhua, "Connecting rod conveyor chain combined persimmon baking device." Chinese patent: 201612173U, 2017(11).

[4] Li Ke, "Automatic kneading and kneading machine of cam-connecting rod persimmon." Chinese patent: 201612173U, 2017(11).

[5] Li Yao, "Research on customer-oriented Collaborative Evaluation system for Industrial Design products." Degree of—Shaanxi University of Technologym, 2007(6)

[6] Liu Yuan, "Research on Advertising originality based on morphological Matrix decomposition elements." Degree of Hubei University Of Technology, 2009(6).

[7] Hong Zhiguo "Calculation of higher-order average Stochastic consistency Index (ri) in Analytic hierarchy process." Journal of Computer engineering and application, 2002(6)

[8] Zhang Zhina, "Research on driving School selection based on Fuzzy Analytic hierarchy process." Journal of Shangqiu Vocational and Technical College2015(4).

Study on Axial Parameters of Stepped Ferrofluid Seals

Guohong Wang[1], Xiaolong Yang[1*] and Ruibo Zhang[1]

[1]School of Mechanical Engineering, Guangxi University of Science and Technology, Liuzhou 545006, China

[*]Corresponding author e-mail: 09116324@bjtu.edu.cn

Abstract. As a new type of intelligent material seal technology, ferrofluid seal has been widely used in high speed, heavy load and large eccentricity occasions in recent years because of its long life, low leakage and high pressure resistance. Because of the limitation of the axial height of the axis, most of the existing studies are focused on the radial parameters, but the specific research on the axial parameters has not yet been carried out. In this study, the theoretical pressure capabilities of height of the distribution of axial teeth and axial slot width was calculated according to the theory of stepped ferrofluid seals by finite element method. And by changing the radial gap, the stepped ferrofluid seal is compared with the ordinary seal, and the calculation results are analysed and discussed. The results show that the distribution mode of the axial teeth has a great influence on the theoretical sealing pressure resistance. And the theoretical sealing pressure resistance reaches the maximum when the distribution mode is evenly distributed. The influence of the width of the axial slot on the theoretical sealing pressure resistance can be neglected, so the technological property and the design should be considered as far as possible. By comparing the stepped seal with the ordinary seal, the practicability of the stepped seal is further verified.

1. Introduction

As a new type of intelligent material, ferrofluid seal has been widely used in modern industry [1]-[4]. At present much more studies of ferrofluid seals with parameter of radial. Fan C et al. proposed a new type of structure for large gap ferrofluid seal, and analysed the influence of radial parameters on the sealing pressure resistance [5]. Yanjuan Z et al. studied the influence of radial teeth number and sealing gap on the pressure resistance of small gap ordinary seal, and obtained the functional relationship between clearance and pressure resistance [6]. Du H et al. theoretically analysed and experimentally verified the relationship between starting moment, viscous resistance moment and pressure resistance of MHD seals at high and low temperatures [7]. However there are few literatures about the research of parameter of radical. In this paper, a converging stepped ferrofluid seal structure was used to study the stepped axial parameters.

2. Ferrofluid seal theory

Generally speaking, the method for computing the pressure resistance of ferrofluid seals is as follows [8]:

$$\Delta P = \mu_0 M_s \sum_{i=1}^{N} (H_{\max}^i - H_{\min}^i) = M_s \sum_{i=1}^{N} (B_{\max}^i - B_{\min}^i) \quad (1)$$

where H_{max}^i are the maximum magnetic field strengths, H_{min}^i are the minimum magnetic field strengths, B_{max}^{i} are the maximum magnetic flux densities and B_{min}^{i} are minimum magnetic flux densities.

The total sealing capability as [8]:

$$\Delta P = \sum_{i}^{N} (P_{ia} + \lambda P_{ir}) \tag{2}$$

In (2), P_{ia} are the pressure capabilities of the ferrofluid seal in the axial gaps formed by the ith pole piece and the stepped shaft. P_{ir} are the pressure resistance of the ferrofluid seal in the radial gaps formed by the ith pole piece and the stepped shaft. If $P_{ia} < P_{ir}$, then $\lambda = 1$, otherwise $\lambda = 0$, P_{ia} and P_{ir} can be calculated by theory (1).

3. Structure design and finite element analysis

The design is illustrated in Figure. 1. Under the working environment of small gap, this design scheme changes the leakage direction of sealed medium on the one hand, and on the other hand, a number of axial and radial gap are formed between the stepped axis and the pole piece, thus improving the overall pressure capability [8]. The dimensions of each part of the sealing structure are shown in Table 1.

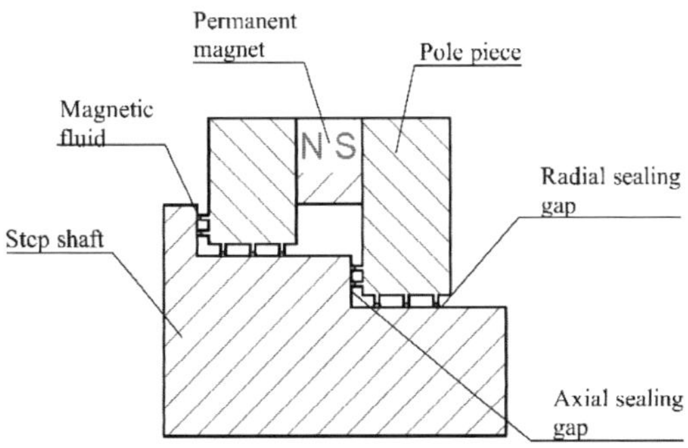

Figure 1. Two-dimensional physical model.

Table 1. Parameter of converging stepped ferrofluid seals.

Item	Value	Item	Value
Inner radius of the 1/2 pole piece (mm)	21/17	Number of radial teeth under the 1/2 pole piece	2
Outer radius of the 1/2 pole piece (mm)	30		
Length of the 1/2 pole piece (mm)	5.3	Width of pole teeth (mm)	0.3
Inner radius of the permanent magnets (mm)	26	Axial gap width (mm)	0.1
Permanent magnets length (mm)	5	Radial gap height (mm)	0.1
Outer radius of the permanent magnets (mm)	30	Slot depth (mm)	0.8
Number of axial teeth under the 1/2 pole piece	3	Slot width (mm)	0.7

The physical model of the ferrofluid seal is created in the pre-processor of ANSYS finite element analysis software by using the ferrofluid seal structure of small gap polymerized stepped ferrofluid seal as shown in Figure. 1. The pole pieces and shaft are made from 2Cr13. The permanent magnetic coercive force and permeability are Hc =1.356 × 10^6A/m and 1.05. And the saturation magnetization of ferrofluid is 30.7 kA/m.

4. Results and discussion

In the post-processor of ANSYS software, a path is defined along the leak path of sealed medium, and the calculated flux density is mapped to the path. In order to verify the influence various factors on the pressure capability of the converging stepped ferrofluid seals, various factors were discussed.

4.1. Effect of the distribution of axial teeth

When the distribution of axial teeth different, the magnetic flux density shown in Figure 2 can be obtained by finite element analysis.

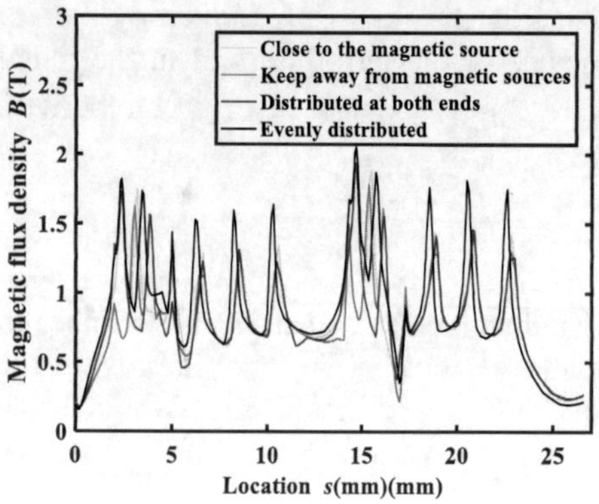

Figure 2. Magnetic field distribution when distribution of axial teeth changes

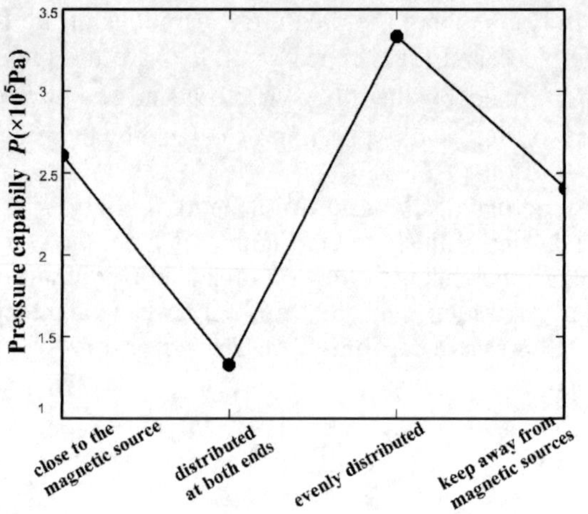

Figure 3. Effect of distribution of axial teeth capacity

As can be seen from Figure 2, the magnetic flux density of the radial teeth varies greatly with the change of the distribution of the axial teeth. When the distribution of the axial teeth is evenly distribute, the radial magnetic flux density reaches the maximum, and the difference of magnetic flux density also reaches the maximum. According to the distribution map of the magnetic force line, it can be found that when the distribution of the axial teeth is different, the leakage phenomenon of the axial teeth changes obviously, and when the distribution of the axial teeth is evenly distributed, the leakage phenomenon of the axial teeth is the lowest, so that the difference of magnetic flux density of the radial teeth increases.

When the distribution of axial teeth are close to the magnetic source, distributed at both ends, evenly distribution and keep away from magnetic source distribution, Figure 3 shows the theoretical comparison of the pressure capability of the stepped ferrofluid seals. It can be seen that when the axial tooth distribution is close to the magnetic source distribution and keep away from magnetic source distribution, the theoretical pressure capacity value of the stepped ferrofluid seal is alike. But greater than the distributed at both ends. When the distribution of the axial teeth is evenly, the theoretical pressure capability of the ferrofluid seal reaches the maximum. This is due to the difference of magnetic leakage phenomenon and magnetic gathering effect. When the distribution of the axial teeth is evenly, the magnetic leakage phenomenon of the axial teeth is the lowest, and the magnetic gathering effect of the radial teeth is the greatest. According to this result, when designing the axial polar teeth, the distribution of the axial polar teeth should be designed to be evenly due to the limitation of the axial height.

4.2. Effect of the axial slot width

When the axial slot width different, the magnetic flux density shown in Figure 4 can be obtained by finite element analysis.

From Figure 4, it can be concluded that the axial flux density increases with the increase of the axial slot width, while the radial flux density remains unchanged. This is due to the effect of magnetic flux leakage phenomenon, the magnetic flux leakage phenomenon in the axial direction reduce when the axial slot width enlarge, and the magnetic flux density distributed on the axial polar teeth enlarge.

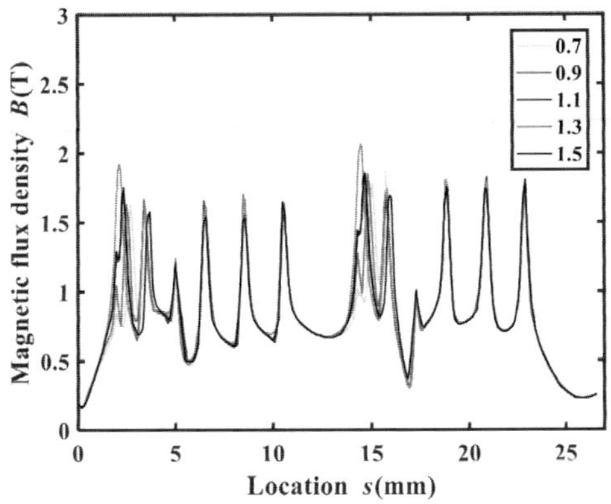

Figure 4. Magnetic field distribution when axial slot width changes

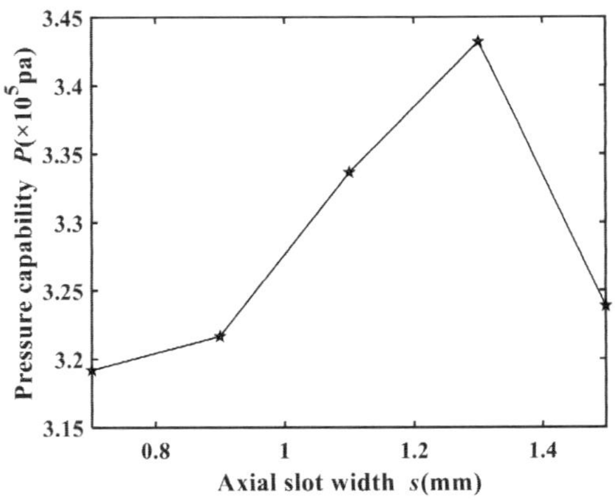

Figure 5. Effect of axial slot width capacity

Figure 5 shows a comparison of theoretical pressure capacity of stepped ferrofluid seals with axial slot width ranging from 0.7 mm to 1.5 mm. It can be seen that the theoretical pressure capability of the stepped ferrofluid seals increases first and then reduces with the axial slot width enlarges, and the fluctuation range of the theoretical pressure capacity value is smaller. This is because when the axial slot width enlarges, the leakage phenomenon reduces with the increase of the width of the axial slot due to the influence of the leakage phenomenon. However, when the width of the slot continues to increase, according to the theory of stepped ferrofluid seal, when the sum of the pressure capacity of the axial teeth is greater than that of the radial teeth, the radial teeth have no effect on the pressure capacity and the theoretical pressure capacity reduces.

4.3. Effect of the axial height of radial gap

When the radial gap different, the magnetic flux density shown in Figure 6 can be obtained by finite element analysis. .

From Figure 6, It can be seen that the flux density in the radial gap reduces with the enlarge of the radial gap height, while the flux density in the axial gap enlarges slowly with the increase of the radial gap height. This is because when the axial gap is unchanged, the increase of the radial gap height will enlarge the reluctance in the radial gap, which will enlarge the reluctance of the whole magnetic circuit, so the total magnetic flux in the magnetic circuit will also reduce, and the magnetic flux in the radial gap will decrease. At the same time, as the axial height remains unchanged and the radial gap height enlarges, the distance between the shaft and the magnetic source enlarges, the magnetic flux leakage in the axial gap reduces, and the magnetic flux in the axial seal gap enlarges slowly.

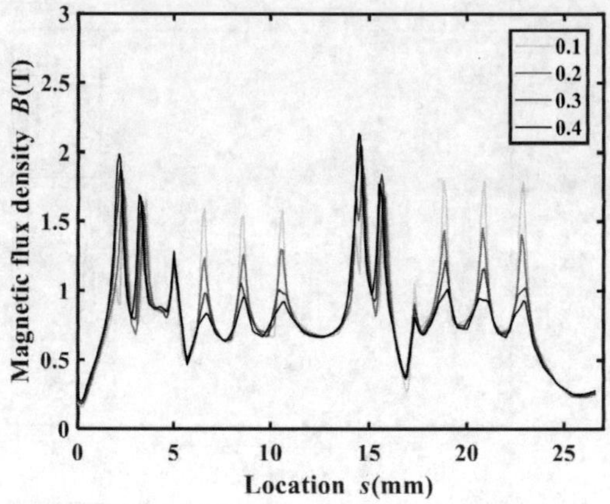

Figure 6. Magnetic field distribution when radial gap changes

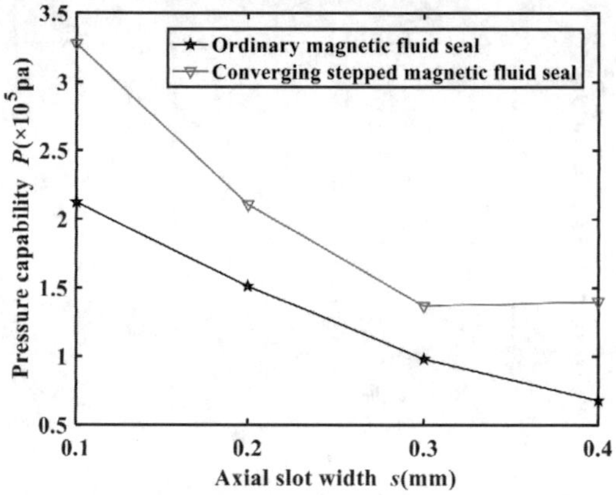

Figure 7. Comparison of theoretical pressure capacity

Figure 7 shows a comparison of theoretical pressure capacity between stepped ferrofluid seals and ordinary ferrofluid seals with radial gap of 0.1-0.4 mm. It can be seen that with the increase of radial gap, the theoretical pressure capacity of stepped ferrofluid seals and ordinary ferrofluid seals reduces, and the trend of decreasing pressure resistance of stepped ferrofluid seals is less than that of ordinary ferrofluid seals. This is because the stepped ferrofluid sealing capacity is related to both its axial and radial gap. And when the radial gap increases, it will only affect the sealing ability in the radial gap, but there is no axial gap in the ordinary ferrofluid seal, so the pressure resistance of the ordinary ferrofluid seal reduces more.

5. Conclusion

In this paper, Converging stepped ferrofluid seals are a new type of ferrofluid seals, the finite element method and numerical study are used to discuss the effect of the axial parameters of the converging stepped ferrofluid seals on the theoretical pressure capacity value. Some conclusions could be briefly listed as follows.

- The distribution mode of the axial teeth has a great influence on the theoretical sealing pressure resistance. And the theoretical sealing pressure resistance reaches the maximum when the distribution mode is evenly distributed.
- The influence of the width of the axial slot on the theoretical sealing pressure resistance can be neglected, so the technological property and the design should be considered as far as possible.
- By comparing the converging stepped seal with the ordinary seal, the practicability of the stepped seal is further verified.

Acknowledgements
The authors gratefully acknowledge the support of the National Nature Science Foundation of Chin (Grant No.51905114), the support of the Science and Technology Project of Guangxi Province (Grant No.2016GXNSFBA380213), the support of the Innovation Project of GuangXi University of Science and Technology Graduate Education (Grant No.GKYC201901) and the support of the Science and Technology Project of Liuzhou (Grant No.2017BC20204).

Reference

[1] Rosensweig R E 1997 Ferrohydrodynamics (New York: Dover Publications)
[2] Sihai Z, Qiang S, Sen L and Lingyu J 2016 The model of seal mechanism for magnetic fluid and related experimental study *Mechanika* (Volume **22**(4)) pp 260–264
[3] Noriyuki H, Hayato Y and Hidenori S 2016 Noncontact gravity compensator with magnetic fluid seals *Journal of Advanced Mechanical Design, Systems, and Manufacturing* (Volume **10**(5))
[4] Yoshinori M and Christopher A D 2017 Miniature magnetic fluid seal working in liquid environments *Journal of Magnetism and Magnetic Materials* pp 285-288
[5] Fan C, Xiaolong Y, Shanghan G and Yanhui C 2017 Numerical study on key parameters of ferrofluid seal with large gap and two magnetic sources *International Journal of Applied Electromagnetics and Mechanics* (Volume **55**) pp 535–543
[6] Yanjuan Z, Yibiao C, Decai L, Zhengmao Y and Yilong Y 2019 Experimental Validation and Numerical Simulation of Static Pressure in Multi-Stage Ferrofluid Seals *IEEE Transactions On Magnetics* (Volume **55**(3))
[7] Du H, Decai L, Jiawei C and Jun Y 2018 Theoretical analysis and experimental study of the characteristics of magnetic fluid seal with a large diameter at high/low temperatures *International Journal of Applied Electromagnetics and Mechanics* pp 531–550
[8] Xiaolong Y, Peng S, Fan C, Fuxiang H, Decai L and Peter J T 2019 Numerical and Experimental Studies of a Novel Converging Stepped Ferrofluid Seal *IEEE Transactions On Magnetics* (Volume **55**(3))

Modeling and Simulation Study of Intake System Characteristics of Wankel Engine for UAV

Xiaqing LIU [1], Minxiang WEI [1] and Rui LIU [1,2,*]

[1]. College of Energy and Power Engineering, Nanjing University of Aeronautics and Astronautics, Nanjing 210016, China
[2]. School of Mechanical and Power Engineering, Nanjing Tech University, Nanjing 211816, China

*Corresponding Author: timliu@njtech.edu.cn

Abstract. Starting from the actual working process of Wankel engine, the intake system model of Wankel engine is constructed based on the mean value engine model. The parameters to be determined in the model are identified by the least-square method based on the engine test data. According to the nonlinear characteristics of the intake system, the structure of the model is improved by introducing the speed correction term, and the problem that the error of original throttle air mass flow model is larger under the larger throttle opening is solved. The results of simulation and steady-state test show that the maximum relative error between simulation and experimental values of throttle air mass flow is less than +3% and that between simulation and experiment values of intake pressure is less than +2%.The parameter identification and structure of the model have high precision, which can calculate the basic air intake flow of Wankel engine accurately and shorten the development cycle of the electric control system of Wankel engine.

1. Introduction

Compared with the traditional piston engine, Wankel engine has the advantages of small volume, low weight, high power density and low noise due to the elimination of crank-connecting rod mechanism and valve mechanism. In recent years, with the upsurge of unmanned aerial vehicle (UAV) research and development, Wankel engine, as a power device, has once again become the focus of the industry. In order to improve the power performance of Wankel engine, reduce the fuel consumption, and increase the cruising range of the engine, it is necessary to develop an electronic fuel injection system for Wankel engine of UAV.

The electronic fuel injection system can realize the real-time control of Wankel engine under different working conditions, thus greatly improving the environmental adaptability and working performance of Wankel engine [1]. Since the electronic fuel injection system requires accurate detection of the air mass flow per cycle to determine the injected fuel mass flow per cycle based on the stoichiometric air/fuel mass ratio, it is necessary to model and analyze the characteristics of the intake system of the Wankel engine based on the structure and actual working process of the engine.

Wankel engine is mainly composed of triangular rotor, cylinder block, eccentric shaft and various seals. The triangular rotor divides the internal space of the cylinder into three independent working chambers. When the engine is working, the triangular rotor in the cylinder rotates around the center of the main bearing and rotates around its own rotating center. The volume of each working chamber changes constantly, and each of them completes four strokes of intake, compression, work and exhaust respectively [2-4], as shown in figure 1. Because of this unique structure and movement mode of Wankel engine, the structure of the intake system is simple, and the direction of intake and exhaust is basically consistent with the direction of gas flow in the engine. Sufficient air intake flow can be provided for Wankel engine over the entire speed range of the engine.

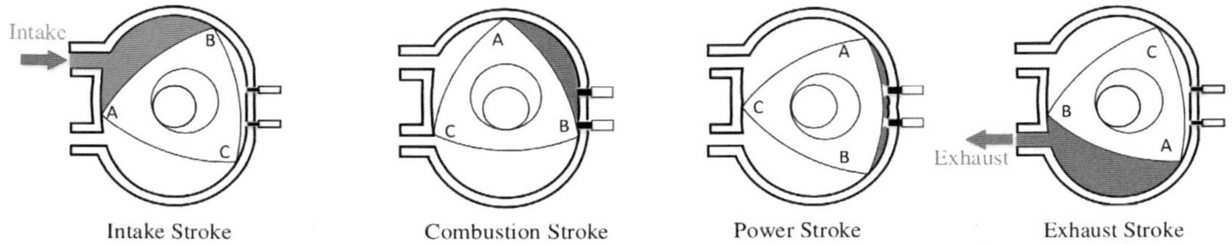

Figure 1. Actual operation cycle of Wankel engine.

For the air/fuel ratio control of piston engine, scholars have been working to establish a control model with high control precision, few undetermined parameters, small amount of calculation, and strong system robustness. A type of simple mathematical engine model for spark ignition engines is the mean value engine model [5-7]. Based on this model, a number of authors have developed air-fuel ratio control strategies and intake air mass flow estimation algorithm for spark ignition engine [8-16]. However, there are few studies on the modeling of the intake characteristics of Wankel engine. In reference [17] and [18], the intake air mass flow model of Wankel engine is established, and the characteristic parameters of the engine intake system are calibrated according to the bench test results. Due to the saturation nonlinearity of air mass flow at throttle, the problem that the original model has a large error under large throttle opening has not been solved in references [17] and [18].

The purpose of this paper is to construct the intake characteristic model of Wankel engine by using the mean value model, and to determine the fixed parameters in the model using the steady test data. The validity of this model is validated by combining the simulation results of MATLAB/Simulink with the test data. Comparison results show that the model has simple structure and high accuracy. It can calculate the basic intake air mass flow of Wankel engine accurately and provide basic data for the development of the electronic controlled system of Wankel engine.

2. The intake system model of Wankel engine.

The mean value engine model is a commonly used control model. The average values of variables in several engine working cycles are used to describe the dynamic process of the engine [5]. A series of state differential equations are used to describe the corresponding system dynamics, so the expression of the model is simple. In addition, the model contains fewer fixed parameters to meet the needs of simulation and control and has a wide range of application. Figure 2 is a schematic diagram of the intake system of Wankel engine.

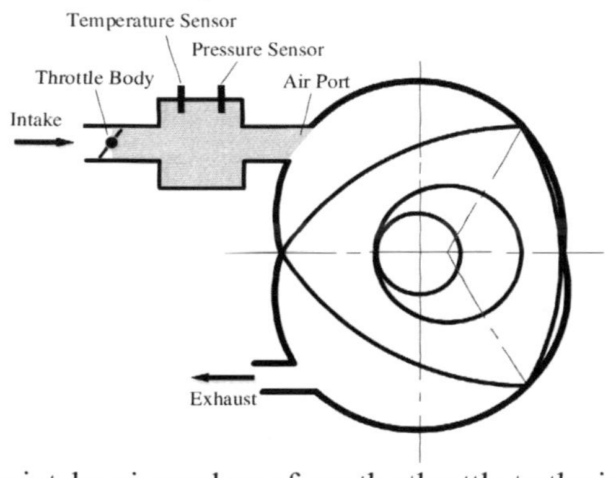

Figure 2. Schematic diagram of intake system of Wankel engine.

The intake pipe volume from the throttle to the intake port is regarded as the control body, as shown

IOP Conf. Series: Materials Science and Engineering **692** (2019) 012045 doi:10.1088/1757-899X/692/1/012045

in figure 3.

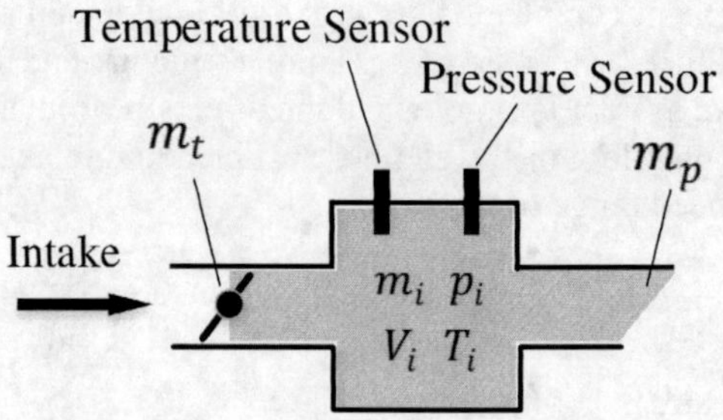

Figure 3. Schematic diagram of Intake manifold.

The air in this control body is regarded as an ideal gas [5]. According to the equation of state of the ideal gas, the air in the control body is given by the expression

$$m_i = \frac{p_i V_i}{R_a T_i} \tag{1}$$

where m_i is the total mass of air in the control body, p_i is the absolute pressure of the air in the control body, V_i is the total volume of the control body, R_a is the air gas constant, T_i is the absolute temperature of the air in the control body.

The derivation on both sides of the antithesis (1) has the form

$$\dot{m}_i = \frac{d}{dt}\left(\frac{p_i V_i}{R_a T_i}\right) = \frac{V_i}{R_a T_i}\dot{p}_i - \frac{V_i p_i}{R_a T_i^2}\dot{T}_i \tag{2}$$

According to literature [7], the change of the absolute pressure of the air in the control body relative to the time is much greater than that of the gas temperature in the control body relative to the time. Therefore, the item on the right side of equation (2) concerning the change of intake temperature can be neglected. In the control body, according to the law of conservation of mass

$$\dot{m}_i = \dot{m}_t - \dot{m}_p \tag{3}$$

where $\dot{m}_t$ is throttle air mass flow, $\dot{m}_p$ is port air mass flow, combined with equation (2), we can deduce

$$\dot{m}_i = \dot{m}_t - \dot{m}_p = \frac{V_i}{R_a T_i}\dot{p}_i \tag{4}$$

$$\dot{p}_i = \frac{R_a T_i}{V_i}(\dot{m}_t - \dot{m}_p) \tag{5}$$

The throttle air mass flow $\dot{m}_t$ is regarded as the flow of a compressible fluid through a converging nozzle [5]

$$\dot{m}_t(\alpha, p_i) = C_{t1}\frac{P_{amb}}{\sqrt{T_{amb}}}\beta_1(\alpha)\beta_2(p_i) + C_{t2} \tag{6}$$

where C_{t1} and C_{t2} are fixed parameters to be calibrated by experiments, P_{amb} is the ambient atmospheric pressure, T_{amb} is the ambient temperature. Ignoring the throttling action of the air filter in the front of the intake pipe, the pressure and temperature upstream of the throttle can be approximately expressed by the ambient atmospheric pressure P_{amb} and the ambient temperature T_{amb} [5]. α is the throttle plate angle, $\beta_1(\alpha)$ is a function of throttle plate angle α, $\beta_2(p_i)$ is a function of the absolute pressure of air in the control body p_i. According to literature [5], functions $\beta_1(\alpha)$ and functions $\beta_2(p_i)$ have the following form

$$\beta_1(\alpha) = 1 - 1.4073\cos(\alpha) + 0.4087\cos^2(\alpha) \tag{7}$$

$$\beta_2(p_i) = \begin{cases} \frac{1}{0.7404}\sqrt{\left(\frac{p_i}{P_{amb}}\right)^{0.4404} - \left(\frac{p_i}{P_{amb}}\right)^{2.3143}} & ,\frac{p_i}{P_{amb}} \geq 0.4125 \\ 1 & ,\frac{p_i}{P_{amb}} < 0.412 \end{cases} \quad (8)$$

According to the speed-density equation, the port air mass flow $\dot{m}_i$ has the form [7]

$$\dot{m}_p = \frac{nV_d p_i \eta_v}{60R_a T_i} \quad (9)$$

where n is the speed of the eccentric shaft of Wankel engine, V_d is engine displacement, η_v is engine volume efficiency, its form can be approximately expressed by literature [10]

$$\eta_v = C_{v0} + C_{v1}n + C_{v2}n^2 + C_{v3}p_i \quad (10)$$

where C_{v0}, C_{v1}, C_{v2}, C_{v3} are all constants fitted by experiments.

Equation (5), (6) and equation (9) constitute a complete model of intake system of Wankel engine. The inputs of the model are throttle plate angle α and speed of the eccentric shaft of Wankel engine n, and the state variable of the model is the absolute pressure of the air in the control body p_i.

3.Identification of undetermined parameters and optimization of model structure

According to the intake system model of Wankel engine, the undetermined parameters in the model need to be identified. The research object of this paper is a single-rotor engine with circumferential intake, whose specific parameters are as shown in table 1. Aiming at the controlled engine, the steady-state bench test was carried out under different throttle opening and rotating speed. The engine test bench is shown in figure 4. In the course of the experiment, the signals such as throttle plate angle α, speed of the eccentric shaft, air mass flow rate $\dot{m}_t$ and the absolute pressure of the air p_i are collected by data acquisition system.

Figure 4. Wankel engine test bench.

Table 1. Main parameters of Wankel engine.

No	Engine Particulars	Value
1	Swept Volume/cm^3	294
2	Compression Ratio	9
3	Generating Radius/mm	69
4	Eccentricity/mm	11.6
5	Maximum Continuous Power/Kw	35.8
6	Maximum Continuous RPM/r/min	7100
7	Cooling	air and liquid cooling systems

3.1 Parameters identification of throttle air mass flow model

The product of the three factors $P_{amb}/\sqrt{T_{amb}}$, function $\beta_1(\alpha)$ and function $\beta_2(p_i)$ in equation (5) is considered as a whole. The relationship between throttle air mass flow obtained from the steady-state test data of the engine and the product $P_{amb}\beta_1(\alpha)\beta_2(p_i)/\sqrt{T_{amb}}$ is shown in figure 5.

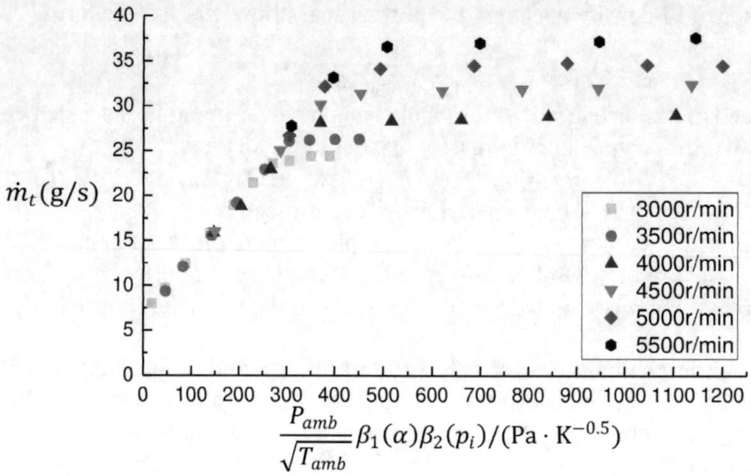

Figure 5. Relationship between throttle air mass flow and $\frac{P_{amb}}{\sqrt{T_{amb}}}\beta_1(\alpha)\beta_2(p_i)$.

According to formula (5), there is a linear relationship between the throttle air mass flow rate $\dot{m}_t$ and the product term $P_{amb}\beta_1(\alpha)\beta_2(p_i)/\sqrt{T_{amb}}$. However, the measured values of the engine steady-state test show (as shown in figure 5) that a single linear relationship cannot completely cover the entire range of throttle opening and intake pressure p_i. When the engine speed is constant and the throttle opening is small, due to the obvious throttle effect of the throttle body, there is a good linear relationship between the throttle air mass flow rate $\dot{m}_t$ and the product term $P_{amb}\beta_1(\alpha)\beta_2(p_i)/\sqrt{T_{amb}}$. When the throttle plate angle α increases to a certain large opening, the throttle effect of the throttle body is weakened, and the pressure drop before and after the throttle body will not change obviously. When the throttle opening continues to increase, there is no obvious effect on the intake pressure p_i , and the throttle air mass flow rate $\dot{m}_t$ tends to be saturated, which is mainly affected by the rotor engine output shaft speed. As can be seen from figure 5, for different engine speeds, the inflection point of the curve is different, that is, when the engine is at different speeds, the value of larger throttle opening is not unique.

From the previous analysis, it can be seen that the error of the throttle air mass flow model is larger under the larger throttle opening, and the throttle air mass flow tends to be saturated with the increase of throttle opening, and it is related to the rotational speed of the engine output shaft. Therefore, the correction term of rotating speed needs to be introduced to improve the overall accuracy of the model:

$$\dot{m}_t(n) = C_{t3}\frac{P_{amb}}{\sqrt{T_{amb}}}\beta_3(n) + C_{t4} \tag{11}$$

where $\dot{m}_t(n)$ is the value of throttle air mass flow under larger throttle opening, which is determined according to the engine output shaft speed. C_{t3} and C_{t4} are the undetermined parameters of the model, $\beta_3(n)$ is the introduced correction term, which can be expressed as follows:

$$\beta_3(n) = \frac{n}{1000} \tag{12}$$

The throttle air mass flow data obtained from the engine steady state test are fitted to the fixed parameters C_{t1}, C_{t2}, C_{t3} and C_{t4} based on the least square method. The identification results are shown in table 2.

Table 2. Undetermined parameters of throttle air mass flow model.

Undetermined Parameters	Value
C_{t1}	65
C_{t2}	6.5
C_{t3}	1
C_{t4}	8

The throttle air mass flow may take the smaller of $\dot{m}_t(\alpha, p_i)$ and $\dot{m}_t(n)$

$$\dot{m}_t = \min(\dot{m}_t(\alpha, p_i), \dot{m}_t(n)) \tag{13}$$

The point in figure 6 is the throttle air mass flow measured by the experiments at different engine speeds and different throttle opening, and the thin solid line is the calculated value of the model whose structure is modified. Compared with the original model, the modified throttle air flow model still has higher precision under larger throttle opening, which solves the problem that the original model has larger error under larger throttle opening.

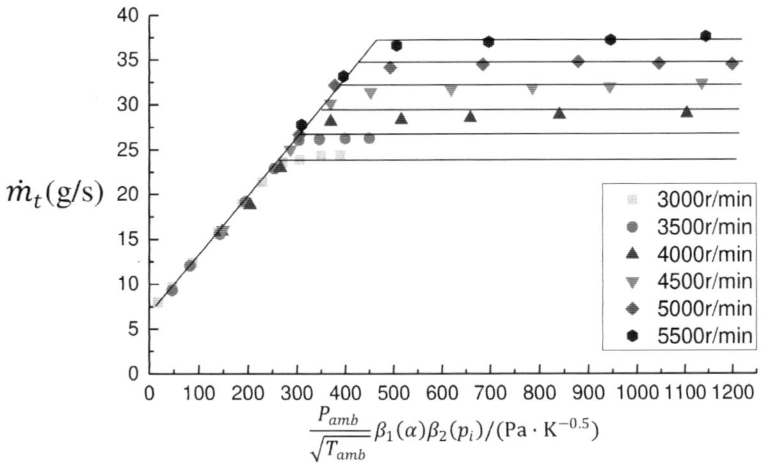

Figure 6. Comparison of measured values of steady-state test and calculated values of modified models.

3.2 Parameters identification of port air mass flow model

The port air mass flow $\dot{m}_p$ can be derived from equations (9) and (10)

$$\dot{m}_p = \frac{nV_d p_i \eta_v}{60R_a T_i} = \frac{nV_d p_i}{60R_a T_i} * (C_{v0} + C_{v1}n + C_{v2}n^2 + C_{v3}p_i) \tag{14}$$

When the engine is working in a steady state, the filling and exhaust effect in the intake pipe is small, and the pressure change in the control body Δp_i is approximately 0. In this case, the port air mass flow $\dot{m}_p$ is approximately equal to the throttle air mass flow $\dot{m}_t$. That is, the measured value of the steady air flow of the engine measured by the flowmeter installed in front of the throttle door. Therefore, the value of $\dot{m}_p$ is the measured value of air mass flow of the engine measured by the flow meter installed in front of the throttle. In equation (14), C_{v0}, C_{v1}, C_{v2} and C_{v3} are the parameters of the model to be determined, which can be regarded as the characteristic parameters of the intake system of a specific engine. For Wankel engines with different intake forms and different rotor numbers, the intake system characteristic parameters need to be calibrated by the engine steady-state test.

In this paper, the single-rotor engine with the form of circumferential surface air intake is taken as the research object. The signals of air mass flow rate and rotational speed n are obtained through the steady-state bench test of the engine, and the least squares method is used to identify the parameters. The identification results are shown in table 3.

Table 3. Undetermined parameters of port air mass flow model.

Undetermined Parameters	Value
C_{v0}	-1.028×10^{2}
C_{v1}	4.339×10^{-2}
C_{v2}	-3.537×10^{-6}
C_{v3}	1.058

4. Model simulation and experimental verification

According to equations (6), (11), (13), (14), Table 2 and Table 3, the simulation model of the intake characteristics of Wankel engine is built in MATLAB/Simulink, as shown in figure 7. The input of the model is the throttle plate angle α and the eccentric shaft rotating speed n, and the state variable of the model is the absolute air pressure in the control body p_i.

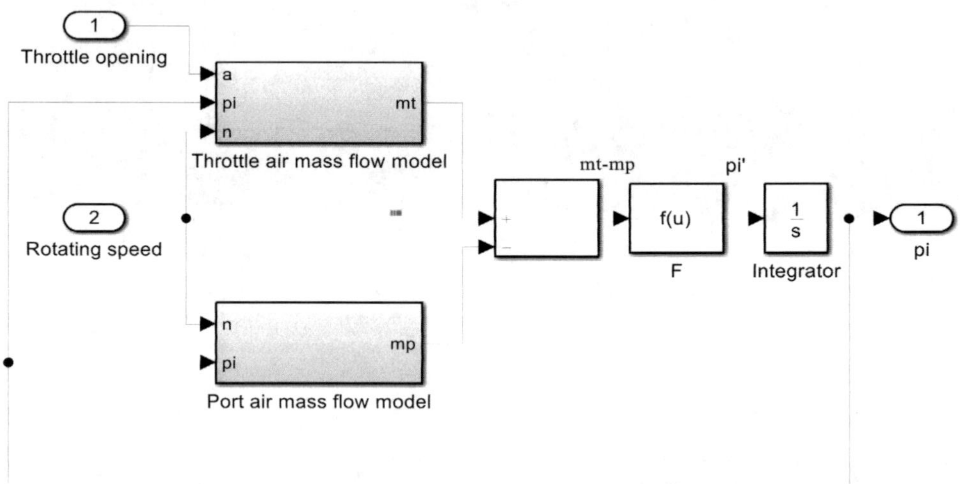

Figure 7. Block diagram of simulation for intake characteristics model of Wankel engine.

In this paper, the structure of the model and the accuracy of parameter identification are verified by a set of partial load tests and a set of full load tests. In the course of the test, the relevant signals are collected by the self-developed data acquisition system. The sequence of throttle plate angle α and eccentric shaft rotating speed n collected by the data acquisition system is taken as the input of the model. The throttle air mass flow value $\dot{m}_t$ and the absolute air pressure p_i calculated by the model simulation are compared with the measured values, and the accuracy of the model is verified according to the comparison results.

For the partial load test, the engine is stably operated at the throttle opening of 10% and rotation speed at 3500 r/min by the dynamometer. We fixed the engine speed at 3500r/min and changed the throttle opening from 10% to 50% to verify the accuracy of the model at low and medium loads and low speeds. The comparison between the simulation results and the test data of throttle air mass flow is shown in figure 8, and the comparison between the simulation results and the test data of absolute air pressure in the control body is shown in figure 9.

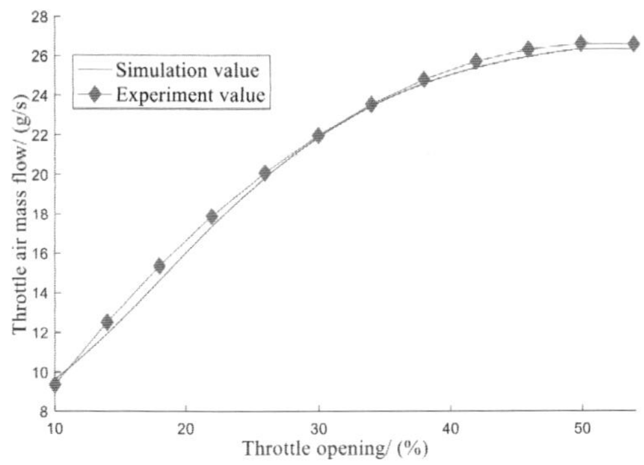

Figure 8. Comparison between throttle air mass flow model simulation and measured values with rotating speed 3500 r/min and throttle opening from 10% to 50%.

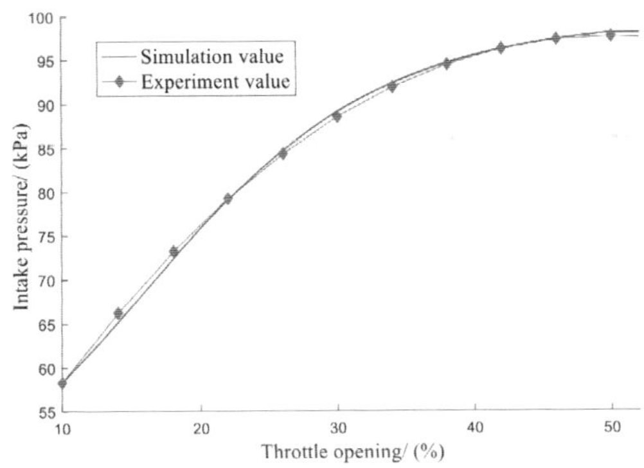

Figure 9. Comparison between intake pressure model simulation and measured values with rotating speed 3500 r/min and throttle opening from 10% to 50%.

For the full load test, the engine is stably operated at the throttle opening of 100% and rotation speed at 5500 r/min by the dynamometer. We fixed the engine throttle opening at 100% and changed the rotation speed from 5500r/min to 7100r/min to verify the accuracy of the model under full load and high rotation speed. Figure 10 shows the comparison between simulation and measured values of throttle air mass flow, and the comparison between the simulation results and t measured values of absolute air pressure in the control body is shown in figure 11.

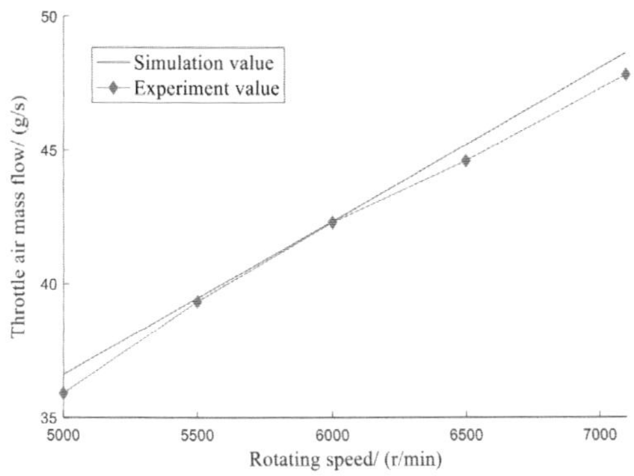

Figure 10. Comparison between throttle air mass flow model simulation and measured values with throttle wide open and rotating speed from 5000 to 7100 r/min.

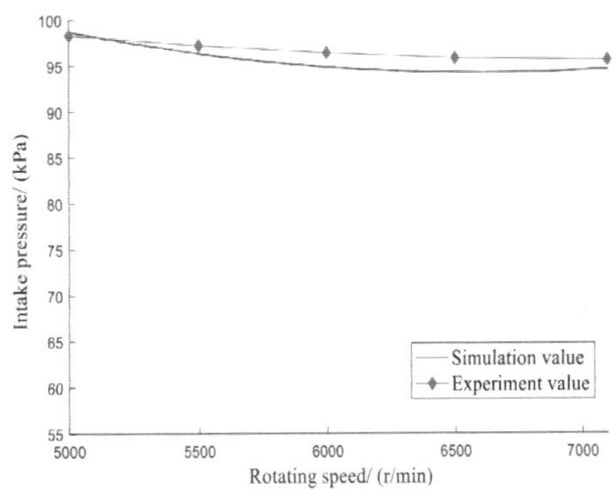

Figure 11. Comparison between intake pressure model simulation and measured values with throttle wide open and rotating speed from 5000 to 7100 r/min.

The comparison between the model simulation and measured values of throttle air mass flow in figure 8 and figure 10 shows that under steady state conditions, the maximum relative error between the model simulation and the measured value does not exceed ±3%.It can be seen from figure 8 that when the throttle opening is small, the throttle effect of throttle body is obvious, and there is an approximate linear relationship between the throttle air mass flow and the throttle plate angle. When the throttle opening increases to a certain large opening, the throttle air mass flow tends to be saturated. Combined with figure 10, it can be found that under the larger throttle opening, the throttle effect of the throttle body is weakened, and the throttle air mass flow is mainly affected by the rotation speed.

Under steady-state operating conditions, the maximum relative error between the model simulation and measured values of intake manifold pressure of Wankel engine is no more than ±2%, as shown in figure 9 and figure 11. Figure 9 shows that when the rotation speed is constant and throttle plate at a smaller angle, as the throttle opening increases, the intake pressure value increases significantly; when the throttle opening increases to a certain larger value, the intake pressure value increases slowly and tends to be stable. When the throttle opening is 100%, the comparison between the intake pressure model simulation and the measured value shows that intake pressure value decreases with the increase of the rotation speed, but compared with the itself, the variation range is not obvious.

Figure 8-figure 11 show that the comparison between simulation and measured values shows that the mean value model of intake system characteristics of Wankel engine established in this paper and the identification of undetermined parameters in the model have high accuracy. Based on this model, the basic intake air mass flow of Wankel engine can be calculated more accurately.

5. Conclusion

Based on the structure and characteristics of air intake system of Wankel engine, the model of Wankel engine intake system is constructed based on the mean value model. According to the engine test data, the least square method is used to identify the undetermined parameters in the model. Because the throttle air mass flow has the nonlinear characteristic, in order to solve this problem, the structure of the model is improved by introducing the speed correction term. The problem that the error of the original throttle air mass flow model is larger under the larger throttle opening is solved. The comparison between the model simulation and experimental values shows that the maximum relative error between the simulation and the measured values of throttle air mass flow does not exceed ±3%, and the maximum relative error between the simulation and the measured values of the intake pressure does not exceed ±2%. The model parameter identification and structure have high accuracy. The intake system model of Wankel engine constructed in this paper contains fewer parameters and has a simple structure and higher precision. Based on this model, the basic intake air mass flow of the engine can be calculated accurately, a lot of calibration work in the development of the engine electronic control system can be reduced, and the development cycle of the rotor engine electronic control system can be shortened.

Acknowledgments

This work was support by the Graduate Innovation Base (Laboratory) Open Fund of Nanjing University of Aeronautics and Astronautics (Grant No. kfjj20180208). The research project was also supported by the Jiangsu Province Key Laboratory of Aerospace Power System (Grant No. CEPE2018003).

References

[1] Wu Chao, Wei Minxiang, Liu Rui. Control Strategy Research on Fuel Injection and Ignition Timing for Rotary Engine[J]. *Small Internal Combustion Engine And Vehicle Technique*,**2019,48(01): 1-7**

[2] Ding Yanci, Guo Pengyan. The technology situation and development tendency of rotary engines at home and abroad[J]. *Automobile Applied Technology*,**2017(16).**

[3] PEI Hailing, ZHOU Naijun, GAO Hongliang. The characteristics and improvement of rotary engines[J]. *Internal Combustion Engines*,**2006(3): 1-3.**

[4] LIU Shengtian, ZHENG Zhong. Structure an-d Application of MAZDA vehicle rotary engine[J]. *Journal of Shandong Jiaotong University*,**2004(02): 22-26.**

[5] Hendricks E, Sorenson S C. Mean value modelling of spark ignition engines[J]. SAE transactions, **1990: 1359-1373.**

[6] Hendricks E, Vesterholm T. The analysis of mean value SI engine models[J]. SAE transactions,**1992: 1396-1414.**

[7] Hendricks E, Chevalier A, Jensen M, et al. Modelling of the intake manifold filling dynamics[J]. SAE transactions,**1996: 122-146.**

[8] ZHOU Dong, SU Tiexiong, LI Chuang. The Research of the Engine Intake Manifold Model's Simulation Accuracy Based on the Mean Value Engine Model[J]. *Science Technology and Engineering*,**2016(32): 238-242.**

[9] CHEN Linlin, WEI Minxiang, YANG Haiqing. Experiment and simulation of the mean value

model on two-stroke gasoline aero-engine[J]. *Journal of Aerospace Power*, 2008,23(12): **2249-2255.**

[10] Chevalier A, Vigild C W, Hendricks E. Predicting the port air mass flow of SI engines in air/fuel ratio control applications[J]. SAE transactions, 2000: 183-210.

[11] Powell J D, Fekete N P, Chang C F. Observer-based air fuel ratio control[J]. *IEEE Control Systems Magazine*, **1998, 18(5): 72-83.**

[12] Jensen P B, Olsen M B, Poulsen J, et al. A new family of nonlinear observers for SI engine air/fuel ratio control[R]. SAE Technical Paper, 1997.

[13] Hendricks E, Vesterholm T, Sorenson S C. Nonlinear, closed loop, SI engine control observers[J]. SAE transactions, **1992: 326-343.**

[14] Tan Derong, Liu Zheng lin, Yan Xinping. Optimum estimate method of intake flowrate for electronic control gasoline engine[J]. *Journal of Traffic and Transportation Engineering*,**2006(02): 39-42.**

[15] FENG Yu, JIAO Xiaohong. Transient cylinder air charge estimation with nonlinear Kalman filter for air-fuel ratio control of Gasoline engines[J]. *Control Theory & Applications*,**2015(4): 16.**

[16] Kerkeni H, Lauber J, Guerra T M. Estimation of Individual In-cynlinder air mass flow via Periodic Observer in Takagi-Sugeno form[C]//2010 IEEE Vehicle Power and Propulsion Conference. IEEE, **2010: 1-6.**

[17] TANG Tingxiao, ZHAO Houliang, CHEN Huaihai. Test model of intake air amount on rotary engine of unmanned aerial vehicle[J]. *Journal of Aerospace Power*,**2010,25(11): 2425-2430.**

[18] MA Panwei, NIU Quanmin, XU Peng. Modeling and Simulation of Rotary Engine Intake System Based on High Airflow Characteristics[J]. *Aeroengine*,**2019(1): 17-21.**

IOP Conf. Series: Materials Science and Engineering **692** (2019) 012046 doi:10.1088/1757-899X/692/1/012046

Inclusive Design of Teaching Aids for Social Science Learning Process for Elementary and Middle School Students based on Creative Mini PC Raspberry Technology

A Suzianti[a], A Mubarak[b], F Edrisy[c], A R Sausan[d], C N Lalita[e] and X Usman[f]

Industrial Engineering Department, Faculty of Engineering Universitas Indonesia, Depok, Indonesia 16424

[a]suzianti@eng.ui.ac.id; [b]mubarak.andri@gmail.com; [c]faizedrisy.21@gmail.com; [d]adhwaranaa@gmail.com; [e]cindylalita@gmail.com; [f]xavusman@gmail.com

Abstract. The purpose of this activity is to educate students and teachers with the inclusive board game invention as a learning tool for memorizing subjects for elementary to high school students. This product answers the problem for the difficulty of students remembering and understanding social science teaching materials. There are two approaches to the Learning Board Game teaching method, the first is visualization of social science teaching materials through interactive videos integrated with images according to subjects and screens and the second is gamification through the elements of challenge and fantasy by answering questions to get rewards cards or marking as a marker gets points. This product was developed with the New Product Development (NPD) Process and uses the Stage-Gate Model. In this model, product design goes through various stages starting from discovery / idea generation to launch with each stage that functions as a quality-gate of product design.

1. Introduction

Student's low-interest in the social science subjects of the archipelago is due to the absence of synergy between educational institutions with methods that support interactive and fun. In addition, until now there are still many instructors who deliver social science material with classical or textual methods [1], whereas instructors only deliver subject matter with the lecture method. Issues regarding social science subjects, especially history have been carried out in Dyah's (2014) research which explains that most social science subjects for students who study these subjects need good memorization lessons.

In the activities discussion, the process of teaching and learning about lessons in the classroom is important to the process of communication between teachers and students, especially in the process of delivering teaching material. However, the communication process may not be effective, not monotonous and not delivered so that the message delivered by the teacher to students can be well received. In this regard, to support the teaching and learning process in the classroom which is more innovative in subjects, which can reduce boredom, a more interactive learning media is needed to facilitate communication between teachers and their students [2].

This research tries to provide solutions for creating products that contain teaching for social science subjects by using New Product Development (NPD). In this study, NPD made a tool to produce attractive teaching aids that did not spend for teachers and students. These teaching aids can be used as interactive learning media to make students happy learning social science.

The stages of implementation generally consist of 7 stages, discussed, namely new product strategies (linking NPD processes with institutional concepts), idea making (looking for product ideas), screening, business analysis, development (developing ideas into tangible forms), testing and commercialization (marketing product) [3]. The final result of the NPD stage is an existence that is able to answer the needs of consumers [4,5]. New product development and new solutions to consumer problems will be called new product development, new product modification, new product development through the company's internal research and development processes (Armstrong and Kotler, 2005). Through this NPD stage, it is expected that the teaching aids on social science lessons that are produced can solve problems for the learning and teaching process in the environment of junior and senior high schools. In addition to NPD, the morphological matrix is used to determine alternative product solutions that we will design and brainstorm using AHP to select products that we will develop.

2. Methodology

2.1 Discovery Stage

In this research, we use the idea generation method to find the right product ideas to solve problems from customers and are feasible to be produced [6].

2.1.1 Idea Generation

2.1.1.1 Fault tree analysis. Our group's fault tree analysis creativity technique is used to determine the factors that make a problem possible. We see that it is difficult for students to understand social science lessons because two different things are boring learning material content, the teacher as the facilitator conveys the material not optimal, and the media delivery of monotonous material. The boring content factors of social science material are due to two factors including:

1) The large amount of material students must memorize makes it difficult for them to understand and be interested in learning social science.

2) Some students have a tendency to prefer physical rather than social subjects such as social science.

Teachers who do not deliver material optimally can be caused by several factors including:

1) One-way teaching methods do not attract students' attention to understand the content of the material delivered by the teacher.

2) The teacher's intonation, gestures, and teaching style are monotonous so students get bored listening to the material for several hours.

3) The teacher leaves class several times and is only given a substitute assignment so that face-to-face meetings between teacher and students are rare.

Factors and media delivery of monotonous material in our opinion is caused by several factors including:

1) Textbooks as the main medium of learning do not attract students' attention to learn in a thick form and contain long unraveled writing so that it does not make it easy for students to read and understand the material to be conveyed [7].

2) Social science teacher only conveys social science teaching material verbally or is illustrated with a presentation without any physical appearance that can attract students' attention

3) Homework assignments or tasks as a way for the teachers to ensure students repeat the material that has been learned in class has not run optimally because the student's absorption of material delivered to the class is not optimal so students cheat on assignments from other friends.

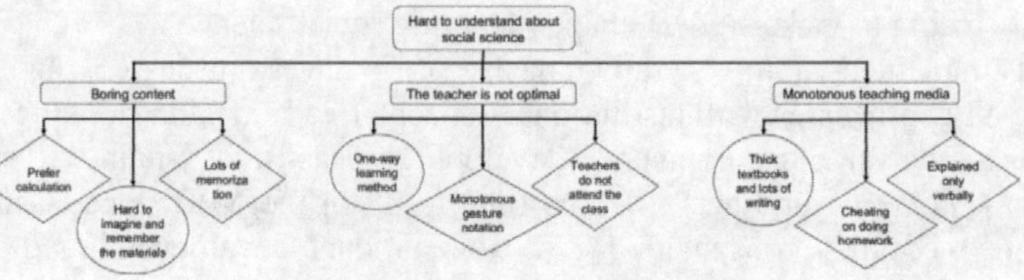

Figure 1. Fault Tree Analysis

2.1.1.2 Morphological matrix. As shown in Figure 1, we have determined that the product that will be designed will work to solve the boring one-way learning media factors so students find it difficult to understand the lesson. To get an idea for a product that will be designed, we use a morphological matrix that can produce several different ideas from the combinations of choices available in Table 1.

Table 1. Morphological Matrix

Activity	Playing	Watching	Discussion	Reading
Inclusive	Color difference	Object surface contour	Sign language	Braille letter
Advantages	Portable	Advanced technology	Unique form	Easy to use

2.1.1.3 Brainstorming. During the brainstorming process, we gathered and each group member gives innovative ideas based on the combination obtained in the previous morphological matrix so we can select the innovative ideas that we get. We select ideas by using rating and weighted tables. Some innovative ideas generated from brainstorming include the following:

- Product 1: Three-dimensional book
- Product 2: Interactive cartoon Indonesian social science
- Product 3: Hologram with characters from an area
- Product 4: Board Game in the form of a map of Indonesia

Our group conducted a product selection process using the AHP method by first rating each of the specified criteria and then calculating according to the product scores available to determine which product was better. The scale used is 1 (very bad) to 5 (very good). The criteria we use in this assessment are:

a. Creativity Aspect: Product uniqueness, whether the innovations made can improve the product image and there is novelty in teaching aids.
b. Technology: Applying appropriate technology to the products that we will design
c. Solve problems: the product solves the root causes of students' difficulties in learning social science appropriately effectively

The following table weights from each criterion:

After weighting each criteria subjectively as shown in Table 2, followed by checking the consistency ratio that is owned, then do a rating calculation for each product. The following table weighs the results of each product evaluation after using the AHP method.

Rating of each product

Table 2. Criteria Weight

Criteria	Weight
Creativity	0,21467689
Technology	0,18729463
Problem Solving	0,69802848

Table 3. Weighted Score Product

Aspect	Creativity	Technology	Problem Solving	TOTAL
Weight	0,21	0,19	0,6	
Three dimension book	0,09	0,16	0,16	0,145499
Interactive cartoon	0,16	0,09	0,09	0,108599
Hologram	0,28	0,47	0,28	0,313429
Map board game	0,47	0,28	0,47	0,432473

As shown in Table 3, the Board Game product in the form of an Indonesian map gets the highest score with a total value of 0.432473.

2.2 FMEA

We use FMEA to identify the risks of developing this board game product as shown in the table below:

Table 4. FMEA

No	Potential failure mode	Potential causes for failure	Potential effects of failure	Current control	Severity	Likelihood	Detection	RPN	Recommendation	Severity	Likelihood	Detection	RPN
1	Product does not sell	The price is too expensive Do not attract customer Disrupted with media	The product fails because it is not profitable	Doing a market research and market testing	5	3	4	60	Innovating the shape and type of board games	3	4	3	36
2	The game is not carried out according to instructions	Teacher does not understand how to play the game Inadequate supporting facilities Students are not serious when playing the game Participants answer the questions fraudulently	Students do not understand the material optimally	The game's rules are easy to obey and manual book	4	4	3	48	A simpler improvement tool and fewer supporting components	3	3	2	18
3	Students got bored playing the board game	Monotonous video Question does not change	The product's life cycle is very short	Make a game series	3	4	3	36	Add application features to the game board	2	3	2	12

As can be seen in Table 4, the biggest risk is the product does not sell in the market because of several factors such as the price is too expensive, not according to needs, and is disrupted with other teaching media so that market research and market testing is important before producing this product [8] and continuous innovation must be done so that the teaching media is not monotonous.

2.3 Idea Decision

Our group determined that the selected social science teaching aids product, namely board games in the form of Indonesian maps. Based on the rating and weighted score assessment, it was concluded that this board game product has a higher value than the other three toy products.

2.4 Use of Empathy Mapping

Empathy mapping is a tool used to determine what is said and done by users is quite easy and determine what users think and feel by observing and observing the behavior and responses of activities and conversations with users. Empathy mapping is divided into four quadrants. After conducting research and surveys, we were able to make an empathy mapping of the use of this Histoboard teaching aids to one of the Social Sciences teachers at SMPN 131 Jakarta. This Empathy mapping is divided into four quadrants, as shown in Figure 2.

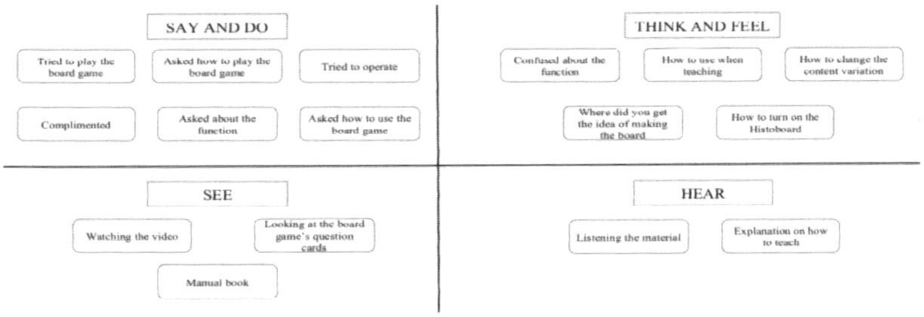

Figure 2. Empathy Mapping

3. Feasibility Study Analysis

3.1 Customer's Needs

Histoboard is a technology-based product in the form of a map of Indonesia which has two modes namely learning mode that can be used by teachers to explain subject matter through instructional education video media, and the second mode is a game mode that can help students in understanding subjects such as social science which has a lot of material to memorize [9,10]. The target customers of this Histoboard product are students with Education levels ranging from Elementary Schools (SD) to High Schools (SMA), which are the generation with a tendency as technology adopters. To find out the conditions, interests and difficulties experienced by students in learning social science, we distributed questionnaires to students from elementary to high school level. The results of the questionnaire can be seen as follows.

IOP Conf. Series: Materials Science and Engineering **692** (2019) 012046 doi:10.1088/1757-899X/692/1/012046

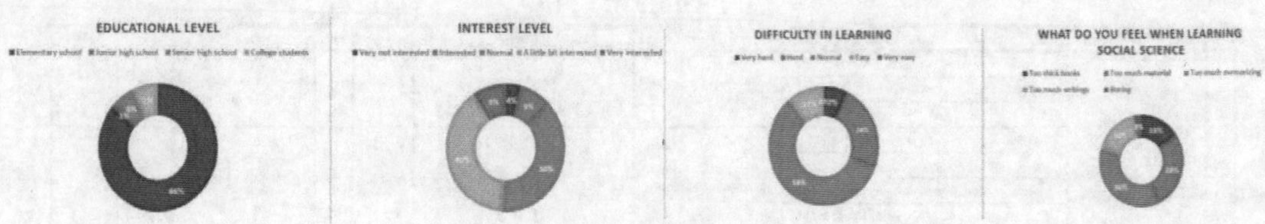

Figure 3. Questionnaires Results

As shown in Figure 3, it resulted that 46% of students who filled out the questionnaire needed media or teaching aids that could help and facilitate them in understanding social science lessons. Next, we provide additional questions to find out what features are needed in a media or teaching aids making it easier for students to understand social science lessons. We provide several choices of methods that can help students in understanding social science lessons, which are educational videos related to the subject matter, questions and practice questions, interactive games, explanations of material in the form of text, explanation of material in the form of images, and at an affordable price as can be seen in Figure 4.

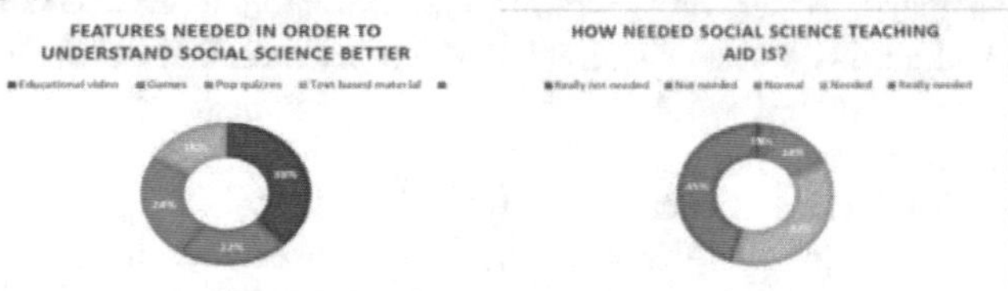

Figure 4. Questionnaires Result

3.1.1 Hierarchy of Consumer's Needs. We formed a hierarchy or levels that will facilitate the design process of technology-based teaching aid products for social studies lessons going forward. The attributes in the questionnaire are presented in the "Level 1" column, while the "Level 2" column is a more specific development of the attributes in the "Level 1" column as can be seen in Table 5.

3.1.2 Level of Importance Based on Consumer's Needs. Because not all the needs and desires of consumers can be translated into a product, it is necessary to prioritize these attributes that are expected by consumers for a teaching aid product for history lessons. Based on the results of the questionnaire, the following level of importance of attributes in accordance with consumer expectations as can be seen in Table 6. The final product is shown in Figure 5.

Table 5. Consumer's Needs Hierarchy

Level 1	Level 2
Features to play educational videos related to learning material	Video series related to teacher's learning syllabus
	Videos will be shown after the teacher's explanation in order to help the students understand better
	Video content that is interesting, creative, interactive, and with short duration
Features with pop quizzes	Pop quizzes related to teacher's learning syllabus
	Answers and quizzes explanations
Features with interactive games	Learning materials in interactive games mode
	Games in groups to improve teamwork and also communication skills between students
	There are some challenges filled with questions related to the learning materials in games
Features with material explanation in texts	Learning materials that will be deliver with PowerPoint presentation
	Material summary that will help the students better
Features with material explanation in pictures	Learning materials that will be deliver with animation
	Using PowerPoint with a visually interesting contents
	Material explanation with pictures about social science
Teaching aid design that is comfortable to use	Teaching aid with easily understandable design
	Unique interface design that could not make students easily bored
	Ergonomics interface design
Light teaching aid	Light teaching aid
	Teaching aid with the right packaging
Affordable price	Price range Rp 500.000-Rp 600.000

Table 6. Level of Importance Based on Consumer's Needs

Consumer's Needs	Level of Importance
Educational videos based on the learning material	5
Pop quizzes	4
Interactive games	3
Material explanation in texts	3
Material explanation in pictures	5
Teaching aid that is comfortable to use	5
Portable teaching aid	4
Affordable price	4

Figure 5. Histoboard

4. Conclusion

From our research, clearly the need to increase student's interest in learning social science is evident and indisputable. The learning method used when teaching social science is known to be very monotonous and the subject itself is hard for the students to understand.

In this paper we get that social science has a boring content, which the students prefer calculation more than memorization, the social science teacher is not yet optimal when teaching because they still use the one-way learning method and do monotonous gesture notation and sometimes they do not attend classes, and social science still uses monotonous teaching media, which has thick textbooks and lots of writing, and usually explained only verbally.

To provide solutions for creating products that contain a new teaching method for social science subjects is by using New Product Development (NPD). Through this NPD stage, the teaching aids on social science lessons that are produced was able to solve the problems for the learning and teaching process in the environment of junior and senior high schools. After conducting idea generation using a number of methods, including Fault Tree Analysis, Morphological Matrix, as well as brainstorming for ideas of early product of learning aids to be made, our group determined that the selected social science teaching aids product, namely board games in the form of Indonesian maps. Based on the rating and weighted score assessment, it was concluded that this board game product has a higher value than the other three toy products.

References

[1] Apriliana, Diah (2014). Pengembangan Pembelajaran Sejarah dengan *Scientific Approach* Melalui *Outdoor Study* di Kelas XI IIS 4 SMA Negeri Ajibarang. Semarang. Indonesian Journal of History Education. **Volume 3**, pp 27-30
[2] Liu, E. Z. F., & Chen, P. K. (2013). The effect of game-based learning on students' learning performance in science learning–A case of "Conveyance Go". Procedia-Social and Behavioral Sciences, **103**, 1044-1051
[3] Booz, Allen, & Hamilton. (1982). New Product Development for the 1980's. New York: Booz, Allen & Hamilton, Inc.
[4] Cooper, RG & Kleinschmidt, EJ (1988). Resource Allocation and the New Product Process. Journal of Industrial Marketing Management, **17**, pp 249-262.
[5] Thomas, RJ (1993). New Product Development – Managing and Forecasting for Strategic Success. John Wiley & Sons.
[6] Twigg, D (1998) Managing Product Development within a design chain. International journal of operations & production management, **18**, 5.
[7] Odenweller, C. M., Hsu, C. T., & DiCarlo, S. E. (1998). Educational card games for understanding gastrointestinal physiology. Advances in Physiology Education, **275**(6), S78
[8] Armstrong Gray & Kotler, Philip (2005). *Marketing An Introduction* (7th Edition). United States Of America, Pearson Education, Inc
[9] Allery, L. (2014). Make use of educational games. Education for Primary Care, 25(1), 65-66.
[10] Cavalho, J. C. Q. D., Beltramini, L. M., & Bossolan, N. R. S. (2018). Using a board game to teach protein synthesis to high school students. Journal of Biological Education, **52**, 1-12.

IOP Conf. Series: Materials Science and Engineering **692** (2019) 012047 doi:10.1088/1757-899X/692/1/012047

Time Series Data Prediction Based on Sequence to Sequence Model

Chao Yang, Zhongwen Guo and Lintao Xian

Department of Computer Science and Technology, Ocean University of China, Qingdao, China

xianlintao@qq.com

Abstract. Home appliance test is an essential part of the R&D process. However, since home appliance test must be carried out under certain circumstances, it makes the development cycle longer. If it is possible to predicting the home appliance test data, it will improve the efficiency of home appliance development. Sequence to sequence model has been proved that it has a strong ability to map the input sequence and output sequence and it has been widely used in machine translation tasks. In order to solve the home appliance test data prediction problem, we firstly tried to apply sequence to sequence model to predict numerically continuous time series data. Our experiments proved that our model can predict the data of relatively long-time steps when inputting data of relatively short time steps. In our experiment, the mean absolute percentage error of the prediction of our model is about 0.1%. Our approach has a strong generalization ability and performs well in different experimental scenarios. Finally, we believe that the sequence to sequence method performs satisfactorily on prediction problems for continuous time series data. With this model, we can obtain more accurate results than traditional methods and it is meaningful for the R&D of home appliances.

1. Introduction

With the development of smart homes, the pattern of the home appliance manufacturing industry has changed from the enterprise productions decides the user's demand to the user's demand decides the original enterprise productions. The home appliance manufacturing industry customizes the products according to the user's needs. Thus, the frequency of research and development of new products is becoming higher and higher. In the process of R&D of home appliances, product test is an indispensable part. Home appliances need to run continuously for hundreds to thousands of hours under certain working conditions, resulting in a long development cycle. Predicting the refrigerator test data shortens the test cycle, which will improve the research and development efficiency of new products.

To address prediction problems, traditionally we use linear regression, lasso regression, polynomial regression, etc. However, these kinds of algorithms depend on choosing the regression terms manually and do not have satisfying performances when the actual formula is extremely complicated. Thanks to the fast development of deep learning, we are able to use artificial neural networks (ANN) to address the regression problems [1]. Because of the character of ANN, researchers no longer need to set regression terms manually, but use an end-to-end approach that the model can automatically learn the features in the data and use it to complete the regression tasks. The simplest ANN, also called Multi-Layer Perception (MLP) [2], has been proven to be a general function approximation that can be used to match the nonlinear part more accurately and have better performances comparing with the previous algorithms.

Unfortunately, when the data have time-series characters, MLP is not able to utilize the time features of the data. In order to take advantage of the time series information of the data, Hihi [3] proposed recurrent neural networks (RNN).

A standard RNN does not only focus on the current input, but also utilize the sequence features. I.e., it is easier for a standard RNN to map sequences to sequences [4,5].

But, when the length of the input sequence is different from the length of the output sequence, it will be challenging for RNN to map the two sequences. In addition, when the length of the sequence is quite long, RNN will be tricky to train and mostly suffer for gradient vanish [6]. Later, a new structure called Long Short-Term Memory (LSTM) [7] became popular. Thanks to the special design of the LSTM module, it can address the gradient vanish problem when facing a long-term sequence and it has a strong ability for continual prediction [8].

The principle of a LSTM is to estimate the conditional probability $p(y_1,...,y_{T'} \mid x_1,...,x_T)$ when given the input sequence $(x_1,...,x_T)$. A standard LSTM formulation is:

$$p(y_1,...,y_{T'} \mid x_1,...,x_T) = \prod_{t=1}^{T'} p(y_t \mid v, y_1,..., y_{t-1}) \tag{1}$$

Thanks to LSTM, it is possible to predict a longer-term sequence. It is almost the best choice when addressing natural language processing problems. To make LSTM module more easily to train and implement, Cho et al. [9] proposed a novel structure called Gated Recurrent Unit (GRU) and it has been proved that GRU is faster for training than LSTM and has almost the same performance.

However, these kinds of structures are still hard to address the problems when the input length and output length are different. To address this problem, Cho et al. [9] developed sequence to sequence model.

The sequence to sequence model, often called seq2seq for short, is inspired by auto-encoder structure [10]. The difference is that instead of using multiple layers of neural network in the encoder and decoder, we use RNN module in the encoder part to change the sequence into an context vector, the context vector, which consist of the information of the input sequence, will be the input or the initial hidden state to send to the decoder part.

Generally, we choose the last time-step output of the encoder as the context vector. But it has been proved to be unreasonable and unsuitable for translation tasks. Graves [11] proposed a novel mechanism called attention which allowed the decoder to concentrate on different part of encoder's hidden state to choose the necessary and important information for translation. This seq2seq with attention structure has been proven to have an excellent performance for sequence prediction tasks such as machine translation [12,13,14], speech recognition [15,16], emotion classification [17], video analysis [18], trajectory prediction [19], image captioning [20], etc.

2. The Model

In this paper, to address the home appliances test data prediction task, instead of the traditional methods, we propose a method based on seq2seq model.

The encoder consists with a two-layer GRU. Comparing with the model that usually uses in machine translation tasks, we abandon word2vec [21], also called embedding layers, in our encoder part because our data do not suffer for sparse input issues. The input will go through all the GRUs and then convert to the context vector.

In the middle of encoder and decoder we use a single fully connected layer network for adjusting the dimension of the data. In addition, it is kind of like attention mechanism that allowed the neural network to choose the information that are more useful for our task automatically.

The decoder consists with a two-layer GRU which has the same structure with the GRU which is used in the encoder. Following the GRU are fully connected layers. We use ReLU as the activation function in the middle layers of the neural network. Generally, we often use softmax at the last layer to force the neural network to choose the word with the maximum possibility. However, the test data of the appliances are continuous. In order to obtain continuous output, we abandon softmax and do not add any activation function in the last layer. Also, we do not add embedding layer in decoder part. The output of the GRUs in the decoder will then go through the fully connected layers and we will obtain the output of the model

An illustration of the structure of our model can be seen in figure 1.

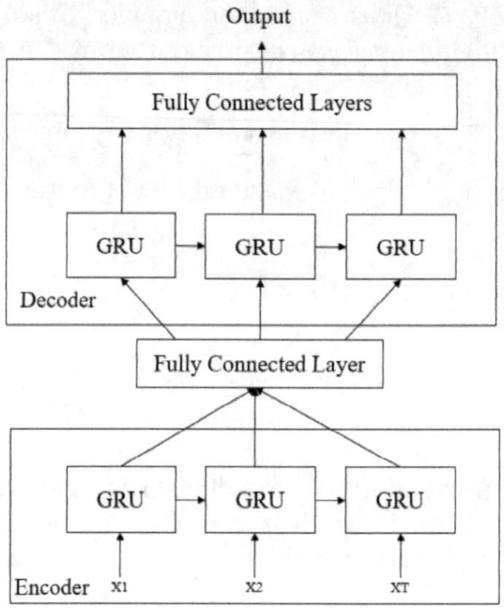

Figure 1. An illustration of the structure of our model

3. Experiments

3.1. Data Description and Processing

In this task, we will try to use the previous few sets of data to predict a large amount of subsequent data. We use refrigerator power consumption test data in our experiment. An overview of the data can be seen in table 1. In refrigerator power consumption test, we use seven sensors to capture the frozen temperature of the seven different position of refrigerator. Our test needs to be done in a certain environment temperature so we have two sensors to capture the environment temperature to make sure the environment temperature meets the requirement. Other sensors are used to capture the status of the refrigerator during the power consumption test.

We use a small length of data as input to predict some essential sensor data. In refrigerator power consumption test, we need to predict the power consumption and the temperatures that are important to the testers.

We collected experimental data from different kinds of refrigerator in different environments to make sure our model is robust and has generalization performance. We accumulated 105,000 time steps of data for training and 10,500 time steps of data for validation.

We have tried different length of the input and output, and to make sure the deviation is small enough and the length of the output is longer enough comparing with the length of the input, we set 20 as the max length of the input and set 500 as the max length of the output. We use normalization to make sure all the data are in similar range to avoid hurting the performance because of the range differences of the data. In this experiment, we have temperatures from 7 sensors, and we do care about mean temperature. So, we just calculate the mean temperature and use it as the input to reduce the dimension of the input data.

Table 1. An overview of our training data

Time Step / Feature	1	2	3	4	5	6
Frozen 1	-17.40	-17.40	-17.30	-17.20	-17.20	-17.20
Frozen 2	-13.20	-13.10	-13.00	-13.00	-13.20	-13.40
Frozen 3	-13.80	-13.80	-13.60	-13.70	-14.10	-14.40
Frozen 4	-16.30	-16.30	-16.20	-16.20	-16.10	-16.10
Frozen 5	-16.30	-16.30	-16.20	-16.20	-16.10	-16.10
Frozen 6	-16.10	-16.10	-16.00	-15.90	-15.90	-15.90
Frozen 7	-17.00	-16.90	-16.90	-16.80	-16.70	-16.70
Environment 1	15.90	15.80	15.90	15.80	15.80	15.80
Environment 2	16.10	16.10	16.10	16.10	16.10	16.10
Voltage	99.80	99.80	99.80	99.80	99.90	99.80
Electricity	1.476	1.442	1.415	1.413	1.424	1.427
Power	77.40	70.60	65.00	65.30	66.00	66.60
Frequency	50.00	50.00	50.00	50.00	50.00	50.00
Consumption	0.1723	0.1729	0.1735	0.1740	0.1746	0.1751
Humid Temp	13.50	13.50	13.50	13.50	13.50	13.50
Power Factor	0.53	0.49	0.46	0.46	0.46	0.46

3.2. Training Details

We tried different kinds of RNN units in our encoder and decoder. We find GRU is easy and efficient to train comparing with LSTM. We use GRUs with two layers and each layer contains 256 cells. Although deep GRUs have been proved its significant performance [22], we want to balance the performance and the depth of our model to make sure it runs faster. We tried different number of layers and it turns out two-layer GRU structure is a good choice.

We reserve all the hidden states of the encoder instead of taking the hidden state of the last time-step to take full advantage of the input data. To make sure that the dimension is suitable for the input of the decoder, we add a single fully connected layer between the encoder and the decoder which is quite different from the traditional seq2seq model. It turns out this layer is quite useful and helps the model predict the output more precisely.

We have considered the fact that in different experiments, the kinds of data that the testers are interested in are different. So, for each kind of data, we trained a single model for it. In our test term, we let multiple models predict the data parallelly. This way has two advantages. Firstly, once the testers change the kinds of data that they are interested in, we just need to fine-tune our model to make it fit the requirement of the testers. Comparing with predicting all the kinds of data, it is cheaper in computation and more convenient to change the output. Secondly, according to our experiments, the model predicts a single kind of data more precisely than predicting multiple kinds of data. This may because the range of the different kinds of data is quite different, thus predicting all the data at the same time is trickier and more imprecise.

We use Adam optimizer and set 1e-6 as the learning rate. We trained our model for a total of 2000 epochs and our model converges.

4. Experiment Result

To evaluate the performance of our model more accurately and reasonably, we use two scores. Firstly, we use mean absolute error (MAE) to assess the deviation numerically. Secondly, we use mean absolute percent error (MAPE) to assess the quality of our prediction.

The specific calculation of MAE and MAPE can be seen in the following equations:

$$MAE(y,\hat{y}) = \frac{1}{n_{samples}} \sum_{i=0}^{n_{samples}-1} |y_i - \hat{y}_i| \tag{2}$$

$$MAPE(y,\hat{y}) = \frac{100}{n_{samples}} * \sum_{i=0}^{n_{samples}-1} |\frac{y_i - \hat{y}_i}{y_i}| \tag{3}$$

In these two equations, y is the original data, $\hat{y}$ is the output of our model. MAE and MAPE are two common indicators to evaluate the regression performance.

Because the numerical range of output may be highly different. Solely MAE or MAPE to assess our output is not reasonable. Therefore, we use both of them to evaluate the performance of our model.

The MAE and MAPE at certain epoch of the training process are presented in table 2 and table 3.

The changes of MAPE and MAPE during the training process are presented in figure 2 and figure 3.

Table 2. Validation mean absolute error loss

	Loss after 500 Epochs	Loss after 1000 Epochs	Loss after 1500 Epochs	Loss after 2000 Epochs
LR	1.185242	1.06315	1.049329	1.047863
MLP	0.011882	0.010381	0.009476	0.010658
RNN	0.006274	0.005227	0.004893	0.004010
Our Model	0.006756	0.003951	0.002512	0.002136

Table 3. Validation mean absolute percentage error loss

	Loss after 500 Epochs	Loss after 1000 Epochs	Loss after 1500 Epochs	Loss after 2000 Epochs
LR	96.6006%	100.7325%	105.3969%	104.8149%
MLP	0.8833%	0.8086%	0.8624%	0.7519%
RNN	0.4256%	0.3387%	0.3008%	0.2471%
Our Model	0.5276%	0.3032%	0.1993%	0.1501%

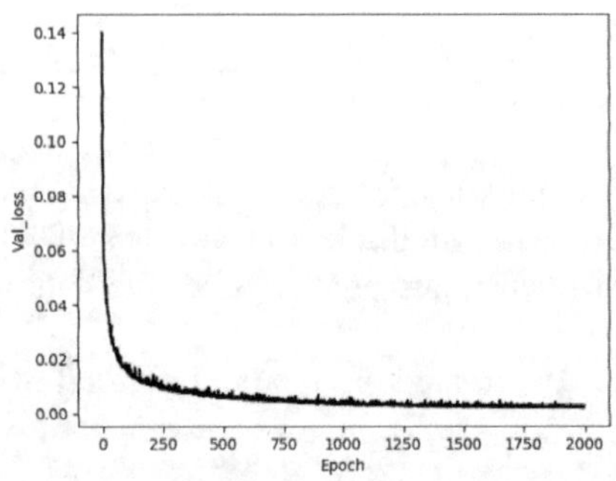

Figure 2. Validation mean absolute error loss during training process

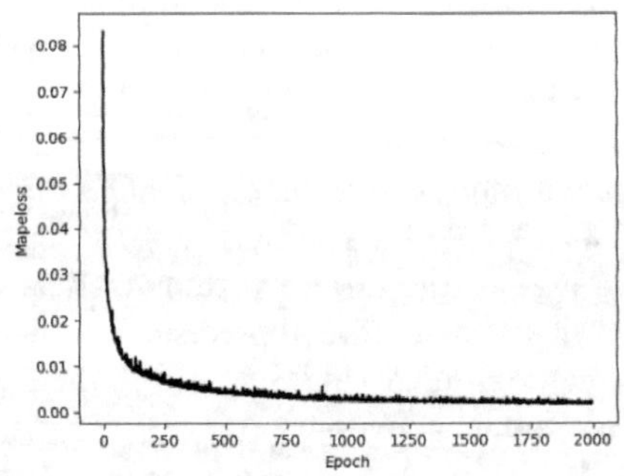

Figure 3. Validation mean absolute percentage error loss during training process

It turns out that based on our seq2seq model, after 2000 epochs, the MAPE is about 0.1% and the MAE is about 0.002, which is precise enough for the usage in the industry field. We can see that comparing with linger regression, multiple layer perception and simple RNN, our model has a much fewer mean absolute error and mean absolute percentage error in validation state. The result suggests that our model has a better performance on predicting the sequence data than the previous methods.

5. Conclusion

In this paper, we showed that sequence to sequence model can learn the mapping from a sequence of a variable-length sequence to another variable-length sequence and it has a pretty good performance for continuous time series data prediction tasks.

We evaluate our model based on refrigerator power consumption test. We use MAE to optimize our model and use both MAE and MAPE to evaluate the performance because of the range distinction of the output data. Our experiment result shows that our model is able to extract and utilize the information efficiently. By using a short length of data, our model can predict a quite long length of data and it is helpful for the testers to reduce the experiment time and increase efficiency.

Because generally the deep learning algorithm is greedy for data, we do believe this time series data prediction method is suitable for predicting the time series data that are influenced by a various of factors. I.e., the input data should have enough dimensions as features to make sure the model has a satisfied performance.

We are excited about the ability of the sequence to sequence model to precisely predict such a long sequence. We believe sequence to sequence model has an excellent generalization and it is suitable for a various kinds of time series prediction tasks.

Acknowledge

We would like to say thank you to all the colleagues in Ocean University of China that have helped us about our research. Chao Yang wants to thank his family for supporting his decision to dive in deep learning and for their support during his career.

References

[1] Kwok T Y and Yeung D Y 1997 Constructive algorithms for structure learning in feedforward neural networks for regression problems *IEEE transactions on neural networks* 8(3) pp 630-645
[2] Gardner M W and Dorling S R 1998 Artificial neural networks (the multilayer perceptron)—a review of applications in the atmospheric sciences *Atmospheric environment* 32(14-15) pp 2627-2636
[3] El Hihi S and Bengio Y 1996 Hierarchical recurrent neural networks for long-term dependencies *In Advances in neural information processing systems* pp 493-499
[4] Lipton Z C, Berkowitz J and Elkan C 2015 A critical review of recurrent neural networks for sequence learning *arXiv preprint arXiv* 1506.00019
[5] Jozefowicz R, Zaremba W and Sutskever I 2015 An empirical exploration of recurrent network architectures *In International Conference on Machine Learning* pp 2342-2350
[6] Hochreiter S 1998 The vanishing gradient problem during learning recurrent neural nets and problem solutions *International Journal of Uncertainty, Fuzziness and Knowledge-Based Systems* 6(02) pp 107-116
[7] Hochreiter S and Schmidhuber Jürgen 1997 Long short-term memory *Neural Computation* 9(8) pp 1735-1780
[8] Gers F A, Schmidhuber Jürgen and Cummins F 2000 Learning to forget: continual prediction with lstm *Neural Computation* 12(10) pp 2451-2471
[9] Cho K, Van Merrienboer B, Gulcehre C, Bahdanau, D, Bougares F and Schwenk H 2014 Learning phrase representations using rnn encoder-decoder for statistical machine translation *arXiv preprint arXiv* 1406.1078
[10] Hinton G E and Zemel R S 1994 Autoencoders, minimum description length and Helmholtz free energy *In Advances in neural information processing systems* pp 3-10
[11] Graves A 2012 Sequence transduction with recurrent neural networks. *Computer Science*

[12] Eriguchi A, Hashimoto K and Tsuruoka Y 2016 Tree-to-sequence attentional neural machine translation *arXiv preprint arXiv* 1603.06075
[13] Zhang J and Zong C 2015 Deep neural networks in machine translation: an overview *IEEE INTELLIGENT SYSTEMS* 30(5) pp 16-25
[14] Singh S P, Kumar A, Darbari H, Singh L, Rastogi A and Jain S 2017 Machine translation using deep learning: An overview. *International Conference on Computer, Communications and Electronics (Comptelix)* pp 162-167
[15] Chiu C C, Sainath T N, Wu Y, Prabhavalkar R, Nguyen P, Chen Z, and Jaitly N 2018 State-of-the-art speech recognition with sequence-to-sequence models *IEEE International Conference on Acoustics, Speech and Signal Processing* pp 4774-4778
[16] Graves A, Mohamed A R and Hinton G 2013 Speech recognition with deep recurrent neural networks *IEEE international conference on acoustics, speech and signal processing* pp 6645-6649
[17] Rana R 2016 Gated recurrent unit (GRU) for emotion classification from noisy speech *arXiv preprint arXiv* 1612.07778
[18] Venugopalan S, Rohrbach M, Donahue J, Mooney R, Darrell T and Saenko K 2015 Sequence to Sequence -- Video to Text *IEEE International Conference on Computer Vision* pp 4534- 4542
[19] Park S H, Kim B, Kang C M, Chung C C, and Choi J W 2018 Sequence-to-sequence prediction of vehicle trajectory via LSTM encoder-decoder architecture *IEEE Intelligent Vehicles Symposium (IV)* pp 1672-1678
[20] Wang C, Yang H, Bartz C and Meinel C 2016 Image captioning with deep bidirectional LSTMs *In Proceedings of the 24th ACM international conference on Multimedia* pp 988-997
[21] Mikolov T, Sutskever I, Chen K, Corrado G S and Dean J 2013 Distributed representations of words and phrases and their compositionality *In Advances in neural information processing systems* pp 3111-3119
[22] Pascanu R, Gulcehre C, Cho K and Bengio Y 2013 How to construct deep recurrent neural networks *arXiv preprint arXiv* 1312.6026

IOP Conf. Series: Materials Science and Engineering **692** (2019) 012048 doi:10.1088/1757-899X/692/1/012048

A Novel Reference Model-Based Neural Network Approach to Temperature Control System

Song Xu[1], Seiji Hashimoto[1*] and Wei Jiang[2]

[1] Division of Electronic and Informatics, Gunma University, 1-5-1 Tenjin Cho, Kiryu, Gunma, Japan.

[2] Department of Electrical Engineering, Yangzhou University, N.196 Huayang West Road, Yangzhou, Jiangsu, China.

[*] hashimotos@gunma-u.ac.jp

Abstract. With the continuous development of deep learning, neural network with its excellent self-learning performance obtains a series of a major breakthrough in target detection, image recognition and so on. In this paper, the temperature control system based on a neural network combined with I-PD compensation is proposed. To improve the neural network self-learning efficiency, the reference model is introduced for providing the teaching signal of neural network. The system simulations are carried out in MATLAB/SIMULINK environment to verify the control efficiency of the proposed reference model based neural network control system. The experiments are carried out on the DSP based temperature control system platform, the results are compared to the conventional I-PD control system to verify the control efficiency.

1. Introduction

In recent years, thermal processing systems that incorporate temperature control are needed in order to achieve high-performance processing. Among the various thermal processing techniques, the PID control has become the most commonly used controller due to its simplicity and applicability, even for multi-point temperature control systems. However, if the thermal processing system has a large delay time, the transient response and the stability could not be guaranteed respected [1]-[9].

Based on the condition that most of the PID controllers cannot learn and adjust its parameter online to suit the variation of the thermal process system, the intelligent control theory has gained more and more attention in process control systems. As the multi-layer neural network with deep learning technology. By introducing the neural networks control into the thermal process system which has the strong nonlinear characteristic, the adaptive system can improve the control performance to the un-modelled dynamics and the bounded disturbances [10]-[14].

In this paper, a neural network approach for a temperature control system with a reference model has been proposed. The control system is driven by using the error between the reference model output and the real system output as the teaching signal of the neural network. By adding the neural network controller into the IPD control system, the control input can be appropriately adjusted. The control performance has been quickly tuned while system response reaches the steady state. In order to evaluate the proposed reference model based neural network control system, the simulations have been carried out in MATLAB/SIMULINK environment and the experiments have been carried out on a DSP based temperature platform.

IOP Conf. Series: Materials Science and Engineering **692** (2019) 012048 doi:10.1088/1757-899X/692/1/012048

2. Configuration of neural network with reference model system

In this section, we describe the configuration of the proposed neural network control system with the reference model is described. Figure 1 shows the block diagram of the reference model based neural network control system.

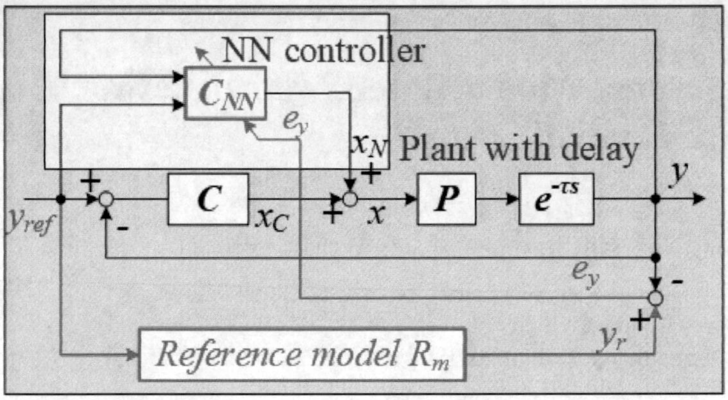

Figure 1. Block diagram of reference model based neural network control system

For simplicity, the controlled object is expressed as a first order plus time delay (FOPTD) model. In Figure 1, y_{ref} is the set reference value of the system, e_y is the teaching signal of the neural network controller, which is calculated by the error between the output y_r of reference model R_m and the output temperature y, C is the conventional PID controller, in this paper I-PD configuration is employed, while the control input x is the sum of the neural network control output x_N and IPD output x_C. Since the control object is the plant with time delay, the reference model R_m can be appropriately designed to provide the ideal temperature output with the same delay. The explanation of the control system can be divided into four main parts.

2.1. Control object with time delay

The control object is the thermal processing system which can be represented as the FOPTD system. With its large delay time, the transfer function of the plant can be expressed as equation (1), where K is the steady state gain of the output temperature against the input signal, T is the time constant of the plant, and the τ is the delay time of the response.

$$P(s) = \frac{K}{Ts+1} e^{-\tau s} \tag{1}$$

2.2. I-PD control

Consider that the neural network controller needs time to learn and training the parameters, and mainly acting on the late time, the conventional I-PD controller is designed for the control of the initial state. The parameter of the I-PD controller is designed based on the Ziegler-Nichols rule (step response method). These values are decided by τ, K and T shown in equation (1). The I-PD parameter K_p, T_i and T_d can be calculated as equations (2), (3) and (4), respectively.

$$K_p = 1.2\frac{T}{\tau} \tag{2}$$

$$T_i = 0.5T \tag{3}$$

$$T_d = 2\tau \tag{4}$$

2.3. Neural network controller

The multi-layer neural network (NN) controller has been introduced to the I-PD control system. In the proposed system, the NN controller has one input layer, two hidden layers and one output layer as shown in Figure 2, where the two hidden layers have 10 neurons.

In this system, the reference value of the system y_{ref} and the output temperature y are set as the input signal of the neural networks. x_N is the output value of the neural networks. The calculation process from the input Nin to the output N_{out} can be shown in Figure 3. Where the W is the weight of neurons, $\boldsymbol{\theta}$ is the offset value of every neuron, σ is the neuron activation function.

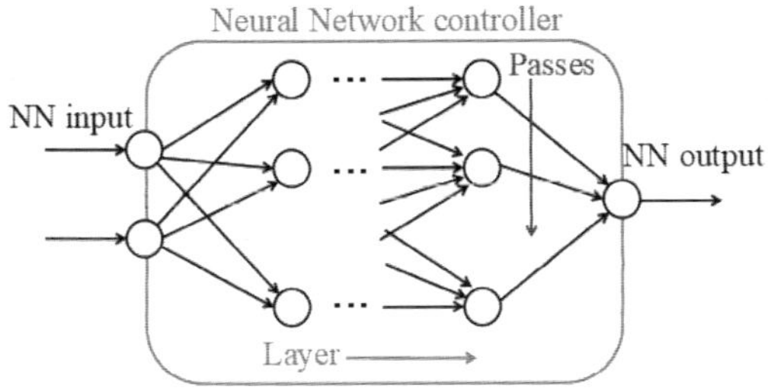

Figure 2. Structure of the multi-layer neural network controller

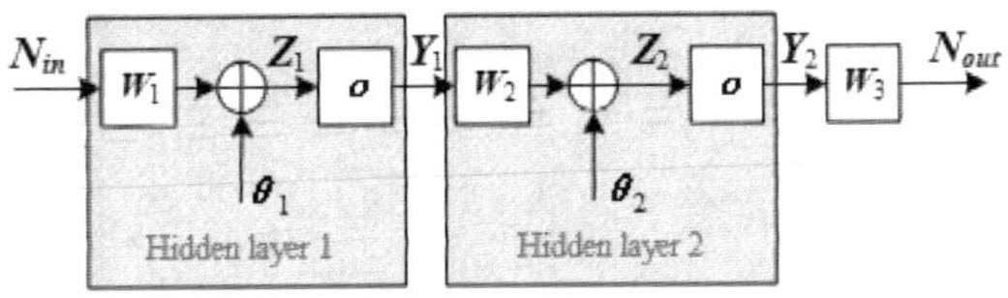

Figure 3. Calculation process of neural network controller

From the calculation process the output of the neural networks can be derived as:

$$NN_{out} = W_3 * \sigma(W_2 * \sigma(W_1 * NN_{in} + \theta_1) + \theta_2) \tag{5}$$

Where $N_{in} = [y_{ref}, y]^T$ is the input vector, W_1, W_2 and W_3 are the weight matrixes, and the $N_{out} = [x_N]$ is the neural network output. With regards to the self-learning, the weight matrix needs to be updated. The learning rule of the weight can be calculated as equation (6), where m is the neural network layer, α is the learning gain, x is the input of each weight represent as N_{in}, Y_1 and Y_2, and δ is the derivation of the weight as equation (7).

$$W(m)^{n+1} = W(m)^n - \alpha \times \delta(m)^n \times x(m)^n \tag{6}$$

$$\delta(m) = \nabla\sigma(z(m)) \times W(m+1) \times \delta(m) \tag{7}$$

In order to improve the self-learning ability of the neural network controller, the *ReLu* neuron activation function as equation (8) has been applied.

$$f(x) = \begin{cases} x & x > 0 \\ 0 & x \le 0 \end{cases} \tag{8}$$

$$f'(x) = \begin{cases} 1 & x > 0 \\ 0 & x \le 0 \end{cases} \tag{9}$$

2.4. Reference model-based NN control system

The Pade approximation method has been introduced to approximate the plant to equation (10), which is easier for the simulation to modify the reference model. The reference model setting is based on the approximated plant model, which is to improve the transient response speed by adding a gain R to the plant time constant the approximated model is shown as equation (11). Approximation of dead time to 1st order transfer function makes the controller realization easy, i.e. the memory to store the output can be saved.

$$P(s) \approx \frac{K}{Ts+1} \bullet \frac{1}{\tau s+1} \tag{10}$$

$$R(s) \approx \frac{1}{T*R*s+1} \bullet \frac{1}{(\tau s+1)^2} \tag{11}$$

3. System simulation

In order to verify the proposed reference model-based NN control method, the simulation has been carried out in MATLAB/SIMULINK environment. In the simulation, the control object transfer is expressed as a FOPTD system as equation (12). Its approximated transfer function becomes equation (13), hence the reference model can be represented as equation (14), wherever in this proposal, R is set as 0.01.

$$P(s) = \frac{2.36}{2626s+1} e^{-524s} \tag{12}$$

$$P(s) \approx \frac{2.36}{2626s+1} \bullet \frac{1}{(524s+1)^2} \tag{13}$$

$$P(s) \approx \frac{1}{26.26s+1} \bullet \frac{1}{(524s+1)^2} \tag{14}$$

The I-PD parameter was decided by the plant parameter as described before. The parameters were K_p = 10.3, T_i = 838, T_d = 209. Moreover, the hyper parameters of the neural network are decided as $\alpha = 1 \times 10^{-9}$, $\beta = 2 \times 10^{-5}$. The initial values of the weight can be set as an optimal random value.

The simulation has been carried out with the following condition. The control performance was evaluated both in a positive direction and a negative direction. The reference value of the temperature was set as a repetitive step signal with an amplitude of 5 deg C, the offset of the reference is 100 deg C. The time response is shown in Figure 4. The control input with the summation of the neural network output and the I-PD output is shown in Figure 5.

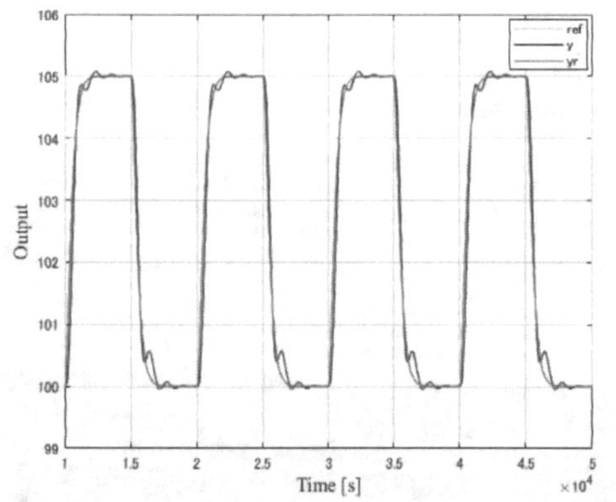

Figure 4. Output of NN controller and I-PD controller

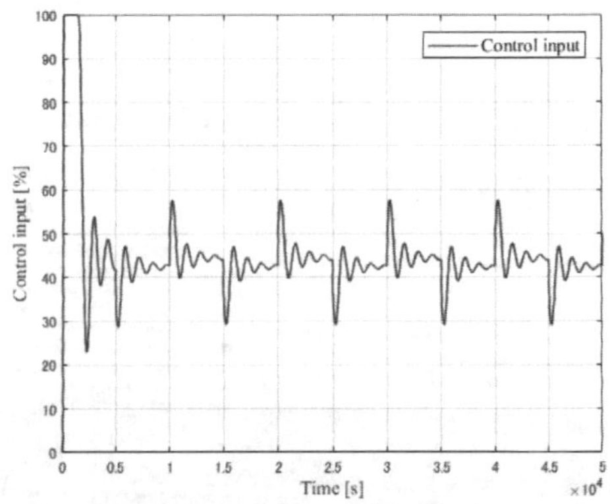

Figure 5. Control input of the system

Also, each positive and negative direction's response need to be analyzed. Focusing on the reference from 100 deg C to 105 deg C as the positive direction and 105 deg C to 100 deg C as the negative direction, each step response is plotted on the same figure as in Figures. 6 and 7, respectively.

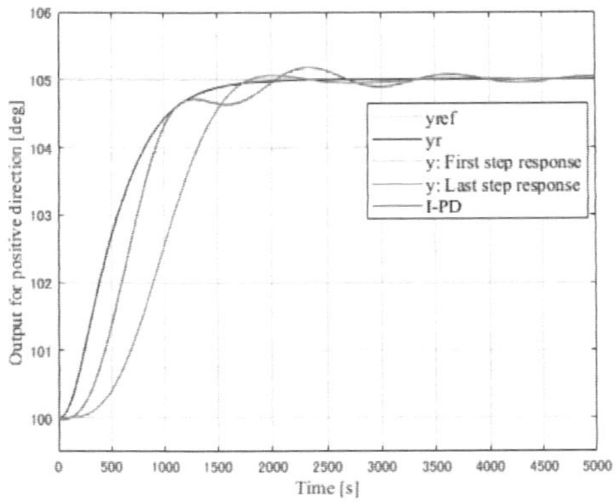

Figure 6. Positive direction control simulation result

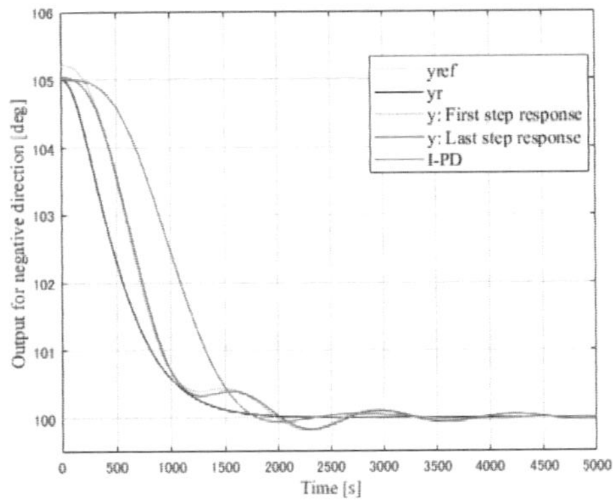

Figure 7. Negative direction control simulation result

From the simulation results, no matter the positive direction control or the negative direction control, the system response with the neural network control in regards to rising time and settling time is improved from the conventional I-PD control. Moreover, since the first step response is almost the same as the last step response, the learning process has been done at the beginning of the first step. The system response follows to the reference model, although there is a little deviation. As a result, the control efficiency has been improved by comparing to the conventional I-PD control.

4. Experimental results

Experiments with the proposed reference model-based neural network control method were carried out based on the platform that equips DSP as the temperature controller as is shown in Figure 8. The system has four coupling channels, each channel has two independent heaters and one temperature sensor. The temperature sensor can transfer temperature 0-400 deg C to 0-10 VDC output voltages. In addition, the heaters are driven by PWM signals. The temperature can be controlled through controlling the duty ratio of the PWM signals.

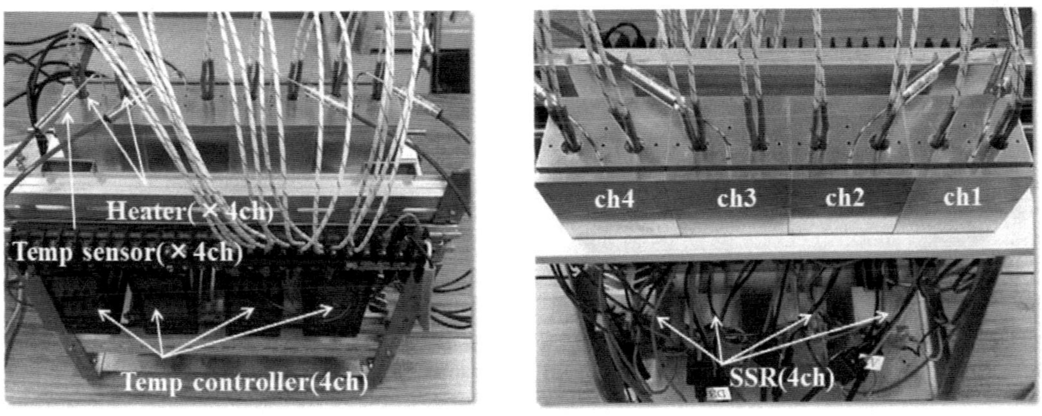

(a) Front view (b) Rear view

Figure 8. Experimental setup

In this proposal, the channel Ch1 is defined as the input channel (heating channel) and Ch4 is defined as the output channel. The step response method was introduced for the system identification. The plant can be identified as equation (15). Based on the identified plant model, the reference model can be obtained as equation (16).

$$P(s) = \frac{3.052}{2161s+1}e^{-615s} \tag{15}$$

$$R_m = \frac{1}{21.61s+1} \bullet \frac{1}{615s+1} \tag{16}$$

The parameters of the I-PD controller are obtained by Ziegler-Nichols method, thus, the parameters K_p =1.38, T_i =1231 and T_d =308. The initial layer offset $\theta_1 = \theta_2 = 0$, the learning gain $\alpha = 1.2\times10^{-12}$, while the training gain $\beta = 2\times10^{-10}$. And the initial weight of the NN controller has been optimized.

In the experiments, to verify the control efficiency of the proposed method, the results were compared to conventional I-PD control system both with the feed-forward gain k_f=0 (slow) and k_f=1 (fast). The experiments were carried out by controlling the temperature of Ch1 from 100 deg C to 105 deg C. The results of output temperature are shown in Figure 9, while the control input is shown in Figure 10.

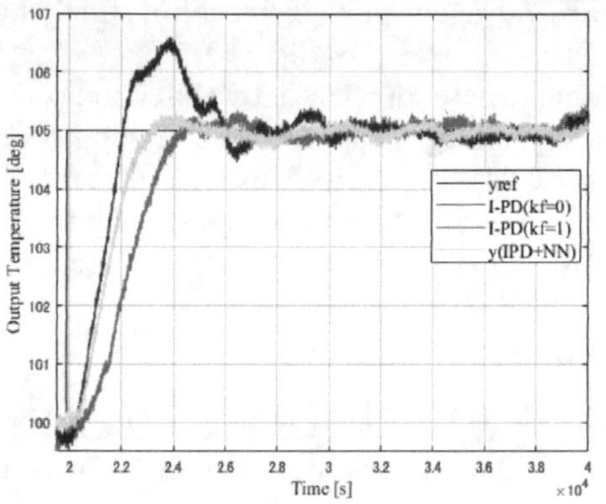

Figure 9. Experiment results of transient response

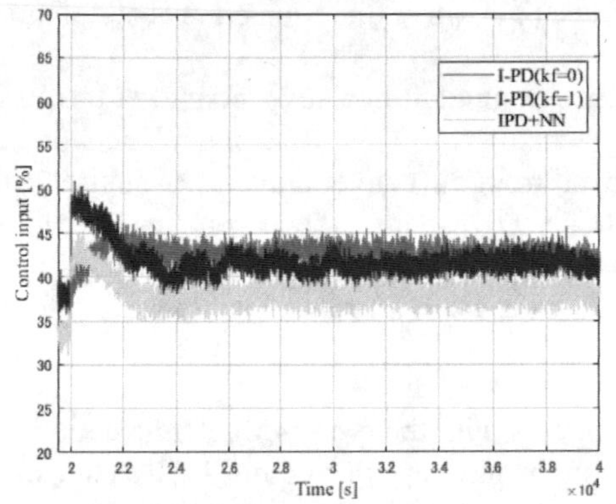

Figure 10. Control input of IPD+NN system and conventional IPD system

From Figure 9, the conventional I-PD controller with feed forward gain k_f=0 has the longest response while no overshoot, and I-PD controller with feed-forward k_f=1 has the fastest transient response, however, have an overshoot as 1.5 deg C. By introducing the NN controller, the transient response has been improved, the system has the transient response as fast as conventional I-PD with feed-forward gain k_f=1, and has no overshoot. Thus, the system response has been improved, and the efficiency of the proposed control method has been successfully evaluated.

5. Conclusion

In this paper, a reference model-based neural network controller for temperature control system has been proposed. The control system is driven by using the error signal between system output and reference model output as the teaching signal of the neural networks. The NN system configuration combined with I-PD control has been described. And then, the proposed design method has been applied to a first order plus time delay plant. Simulation has been carried out in MATLAB, and the results have been quantitatively compared to the convention I-PD control method with the summation of squared error. The experiments have been carried out on a DSP based temperature system platform, the results also have been compared to the conventional I-PD control system to evaluated the improvement taken in the transient response. The control efficiency of the proposed method has been successfully evaluated.

References

[1] N. Suda, PID control, *Asakura Publishing*, 1992.
[2] M. Oshima, Process Control Systems, *Corona Publishing*, 2003.
[3] G. C. Goodwin, S. F. Fraebe and M. E. Salgado, Control System Design, *Prentice Hall*, 2001.
[4] M. Morari and E. Zafiriou, Robust Process Control, *Prentice Hall,* 1992.

[5] J. M. Maciejowski, Multivariable Feedback Design, *Addison Wesley*, 1989.
[6] N. Abe, A. Kojima, Control in Time-delay and Distributed Parameter Systems, *Corona Publishing*, 2007.
[7] Nanno, M. Tanaka, N. Matsunaga and S. Kawaji, "On Performance of the Gradient Temperature Control Method for Uniform Heating", *IEEJ Trans. EIS,* **124, 8,** pp. 1606-1612, 2004.
[8] N. Matsunaga, S. Kawaji, M. Tanaka and I. Nanno, "A Novel Approach of Thermal Process Control for Uniform Temperature", *16th IFAC World Congress*, **38, 1**, pp. 111-116, 2005.
[9] T. Tomaru and Y. Mori, "Design Method of Minimum-Phase State Decoupling Control with Feedforward Compensation", *SICE Annual Conference* 2007, **1C07-1,** 2007.
[10] D. Ma, T. Zhou, S. Qi, M.A. Shahzad, Z. Xiao, "Supercritical water heat transfer coefficient prediction analysis based on BP neural network", *Nucl. Eng.Des.* **320**, 400-408, 2017.
[11] M. S. A. Oliveira, A. C. M. Sousa, "Neural network analysis of experimental data for air/water spray cooling", *J. Mater. Process. Technol.* **113**, 439-445, 2007.
[12] B. X. Wang, D. H. Zhang, J. Wang, M. Yu, N. Zhou, G. M. Cao, "Application of neural network to prediction of plate finish cooling temperature", *J. Central South Univ. Technol.* 15, 136-140, 2008.
[13] L. Sudha, R. Dillibabu, S. S. Srinivas, A. Annamalai, "Optimization of process parameters in feed manufacturing using artificial neural network", *Comput. Electron. Agric.* **120**, 1-6, 2016.
[14] T. Wen, Z.-Y. Liu, W. Di, G.-D. Wang, Artificial neural network modeling of microstructure during C-Mn and HSLA plate rolling, *J. Iron. Steel Res. Int.* **16**, 80-83, 2009.

IOP Conf. Series: Materials Science and Engineering **692** (2019) 012049 doi:10.1088/1757-899X/692/1/012049

Hybrid Non-Local Network for Abstractive Summarization

Zepeng Hao[1, *], Yupu Guo[1], Cheng Gong[1] and Honghui Chen[1]

[1] Science and Technology on Information Systems Engineering Laboratory, National University of Defense Technology, Changsha, China, 410073.

[*]hao3184045@gmail.com

Abstract. In the field of abstractive automatic summarization, the Seq2Seq model constructed by the RNN network is widely used and has achieved perfect results. However, due to the Seq2Seq model and the internal structure of the RNN network, only short-term and local information can be referenced in the process of generating summary, resulting the generated summary contains less information, and may even happen that the generated summary does not match the corresponding document. There have been many previous attempts to improve the model by using the attention mechanism, but only to use long-term information within a certain range. In our paper, we present a abstractive summarization model that enhances the global feature, which called Hybrid Non-Local Network model for single document summarization task, consisting of a non-local block and a traditional Seq2Seq module. Our model uses the non-local network block to transform the Seq2Seq model to better capture global feature. For example, the relationship between words and words, sentences and sentences in a large span of the original document, making the generated summary of higher quality, more informative, more accurate and rich representation of the original document. The experiments in our paper are performed on the CNN/DailyMail corpus. The proposed model was improved by 1.37, 0.47, and 1.16 on R-1 score, R-2 score, and R-L score, separately, compared to the baseline model.

1. Introduction

With the rapid growth in the amount of data, the demand for the retrieval, understanding and analysis of a large amount of information is constantly increasing, which continues to promote the development of the automatic text summarization field, and the automatic text summarization model is constantly improving, in order to generate more informative, smoother and more similar to human-generated summaries. The automatic text summarization methods can be separated into two categories: the extractive summarization method, and the abstractive summarization method. The extractive summarization method is to judge, score and sort the importance of words and sentences in the source document, and then extract words or sentences from the source document according to the order of importance to form the summary. The absrtactive summarization method is to encode and understand the original document through the model, and then generate words and phrases by decoding to form the summary.

The extractive summarization method[1] is very mature, and the effect obtained on each dataset is remarkable. Most summarization systems also use the extractive summarization method, but with the continuous development of the deep learning network structure and the improvement of computing power, the abstractive summarization method slowly showing the advantages: smoother and closer to human-generated summaries. At present, most of the abstractive summarization models are built on the Seq2Seq[2][3] structure with attention mechanism. The attention mechanism[4][5] can help the model to collect and utilize global information better, but the scope of the attention mechanism is relatively limited. The attention mechanism cannot capture the relationship between words or sentences that are far apart in the source document. This problem will result that the generated summary contains less information, and may even happen that the generated summary does not match the corresponding document. In the recent work in the field of imagery, the Facebook team proposed a non-local network block[6] to capture long-range dependencies in images, and non-local network block can calculate the response at a position as a weighted sum of the features at all positions. This block captures more global features and achieves significant results on image recognition and image restoration tasks[7].

Inspired by the Facebook team, we try to migrate the application of the non-local network block to the automatic text summarization field, and proposed our model in this paper: Hybrid Non-Local Network model. Our model is mainly composed of two conponents, namely the non-local network block and the Seq2Seq module. We mainly improve the decoder in the Seq2Seq module through the non-local network bolck. In the non-local network block, by analyzing and mining the information generated by the encoder, a richer global information is obtained, so that richer reference information is given when the decoder generates the summary. In the Seq2Seq module, encoding is performed according to the feature representation of the document, and then the summary is generated according to the encoded information and the information provided by the non-local network block. This model ensures that the generated summary is more informative, smoother, and closer to human-generated summaries.

We test our model on the CNN/DailyMail dataset[8]. Through experiments, we found that our model performance is 1.37, 0.47 and 1.16 higher than the state-of-the-art baseline at R-1 score, R-2 score, and R-L score, separately.

In our paper, we summarize our significant contributions follow:

- We present a hybrid model consisting of the non-local network block and the Seq2Seq module called Hybrid Non-Local Network model.
- We migrate non-local network block from the image domain to the text domain. The non-local network block can better capture the global features, especially the relationships between words or sentences that span a large range in the source document, so that the information obtained by the model more abundant, the generated summary more informative and smoother.
- We test our model under conditions of different fixed-length summaries and find that our model is better at generating longer summaries.

2. Approach

In this paper, the model we proposed consists of two components: non-local network block and Seq2Seq module. The structure of the whole model is shown in the figure 1. Next we discuss the details of the non-local network block and the Seq2Seq module.

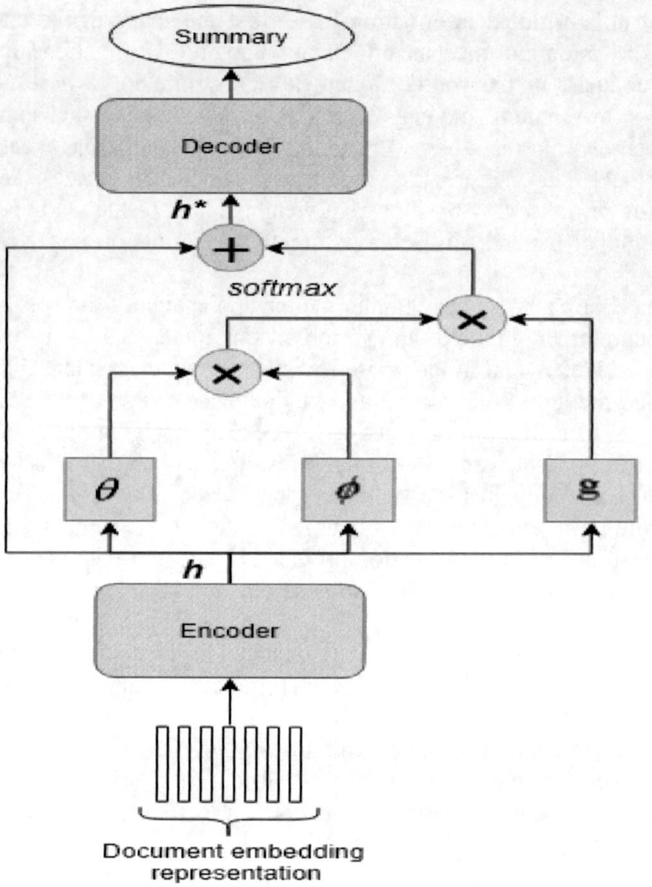

Figure 1. The framework of our model.

2.1. Non-Local Network block
We migrate the non-local network block from the image domain to the text domain. Building the non-local network block requires three steps. First, define the non-local network according to the non-local operation, then select the specific instantiations, and finally define and construct the non-local network block of the text domain.

2.1.1. Non-Local Network. According to the description of non-local operations, we define the expression of the non-local network as follows:

$$y_i = \frac{1}{C(x)} \sum_{\forall j} f(x_i, x_j) g(x_j) \tag{1}$$

In the above formula, i is the index of the computed output position, j is the index of all positions that are likely to be related with i, x is the input information (here is the text representation), y is the same output information as the x dimension, f is a scalar function that computes the relationship between i and all j, the g function calculates the representation of the input information at j position, and $C(x)$ is the function used for normalization. In the traditional Seq2Seq model, only the adjacent relationship with the corresponding position is considered. The addition of the attention mechanism can make the model consider a large range of relationships, but there is no way to fully consider all positions. It can be seen from the definition of non-local network that the non-local network considers the relationship between all locations and corresponding location, which enhances the capture of global features by the model and enriches the global information.

2.1.2. Instantiations. In order to make non-local network work, we introduce several versions of f and g functions. Based on previous research experience, we set the g function to consist of a linear embedding function. The g function is represented by the following formula:

$$g(x_j) = W_g x_j \quad (2)$$

Where W_g is a learnable parameter matrix. Next we describe several alternative f functions.

Dot product. The f function can be defined as a dot product to calculate the similarity at different positions:

$$f(x_i, x_j) = \theta(x_i)^T \phi(x_j) \quad (3)$$

Here we use the embedding mode, the corresponding regularization factor $C(x) = N$, where N is the number of positions of the input information x.

Gaussian. Here the f function is defined as a Gaussian function to compute the similarity:

$$f(x_i, x_j) = e^{x_i^T x_j} \quad (4)$$

Which is equivalent to the improvement of the dot product function, and the calculation of the Euclidean distance is applied. The normalization function corresponding to the f function should be set as $C(x) = \sum_{\forall j} f(x_i, x_j)$.

Embedded Gaussian. Embedded Gaussian function is an extension of the Gaussian function. Here we design the f function as follows:

$$f(x_i, x_j) = e^{\theta(x_i)^T \phi(x_j)} \quad (5)$$

In the above formula, $\theta(x_i) = W_\theta x_i$ and $\phi(x_j) = W_\phi x_j$ are both composed of a learnable embedded matrix. The regularization factor corresponding to the f function is $C(x) = \sum_{\forall j} f(x_i, x_j)$.

Since we move the non-local network to the text domain, the input of the non-local network may be embedded text representations, so we choose the embedded Gaussian function and its corresponding nomalization factor to be applied to the non-local network.The non-local network can be flexibly constructed into block, adding and improving existing model structures to enrich the global features captured by the model.

2.1.3. Non-Local block. In order to make the non-local network better fit into the existing model, we design the non-local network block as follows:

$$z_i = W_z y_i + x_i \quad (6)$$

Where y_i is the output of the non-local network, and $+x_i$ allows the non-local network block to be flexibly added to the pre-trained model.

2.2. Seq2Seq module

Next we introduce the Seq2Seq model that combines non-local network block. We use non-local network block to improve the decoder and capture more global features, making the information referenced when generating the summary more abundant. Next, I will divide the model into two components (encoder and decoder) to discuss.

2.2.1. Encoder. Firstly, we use the Word2vec method to get the embedding feature representation of each word in the original document. And the embedding representation of each token as input through the Bi-LSTM (bi-directional LSTM) structure. This structure can concatenate the hidden states from the previous and next word at each time step, capturing semantic relationships from both sides of each word, resulting in a series of hidden state h_i . The number of neurons in the LSTM structure in each direction is half the number of neurons in the output representation, then in the same number of neurons, the both sides states can be connected.

2.2.2. Decoder. In this part, we improve the decoder by adding non-local network block to capture more global features, solve the problem that the attention mechanism obtains the limitation of the relationship, and make the generated summary information more abundant.

First, we use the sequence of hidden states h_i obtained by the encoder as the input of the non-local network block, and capture more potential global features through the non-local network block, so that the content vector information generated in the next step is more abundant.

$$h_i^* = W_z y_i + h_i \tag{7}$$

$$y_i = \frac{1}{\sum_{\forall j} e^{(W_\theta h_i)^{\mathrm{T}} W_\phi h_j}} \sum_{\forall j} e^{(W_\theta h_i)^{\mathrm{T}} W_\phi h_j} W_g h_j \tag{8}$$

Next, we input the embedding representation of the previous word into the LSTM to get the hidden state of the decoder s_t, and then calculate the attention distribution a^t according to the output of the non-local network block h_i^* and the hidden state of the decoder s_t :

$$e_i^t = v^T \tanh\left(W_h h_i^* + W_s s_t + b_a\right) \tag{9}$$

$$a^t = soft\max\left(e^t\right) \tag{10}$$

In the above formula, v, W_h, W_s and b_a are learnable parameters. The attention distribution depends on the relationship between the hidden state of the decoder and the hidden state of the encoder with enhanced global features. Next, we can compute the content vector c_t by the following formula:

$$c_t = \sum_i a_i^t h_i^* \tag{11}$$

We incorporate the content vector c_t with the hidden state of the decoder s_t and then calculate the probability distribution of the vocabulary P_v through a two-layer linear network:

$$P_v = soft\max\left(W'\left(W[s_t, c_t] + b\right) + b'\right) \tag{12}$$

where W, W', b and b' are learnable parameters. Simultaneously, we add the pointer-generator network[9][10] to the decoder to solve the out-of-vocabulary (OOV) problem. Then the probability distribution of the final predictive word is as follows:

$$P(w) = p_g P_v(w) + (1 - p_g)\sum_{i:w_i = w} a_i^t \tag{13}$$

p_g is an influence factor, it decides whether to sample a word from the probability distribution or copy a word from the original document. p_g is calculated by the content vector c_t and the hidden state of the decoder s_t :

$$p_g = \sigma\left(w_c^T c_t + w_s^T s_t + b_p\right) \tag{14}$$

Where w_c, w_s and b_p are parameters obtained through training and σ is the sigmoid function. Next, we use the sum of the negative log likelihood of the target word w_t^* in each time step t in the whole sequence as the TotalLoss. In the training process, we minimize the TotalLoss by updating the model parameters λ :

$$\min_{\lambda} \sum_{t=0}^{T} -\log P\left(w_t^*\right) \tag{15}$$

At the same time, we added the beam-search mechanism to our model to solve the optimization problems in model training and testing, so that we can better train and test our model. Finally, we can get the generated summary through the model.

3. Experiments

In this section, we describe the dataset and evaluation metrics for the automatic text abstractive summarization task. Also introduced the baseline models compared to our model, research questions as well as the details of the parameters used in the experiment.

3.1. Datasets and Evaluation metrics

Here, we apply the CNN/DailyMail corpus that is commonly used in the domain. The dataset has 286,817 training data pairs, 13,368 validation data pairs and 11,487 test data pairs. At the same time, the source document in the training set have an average of 766 words, with an average of 29.74 sentences, while the summary has an average of 53 words with an average of 3.72 sentences. The uniqueness of this training set is that there are many long documents, and the summaries are mostly composed of multiple sentences in an order, which brings certain challenges to the abstractive summarization model. We use the ROUGE metrics to evaluate the summary generated by the model: ROUGE-1 and ROUGE-2 mainly reflect the amount of information, and ROUGE-L mainly reflects fluency.

3.2. Baseline Models and Research questions

We select two abstractive summarization models to be compared to our hybrid non-local network model, which are ABS and PGC, respectively. ABS is a traditional Seq2Seq model with attention mechanism. PGC is a Seq2Seq model with pointer-generator coverage network. In the experiments we mainly discuss two research questions: (**RQ1**) Can our model perform better than baselines on dataset? (**RQ2**) When the generated summary length is fixed to 75 bytes, 175 bytes, and full length, what is the quality of summaries generated by models?

3.3. Parameters

In the experiment we apply the 128-dimensional word2vec embedded representation trained on the dataset as input, the hidden states of our model is set to 256 dimensions, for the pointer generation network we apply a 50k-word vocabulary.

4. Results and Analysis

4.1. Model performance

In order to answer **RQ1**, we test and evaluate the summaries generated by models through the ROUGE metrics. The evaluation results are shown in Table 1. It can be seen from the results that the PGC model is significantly higher than the ABS model in the three metrics, indicating that the summary generated by the PGC model is better than the summary generated by the ABS model in terms of information volume and fluency.

Our model is obviously higher than the PGC model on R-1 score and R-L score, and slightly higher than the PGC model on the R-2 score, indicating that the summary generated by our model smoother and more informative. Experiments show that the non-local network block can enhance the model's capture of global features and enable the model to generate a more informative summary.

Table 1. The score on the evaluation metrics when the generated summary length is fixed to the full length. (%)

Method	R-1 Score	R-2 Score	R-L Score
ABS Model	36.41	15.43	32.79
PGC Model	38.89	16.90	35.32
Hybrid Model	**40.26**	**17.37**	**36.48**

4.2. Different length summaries

To answer **RQ2**, we use the model to generate summaries of full length, 75 bytes, and 175 bytes. These summaries are evaluated by the ROUGE metrics, and the results of the evaluation are shown in Tables 1, 2, and 3, separately. As can be seen from the evaluation results, the performance of our model is similar to the PGC model when generating the 75-byte length summary. However, when generating the summary of 175 bytes in length, the metrics of R-1 score, R-2 score and R-L score are significantly higher than the other two models. This shows that our Hybrid Non-Local Network model can capture more global features, and can make the summary more informative when generating the longer summary. At the same time, we can see from Figure 2 that for our model itself, the model performs better when generating the longer summary, indicating that our model is more suitable for generating long summaries.

Table 2. The score on the evaluation metrics when the generated summary length is fixed to the 75 bytes. (%)

Method	R-1 Score	R-2 Score	R-L Score
ABS Model	13,74	6.20	12.88
PGC Model	21.40	9.66	20.44
Hybrid Model	**21.60**	**9.80**	**20.46**

Table 3. The score on the evaluation metrics when the generated summary length is fixed to the 175 bytes. (%)

Method	R-1 Score	R-2 Score	R-L Score
ABS Model	26.77	11.68	24.52
PGC Model	32.58	14.16	29.97
Hybrid Model	**32.95**	**14.41**	**30.28**

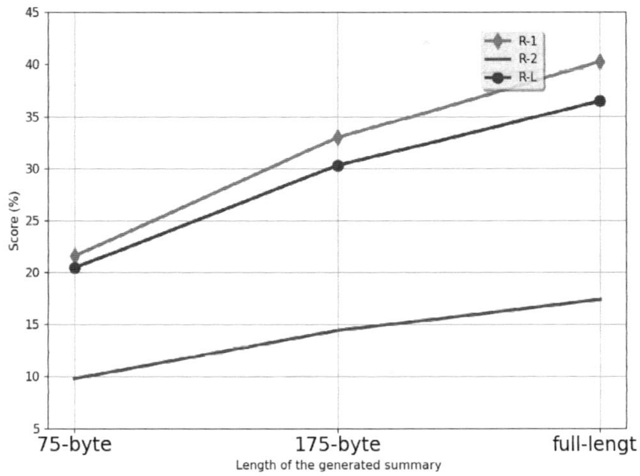

Figure 2. The score of our model on the evaluation metrics when the generated summary length changes.

5. Conclusions

In our paper, we present the Hybrid Non-Local Network model for single-document absrtactive summarization. The model improves the traditional Seq2Seq model through non-local network block, so that the hybrid model can capture more global features and generate higher-quality, more informative summaries. The experimental results demonstrate that our model performs better on the ROUGE evaluation metircs than the baseline models. Our model is more suitable for generating long summaries.

References

[1] Moratanch N, Chitrakala S. A survey on extractive text summarization[C]// International Conference on Computer. 2017.
[2] Nallapati R, Zhou B, Gulcehre C, et al. Abstractive text summarization using sequence-to-sequence rnns and beyond[J]. arXiv preprint arXiv:1602.06023, 2016.
[3] Kim S, Kang S G. TL; DR: Improving Abstractive Summarization Using LSTMs[J].
[4] Rush, A.M., Chopra, S., and Weston, J. A neural attention model for abstractive sentence summarization. In EMNLP, 2015.
[5] Vaswani A, Shazeer N, Parmar N, et al. Attention is all you need[C]//Advances in Neural Information Processing Systems. 2017: 5998-6008.
[6] Wang X, Girshick R, Gupta A, et al. Non-local Neural Networks[J]. 2017.
[7] Liu D, Wen B, Fan Y, et al. Non-Local Recurrent Network for Image Restoration[J]. 2018.
[8] Hermann K M, Kocisky T, Grefenstette E, et al. Teaching machines to read and comprehend[C]//Advances in Neural Information Processing Systems. 2015: 1693-1701.
[9] See A, Liu P J, Manning C D. Get to the point: Summarization with pointer-generator networks[J]. arXiv preprint arXiv:1704.04368, 2017.
[10] Jiang X, Hu P, Hou L, et al. Improving Pointer-Generator Network with Keywords Information for Chinese Abstractive Summarization[J]. 2018.

IOP Conf. Series: Materials Science and Engineering **692** (2019) 012050 doi:10.1088/1757-899X/692/1/012050

An End-to-End Named Entity Recognition Model for Chinese

Cheng Gong[1,*], Jiuyang Tang[1] and Zhen Li[1]

[1] Science and Technology on Information Systems Engineering Laboratory, National University of Defence Technology, Changsha, China, 410072.

[*] gcclgl95@126.com

Abstract. Named entity recognition is an important basic task in natural language processing. This paper proposes a named entity recognition method for end-to-end and efficient deep-loop neural networks. Using BERT to train sub-vectors as raw input enables the model to obtain more comprehensive text information, and at the same time, the BiLSTM network is focused on the attention mechanism, so that the network pays more attention to the key information in the text and ignores the redundant information to improve the recognition efficiency of the model. Finally, the relationship between any two tags is captured by the CRF layer, and the entire sentence is decoded and predicted. Experiments show that the method performs well on MSRA corpus.

1. Introduction

With the rapid update of the Internet, the traditional search method cannot meet the user's demand for information and knowledge. The main reason behind this is that the current search engine and other systems can only provide users with fragmented knowledge, such as a separate URL, a piece of text, a picture, or a video, they do not provide the connection between different knowledge, it is difficult to build a complete knowledge system for users. To this end, a knowledge map led by Google knowledge graph came into being.

In the process of building knowledge maps, in addition to structured knowledge, a large amount of knowledge is hidden in unstructured text. Named Entity Recognition (NER), refers to the process of automatically identifying the naming and referencing items of an entity from unstructured text and indicating its category, which is an urgent problem to be solved. A named entity is a research subject identified by a named entity. It generally consists of three major categories (entity class, time class, and numeric class) and seven subclasses (person name, place name, institution name, time, percentage, currency, and data). Judging whether a named entity is correctly identified includes two aspects: whether the boundary of the entity is correctly classified; whether the Entity type is correctly labeled.

In order to reduce the dependence on linguistic knowledge and construct complex feature engineering, the research of named entity recognition has been extended from traditional statistical methods to deep learning methods. By constructing neural networks to learn the potential relationships within the text to extract relevant entity references. Different from English named entity recognition, Chinese named entity recognition has the following nodus:

- Compared with English entities, Chinese entities lack obvious boundaries and inherent definite articles, and also there is no obvious morphological spelling change in Chinese.
- The rapid development of the Internet has led to the emergence of many emerging entities, making identification more difficult.
- Chinese entities do not have a fixed format, it will appear a lot of noise when identifying.

Therefore, how to extract more effective features in Chinese text in a complex language environment has important research value. In this regard, this paper proposes to add a self-attention mechanism based on the existing BiLSTM+CRF sequence labeling model, which can reduce the dependence on external information and capture the internal correlation of data or features, at the same time, this paper also uses BERT [1] to train the word vector as the original input, which has a certain improvement on the recognition result.

2. Related Work

In the early NER research, it was a mainstream method to manually construct finite rules and then search for strings that match these rules from the text. The most representative of these is the DL-CoTrain method proposed by Collins et al. [2], which pre-defines the seed rule set Decision List, and then performs unsupervised training iterations on the set according to the corpus to get more rules and the final rule set, the method have a classification accuracy of more than 91% for three categories of named entities (person, place, and institution names). Although the rule-based method can obtain a better recognition effect on a specific corpus, it requires a large number of rules, and it is too feasible to manually formulate these rules.

With the rise of machine learning in the NLP field, NER research has gradually turned to machine learning. Classic machine learning classification models such as HMM [3]-[4], ME [5], CRF [6], and SVM [7] have been successfully used to serialize annotations of named entities, and have achieved good results. In recent years, people's attention has shifted from the field of machine learning to the field of deep learning based on neural network models. Lample et al. [8] proposed LSTM and two neural network models based on transformation. At the same time, the features were obtained from the labeled corpus and the unlabeled corpus, and the best NER effects were obtained in all four languages. At present, the study of entity extraction based on neural network structure mainly focuses on using the attention mechanism to improve the model effect and model construction for a small number of labeled training data. Based on the RNN+CRF model structure, Rei et al. [9] use the attention mechanism to improve the splicing process of word vector and character vector, so that the model can dynamically use word vector and character vector information. In this paper, we add the attention mechanism to the LSTM output hidden layer.

3. Model

In this paper, the named entity problem is solved as a sequence labeling problem. The model is mainly divided into the following layers: word embedding layer, BiLSTM layer, Multi-headattention layer, CRF layer.

3.1. The word embedding layer

Compared with English, Chinese text lacks a separator like a space character to indicate the boundary identifier of the word boundary. Therefore, Chinese named entity recognition tasks are often accompanied by Chinese word segmentation, but incorrect word segmentation will lead to errors in subsequent named entity recognition tasks. This will have an important impact on the efficiency of Chinese named entity recognition, and the use of words as the basic unit will lose the displayed vocabulary information, which will also have an impact on the efficiency of the task.

Therefore, this paper uses the BERT [1] model to train Chinese text, generate dynamic word representation to improve the efficiency of entity recognition, the BERT model is released by google, it follows the structure of GPT model, using Transfomer encoder as the main model structure, it can continually overlaps the attentional mechanism layer and the normal nonlinear layer to obtain the final textual representation after we input the text.

This paper uses the BERT-BaseChinese model provided by Google to get the dynamic word vector, and input it into the neural network as an embedded layer.

3.2. BiLSTM layer

In recent years, LSTM (Long Short-Term Memory) network has been widely used in Long sequence dependent learning. The traditional RNN network has the problem of gradient disappearance, and it is difficult to process long sequence data, but LSTM network can solve this problem well. The main cell structure is shown as figure 1:

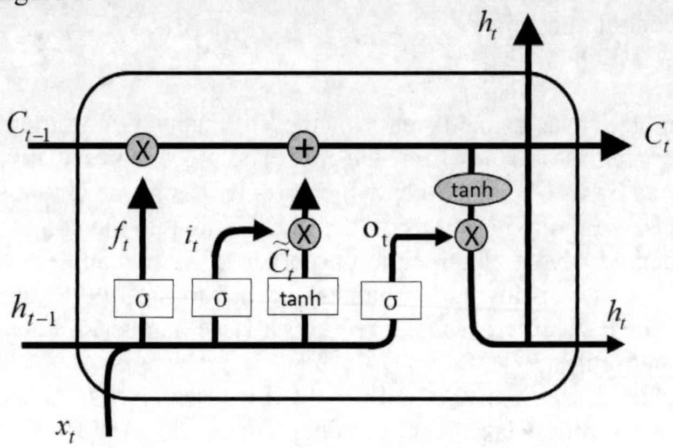

Figure 1. The main cell structure of LSTM

As can be seen from the above figure, there is an additional hidden state at each sequence index position t. This hidden state is generally called cell state. At the same time, compared to the RNN network, the LSTM network adds three door structures at each sequence index position, namely the forgetting gate, the input gate and the output gate.

- Input gate: Determine what new information is stored in the cell state (C_t);
- Forgotten gate: Decide which information to discard from (C_t);
- Output gate: Decide which information to output.

The formula for this network is as follows:

$$\begin{cases} i_t = \sigma(W_t h_{t-1} + U_i x_t + b_i) \\ \overline{C}_t = \tanh(W_c h_{t-1} + U_c x_t + b_c) \\ C_t = f_t \otimes C_{t-1} + i_t \otimes \overline{C}_t \\ f_t = \sigma(W_f h_{t-1} + U_f x_t + b_f) \\ o_t = \sigma(W_o h_{t-1} + U_o x_t + b_o) \\ h_t = o_t \otimes \tanh(C_t) \end{cases} \tag{1}$$

In the formula: X_t represents the input at time t; W_t, W_c, W_f, W_o, U_i, U_f, U_c, U_o respectively represent weight matrix; b_i, b_f, b_c, b_o represent offset items; $\otimes$ is a dot product; σ is the Sigmoid activation function. One-way LSTM can only use one side of text information, so this article uses bidirectional LSTMs to take advantage of contextual information. At the same time, this article connects the output of all previous layers and the initial input of the neural network as the input of the next layer. This connection method can maximize the information flow, and there will be no problems such as gradient disappearance and gradient explosion with the increase of neural network layers.

3.3. Self-attention layer

Although BiLSTM can fully consider the global information, it can not highlight the role of local key information. For example, “Mayor Li Ming inspected the school” , in the name entity recognition, the word “school” is of little significance to identify “Li Ming” . The term "mayor" is important for identification. Therefore, this paper adds attention mechanism in the BiLSTM hidden layer, which makes the model focus on finding related information, removing redundant and redundant information, and configuring different weights to improve the quality of named entity identification.

Attention mechanism is a selection mechanism used to allocate limited information processing capabilities. It is characterized by selective attention to certain important information and correspondingly ignoring other information received at the same time. The performance corresponding to text processing is to assign higher weights to important words and smaller weights to other words. The nature of the attention function $attention(Q,K,V)$ can be described as a query to a series of key values. The calculation of the attention mechanism is given below:

$$attention(Q,K,V) = soft\max(\frac{QK^T}{\sqrt{d_K}})V \tag{2}$$

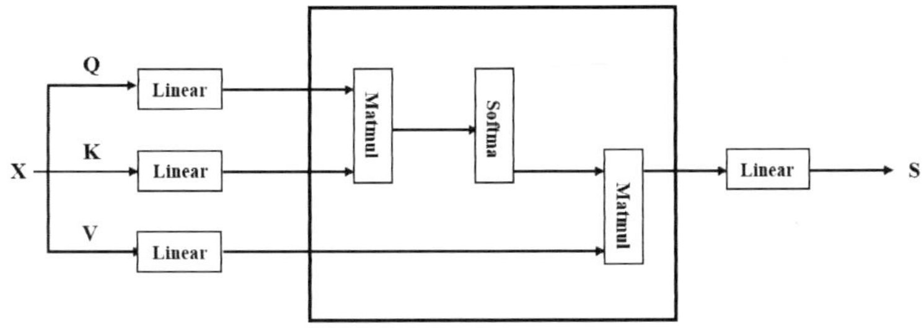

Figure 2. Flow chart of self-attention mechanism

The attention layer essentially encodes the sequence Q of $nn \times d_K$ into a new sequence of $n \times d_V$. The self-attention mechanism is a special case in the attention mechanism which is shown in figure 2. It only needs one sequence to calculate its representation. So the Q, K and V are the same sequence, attention(X, X, X), where X is the input sequence. The self-attention mechanism is characterized by ignoring the distance between each word, directly calculating its dependencies, learning the internal structure of the sentence, and obtaining long-term dependence.

3.4. CRF layer

The existence of the CRF layer is based on the fact that the tags are not isolated from each other, but have a certain correlation. The CRF layer can find the optimal output sequence while considering the dependencies between consecutive tags, and also prevent the occurrence of an illegal sequence. The final label prediction result consists of two parts, one is the output network P from the previous network, and the second is the transition probability T between the labels output by the CRF layer. The score of the sequence is:

$$S = \sum_{t=1}^{n} P_{t,y_t} + \sum_{t=0}^{n} T_{y_t,y_{t+1}} \tag{3}$$

Where $T_{y_t,y_{t+1}}$ is the transfer feature matrix used to store the probability of transition between all tags

3.5. Overall model

The model proposed in this paper is shown in the following figure 3. It is mainly divided into four parts, embedded module, BiLSTM module, Self-attention module, CRF module. The input is the BERT word vector trained by BERT-Base Chinese model provided by Google.

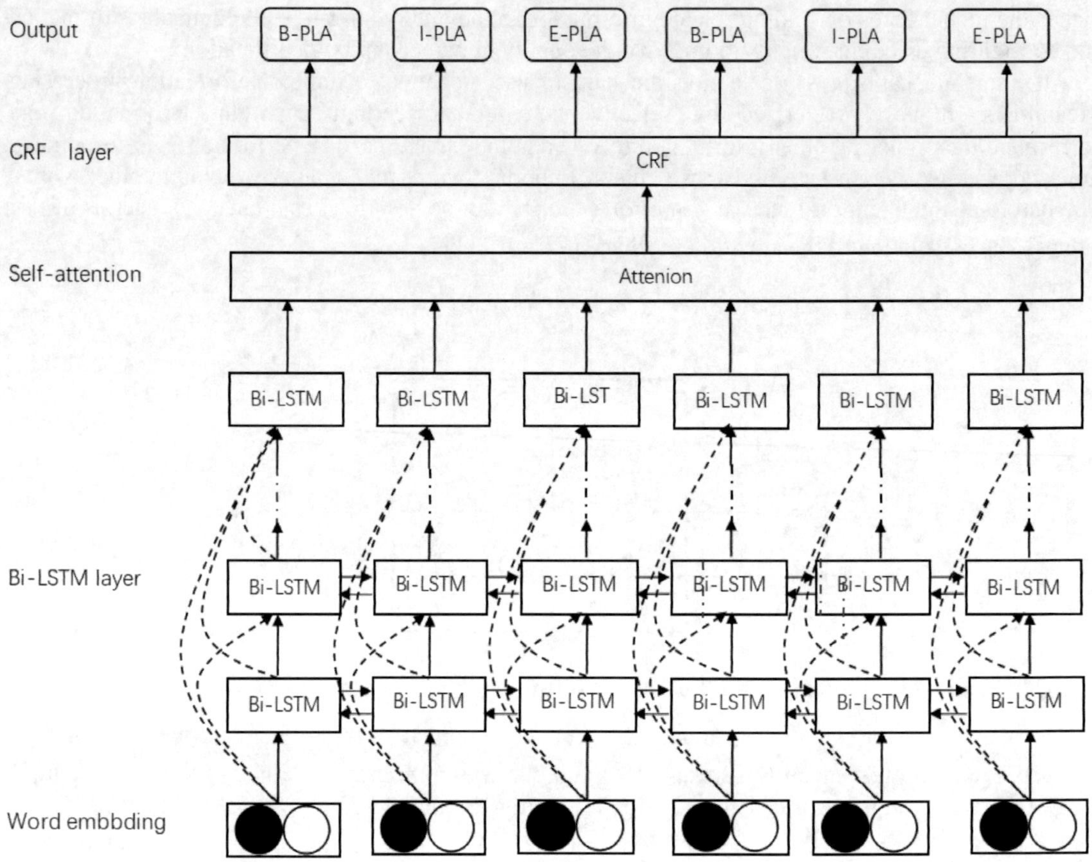

Figure 3. The overall model structures

For more efficient identification, this article uses multiple layers of BiLSTM, we input the word vector as the original input to the first layer of BiLSTM, and splicing the output of the first layer of BiLSTM with the original input as the input of the second layer. According to this method, after multiple layers of extraction, the final output is a deep meaning, syntactic and semantic feature representation, and then use the zoom point product attention to learn the structural features inside the sentence, capturing long-term dependencies:

$$S = Attention(h, h, h) \tag{4}$$

Finally, the feature vector is input to the CRF layer to obtain a prediction label.

4. Experimental design and results analysis

4.1. The data set

The data set used in this paper is SIGHAN 2006 Bakeoff - 3 Chinese evaluation data set MSRA, which mainly contains three types of entities, namely person name, place name, organization name, and its specific parameters are included in table 1:

Table 1. MSRA data set information statistics

Type	Train	Test	validation
Number of sentences	20.8K	2.3K	4.6K
Number of characters	1000.0K	112.1K	223.8K

4.1.1. tagging scheme and evaluating indicator. This article uses the BIOES tagging scheme instead of BIO, because the BIOES policy can more clearly define the boundaries of the entity, in the scheme B represents the starting character of the entity and I represents the intermediate character of the entity. O represents a non-entity character, E represents the end character of the entity, S represents an entity having only one character.The experiment uses the Precision value P, the recall value R and the F_1 value to evaluate the recognition results. The F1 value can accurately evaluate the model performance. The calculation formula is as follows:

$$P = \frac{\text{Number of entities correctlyidentified}}{\text{Number of entities identified}} \times 100\% \tag{5}$$

$$R = \frac{\text{Number of entities correctlyidentified}}{\text{Number of entities in the sample}} \times 100\% \tag{6}$$

$$F_1 = \frac{2 \times R \times P}{P + R} \tag{7}$$

4.2. Parameter settings

The input of BiLSTM is a tensor of $100 \times \max\_sen\_len\_\times 200$, where the first dimension represents batch size, the second dimension represents the largest sentence length in the batch size sentence, is a variable, and the third dimension represents the dimension of the hidden layer.

We set **Dropout** to 0.5 and use it in the softmax layer to prevent overfitting. To speed up the convergence of the gradient descent, we use a Gaussian distribution whose mean is zero and variance is one to initialize the model parameters.The optimization function is (the stochastic gradient decent, SGD) algorithm, the learning rate is set to 0.015, the step size is set to 0.05, and the d_k in self-attention is set to 4. We find that the model can achieve the highest efficiency when the BiLSTM layer number is 6 in the same conditions after testing.

4.3. Results analysis

The following table 2 is the comparison between the model in this paper and other model:

Table 2. The results of the model on the MSRA dataset

Number	Model	Precision	Recall	F1
1	Chen et al	91.22	81.71	86.2
2	zhang et al	92.2	90.18	91.18
3	Zhou et al	91.86	88.75	90.28
4	BiLSTM+CRF(None)	86.21	83.54	84.84
5	BiLSTM+CRF(Seg)	88.16	85.42	86.77
6	BiLSTM+CRF(Self-All)	91.13	90.05	90.59
7	BiLSTM+CRF(Self-All+BERT)	94.57	94.23	94.4

The first three items use external knowledge and artificial features. The fifth item is the model with word segmentation Simultaneously. We train the input word vectors by Word2vec and BERT respectively. The sixth and seventh items in the table above are the experimental results. From the table, we can see that the model with self-attention has higher F1 value than other models, which shows that the self-attention mechanism has greatly improved the performance of the model. Among them, the use of Word2vec can improve the performance of the model. The P value of the self-attention model with the original input of the word vector trained by Word2vec is slightly lower than that of Chen et al., because Chen uses the artificially designed features. From the results of the sixth and seventh experiments, it can be concluded that BERT, as a dynamic vector, greatly improves the efficiency of named entity tasks compared with Word2vec.

5. Conclusion

This paper proposes an end-to-end, high-efficiency model for Chinese named entity tasks. We evaluate it on the MASA dataset, and the model performs more efficient than the previous LSTM model. We use BERT training word vector as the original input to understand the meaning of Chinese phrases more comprehensively, and can improve the problem of Chinese new words and unclear boundaries. At the same time, this paper adds self-attention mechanism to the model, so that the model can be given The weighting method focuses on the key information in the sentence, ignoring other redundant information, and has certain benefits for improving the performance of the model. Our model does not use any artifacts or domain-specific features, so the model can be easily transferred to other areas.

Reference

[1] Devlin J, Chang M W, Lee K, et al. BERT: Pre-training of Deep Bidirectional Transformers for Language Understanding[J]. 2018.
[2] Collins M, Singer Y. Unsupervised models for named entity classification[C]// Proceedings of the Joint SIGDAT Conference on Empirical Methods in Natural Language Processing and Very Large Corpora, 1999:100—110.
[3] Bikel D M , Miller S , Schwartz R , et al. Nymble: a High-Performance Learning Name-finder[J]. Anlp, 1998:194—201.
[4] Bikel D M . An algorithm that learns what's in a name[J]. Machine Learning, 1999, 34.
[5] Borthwick A . A Maximum Entropy Approach to Named Entity Recognition[J]. Ph. D. Thesis New York University, 1999.
[6] McCallum A, Li W. Early results for named entity recognition with conditional random fields, feature induction and web-enhanced lexicons[C]// Proceedings of the Seventh Conference on Natural Language Learning at HLT-NAACL, Stroudsburg: Association for Computational Linguistics, 2003,4: 188-191.
[7] Isozaki H, Kazawa H. Efficient support vector classifiers for named entity recognition[C]// Proceedings of the 19th International Conference on Computational Linguistics. Stroudsburg:Association for Computational Linguistics, 2002, 1: 1-7.
[8] Lample G, Ballesteros M, Subramanian S, et al. Neural architectures for named entity recognition[C]// Proceedings of the 2016 Conference of the North American Chapter of the Association for Computational Linguistics: Human Language Technologies. Stroudsburg: Association for Computational Linguistics, 2016: 260-270.
[9] Rei M, Crichton G K O, Pyysalo S. Attending to Characters in Neural Sequence Labeling Models [C]. In COLING 2016, 26th International Conference on Computational Linguistics, Proceedings of the Conference: Technical Papers, December 11-16, 2016, Osaka, Japan. 2016: 309–318.

IOP Conf. Series: Materials Science and Engineering **692** (2019) 012051 doi:10.1088/1757-899X/692/1/012051

Study on Vessel Maintenance Support System Modelling Based on DODAF

Wang Chaowei[1], Ding Lvhui[1], Wu Qian[2,*], Zhang Jie[1] and Li Peichang[1]

[1]China Ship Develop and Design Center, Wuhan 430064, China
[2]China National Institute of Standardization, Beijing 100191,China

[*] Corresponding author's e-mail address: ustbwuqian@163.com

Abstract. The vessel maintenance support system, which consist of aboard and onboard two parts, is different from the other equipments, with the characteristics of tasks varied, complex structures and onerous business. It is a virtual system rather than a material system. So the traditional modelling method which is used to describe the material objects has some defects of long communication period, weak reusability and poor expansion in describing support system in design process, which apparently cannot meet the current demands. It is imperative to introducing standardized design and modelling approach to establish the vessel support system. The study firstly analyses the function, organization, resource and operation of vessel support system. The support system modelling method is introduced based on the analysis of different view products in DODAF framework. The support system model, which is consists of organization model, activity model and information exchange matrix is given based on active based model. It can provide guidance in support system design, support concept generation under new situation, even in integrated joint operation system design in the future.

1. Introduction

Vessel is one of the most important equipments in integrated joint operationwiththe characteristics of complex structure, long developing period, high cost and wide range of service cycle and user. The vessel maintenance support capability determines the whole equipment efficiency as well as its performance. Whereas the maintenance support capability of vessel is more complicated and tremendous than other ordnances, such as aircraft, tank and armored car. The traditional modelling method with words and diagrams or chats doesn't meet the modernized equipment's design and operation requirements. Nowadays, the design project of support system involves many relevant stakeholders. A new modelling approach must be found to establish the relationship of all the fundamental data and sub-models and integrate all the joint developers effectively on the concurrent activities during the design phrase. This association work can guarantee consistency, veracity and timeliness the modelling

The paper presents an approachof vessel maintenance support system modelling based on DODAF with the principles of system architecture building, which takes organization, personnel, activity and information into consideration. Firstly, the DODAF products are introduced. Then the ABM modelling method with DODAF operational views is given with the reviewing of functions. The relationship among the views and the modelling steps in ABM are also analysed. The make-up factors of vessel maintenance support system are introduced by the way of three-dimensional model. The reflection between the support factors and DODAF products are analysed. Finally, the model of maintenance support system. It shows thatmaintenance support system modelling based on ABM with DODAF is more efficient and convenient. The model can be used repeat and is able to expand, which aid a larger amount of developer distributed to work more coordinately.

2. DODAF products applied in system modelling

2.1. DODAF Framework

2.1.1. DODAF introduction. The Department of Defence architecture framework (DODAF) is developed by DoD of USA based on C4ISR system (Command, Control, Communication, Computer, Intelligence, Surveillance and Reconnaissance) according to the military experience in combat. Nowadaysit applies in more and more engineering system. The DODAF 1.5 and 2.0 is published gradually, which emphasises the data centred design opinion [1][2].

*2.1.2. DODAF products.*The DODAF 1.5 products consists of four parts: all view(AV), operation view (OV), system view (SV) and technical system view (TV). The AVs shows the reviewing and environments of the whole system. The OVs describes the types, activity, organization and other related information within businesses that system deals with. The SVs determines the requirements of function and performance of system. The TVs is usedto expatiate the technical standards and guidance which support the system during the design. It should be noticed that not all the products will be used in system establishment. It varies according to the specific demands of user or designers' purpose[3].

*2.1.3 Modelling procedures with DODAF.*The modelling progress in which a system with clear characteristics and meet certain requirements has to be guided under a certain means or principles due to the complexity of both the system in expectation but also the modelling progress itself. Firstly, the users should be aware that it is impossible to build a full-fledged model consisting of all elements of the system. all the project participates must acknowledge the purposes and applied situations of model and confirm the key points in analysis and modelling to make the project more efficiently. Secondly, the application background, operation scenario, geographic area and management factors which would influence the above aspects should be determined. Thirdly, the characteristics and features of the system have to be confirm in case of a useless modelling if ignoring the details. Similarly, the model can be huge and hard to recycle if mangy needless characteristics are included. Then cut off the unnecessary view products, which may decide the amount of work to a certain extent. Furthermore, all the useful data should be gathering as soon as possible,which is a continuous work started in early phrase. The data may be updated with the development, Finally, the DODAF product would be developed as well as the related documents.The verification of model can be conducted according to the user's demands[4].

3. Vessel Maintenance Support System Modellingbased on ABM with DODAF products

3.1. ABM method.

The total products in DODAF is 26 views. If the system is going to be built with all the views, it would take a large amount of personnel, time and cost. As the development of DODAF, the researchers make several inductions of modelling methods. ABM is one of the methods proposed by Steven.J.Ring. Compared to the traditional modelling, which focuses on the products, ABM is a data-centred modelling method base on the three kinds of views. ABM supports the cross of different views and automatically data generation. Once the relationships among some certain views are confirmed, the remaining views can automatically generate by software tools. The system structure can be divided into three objects: entity, relation and attribute. The entity is used to store and process the data in system. The relation is the link among the entities. And the attribute is the features and characteristic of entity and relation[5].

3.2. Modelling steps based on ABM.

According to 3.1, the modelling steps based on DODAF is shown in figure.1.The first step is to establish the operational activity views (OV-5), which includes the logic relationship among activities, and the inputs and outputs among activities and environments. The second step is to establish nodes in OV-2. The whole OV-2 consists of two parts, nodes and requirement lines. The requirement lines can automatically generate. So the nodes should be determined at first. The third step is to establish roles and units in OV-4. The OV-4 is mainly used to describe the structure relation among entries in system, which includes manage, command, report and other relation. The former three steps separately describe activities, nodes and roles. And the fourth step in modelling is to relate them manually. The fifth step is to generate the relationship among activities, nodes and roles. Thus, the activities and role would in the corresponding nodes in OV-2. So would the other two views. The sixth step is to generate the information exchange requirements in OV-2. This step can be also accomplished by software automatically. Essentially, the information exchange is conducted by the activities in the nodes. The seventh step is to generate the requirements line in OV-2, which is to map the information to the exchange line. The eighth step is to generate the information exchange matrix, which is OV-3. OV-3 describes the reason for information exchanging, direction and way of information exchanging and the specific information. Thus, the ABM provides a certain flow and standard for products selecting in DODAF.

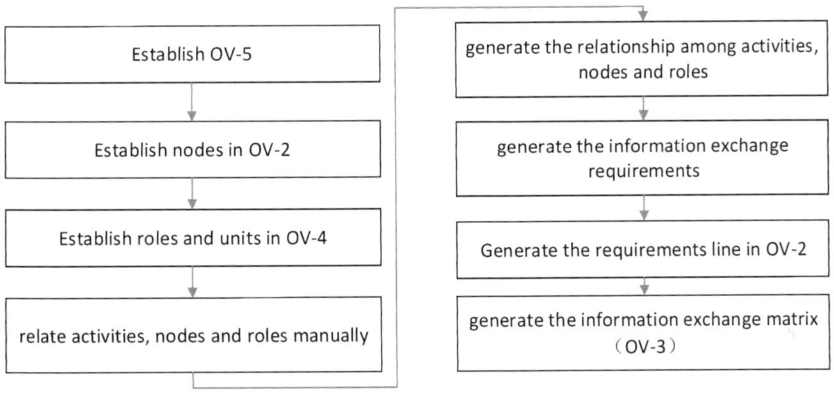

Figure 1. Modelling steps based on DODAF

4. Vessel Maintenance Support System Modelling Based on ABM

4.1. Vessel Maintenance Support System.

VesselMaintenance support system refers to the combination of support resources and management factors which vessel requires during maintenance, including on-board and base support activities.The support resourcesconsist of all the personnel, material, information and other kinds of resources. The supportresources are distributed in every node in support system. And support functions of system can be executed through the support activities operated by the corresponding organization. As shown in figure 2.

IOP Conf. Series: Materials Science and Engineering **692** (2019) 012051 doi:10.1088/1757-899X/692/1/012051

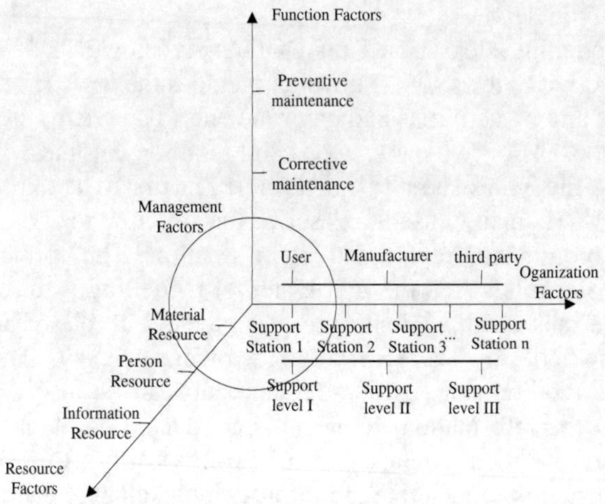

Figure 2. Three-dimensional Model of Support System

4.2. OV-5 of Vessel Maintenance Support System.

The activities within vesselmaintenance support system are integrated into the functions of system. The activities are hierarchy, in which the maintenance function includes corrective maintenance activities and preventive maintenance activities. The two kinds of maintenance consist of various maintenance procedures. The modelling method of IDEF0 is applied to establish OV-5 of Vessel Maintenance Support System. The five factors have to be described clearly in IDEF0, as shown in figure 3. The first is activity, which describe the specific progresses of maintenance, the activities are arranged in a certain logic way and can be decomposed. The second is the input of activity, which can be actual martial or virtual data or information. The actual martials compose of spares, oil and other consumption. The virtual data can be administrationorder or intelligence. The third is output of activity, which is the recovery equipment with supporting and records. The forth is mechanism ofactivity, which support the activity. The mechanisms can be specific person or a kind of organization, and even the support nodes. In the paper. Support tools, facilities and equipments are regarded as the mechanisms. The fifth is control of activity, which is mainly the regulations and laws support system should obey. The technical manual is one of the typical controls. Other factors which may influence the system can also be control, such as battle field environment and support budgets.

4.3. OV-2 of Vessel Maintenance Support System.

The support node is the combination of support organizations, which may include one or more organizations. The amount of nodes varies with the requirements of support. The rule of nodes sort is the function of organizations rather than the geographic distribution. The OV-2 of Vessel Maintenance Support System includes three parts. The first one is support nodes, which is the basic factor that generate, apply and process information. The node can be a concrete organization (e.g. strategy office of DoD, manufacture), or a node in functional or logic aspect (e.g. logistic node, intelligence node). It should be noted that the OV-2 should be established according to the actually executed mission. The specific location or buildings should not be described as a node. The second is the information lines, which describe the information exchange requirements among nodes. The information is generated or applied to accomplish the support activities. The third one is business activities. All the nodes should a certain set of business. The OV-2 of Vessel Maintenance Support System is shown in figure.4. The model of OV-2 is established by the method of UML. Additionally, the activities and information lines can be generated automatically with the modelling tools (e.g. IBM SA) once the connection among three views is built. The repair level of vessel can be divided into three levels, the operational level (O), the immediate level (I) and the depot level (D), with the growth of maintenance capability. The operational level in vessel maintenance support system refers to the on-board and dock maintenance.

The immediate level refers to the maintenance at the repair ship or institute. The depot level refers to the repair factory of navy or civil use.

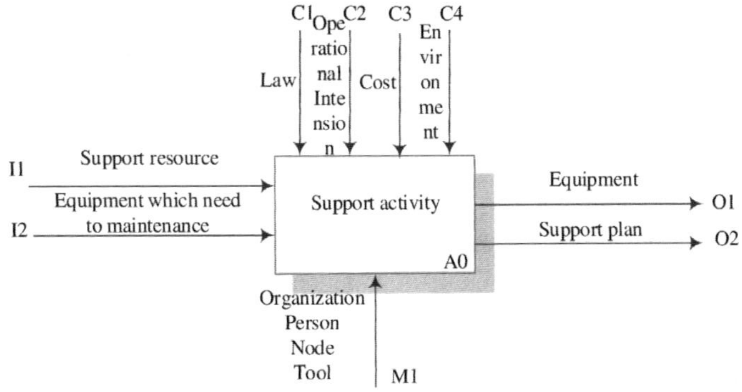

Figure 3. OV-5Modelling based onIDEF0

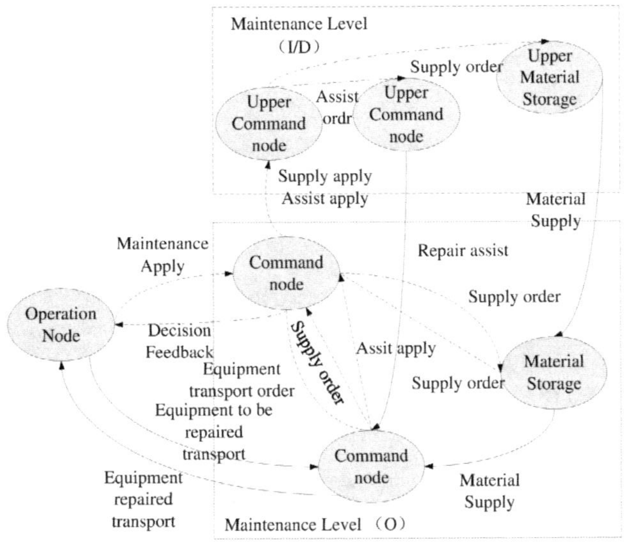

Figure.4 OV-2 of Vessel Maintenance Support System

4.4. OV-4 of Vessel Maintenance Support System.

The main purpose of OV-4 is to describe the roles and authorities to realise the objections and functions of support system. In other words, it is to illustrate the activities conducted by whom. The logic links and connections among the organizations, which provide foundation of normal work of system should be established. The organizations in the support system have the following characteristics: hierarchy, dispersiveness, coordinated and autonomy. The hierarchy means that one support organization can be divided into several sub- organizations, which can be also divided. The organizations are multilevel. The dispersiveness refers to that the organizations are disperse everywhere. It is necessary to build material and information exchange channel among the organizations. The coordinated means that the sub-organizationsconstitute an entirety one to accomplish the specific mission. The autonomy refers to that though sub-organizationsare subject to the upper ones, they are autonomy under some situation in certain period and can make decision about affairs of themselves. The figure.5 shows a template in common use for OV-4. The lines that are used to link organizationsin the template is divided into two parts. The full lines refer to the relation of command, and the dotted lines refer to the relation of coordinate, cooperate or other relation.

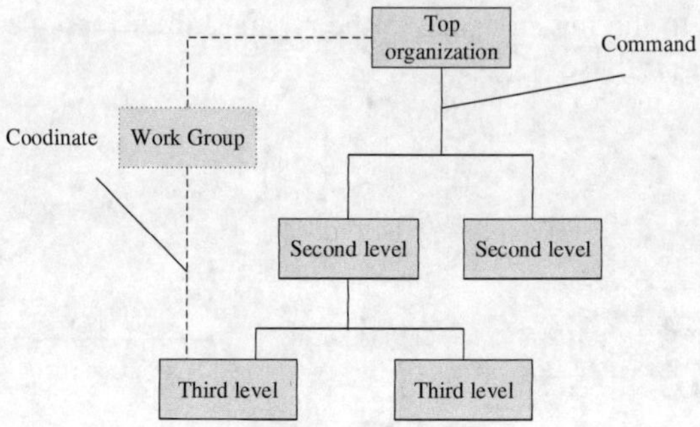

Figure 5. OV-4template

4.5. OV-3 of Vessel Maintenance Support System.

The OV-3 is a matrix (or called chat) which describe the information exchange's reason, flow direction, category, specific content, method and other required factors among nodes. The information in OV-3 is a generalized concept, which can be maintenance plan, material apply order or spares. In DODAF, there is no concrete describing items about OV-3. The user can define it according to the project, such as the period, language, confidentiality.Not all the details of information exchange need to be record. The content, input nodes, output nodes, and attribute of activities should be emphasized. The same content should be separated if the receive node is different. The types of information (e.g. command, intelligence and so on) also have an effect on the line number of matrix Generally, the OV-3 can be generated automatically by DODAF modelling software.

5. Conclusion

This study presented a method of vessel maintenance support system based on DODAF. The DODAF views products are introduced, then the procedures of ABM modeling method are present to give ways to build the support system in a certain reasonable way. The concept and connotation of vessel maintenance support system is introduced. Then the views products of OV-2, OV-3,OV4,OV-5 is built. The reflection between DODAF models and support system's function, organization and resource is established. The method can improve the recycle rate of modelling and the work efficiency. The simulation can be conducted based on the DODAF model. It can aid the developer to make the decision and optimization at the early stage in design phase.

References

[1] Department of Defense Architecture Framework Working Group.DoD Architectural Framework, Version 1.0, 2003

[2] Department of Defense Architecture Framework Working Group.Department of Defense Architecture Framework Version2.0.2008

[3] Kelly Griendling,Dimitri N. Mavris. Development of a DODAF-based executable architecting approach to analyze system-of-systems alternatives[C]. 2010 4th Systems Conference

[4] A. Biswas J. Hayden, M. S Phillips, K. B. Bhasin, C. Putt, T. Sartwell Applying DODAF to NASA Orion Mission Communication[C],Aerospace Conference, 2008

[5] Xing Pan,Baoshi Yin,Jianmi Hu Modeling and Simulation for SoS Based on The DODAF Framework[C] The 5th Asia-Pacific System Engineering Conference,2011

IOP Conf. Series: Materials Science and Engineering **692** (2019) 012052 doi:10.1088/1757-899X/692/1/012052

Study on Evaluation Model of Equipment Transportation Efficiency Based on Analytic Hierarchy Process

Wu Qian[1], Wang Chaowei[2], Pei Fei[3,*], Xu Yingcheng[1] and Lu Xiaowei[1]

[1]China National Institute of Standardization, Beijing 100191, China
[2]China Ship Develop and Design Center, Wuhan 430064, China
[3]China standard promoting quality science and technology(beijing)co., ltd, Beijing 100191, China

[*] Corresponding author's e-mail address: Peifei@cnis.ac.cn

Abstract. On the basis of factors influencing the evaluation of equipment transportation efficiency, such as safety, timeliness, universality and economy of system analysis, and in combination with the basic theory of Analytic Hierarchy Process (AHP), this paper sorted out the main steps for evaluation of equipment transportation efficiency, established the hierarchical chart for evaluation index of equipment transportation efficiency, proposed a comprehensive evaluation model of equipment transportation efficiency based on AHP, and verified the feasibility of this model through the specific cases. The comprehensive evaluation model of equipment transportation efficiency based on AHP proposed by this paper provides a method for the decision makers to select the optimal design scheme for equipment transportability, and lays a foundation for related studies of equipment transportability.

1. Introduction

Transportation efficiency is one of important indexes to evaluate the equipment readiness performance, and is also one of main factors affecting the outcome of a war [1]. The equipment transportation efficiency is closely related to the reliability of transportability design scheme, expenses and expenditures, universality of transportation equipment, transportation time, etc. Therefore, the transportation efficiency evaluation of equipment transportability design scheme pertains to a multi-attribute and complicated decision-making problem analysis. And AHP can analyze the multiple influencing factors of the complicated decision-making problem and their interaction well, decompose the decision-making problems level upon level through a combination of qualitative and quantitative methods, and determine the optimal scheme. Therefore, it is widely used in all fields. Dongzhi Wang, et al. (2019)[2], Yu Lu, et al. (2019) [3], Yang Wang, et al. (2019) [4], Chen Li, el al. (2019) [5], and Yanfeng Zhang (2019) [6] respectively applied the AHP to the evaluation of groundwater quality, weight determination of road in the driving condition, comprehensive monitoring of transmission section, power grid dispatching evaluation, multi-attribute threat evaluation of equipment, etc. Guoquan Yan, et al (2017) [7] applied the AHP to the risk analysis and assessment of transportation and construction supervision, and established a comprehensive evaluation model to determine the project risk. Jiayong Zhu, et al. (2019) [8] used the AHP to analyze the influence relationship among overall control, resource input, coordination workload, procedure complexity, cost controllability and quality safety risk level of port construction project, and established an analytic hierarchy model, thereby providing a reasonable and reliable basis for the project undertaker to make decision.

On the basis of the above research results and in combination with the factors influencing the evaluation of equipment transportation efficiency, this paper established the evaluation model of equipment transportation efficiency based on AHP, thereby providing a scientific method for selection of optimal equipment transportability design scheme.

2. Factors Influencing Equipment Transportation Efficiency

There are many factors influencing the equipment transportability efficiency, mainly including the safety, economy, universality and timeliness [9, 10].

2.1. Safety

Safety of equipment transportation process has a direct influence on smooth assembly or prompt use of equipment to be transported after its arrival at the destination, and is one of important factors influencing the equipment transportation efficiency. Safety factor is usually expressed by the reliability of equipment transportability design scheme to accomplish the equipment transportation task.

2.2. Economy

Equipment transportation involves packaging, loading and unloading, storage, transportation and other activities, each link of which will generate the expenses on personnel, maintenance, energy consumption and leasing. How to achieve the maximum benefit at the lowest cost is also a key consideration in the equipment transportability design scheme.

2.3. Universality

The more the kinds, quantities and volumes of special resources consumed in equipment transportation are, the more the time and expenditures consumed in the transportation process are. Therefore, the reduction of ratio of special transportation resources in the total transportation resources during the equipment transportation is also one of main methods to improve the equipment transportation efficiency.

2.4. Timeliness

Reasonable and efficient utilization of existing personnel and equipment resources to deliver the equipment to be transported to the destination on time and in good condition is a precondition to ensure the equipment readiness. Under the equal conditions, the higher the timeliness is, the better the equipment transportability design scheme is.

3. Evaluation Model of Equipment Transportation Efficiency Based on AHP

3.1. Basic Theory of AHP

AHP refers to establishing an index hierarchy and judgment matrix by applying the network system theory and multi-objective comprehensive evaluation method and analyzing the influencing factors, calculating the index weight with the mathematical method, and conducting the consistency check to obtain the weight of index in each hierarchy.

3.1.1. Establishment of progressive hierarchy. AHP divides the index into different hierarchies, generally including the objective level, criterion level and scheme level. The objective level mainly refers to the predetermined objective of the problem, the criterion level generally refers to the relevant criteria affecting the objective achievement and can also be further subdivided into sub-criterion level as required, and the scheme level refers to the schemes related to the objective achievement, specifically as shown in Figure 1.

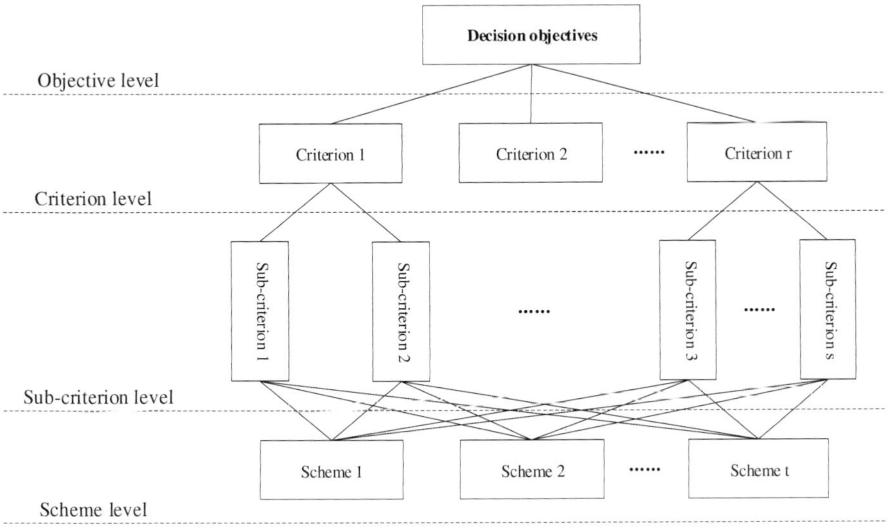

Figure 1. Index Hierarchy

3.1.2. Construction of judgment matrix. The judgment matrix is constructed according to the hierarchical structure. Each element having downward subordination is taken as the first element of judgment matrix, and elements attached to it are successively arranged in the first row and first column of the climate. Associated elements are compared pairwise, and assigned 1 to 9 according to the importance. Refer to Table 1 for the scale value of importance.

Table 1. Scale Value of Importance

Scale value of importance	**Degree to which one factor is more important than the other one by comparing them**
1	Equally important
3	Slightly important
5	Obviously important
7	Strongly important
9	Absolutely important
2,4,6,8	Median of the above adjacent judgment
Inverse	If A_i is compared with A_j to obtain A_{ij}, A_j and A_i are compared to obtain $A_{ji} = \frac{1}{A_{ij}}$

Thus, the judgment matrix $A = (A_{ij})_{n \times n}$ can be obtained, as shown in Table 2:

Table 2. Elements of Judgment Matrix A

A_{ij}	A_1	A_2	......	A_n
A_1	1	A_{12}	......	A_{1n}
A_2	A_{21}	1	......	A_{2n}
......	......	......	......	......
A_n	A_{n1}	A_{n2}	......	1

3.1.3. Calculation of λ_{max} and its corresponding eigenvector. The maximum eigenvalue λ_{max} of the above judgment matrix is calculated by sum product, and the eigenvector corresponding to the maximum eigenvalue is the weight vector of index in each level. Specific steps are as follows:

- Normalizing the judgment matrix array

$$\overline{A_{ij}} = \frac{A_{ij}}{\sum_{k=1}^{n} A_{kj}} \tag{1}$$

Where, i, j=1, 2, …, n.

- Adding the rows of normalized judgment matrix

$$\overline{W_i} = \sum_{j=1}^{n} \overline{A_{ij}} \tag{2}$$

Where, i, j=1, 2, …, n.

- Normalizing the vector $\overline{W}$ to obtain the eigenvector $W = [W_1, W_2, \dots\dots, W_n]^T$

$$W_i = \frac{\overline{W_i}}{\sum_{j=1}^{n} \overline{W_j}} \tag{3}$$

Where, i, j=1, 2, …, n.

- Calculating the maximum eigenvalue λ_{max}

$$\lambda_{max} = \sum_{i=1}^{n} \frac{(AW)_i}{nW_i} = \frac{\sum_{i=1}^{n} \frac{(AW)_i}{W_i}}{n} \tag{4}$$

Where, i=1, 2, …, n.

3.1.4. Consistency check. During the pairwise comparison for importance of indexes, there may be some errors, and thus, the consistency cannot be satisfied. In order to ensure the rationality of determining the weight with the AHP, it is necessary to check the consistency of the judgment matrix.

- Calculating the consistency index C.I.

$$\text{C. I.} = \frac{\lambda_{max} - n}{n-1} \tag{5}$$

- Determining the corresponding average random consistency index R.I. through table look-up, as shown in Table 3.

Table 3. R.I. Repeatedly Calculated for 1,000 Times

n	1	2	3	4	5	6	7	8	9	10	11	12
R.I.	0.00	0.00	0.52	0.89	1.12	1.26	1.36	1.41	1.46	1.49	1.52	1.54

- Calculating the consistency ratio C.R.

$$\text{C. R.} = \frac{\text{C.I.}}{\text{R.I.}} \tag{6}$$

Generally, when C.R.<0.1, the consistency of the judgment matrix is considered acceptable. Otherwise, the judgment matrix does not meet the consistency requirement and needs to be revised again.

3.2. Comprehensive Evaluation Model of Transportation Efficiency

In this paper, the comprehensive weighted evaluation method is adopted to determine the transportation efficiency of equipment transportability design scheme on the basis of obtaining the weight vectors of indexes. Assuming that there are n alternative equipment transportability design schemes and m evaluation indexes, a decision-making matrix B is established by combining the alternative transportability design scheme with the transportation efficiency evaluation indexes, and can be expressed as:

$$B=(b_{ij})_{m\times n} \tag{7}$$

Where, $i = 1, 2, \dots, m$; $j = 1, 2, \dots, n$; b_{ij} refers to the attribute value when the j-th equipment transportation scheme corresponds to the i-th evaluation index.

The transportation efficiency of alternative scheme can be calculated in accordance with the weight vectors of the indexes and the decision-making matrix B, and expressed by Z:

$$Z_j = \sum_{j=1}^{n} w_j\, b_{ij} \tag{8}$$

Where, $i = 1, 2, \dots, m$; $j = 1, 2, \dots, n$.

The higher the multi-factor comprehensive evaluation value of equipment transportation efficiency is, the better the comprehensive consideration result of this transportation scheme is, and is expressed by Z_j^* which is specifically calculated as follows:

$$Z_j^* = \max\{Z_j | j = 1,2, \ldots, n\} \tag{9}$$

So it can be obtained that, the equipment transportation scheme j is the optimal transportation scheme.

4. Case Analysis

4.1. Basic Information

The hierarchical chart for evaluation index of transportation efficiency of newly developed equipment is established in combination with the analysis on influencing factors of equipment transportation efficiency [9, 10], as shown in Figure 2.

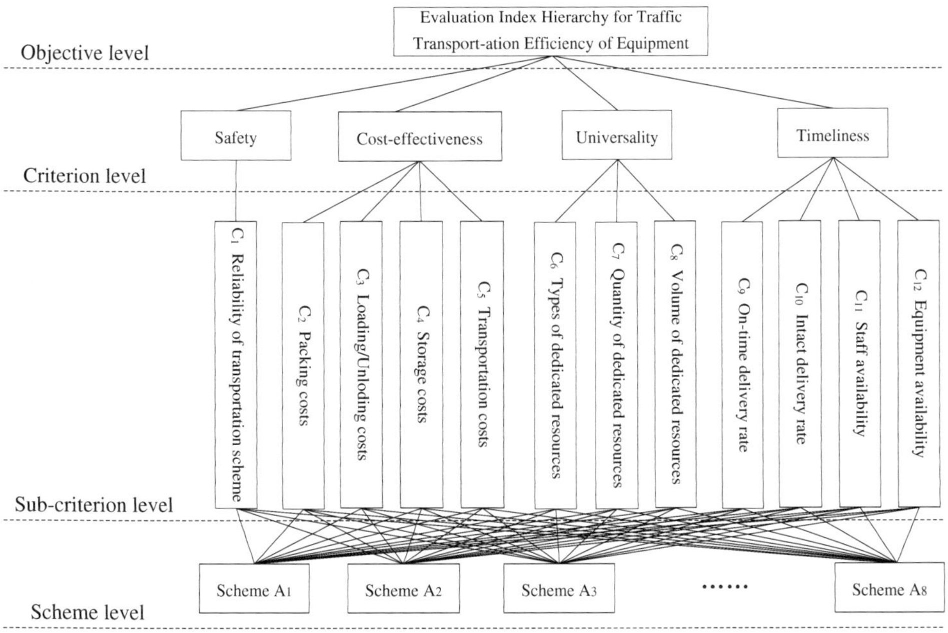

Figure 2. Evaluation Index Hierarchy for Traffic Transport-ation Efficiency of Equipment

Ten exports are organized to grade the transportation efficiency indexes of all schemes according to the basic information table for alternative scheme of equipment (refer to Table 4), and to discretize them according to the ceiling principle, to obtain the decision table of transportation efficiency evaluation (refer to Table 5).

Table 4. Basic Information Table of Equipment Transportability Design Scheme

Scheme	Safety (%)	Economy (RMB 10,000)	Universality (%)	Timeliness (%)
A_1	97	10	70	80
A_2	95	15	80	80
A_3	96	11	60	90
A_4	95	12	70	80
A_5	94	13	80	70
A_6	97	12	80	80

A_7	98	13	70	80
A_8	97	9	70	80

Table 5. Decision Table after Discretization

Scheme	C_1	C_2	C_3	C_4	C_5	C_6	C_7	C_8	C_9	C_{10}	C_{11}	C_{12}
A_1	5	5	4	3	3	4	2	4	4	3	3	4
A_2	5	4	4	1	2	4	3	4	5	3	3	4
A_3	4	3	5	4	4	5	5	5	4	5	5	4
A_4	3	4	4	3	4	3	4	3	3	4	3	4
A_5	5	4	3	1	2	5	4	4	5	4	4	3
A_6	1	5	5	5	5	2	2	2	2	2	3	3
A_7	2	3	5	3	5	1	4	3	3	3	3	4
A_8	3	4	4	3	5	5	3	4	4	5	4	5

4.2. Weight Calculation of Index

In this case, detailed calculation is conducted with the sum product by taking the index of the criterion level in the hierarchical chart for evaluation index of equipment transportation efficiency as an example, with the specific process as follows:

4.2.1. Establishing the judgment matrix and normalizing it. Determining and establishing the judgment matrix according to the ratio scales of 1-9 in Table 1, comparing the matrix $A = (a_{ij})_{n\times n}$ according to the importance, and normalizing each column of the judgment matrix to obtain:

$$A = \begin{bmatrix} 1 & 1/2 & 5 & 4 \\ 2 & 1 & 6 & 5 \\ 1/5 & 1/6 & 1 & 1/3 \\ 1/4 & 1/5 & 3 & 1 \end{bmatrix}$$

So:

$$\bar{A} = \begin{bmatrix} 0.2899 & 0.2679 & 0.3333 & 0.3871 \\ 0.5797 & 0.5357 & 0.4000 & 0.4839 \\ 0.0580 & 0.0893 & 0.0667 & 0.0323 \\ 0.0725 & 0.1071 & 0.2000 & 0.0968 \end{bmatrix}$$

4.2.2. Adding the rows of normalized judgment matrix. Adding the normalized judgment matrix by rows to obtain the vector $\bar{W} = (\overline{w_1}, \overline{w_2}, \dots\dots, \overline{w_n})^\tau$ as follows:

$$\bar{W} = \begin{bmatrix} 1.2781 \\ 1.9993 \\ 0.2462 \\ 0.4764 \end{bmatrix}$$

4.2.3. Solving the eigenvector. Normalizing the vector $\bar{W} = (\overline{w_1}, \overline{w_2}, \dots\dots, \overline{w_n})^\tau$ to obtain the solved eigenvector $W = (w_1, w_2, \dots\dots, w_n)^\tau$, namely, the weight vector of index in the criterion level, as follows:

$$W = \begin{bmatrix} 0.3195 \\ 0.4998 \\ 0.0615 \\ 0.1191 \end{bmatrix}$$

So:

$$BW = \begin{bmatrix} 1.3536 \\ 2.1036 \\ 0.2485 \\ 0.4836 \end{bmatrix}$$

4.2.4. Solving the maximum eigenvalue. Calculating the maximum eigenvalue λ_{max} of the judgment matrix. The following equation is obtained by BW and w_i :

$$\lambda_{\max} = \sum_{i=1}^{n} \frac{(BW)_i}{nw_i} = \frac{1.3536}{4 \times 0.3195} + \frac{2.1036}{4 \times 0.4998} + \frac{0.2485}{4 \times 0.0615} + \frac{0.4836}{4 \times 0.1191} = 4.1355$$

4.2.5. Consistency check. The consistency check index C.I. could be obtained according to the maximum eigenvalue of the judgment matrix:

$$C.I. = \frac{\lambda_{\max} - n}{n-1} = \frac{4.1355 - 4}{4 - 1} = 0.0452$$

Then, based on Table 3, the consistency index RI could be obtained, and the consistency ratio C.R. is calculated as follows:

$$C.R. = \frac{C.I.}{R.I.} = \frac{0.0452}{0.89} = 0.0508 < 0.1$$

Accordingly, it could be determined that, the judgment matrix has satisfied consistency, and thus, the relative weights of four indexes in the criterion level are determined. Similarly, repeating the above processes, the weights of all indexes in the sub-index level could be determined, and finally, the weight relationship among the indexes is determined, as shown in Table 6 below:

Table 6. Subjective Weights of Indexes

Attribute	**Weight**	**Attribute**	**Weight**	**Attribute**	**Weight**	**Attribute**	**Weight**
C_1	0.1023	C_4	0.1999	C_7	0.0205	C_{10}	0.0244
C_2	0.1757	C_5	0.1749	C_8	0.0266	C_{11}	0.0422
C_3	0.0415	C_6	0.1250	C_9	0.0145	C_{12}	0.0525

4.3. Transportation Efficiency Evaluation

The decision tables and weight vectors of discretized indexes could be respectively obtained according to Table 5 and Table 6, and in combination with Formula (8), the equipment transportation efficiency of each transportability design scheme is comprehensively evaluated, and the evaluation result is shown in Table 7.

Table 7. Comprehensive Evaluation Result of Equipment Transportability Design Scheme

Alternative scheme	**Comprehensive evaluation value of transportation efficiency**	**Ranking of evaluation result**
A_1	3.7956	3
A_2	3.0802	8
A_3	4.1045	1
A_4	3.4895	4
A_5	3.1983	6
A_6	3.7684	4
A_7	3.1535	7
A_8	4.0541	2

The comprehensive evaluation result for transportation efficiency of equipment transportability design scheme obtained from the above table shows that, the Scheme A_3 is the optimal design scheme, and the Scheme A_8 is the suboptimal design scheme. The comprehensive evaluation results of transportation efficiency for these two schemes are significantly better than those of other six schemes, and there is little difference in the evaluation results of the two schemes. Therefore, the decision makers are suggested to select one of the two schemes A_3 and A_8 as the transportability design scheme of this equipment.

5. Summary

As the recognition of importance to the equipment transportation efficiency evaluation is gradually increased, the research in this field in China is also gradually deepened. The comprehensive evaluation model of equipment transportation efficiency based on AHP proposed by this paper is imperfect, but combines the qualitative and quantitative methods to a certain extent. This concept is introduced into the field of equipment transportation efficiency evaluation, thereby providing a more scientific method for the decision makers to select the optimal equipment transportability design scheme, and having certain reference significance for follow-up study on equipment transportability analysis.

Acknowledgements

We would like to acknowledge that this research are supported and funded by the National Science Foundation of China under Grant No. 71571191, 91646122 and 91746202, the Basic Scientific Research Business Projects 552018Y-5927, the National Key R&D Program of China under Grant No.2016YFF0202604, 2017YFF0209603, 2017YFF0209604, 2017YFF0206504.

References

[1] Qian Wu. Study on Transportability Analysis Method of Unmanned Aerial Vehicle System [D]. Beijing: Beihang University, 2014.

[2] Dongzhi Wang, Jianhui Liang. Evaluation on Groundwater Quality Based on Analytic Hierarchy Process - Taking Aksu as an Example [J]. Journal of Anhui Agricultural Sciences, 2019, 47 (8): 80-85.

[3] Yu Lu. Weight Determination of Road in Driving Condition Based on Analytic Hierarchy Process [J]. Test, 2019 (2): 114-117.

[4] Yang Wang, Biao Xu, Chaofan Zhou, et al. Comprehensive Monitoring Method of Transmission Section Based on Analytic Hierarchy Process [J]. China Power, 2019, 52 (4): 89-95.

[5] Cheng Li, Zili Yin, Xiaohui Wang, et al. Dispatching Evaluation of Power Distribution Network Based on Analytic Hierarchy Process and Entropy Weight Method [J]. Proceedings of the CSU-EPSA, 2019: 1-8.

[6] Yanfeng Zhang, Jianshu Liu, Shifeng Zhang. Multi-attribute Threat Evaluation of Objective Based on Analytic Hierarchy Process and Entropy Method [J]. Journal of Projectiles, Rockets, Missiles and Guidance, 2019, 2:1-6.

[7] Guoquan Yan, Hongwei Chen. Study and Application of Analytic Hierarchy Process in Risk Analysis and Assessment of Construction Supervision of Waterway Engineering [J]. Management and Administration, 2017, 24 (12): 230-235.

[8] Jiayong Zhu, Hang Su. Discussion of Port Construction Mode Based on Analytic Hierarchy Process [J]. Port & Waterway Engineering, 2019, 554 (4): 89-94.

[9] Qian Wu, Ma Lin, Chaowei Wang. Study on Evaluation Index System of Equipment System Transportability [C], WCEAM2013, 2015: 1539-1547.

[10] Lei Zhang, Yunpeng Chen, Minglu Guan. Study on Evaluation Model of Military Air Transportation Efficiency in Wartime [J]. Fire Control & Command Control, 2013, 38 (3): 69-75.

IOP Conf. Series: Materials Science and Engineering **692** (2019) 012053 doi:10.1088/1757-899X/692/1/012053

Design of High-Efficiency Refrigerator Test System for Industrial Internet of Things

Lintao Xian, Wei Jing and Chao Yang

Department of Computer Science and Technology, Ocean University of China, Qingdao, China

744860372@qq.com

Abstract. Smart homes are developing rapidly, and refrigerators are an indispensable part of "smart homes". Therefore, the research efficiency and quality assurance of refrigerators are highly valued by home appliance companies. Refrigerator test is required in both research and production processes. At present, the quality test of refrigerator has a high degree of manual participation, such as the selection of the stable operating range of refrigerator and the conversion of test stage require manual determination, which cause low test efficiency and high electric energy consumption. In order to solve these problems, this paper design a High-Efficiency Refrigerator Test System (HERTS) for Industrial Internet of Things (IIoT). The core test data intelligent analysis module realizes timely determination of test data and automatic transition of test stage. The proposed HERTS increases the efficiency of automate refrigerator test and greatly reduces the power consumption.

1. Introduction

The development of IIoT is the key of transformation and upgrade of manufacturing industries. With the proposal and development of industry 4.0 [1,2] and "Made in China 2025" [3], higher requirements are proposed for the manufacturing industry. The manufacturing industry develops from the original automation to intelligent and unmanned [4]. The rapid development of smart homes is prompting home appliance manufacturers to continuously develop new products to seize the market [5,6]. Refrigerators are an indispensable part of "smart homes" [7]. Therefore, the research efficiency and quality assurance of refrigerators have been highly valued by home appliance companies [8]. In the research and production process of refrigerators, test is required. However, the current refrigerator test system [9,10] only realizes semi-automatic testing of refrigerators, and only realizes basic functions such as sensor data acquisition, data display, and data management. With the development of intelligent factory oriented to robot, the refrigerator test industry is faced with inefficient problems.

The selection of refrigerator operation stability range and the transformation of test stage need to be judged manually. GBT8059[11] as China's refrigerator product test standard stipulates that in the test of refrigerator energy consumption, the data of refrigerator which continuous operation stable for 6 hours or more should be chose as the test data. However, the selection of stable data requires the tester to check the collected refrigerator data periodically [12]. And the tester needs select the possible stable interval to calculate whether it meets the criteria for stable operation of the refrigerator. According to the freezing capacity test of the refrigerator, each refrigerator compartment should be loaded with the freezing load after the stable operation state. At the same time, the time difference of refrigeration load from 25°C and -18°C should be recorded. The temperature rise test stipulates that the refrigerator power should be cut off after stable operation, and the time difference of the load package from -18℃ to -9℃ should be recorded. But the switching of different test phases requires manual determination and operation. At present, refrigerator product test are over-reliant on manual, which waste manpower, and manual operations are not timely enough to increasing the test cycle [13].

In daily life, the refrigerator will work in different temperature and humidity environments [14], which is the maximum simulation of the actual working environment of the refrigerator, so the laboratory test environment will be maintained under specific conditions. Inefficient refrigerator testing increases the test cycle and wastes a lot of energy.

To address the problems faced by the current refrigerator test industry mentioned above, this paper presents the design of High-efficiency refrigerator test system. The method of judging and marking the stable state of refrigerator product test data based on feature selection and the method of intelligent test switching are presented.

2. HTRES Design and Implementation

The refrigerator test system can be divided into three parts: the refrigerator to be tested, the intelligent sensing device, and the refrigerator test software. The connection structure is shown in figure 1. Sensors are arranged on the refrigerator to be detected to sense the running state of the refrigerator, and the refrigerator test software collects sensor data and performs analysis processing.

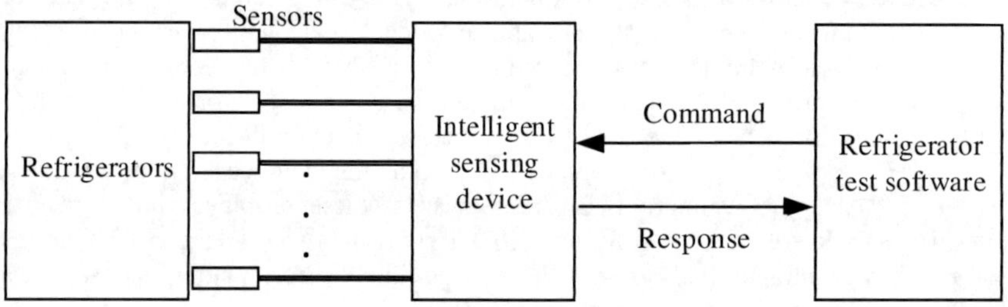

Figure 1. Framework of refrigerator test system.

2.1. Function module of refrigerator test software

The functions module of the refrigerator test software include sensor data acquisition module, nameplate information input module, test data management module, data storage module, data curve display module, test data intelligent analysis module, test report generation module. Functional module workflow is shown in figure 2.

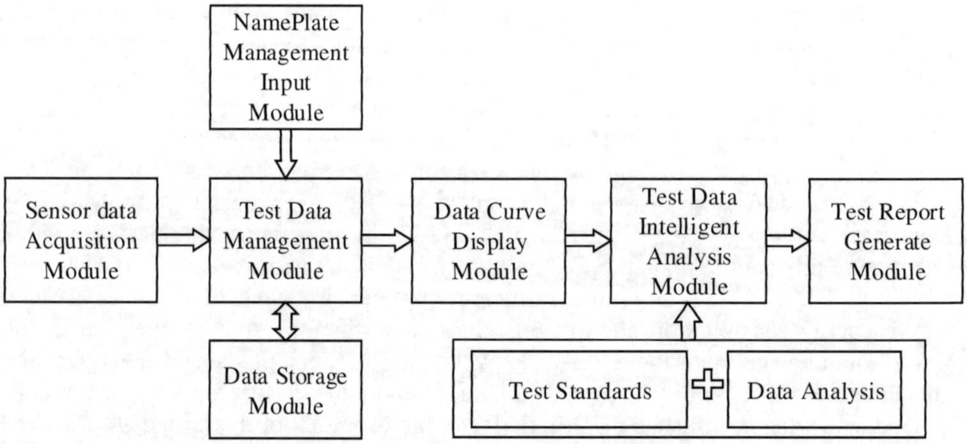

Figure 2. Functional module workflow

Sensor data acquisition module can automatically collect data from the sensor. Nameplate information input module can be used to input refrigerator test tasks information and refrigerator attribute information. Test data management module manage the data transferred by the sensor data acquisition module and nameplate information input module and call other modules to process the data. Data storage module can store sensor data and nameplate information into a database. Data curve display module enables the sensor data to be plotted into an intuitive curve according to time. Test data intelligent analysis module can be used to analyze and judge the collected sensor values according to the standard of refrigerator test. Test report generation module can generate test report according to test data intelligent analysis results and push it to refrigerator tester.

2.2. *Intelligent Analysis Module*

Aiming at the problem of more manual participation in test data analysis. We analyse the sensor data according to the refrigerator product test standard. Then we design the fast location stability interval algorithm based on eigenvalues and the intelligent switching test method, which are used to realize the function of the test data intelligent analysis module. As a result, the sensor data of a refrigerator test can be analysed quickly and intelligently, and the test results of a refrigerator test are obtained in time. The cycle of the refrigerator operating is defined as the time for the temperature of the refrigerator from peak (or trough) to peak (or trough), or the time for compressor power starts running until it stops running, and then the next time it starts running, it is recorded as Tcycle. The standard specifies the energy consumption test for the stable operating condition. It is necessary to select a continuous operating interval. This interval includes 3*n cycle (n=1, 2, 3 ...), and the interval duration is greater than or equal to 6 hours, less than or equal to 30* Tcycle. The interval is divided into three segments according to the number of cycles, interval A, interval B, and interval C. The average of the temperature calculations of the three intervals T_A, T_B, T_C, the average of the power calculations of the three intervals P_A, P_B, P_C, the length of time of the three intervals t_A, t_B and t_C. Meet the requirements (1)(2)(3)(4) formula.

$$T_{\max(A,B,C)} - T_{\min(A,B,C)} < 0.025K \tag{1}$$

$$\frac{ABS(T_C - T_A)}{(\frac{t_A}{2} + t_B + \frac{t_C}{2})} < 0.025K/h \tag{2}$$

$$\frac{P_{\max(A,B,C)} - P_{\min(A,B,C)}}{P_{av(A,B,C)}} < 1\% \tag{3}$$

$$\frac{ABS(P_C - P_A)}{(\frac{t_A}{2} + t_B + \frac{t_C}{2}) \times P_{av(A,B,C)}} < 0.25\%/h \tag{4}$$

$ABS()$ is calculating the average value, $Pmax()$ is calculating the maximum value of power, $Pmin()$ is calculating the minimum value of power, $Tmax()$ is calculating the maximum value of temperature, and $Tmin()$ is calculating the minimum value of temperature.

IOP Conf. Series: Materials Science and Engineering **692** (2019) 012053 doi:10.1088/1757-899X/692/1/012053

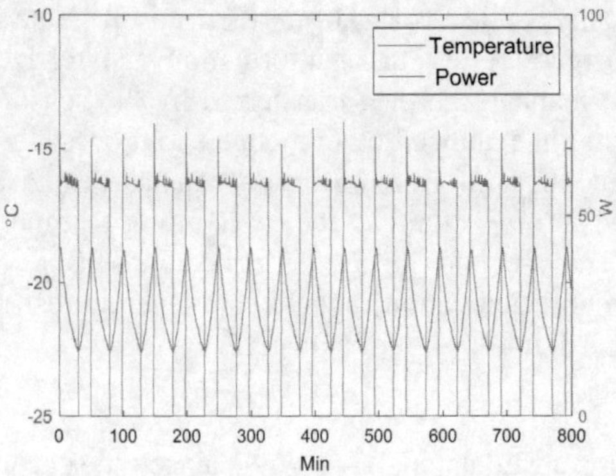

Figure 3. Stable operating condition data

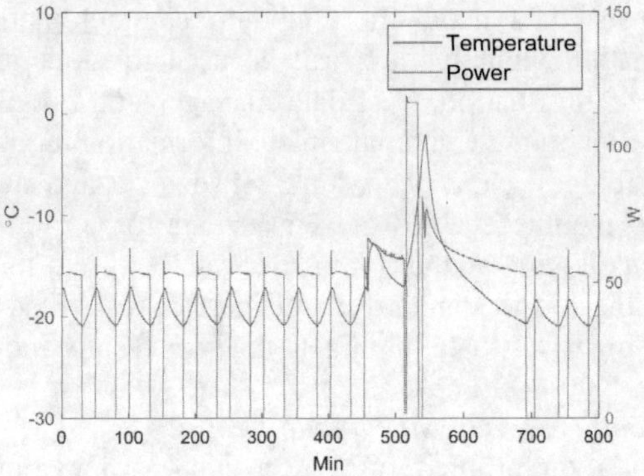

Figure 4. Unstable operating condition data

The manual method stead state interval selection is whether the power and temperature of each operating cycle are similar. Figure 3 shown stable operating condition data, and figure 4 shown unstable operating condition data. Therefore, each cycle can be used as a whole to judge the stability of data between cycles. Extract the characteristic value set of the refrigerator operating data curve F .

$$F = \{(t_1, T_{w1}, T_{t1}, t_{r1}, t_{h1}, P_{a1}), (t_2, T_{w2}, T_{t2}, t_{r2}, t_{h2}, P_{a2}), ..., (t_i, T_{wi}, T_{ti}, t_{ri}, t_{hi}, P_{ai})\} \quad (5)$$

Where the T_w is temperature peak, T_t is temperature trough, t_r is start-up time (compressor operation), t_h is downtime (compressor standby), P_a is power average of each cycle. In statistics, the standard deviation can determine the stability of the data. Performing standard deviation calculation for each characteristic value, and then finally sum.

$$\sigma_j = \frac{1}{n}\left(\sum_{i=1}^{n}(T_{wi} - \mu_{Tw})^2 + \sum_{i=1}^{n}(T_{ti} - \mu_{Tt})^2 + \sum_{i=1}^{n}(t_{ri} - \mu_{tr})^2 + \sum_{i=1}^{n}(t_{hi} - \mu_{th})^2 + \sum_{i=1}^{n}(P_{ai} - \mu_{Pa})^2\right) \quad (6)$$

Among them, $\left\lceil 2/T\text{cycle} \right\rceil \le n \le 30$. The interval with the smallest standard deviation is locked depending on $min(\sigma_i)$ instead of manual selection the stead state interval.

Refrigerator test software obtains the test information through the Nameplate entry module, data acquisition module to provide data collection, the stability judgment method locates the refrigerator operation stability interval, and we design the test analysis algorithm to realize the unmanned test of the refrigerator. Test analysis algorithm steps are described below.

Step 1 Input the sensor data marked by the label.

Step 2 Extract sensor data feature values

Step 3 Find the standard deviation of the feature point values separately and obtain the minimum standard deviation interval.

Step 4 Determine whether the interval compliances with the stead state standards, if yes, jump to step 5; if no jump to step 2 and continue to run the test without meeting the standard; if the interval is more than 108h, jump to step 9

Step 5 Judging the type of test (energy consumption test, freezing capacity test, temperature rise test, etc.)

Step 6 If the energy consumption test is carried out, calculate the power consumption of the stable interval

Step 7 If the freezing capacity test is carried out, record the time difference between 25 ° C and -18 ° C of the load temperature. If the time exceeds 24 h, skip to step 9.

Step 8 If the temperature rises test is carried out, cut off the power supply, and record the time difference of the load package from -18 ° C to -9 ° C

Step 9 Submit a test report

IOP Conf. Series: Materials Science and Engineering **692** (2019) 012053 doi:10.1088/1757-899X/692/1/012053

3. HTRES Power economy

In the refrigerator test, the energy consumption for running the refrigerator is low, but the test is carried out under specific working conditions, and large amounts of electricity is needed to maintain the working conditions. According to the actual statistics of a household appliance enterprise in Qingdao, it is usually applied 7.5 kW·h to maintain the test working condition. Table 1 shows the energy consumption statistics of three types of refrigerator tests conducted manually.

Table1. Manual refrigerator test energy consumption

Test type	Average test time (h)	Working condition energy consumption(kW·h)
Freezing capacity test	48	360
Energy consumption test	30	225
Temperature rise test	28	210

Instead of manual operation, the HERTS automatically selects the stability interval and switches the test stages, which improves the test efficiency, shortens the test period and effectively saves electricity consumption. Tables 2 shows the energy consumption statistics of the three types of refrigerator tests conducted by high-efficiency refrigerator test system.

Table2. HTRES test energy consumption

Test type	HTRES average test time(h)	Reduce time(h)	Working condition energy consumption(kW·h)
Freezing capacity test	32	16	240
Energy consumption test	20	10	150
Temperature rise test	19.5	8.5	146.25

From table 1 and table 2, it's obvious to find that the energy saving for each refrigerator test of the laboratory. The freezing capacity test save 120 kW·h, the energy consumption test save 75 kW·h, and the temperature rise test save 63.75 kW·h. If the three types of refrigerator tests do 50 times annually, 0.129 million kW·h of energy consumption can be saved.

4. Conclusion

In this paper, we present the high-efficiency refrigerator test system, which leads to the realization of the unmanned operation of refrigerator test under the working environment of robot. The high-efficiency refrigerator test system achieve unmanned operation, improve the efficiency of refrigerator test and save energy consumption. It has been applied by household appliance enterprise in Qingdao. In a typical application scenario as shown in figure 5. It realizes the timely selection stead state interval and transition of test stage. It improve the production efficiency.

IOP Conf. Series: Materials Science and Engineering **692** (2019) 012053 doi:10.1088/1757-899X/692/1/012053

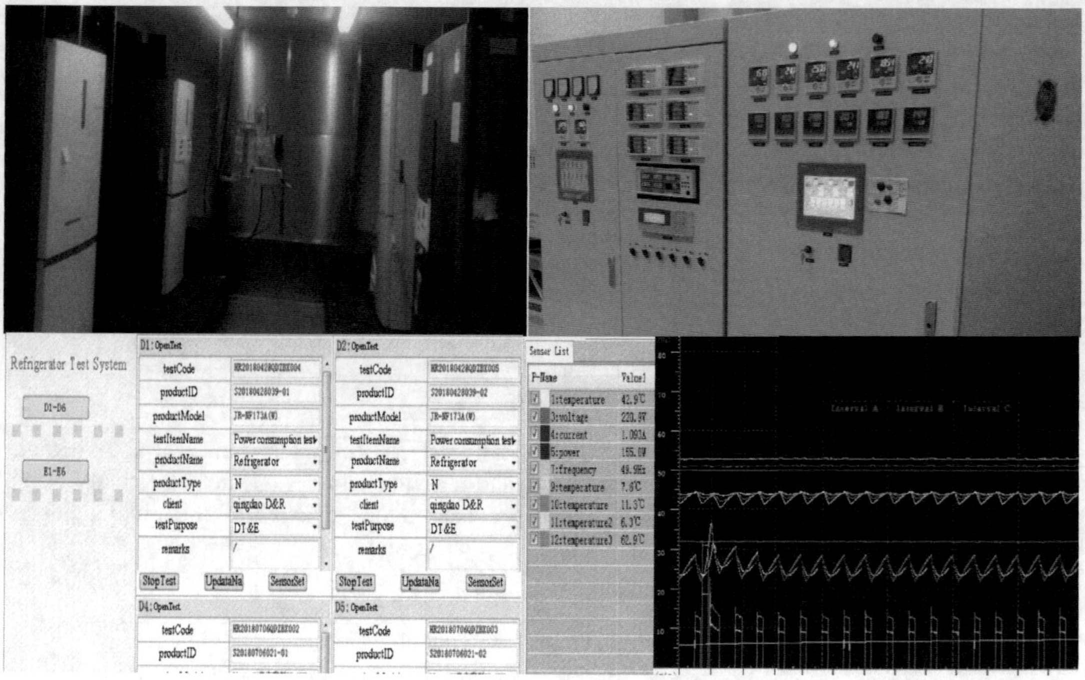

Figure 5. The test scenario of HERTS.

Acknowledge

We thanks to all the colleagues in the network new technology laboratory for their support of us research. Thanks for the Refrigerator tester in Haier household appliance who explained the refrigerator testing process. We gratefully acknowledge financial support from China Scholarship Council.

References

[1] Chen B, Wan J, Shu L, Li P, Mukherjee M and Yin B 2017 Smart factory of industry 4.0: Key technologies, application case, and challenges *IEEE Access* pp 6505-6519

[2] Wang S, Wan J, Zhang D, Li D and Zhang C 2016 Towards smart factory for industry 4.0: a self-organized multi-agent system with big data based feedback and coordination *Computer Networks* pp 158-168

[3] Li L 2018 China's manufacturing locus in 2025: With a comparison of "Made-in-China 2025" and "Industry 4.0" *Technological Forecasting and Social Change* pp 66-74

[4] Zhou K, Liu T and Zhou L 2015 Industry 4.0: Towards future industrial opportunities and challenges *International conference on fuzzy systems and knowledge discovery (FSKD)* pp2147-2152

[5] Hui T, Sherratt R and Sánchez D 2017 Major requirements for building Smart Homes in Smart Cities based on Internet of Things technologies *Future Generation Computer Systems* 76 pp 358-369

[6] Fensel A, Tomic D and Koller A 2017 Contributing to appliances' energy efficiency with Internet of Things, smart data and user engagement *Future Generation Computer Systems* pp 329-338

[7] Floarea A and Sgârciu V 2016 Smart refrigerator: A next generation refrigerator connected to the IoT *International Conference on Electronics, Computers and Artificial Intelligence (ECAI)* pp 1-6

[8] Singh D and Jain P 2016 IoT based smart refrigerator system *International Journal of Advanced Research in Electronics and Communication Engineering (IJARECE)* pp 2080-2084

[9] Guo Z, Liu C, Wang X, Ma H, Jiang Y, Zheng, B and He B 2013 CVIS: Complex virtual instruments system architecture *IEEE International Instrumentation and Measurement Technology Conference (I2MTC)* pp 566-571
[10] Hu K, Guo Z, Ma G, Zhou W and Sun Z 2018 Complex Virtual Instrument Model for Ocean Sensor Networks *IEEE Access* pp 21934-21944
[11] National Household Appliance Standardization Technical Committee 2016 Household and similar refrigerating appliances *Beijing:Standard Press of China* GB/T 8059-2016
[12] Prapulla S, Shobha G and Thanuja T 2015 Smart refrigerator using internet of things *Journal of Multidisciplinary Engineering Science and Technology* pp 1795-1801
[13] Khan M, Silva B and Han K 2016 Internet of things based energy aware smart home control system *IEEE Access* pp 7556-7566
[14] Wu H and Chuang Y 2017 Low-cost smart refrigerator *IEEE International Conference on Edge Computing (EDGE)* pp 228-231

AUTHOR INDEX

AUTHOR INDEX